THE RIVER RETURNS

McGill-Queen's University Press
Montreal & Kingston | London | Ithaca

The River Returns

AN ENVIRONMENTAL HISTORY OF THE BOW

Christopher Armstrong,
Matthew Evenden,
and H.V. Nelles

ISBN 978-0-7735-3584-8

Legal deposit fourth quarter 2009
Bibliothèque nationale du Québec

Printed in Canada on acid-free paper

This book has been published with the help of a grant from the Canadian Federation for the Humanities and Social Sciences, through the Aid to Scholarly Publications Programme, using funds provided by the Social Sciences and Humanities Research Council of Canada.

McGill-Queen's University Press acknowledges the support of the Canada Council for the Arts for our publishing program. We also acknowledge the financial support of the Government of Canada through the Book Publishing Industry Development Program (BPIDP) for our publishing activities.

Library and Archives Canada Cataloguing in Publication

Armstrong, Christopher, 1942–
The river returns : an environmental history of the Bow / Christopher Armstrong, Matthew Evenden, and H.V. Nelles.

Includes bibliographical references and index.
ISBN 978-0-7735-3584-8

1. Bow River (Alta.) – Environmental conditions. 2. Human ecology – Bow River (Alta.) – History. 3. Bow River (Alta.) – History. 4. Bow River Valley (Alta.) – History. I. Evenden, Matthew Dominic, 1971– II. Nelles, H. V. (Henry Vivian), 1942– III. Title.

FC3695.B67A74 2009 971.23'3 C2009-902501-9

Set in 11/14 Minion with Stone Informal and Stone Sans
Book design & typesetting by Garet Markvoort, zijn digital

To

Emily,

Jen,

Geoff,

and

Maggie

Contents

Preface

This book tells the story of a river, the Bow in Alberta. It is a story about how the Bow flowed into the lives of the people who came to live in its valley, what happened to it as a result, and how after long experience people came to terms with the river. It is an incomplete story because the ultimate compact that people have made with their river involves trade-offs and adjustments that mainly lie ahead. Still, it is a history worth knowing as this generation and the next confront the challenges of adapting the demands they make on rivers.

It is also a somewhat one-sided story as only one of the two major players has a voice and agency – the human population. The river, like all others, faithfully flows according to observable principles and responds to natural and human disturbance in accordance with fathomable ways. Only over time were those mechanisms gradually deduced and incompletely understood. Rivers, of course, continue to surprise, but this characteristic does not mean that they express intentionality or voice. Humans act and react; the river simply flows. But if in this narrative the river is more acted upon than acting, it nevertheless does eventually assert itself in an essentially human discourse.

This is a story, too, of a particular river, the Bow. We make no claim that the Bow is a typical river; no river is. Indeed, we have chosen to study this river because of its remarkable qualities: among other things, its beauty, cultural variation along its length, and the compact cohesiveness of its valley. While the story of the Bow is not exactly like that of other rivers, it is largely comparable to other European and North American examples, making allowances for regional particularism. The human relationship with the river encountered in these pages would be broadly similar had another river or many rivers been chosen. The instrumentalist approach to river use and management, so influential in shaping the Bow, is virtually universal.

Telling the story of the Bow has the advantage not only of raising these larger generalizations but also of doing so in a specific context, where we are able to explore that history in greater detail, affording a glimpse of the intense relationships humans have with rivers. Moreover, the Bow is a river fondly and intimately known to the people who have grown up on its banks, but also known to millions of tourists and visitors and many more who know it only through iconographic images. In that sense the Bow is both a local and a world river, and its story could be told, *mutatis mutandis*, in many other places.

Our narrative will unfold topically after we discover our river, place its history in context, and introduce our environmental history approach. While an effort has been made to set topics within a chronological framework, inevitably our structure necessitates a good deal of doubling back and forging ahead. Our approach does, however, allow us to explore in a sustained manner the interaction of nature and culture along several axes of engagement. This river has been, and still is, a source of power and wealth for those able to command it, but the powers, the objectives, and the beneficiaries have changed over time. This is the story of a struggle, amidst changing social, economic, and technological conditions, to control the river for private and public benefit during a time when the compliant river seemed to accommodate everyone. Eventually, however, a countervailing ethics and politics set limits on use.

Our title refers to the classic Hollywood western filmed along the banks of the Bow in the summer of 1953, *The River of No Return*. Marilyn Monroe, Robert Mitchum, and Rory Calhoun starred in this picture, but so too did the Bow River; though it must be confessed that under Otto Preminger's direction the Athabasca and Maligne Rivers in Jasper National Park stood in from time to time. On a makeshift raft (echoes of Huckleberry Finn) Monroe, Mitchum, and a young boy escape an Indian uprising down a treacherous river in pursuit of the gambler villain Calhoun, who has betrayed them. The river not only provided the majestic Cinemascope scenery within which big dreams might be projected but also played a role on its own. It advanced the plot from one adventure to the next, swept the raft of Hollywood stars into life-threatening rapids, and carried them gently in interludes through calm reaches. In the conventions of the genre the hero tames the wild woman as he commands an even wilder river. Ultimately the river delivers its cinematic cargo to the climax in fictional Council City, where manhood is achieved out of the barrel of a gun.

The title *The River of No Return* dramatically states the essentially one-way relationship between people and rivers that prevailed for much of the twentieth century. The river was a utility; it carried things downstream. By contrast, in this book, *The River Returns*, we want to stress the circular relationship between the inhabitants of the Bow valley and the river. What is carried downstream comes back. Acted upon, the river invariably returns the consequences of those actions in ways that cannot be avoided.

In our telling the environmental history of the Bow is not an "end of nature" fable. Nor is it a story to give comfort to "river management" hubris. Ours is a narrative of the middle ground, of nature certainly modified through human use but also through lived experience more effectively understood. It is an account of people putting themselves back into a natural world and learning, however imperfectly, to live with it.

A Note on First Nations Names

In this book we have used the names preferred by First Nations today and used in their own traditions, as elaborated below. The exceptions are when we quote from documents, cite sources, or reprint maps.

First Nation	Formerly Known As
Kainai	Blood
Nakoda	Stoney
Piikani	Peigan
Shoshoni	Snake
Siksika	Blackfoot
Tsuu T'ina	Sarcee

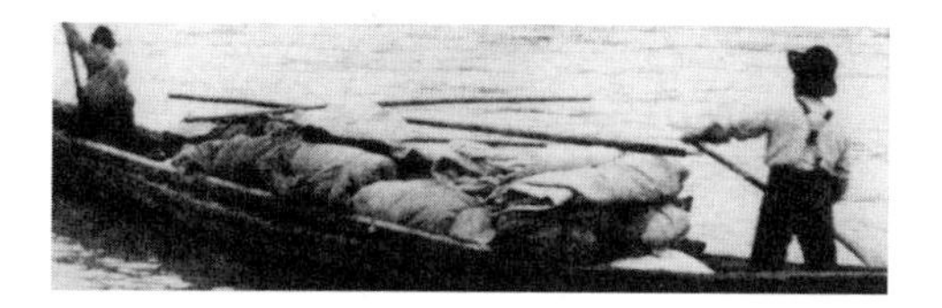

Abbreviations

Archives of Alberta	AA
Alberta Water Resources Commission	AWRC
biochemical oxygen demand	BOD
City of Calgary Archives	CCA
cubic feet per second	CFS
Canadian National Parks Association	CNPA
Canadian Parks and Wildlife Society	CPAWS
Canadian Pacific Railway	CPR
Eastern Irrigation District	EID
Eastern Rockies Forest Conservation Board	ERFCB
Glenbow Archives	GA
Library and Archives Canada	LAC
Prairie Farm Rehabilitation Administration	PFRA
Prairie Provinces Water Board	PPRB
South Saskatchewan River Basin Study	SSRBS
South Saskatchewan River Project	SSRP
University of Alberta Archives	UAA
United Farmers of Alberta	UFA

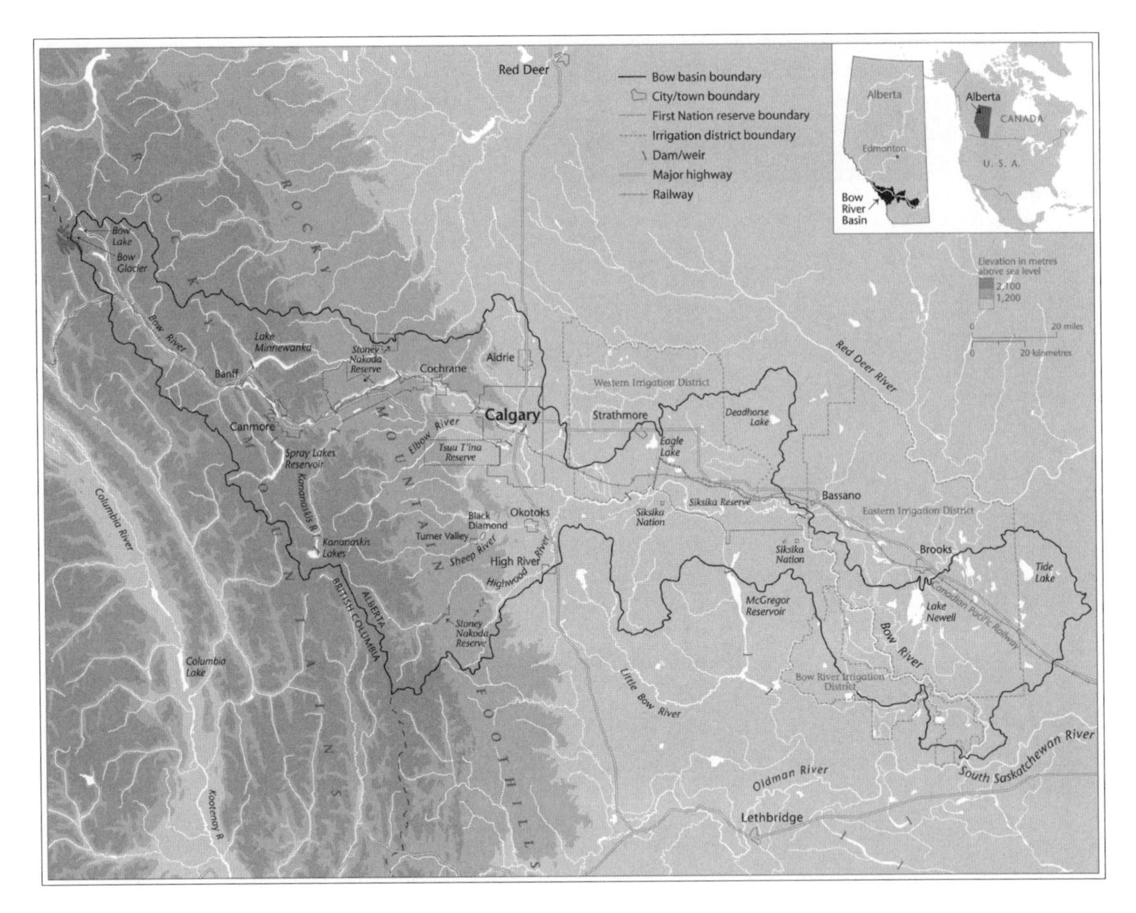

Map 1 The contemporary Bow basin

THE RIVER RETURNS

1 · Discovery

Discovery is a relative concept. It begs the question, by whom? Eurocentric tradition accords the honour of discovering the Bow River to the old fur trader James Gaddy and his seventeen-year-old companion, the Hudson's Bay Company explorer David Thompson, who camped on the Bow with a group of Piikani over the winter of 1787–88.[1] But obviously, different groups of indigenous peoples had hunted, fished, and sought shelter in its valley for thousands of years before that. The Bow was probably discovered or first seen by humans sometime after glacial ice in the valley retreated in the Pleistocene Era.

Before the last glacial advance a river like the Bow had run out of the Rockies onto the plains but by a different route, through what is now Lake Minnewanka. However, during the last ice age, when the Laurentide Glacier buried what would become the prairies and the Cordilleran Glacier ground away at the Rocky Mountains and its valleys, that river ceased, its water locked in ice thousands of metres thick. At the end of the Pleistocene, about thirteen thousand years ago, the present Bow River emerged from the ice.

With global warming the glaciers began their last retreat. Meltwater became impounded in massive lakes that eventually burst their barriers. The shifting of the ice and the force of the meltwater at the end of the Pleistocene left a considerable imprint on the landscape. Silt and rock material dropped by the glaciers formed eskers and moraines. On the open prairie, large boulders, called erratics, fell from the ice and stood out against a flat horizon. The Bow Valley was worked and reworked in this process. The erosive power of melting water transporting coarse material cut away at existing banks and deepened and widened the valley. In places the river dropped its load and formed terraces high above the modern river course.

Even as the ice sheet dwindled and retreated to the east, its remnants modified the climate, leaving the exposed lands of the Bow and surrounding areas within a relatively cold and dry climate that was only slowly colonized by vegetation and animals. Some of this land would have remained as permafrost. With time, however, the winds carried light, mobile seeds from the south, the land thawed, and plants began to take root. The landscape may have appeared like tundra at first, dotted with sage and grasses. Later, poplar forests became established, and later still, conifers began to grow in the upper valley and grasslands in the south and east. Around nine thousand years ago, the plant geography of the Bow Valley started to approximate its contemporary appearance. Bison and other large mammals moved north.[2]

People entered this land of melting ice as early as eleven thousand years ago. At the Vermilion Lakes, in the upper basin, radiocarbon dating suggests that humans occupied the area not long after the ice retreated from the corridor separating the Cordilleran and Laurentide Glaciers. On a river terrace at the base of Mount Edith Cavell, hunter-gatherers camped, hunted mountain sheep, elk, and deer, and broke and carved their bones into tools. Although no stone tools remain in this earliest of Bow basin archaeological sites, it is possible that these were so-called Clovis peoples, a term used to identify some of the earliest occupants of North America, whose archaeological remains have been dated to about 11,000 BP.[3] Once these hunter-gatherers moved on, the river preserved the site by flooding silt and gravels over top. As time passed, other groups of hunter-gatherers passed through, camping on the same exposed and sunny site. Flooding preserved the remnants of these camps, too, as well as several arrowheads that suggest the period and provenance of different hunting groups. A trace layer of ash, from the volcanic eruption on Mount Mazama, Oregon, (6,800 BP) also provides a more definite marker of time's arrow. For these earliest peoples, the river encampment served as one point in a seasonal cycle from which to hunt and possibly gather plants and berries. It was an early and enduring site of refuge in a dynamic landscape.[4] At the dawn of our Holocene epoch, humans discovered the Bow and learned to live off the bounty of its valley.

A succession of peoples embodying different cultures would over the millennia discover the river and occupy its valley.[5] Nor would they remain passive elements of the environment. Scientific analysis of pond sediments provides a record not only of climate and vegetation change but also of the continuous presence of fire in the landscape. Some of this would have been the so-called first fire, sparked by lighting. But some would also have been "anthropogenic fire," combustion deployed by humans to alter the ecology in order to enhance hunting.[6] In subtly different ways peoples adjusted to the opportunities their changing cultures created in the valley and altered the environment, primarily with the tool of fire, to improve their lot.

This churning of peoples and blending of cultures would greatly accelerate in the nineteenth century, when Euro-American intrusions initiated a more intensive epoch of human-environmental interaction. For these newcomers,

Illus. 1.1 William Pearce, surveyor, federal government land commissioner, park promoter, and irrigation evangelist, about 1910. (Glenbow Archives, NA 813-29)

discovery of the river also involved imagining what it might be made to do and how its undeveloped potential could be exploited. For them, nature existed to be used or improved to its highest form. William Pearce, the authoritarian federal government agent of western lands in the late nineteenth century, whom we will meet several times in these pages, typified this position. Seen from his perspective, the semi-arid landscape of the southern prairies could not sustain conventional agriculture. It could barely support free-range cattle ranching. He would spend much of his professional life gruffly trying to keep people off the land. On the other hand, he argued, the land could be improved with water. With elaborate irrigation works of weirs, channels, and ditches, the river could make a dry land flourish with crops, trees, and prosperous farms. But that would take capital, labour, and organization beyond the reach of the individual homesteader. Others, like Pearce, would discover similarly capital-intensive hydroelectric and industrial possibilities inherent in the river.

But beyond narrow conceptions of utility, Pearce imagined other higher "uses" of the river. Its gentle curves, its cooling breezes, and – rare for the bald prairie – its treed banks created the conditions for culture-enhancing parks. Improving the river with parks ennobled the society, making a good life beyond mere existence possible. But these latent parks also required organization, imagination, and investment. In the late nineteenth century Pearce prophesied, and did a good deal personally to achieve, these goals. His discovery of the river embodied the two principal conceptions of use – economic and aesthetic, or use for production/consumption versus use for recreation/pleasure – that would dominate twentieth-century practice.

But the verb "to discover" has other meanings besides the conventional "to obtain sight or knowledge of for the first time" or to explore. The dictionary reminds us that originally it meant "to remove the covering from" and, by extension, to disclose to view, expose, make known, reveal the identity of, and exhibit. In these other senses the "discovery" of the Bow River can be said to have been quite recent, and it is ongoing. Only within the last generation has a covering over the eyes of its human inhabitants been dramatically pulled back to reveal the river in an entirely new light.

With a mounting sense of alarm over the past three decades, books, pamphlets, government task forces, and the mass media have disclosed the importance of the river to the economy and society, how it shapes human destiny, and how generations of use and abuse have fundamentally changed its character. The redoubtable Ralph Klein might be the best exemplar of this phenomenon. He discovered the Bow River paddling on it in his canoe and fishing on its banks. As a crusading journalist, he drew public attention to its deteriorating water quality in the early 1980s. But later, as mayor of Calgary, then minister of the environment, and ultimately premier of the province, he assumed responsibility for cleaning it up while at the same time promoting provincial economic growth. Klein's career trajectory closely mapped a consciousness-raising in the broader community. Within the last decade or so, many people have purported to speak for the river, to study with urgency the myriad of ways it flows through their lives. Thus the most recent human occupants of the Bow valley have only recently discovered their culture in the river and the river in their culture.

We also make our own personal discoveries. Like most Canadians, we discovered the Bow River first as tourists. Christopher Armstrong travelled west on the Canadian Pacific Railway as a fourteen-year-old with a group of other high school students in the summer of 1956 in a colonists' car. He visited the Calgary stampede and then camped at Canmore, where he climbed those alpine meadows on the north side of the river now covered with million-dollar homes. His group trucked up to the Spray Lakes on a very rough-and-ready road across a floating bridge and explored the foot of Mount Assiniboine on horseback. H.V. Nelles first encountered the Bow River on a cross-country hitchhiking expedition in 1964 in Calgary, on a visit to the zoo, and then a short time later above Banff near Bow Lake, where he camped on the river-bank, no doubt illegally. Matthew Evenden undoubtedly saw the Bow River first at the age of three in 1974, while driving along the Trans-Canada Highway stuffed into the backseat of a Meteor, heading east across the Rockies with his family. And we three have since returned to the Bow many times variously to visit relatives, vacation, hike, trail-ride, ski, soak in the hot springs, dine, and get married in its valley. Each time we have reconnected with the river.

The Bow River insinuated itself into our lives as historians as well. Armstrong and Nelles, studying hydroelectric development and utilities regulation for their book *Monopoly's Moment*, discovered the rich history of the river and the depth of records among the collections of the Glenbow Archives in the 1970s. Evenden's research paper in the 1990s on the history of irrigation in Canada drew him to Calgary to study the Bow River irrigation districts. Independently, as we pursued other projects, we uncovered new aspects of the Bow River.

More than nostalgia and coincidence, however, led the three of us to concentrate our collective energies on writing an environmental history of the Bow River. There were reasons beyond the chance convergence of different research agendas on one place that drove this inquiry forward. Why study a river, and this river in particular? Rivers are dynamic historical subjects that have been changed by and have changed human societies. They give coherence to life and define a sense of place; they connect separate spheres of activity and feedback consequences. Historians over several generations have framed their narratives around rivers. As metaphors, rivers speak to the deeper meanings of life. They are of undeniable importance in history and thus deserve a history of their own. A river is an ideal subject through which to explore the native terrain of the environmental historian, the reciprocal role of nature in history and of history in nature. A word or two about each of these points before turning to the Bow River in particular.

River Gods

The ancient Greeks inhabited a world crowded with about sixty specifically named river gods; admittedly lesser gods in their pantheon, they reputedly took mistresses from the human population and thus fathered many Greeks. The Romans made do with considerably fewer river deities, but one of them, Tibernius, played a major role in their founding myth, giving Romulus and Remus over to the care of Lupus. In the Christian era four rivers divided paradise, but rivers as gods were necessarily dethroned, being demoted to mere decorative motifs and tropes. They can be seen disporting themselves in Renaissance fountains, life-giving streams gushing from their gaping mouths or vessels cradled in their arms. Representations of river gods, too, acknowledged the importance of river commerce – witness, for example, the glorious Customs House in Dublin faced with four fearsome but beneficent river deities.

River gods lived because rivers were important in life. Humans have historically favoured riparian locations for their communities because rivers have been the source of power, commerce, communication, protection and safety, food, sanitation, pleasure, and beauty.[7] With abundant water, civilization of a very high order is possible. Water from rivers nourishes culture and agriculture; it slakes thirst, bathes, cleans, boils food, washes laundry, soothes the soul, puts out fires, disposes of filth, baptizes, and in some cultures carries away the ashes of the dead.

Apart from their practical utility, as inspiring objects of exquisite beauty, and as formidable obstacles, poets and artists across the centuries have utilized rivers as metaphors to give meaning to life. Rivers have symbolized the passage of time, eternity, life, death, separation, fate, and danger. Crossing the river, whether the Styx, the Rubicon, the Jordan, the Delaware, or the Mississippi, signalled major changes in condition or decisive action that changed the course of history. The luminous quality of rivers – their elegant curves, mirrored depths, and sublime gorges and cascades – has challenged some of the greatest artists of the Western canon, most notably in different eras: Poussin, Monet, and Seurat in France; Constable and Turner in England. In the United States whole stylistic movements of artists – the Hudson River and Brandywine River schools, for example – have organized themselves around the river motif. In Canada the list of artists who have taken rivers as their subject reads like a roll call of Canadian art history. To write of rivers, then, is to join, in however small a way, in this cultural conversation between writers and artists across the centuries.

At the vernacular level, rivers lend coherence to life. Rivers and their valleys define meaningful places. River valleys establish boundaries and connections, linking separate places in a space than can be perceived. The valley is more than a place; it is quite commonly home, a community of shared experience. People proudly name their businesses and institutions after their rivers and valleys. In folk and popular melodies we sing of our river homes: the Red, the Mississippi, the Shenandoah. Rivers in song signify persistence, strength, a distant desired state to which wanderers long to return.

In Simon Schama's *Landscape and Memory*, rivers form but a part of a breathless tour of aquatic symbolism in mainly European myth, architecture, and art. With daunting erudition Schama demonstrates the remarkable descent and striking continuity of ideas about water and rivers from Egyptian times to the Victorian era, emphasizing the way in which the pagan flowed into the Christian. From antiquity to modern times, water and rivers have been the source of life, symbols of fecundity, the genesis of civilization. Upstream, rivers metaphorically lead ever higher towards sources, origins, springs, through moist clefts, and ultimately to the gods and final things. The relentless, unending, and mysterious flow of rivers has always exerted a spiritual force counterpoised against the intriguing potential of their physical power. Downstream, the ceaseless flow leads forward to destinies foretold and the passage of time.

To control water is to possess power. Those possessing power are obliged in turn to shower their people with a fountain of blessings. But there is a second thematic tradition that Schama uncovers, a circular counterpoint to the linearity of flow. Rivers connect us with cycles of rising and falling, flooding and drought, diurnal, seasonal, and dynastic destruction, renewal and regeneration. The river as blood sustains life by nourishing and cleansing in circulation. There is a circle of eternity in the river.[8]

The Canadian writer who comes closest to the tradition of seeing deeper things in rivers is Hugh MacLennan. In 1974 MacLennan composed a powerful and anguished personal meditation under the title "Thinking Like a River," as a preface to a beautiful book called *Rivers of Canada*. He had begun thinking about Canadian rivers in the late 1950s, following the death of his first wife, the exhausting effort of writing *The Watch that Ends the Night*, and the renewed joy of his remarriage. *Maclean's* magazine commissioned him to write some essays on Canadian rivers, which he gathered in a small book, *Seven Rivers of Canada*, in 1959. Fifteen years later, revised and expanded, these essays reappeared as the text of the *Rivers of Canada*.

River exploration, MacLennan wrote in 1974, "took me out of the small subjective world of myself and my characters into an enormous landscape which I thought at the time was eternal." By his own admission, his encounters with Canada's rivers brought him face to face with the big problems of time and space for the first time. "Space became a living personality to me and rivers the pulse of its life." In rivers MacLennan inevitably encountered time and with time the cycle of life, death, and renewal. "The analogy between rivers and lives has been overworked," he concluded, barely dodging the cliché, "but only because it is unavoidable."[9]

But MacLennan went further: rivers "moulded lives," he insisted, making people in their own image. He illustrated the influence of rivers on people by quoting what he termed the "bravura" description of the St Lawrence River that opens his 1945 novel, *Two Solitudes*, one of the most famous passages in Canadian literature. For MacLennan, the important point was not that the river faithfully reflected the social characteristics of the population on its banks but, rather, that it made them. The river was more than metaphor; it was an actor and a creator.[10] In this, MacLennan was following a long-standing Western cultural tradition of seeing nature as an influence in the shaping of human character.[11]

More than ten years later, when MacLennan returned to his text for Macmillan's coffee-table book, rapid social and economic change made it seem to him as if *Seven Rivers* had been written in another era.[12] In a gloomy mood, profoundly alienated by the excesses and uprisings of the sixties, he sought reassurance from his rivers. In such confusing times, he confessed in his prefatory essay, "I wanted to return to the rivers, and please don't laugh at this: I wanted to think like a river even though a river doesn't think. Because every river on this earth, some of them against incredible obstacles, ultimately finds its way through the labyrinth to the universal sea." Looking in the mirror at the ravages of time on his own face and recalling dead friends, MacLennan came to the grim realization: "It was the fate of my generation to have been born in the death throes of a civilization that had supported the west for two thousand years. It was our tragedy that hardly any of us understood what this had meant."[13] But once again writing about rivers carried him beyond despair to affirmation. For the aging novelist, thinking like a river meant finding reassuring certainties in a relativistic universe.

River Histories

Historians in North America over several generations have also used rivers as metaphors to frame their narratives. Rivers have been central organizing metaphors in both Canadian and American historiography. Whiggish accounts of the advance of civilization as well as Declensionist tales of decline and fall have been carried forward on the river's current.

During the 1930s it could be said that Canadian historical writing achieved a measure of distinctiveness and maturity in examining the material influences of rivers on national development. In his classic book *The Fur Trade*, first published in 1930, and his essays on transportation as a factor in Canadian history, Harold Innis developed a new understanding of the logic of Canadian history. His famous epigram "The present Dominion emerged not in spite of geography but because of it" neatly sums up this point of view.[14] Or, as he observed less pithily on another occasion, "It is no mere accident that the present Dominion coincides roughly with the fur-trading areas of northern North America."[15] Innis's staples interpretation of Canadian history stressed "the overwhelming significance of waterways and especially the St. Lawrence."[16] Geography endowed the regions with a host of natural products. The rivers connected the hinterland with cheap transportation to overseas markets. On this discrepancy between the centre and the margin Canada was born, its shape and allegiance determined by its natural transportation system. And so, too, a distinctive Canadian historiography took shape, rejecting the earlier Liberal nationalist view of Canada as essentially a political creation. According to this Laurentian interpretation, Canada survived on the North American continent because it had a natural foundation. Environment and economy combined to determine its destiny.

The historian Donald Creighton elevated Innis's view of the St Lawrence into new poetic realms, giving it music, drama, and a brooding sense of destiny denied in *The Commercial Empire of the St. Lawrence*, published in 1937. He extended Innis's staples interpretation of Canadian history based upon the transportation capabilities, deep continental penetration, and metropolitan linkages centred upon the St Lawrence.[17] But Creighton, a romantic historian, went further. For him, the river itself was a protagonist. Natural history, in the form of a river, foretold the human history to follow. In a famous, much quoted passage at the beginning of the book, Creighton's theme emerges in a tumbling torrent before settling into sweeping cadences: "The river meant movement, transport, the ceaseless passage west and east, the long procession of river-craft – canoes, bateaux, timber rafts and steamboats – which followed each other into history. It seemed the destined pathway of North American trade; and from the river there rose, like an exhalation, the dream of western commercial empire."

If the river foretold the possibilities of a vast transportation system tapping the natural riches of the continent, the next paragraph introduces the "root defect" in nature's architecture that "stood between the design and fulfilment." The pull of the river diminished in strength in the vast expanse of lakes in the western interior. The low-lying southern boundary between the St Lawrence and the Hudson exposed a strategic weakness. The rapids of the St Lawrence and the falls at Niagara required costly remediation works for the system to function efficiently. But those with faith in the religion of the river could see its possibilities through these flaws: "The dream of the commercial empire of the St. Lawrence runs like an obsession through the whole of Canadian history; and men followed each other through life, planning and toiling to achieve it. The river was not only a great actuality; it was the central truth of a religion. Men lived by it, and once consoled and inspired by its promises, its whispered suggestion, and its shouted commands; and it was a force in history, not merely because of its accomplishments, but because of its shining, ever-receding possibilities."[18]

In this Innis-Creighton "commercial empire" interpretation, rivers – particularly the St Lawrence and the Saskatchewan – drove history forward, determined the boundaries of the country, and gave its politics and economy a logic and a fundamental coherence. Theirs was an environmental interpretation of history in the old sense of the word, in the same way that Frederick Jackson Turner's frontier theory was an environmental interpretation. The environment conditioned human history, made certain things possible, and denied others; it set boundaries, creating situations in which human character and political institutions were determined. The Laurentian interpretation would influence the writing of Canadian academic history until the 1960s.

This phase of writing about rivers in Canada, the commercial empire school, connected loosely to a softer-edged, US approach to the writing of popular American history that emerged about the same time. In the depths of the Great Depression, in the dark shadow of social dislocation and possible political upheaval, the Canadian-born, New York–based historian and editor Constance Lindsay Skinner conceived of a grand series of books intended to "kindle the imagination and to reveal American folk to one another."[19] By "folk" she meant ordinary people, the bedrock of the American republic. Skinner argued, "We began to be Americans on the rivers." She continued:

> By the rivers the explorers and the fur traders entered America. The pioneers, who followed them, built their homes and raised their grain and stock generally at, or near, the mouths of rivers. As their numbers increased they spread up the valleys, keeping close to the streams, since water is an indispensable element of the sustenance of the soil and all animal life. The rivers were the only highways of communication and commerce between solitary hamlets. Settlement expanded from rivers. To repeat, the first foreigners on these shores began their transition from Europeans to Americans as River Folk.[20]

Thus was born the great Rivers of America series of popular histories and with it a particular approach to river history. It began with morale-boosting, populist purpose, a private-enterprise, proto-WPA project to employ authors, artists, and editors but also to raise the consciousness and the spirit of Americans. Skinner believed that "the average American" had been alienated from this folk history "because the epic material of America has been formulated by scholastics instead of by the artists." Her illustrated series, written by popular writers, would correct this estrangement by revealing the American through the "magic in rivers," by showing how they symbolized life and American triumph. In short, she believed, "The American nation came to birth upon rivers," and that is the story she wanted her series Rivers of America to tell. Rivers would carry America safely out of its Great Depression.

By any measure, the Rivers of America series was an astonishing success. The twenty-four volumes Skinner projected when she began the series in 1937 grew to sixty-five by the time the series formally concluded in 1976. It continued on through three more editors after Skinner died in 1939. The authors were for the most part non-fiction writers, journalists, and magazine writers, rather than academic historians. They adopted a breezy, anecdotal approach. All the volumes were illustrated with black and white drawings by well-known artists. Some volumes attracted an enormous following. *The Hudson*, for example, was reprinted eleven times. *The Upper Mississippi* and *The James* were revised and reissued. The series ran out of rivers before it ran out of readers. Eventually the series editors even commissioned books on Canadian rivers. Henry Beston, a Maine writer, wrote *The St. Lawrence*, illustrated by A.Y. Jackson; Bruce Hutchison did *The Fraser*, illustrated by Richard Bennet; Marjorie Wilkins Campbell contributed *The Saskatchewan*, with illustrations by Illingworth Kerr; and Leslie Roberts wrote *The Mackenzie*, which featured Thoreau Macdonald's drawings.[21]

Through Rinehart's series, American and Canadian general readers learned their regional history. Only with difficulty can so many books and authors be squeezed into one interpretive framework, but it might be said that Skinner's authors regarded rivers as "cradles of civilization"[22] and used watersheds as organizing frameworks for telling the story of progress. Rivers opened territory, carried peoples to their destiny, and were the stages on which the great conflicts of American history were played out. "My river has everything a person could ask for," crowed the author of *The Kennebec*. "Great events and greater men."[23] Most books took the form of an epic of discovery, settlement, and trial through conflict to the rise of commercial civilization and major economic and cultural achievement. They typically began with geology and indigenous people and ended with painters, poets, and artists. Along the way the emphasis fell upon characters, great moments, and dramatic episodes. At some point each volume featured a journey downstream or upstream.[24] In most cases, rivers were portrayed as highways up which culture flowed, though there were some exceptions. *The Tennessee* bucked the trend: "Tawny and unsubdued, an Indian among rivers, the old Tennessee threw back men's improvements in his face and went on its own way, which was not the way of the white man."[25] The author of *The Missouri* placed his initial emphasis upon the river itself: "There are streams that have no story except that of the people on their banks, but the Missouri River is a story in itself – and no idyll or ecologue either, but a heroic poem, an epic. It is a thoroughly masculine river, a burly, husky bulldozer of a stream, which has taken on the biggest job of moving dirt in North America."[26]

Treatment of the Canadian rivers differed slightly from the march of progress school. The St Lawrence focused on the folklore of the region and its abundant wildlife. The Mackenzie remained a wilderness river from beginning to end. Campbell's Saskatchewan perhaps came closest to the American triumphalist trope. Although Hutchison depicted the Fraser in a "naked, brutal and ceaseless … war with all living creatures,"[27] he ended his story with a surprisingly upbeat advertisement for the proposed hydroelectric developments that would, he anticipated, soon tame this beast.

The success of the Rivers of America series spawned Canadian imitators. In 1945 McClelland and Stewart published Mable Dunham's *Grand River*, and in 1959 William Toye made his debut with Oxford University Press with *The St. Lawrence.*[28] River history is a remarkably durable genre that continues to attract readers on both sides of the Atlantic. The American writer Ted Lewis has recently published a history of the Hudson River crafted in the Rivers of America mode. The prolific Peter Ackroyd has followed up his biography of the city of London with an equally formidable volume, *Thames: Sacred River.* His forty-five chapters of topically arranged vignettes suggest the range of themes rivers give rise to, especially in his case of a metaphorical nature.[29] It could be argued that Peter Truman's Great Canadian Rivers television series on the Outdoor Life Network keeps alive the Rivers of America spirit, but it does so under a lush photogenic overlay of nature worship.[30]

The Rivers of America series not only ran out of rivers in the 1970s; it lost its faith. In the earlier volumes from time to time a tone of regret or sense of loss would creep into the text, only to be banished by hope. Again, *The Kennebec* provides a good example: "Greed has fouled the Kennebec," its author, Robert Coffin, conceded in 1937. "But the promise of life is still there."[31] By the 1970s this confidence had begun to wane. A 1974 reprint of 1945's *The Columbia* bore a preface lamenting the damage done to the salmon by the dams and nuclear reactors. Nevertheless, the author concluded with unconvincing optimism, noting that the Hanford Atomic Works and the many hydroelectric dams across the mainstem "compose the perfect symbol of the day, the essence of modernity … no other river in America is as true to the spirit of the times."[32]

Frank Waters, the aptly named author of the book on the Colorado River, did not return in such an optimistic spirit in 1974 after revisiting the river he had first written about in 1946. In the intervening years huge dams and irrigation projects had reduced the river "to not much more than a cement lined ditch." In the forties he had found the Colorado "a virtual wilderness." Over a mere thirty years "a savage onslaught against nature" had been waged. Waters lamented that in the earlier edition he had not been attuned to this "impending tragedy." His new preface contained not only a damning assessment of "the march of progress" but also a devastating critique of the Whiggish premises of the Rivers of America series itself:

> In our march of conquest westward from the Atlantic, we Anglo-Americans viewed the earth as an inanimate treasure house existing solely to be exploited for our material gain. With a calculated program of genocide we exterminated almost all Indians and wantonly wiped out all the animals and birds save a few remaining species. We leveled whole forests under the axe, plowed under the grasslands, dammed and drained the rivers, gutted the mountains for gold and silver, divided and sold the land itself. But that which was destroyed and stolen from nature was at the same time paid for by loss of soul, our alienation not only from the earth, but from the dark maternal unconscious, its psychic counterpart.[33]

As this bleak prefatory essay suggests, it was no longer possible to write about rivers in the 1970s with the same confident spirit of a generation earlier. So much had changed, both on the rivers and in the culture. In the first instance the scale of development along rivers had changed. But public attitudes towards the exploitation of natural resources had changed even more profoundly. After Rachel Carson's *Silent Spring* (1962) and other alarming warning cries, the American environment no longer appeared eternal, inexhaustible, or its possibilities limitless. Rather, the environment seemed fragile, desecrated, perhaps in some cases doomed. The rise of the environmental movement in the United States and Canada cast an entirely different light on river history.[34] And when rivers actually caught fire, as an oil slick on the Cuyahoga River did just southeast of downtown Cleveland, Ohio, at noon on 22 June 1969, it was no longer possible to ignore how profoundly nature had been damaged by industrial and human pollution. Then came the Love Canal. Finally, the resurgence of the political left in the United States over foreign policy, race, and social issues introduced a new, piquant critical spirit into historical scholarship. The darker side of history needed to be examined, rather than smothered under a prevailing consensus. Political dissent made power and its unequal distribution central questions.

Meanwhile, a political economy approach to rivers had evolved within Canadian historiography. With some of the nationalist and metahistorical elements stripped away from the Laurentian interpretation, this body of literature narrowed its focus to the economics and politics of development. In the case of hydraulic development, the essential questions were what possibilities did technology and natural resources create, how did entrepreneurs emerge, how did the state and private interests combine to exploit those opportunities, and how were the benefits distributed? Hugh Aitken's *Welland Canal Company* and John Dales's *Hydroelectricity and Industrial Development* pioneered this approach.[35] Sometime later Nelles's *Politics of Development* extended this treatment to early twentieth-century Ontario, emphasizing the role of the state in resource exploitation and, in particular, hydroelectric development.[36] Armstrong's *Politics of Federalism* showed how the struggle to control resource exploitation, among other things, influenced Ontario's relations with the federal government.[37] This literature, which focused upon deep structural elements of the polity and the economy, was at its best in revealing the close linkages between the state and business and the way in which the power of the rivers had been channelled through the social and economic system.

River history acquired a harder edge in subsequent years. Later commentators would criticize the functionalist bias of both the commercial empire and political economy schools, which took nature merely as an industrial input. Rivers were there to be tamed, diverted, developed. Natural resources were seen as commodities in waiting. These frameworks minimized the social and environmental consequences of deforestation, mining, monoculture, irrigation, and hydroelectric development. The consequences were not part of the picture. To a later generation these "costs" of development had to be more accurately weighted in the historical balance.

In the gap between the functionalist assumptions of the commercial empire and political economy approaches and the critical spirit of the environmental movement, a new branch of historiography emerged in the 1980s which called itself "environmental history." Influenced by the writing of historians such as William McNeill, Alfred Crosby, and, in the case of rivers, Donald Worster, this group sought to restore some balance by putting nature back in history.[38]

A decade after MacLennan wrote his elegy "Thinking Like a River," Worster published an essay with the same title. As a young scholar he had become frustrated with the American historiography of his time: "There was no nature in their history – no sense of the presence and influence of the land on past human experience, no soil, no countryside, no smell of fungus, no sound of spring peepers trilling from the marsh at dusk. Historians seemed to have forgotten completely that, until very recently, almost all people lived as intimately with other species and with the wind and weather as they did with their own kind." Worster went on to become one of the founders of the emerging field of environmental history or, as he defined it, "the interdisciplinary study of the relations of culture, technology and nature through time."[39] His title is a conscious gesture to his inspiration, Aldo Leopold, the most influential conservationist of the preceding generation. Leopold had included an essay, "Thinking Like a Mountain," in his famous posthumously published collection *A Sand County Almanac*.[40] In an effort to illustrate the necessary connection between phenomena in nature, he showed how the elimination of wolves led to the proliferation of deer, which in turn deforested the slopes, which in turn led to the erosion of the mountain. Leopold wanted us to think like a mountain about the elimination of wolves.

Worster got more than his title from Leopold. Following his predecessor, he argued that humans and rivers think in different ways. Rivers "think" round; humans looking at them think straight. Drawing upon Leopold's observations on the circularity of natural systems in his essay "Thinking Like a Mountain" and his book *Round River*,[41] Worster noted that a river is but one part of a cycle through which water passes. The water in the river will flow past, but it will also go around and come back again. Changing one part of the cycle influences the rest of the system and all that will come after. "In water we see all of nature reflected. And in our use of that water, that nature, we see much of our past and future mirrored."[42]

Humans, on the other hand, have taken a narrowly presentist economic view of rivers: What work can they be made to perform? How can they be improved for navigation? What will they carry away? How can they be diverted for irrigation or dammed for power? Summing up two centuries of the exploitation of western rivers for power production and irrigation, Worster concludes: "The western river thus ends up becoming an assembly line, rolling unceasingly toward the goal of unlimited production." Water-taking and diversion on a massive scale to sustain energy-intensive industrial farms and thirsty cities in an arid landscape, by interfering with one phase of the cycle of water, have long-term detrimental consequences for the present and future environment. In this fragile landscape above all, humans must learn to think in the round, like a river, considering the implications of intervention on the whole water cycle. Doing so would help to break this habit of instrumental, linear thinking, which not only devastated the rivers of the west but also created a self-destructive society in their valleys.

Worster published this essay in 1984, just before the appearance of his definitive historical critique of the "hydraulic society" created by irrigated agriculture, *Rivers of Empire*.[43] In that book he reinterpreted the history of the American West in terms of mastery over water. Within this "hydraulic society" the rivers had been diverted and their wealth extracted, but the benefits had been unequally distributed. Moreover, the resulting social system remained precariously dependent upon a diminishing supply of water. After Worster, we have entered a new era in which rivers once again form a central subject of historical analysis.[44] His main focus is on the power relations resulting from control over water, but also on the power exerted by the consequences of environmental manipulation. His work has been a guiding influence within the emerging field of environmental history.[45]

Environmentalism has been extremely influential as well in shaping the bleaker, often apocalyptic vision in what might be termed the Paradise Lost school of rivers. The Whig approach to history characteristic of the Rivers of America genre has in some sense been turned on its head in the face of major environmental catastrophes on some rivers. In these more recent books the emphasis has shifted from what rivers have allowed people to do to what people have done to rivers and the devastation that has resulted. Philip Fradkin perhaps began this trend with his 1968 jeremiad, *A River No More: The Colorado River and the West*, a new edition of which appeared in 1995. Blaine Harden, in an angst-ridden personal essay, did the same thing for the Columbia in 1996: *A River Lost: The Life and Death of the Columbia*.[46] Fred Pearce punctuated the "hydro hubris" of much of the earlier writing in *The Dammed: Rivers, Dams and the Coming World Water Crisis*.[47]

An excellent example of this kind of extremely sensitive and scientifically informed criticism in a Canadian context is to be found in Richard C. Bocking's recent remake of the Fraser, *Mighty River*.[48] As he journeys downstream, Bocking catalogues the effects of extensive logging, habitat destruction, diking, draining, paving, ploughing, and diversion on the much transformed river. This Paradise Lost approach, a poignant and extremely powerful literature, often speaks directly to contemporary struggles over competing uses, conveying an overall tone of regret at irreversible change and devastation, but it occasionally admits of some hope that what little remains may be saved. In an ironic turn of events that puts the Rivers of America ideology into reverse, declensionist narratives about rivers are the most effective means of conveying environmental ideologies to a popular audience.

Sometimes, too, these books point out the ultimate futility of human attempts to control natural forces. John McPhee is a gentle and engaging master of this genre, as can be seen in his three essays on Mississippi flood control, Icelandic volcano management, and mountain erosion in Los Angeles, collected in *The Control of Nature*.[49] This approach shares much with what might be termed the "rogue nature" school of river history. Seen from this perspective, nature is a protean, untameable force that unexpectedly throws off feeble human traces to work its will. John M. Barry's bestseller *Rising Tide: The Great Mississippi Flood of 1927 and How It Changed America* is perhaps the best-known of the US examples.[50] In Canada we have Jack Bumsted's *Floods of the Centuries*.[51]

In the recent literature, nature and culture have often been put at odds and rendered as irreconcilable opposing alternatives: development destroys nature. Exhibit A of a long list of rivers lost has been the Columbia, whose salmon run, scenery, indigenous places, and riverine ecology have been drowned by sixteen mainstem dams to improve navigation and generate hydroelectric power. That is why Richard White chose that site to attempt a reconciliation of these traditional dualities in his influential little book *The Organic Machine: Remaking the Columbia River*. Rather than oppose nature and culture, he directs attention instead to the way in which they become inextricably intertwined. "Nature, at once a cultural construct and a set of actual things outside of us and not wholly contained by our constructions, needs to be put into human history," he begins. "We might want to look for the natural in the dams and the unnatural in the salmon. The boundaries between the human and the natural have existed only to be crossed on the river."[52]

Through the seeming oxymoronic metaphor, the organic machine, he follows the poet Ralph Waldo Emerson and the historian of technology Lewis Mumford in looking for the natural in human intervention and the contrived in natural systems. The Columbia has been put irredeemably to work, White observes, but it has not been destroyed. Rather, it has become a part-human, part-natural creation. "It is important that we get our metaphors right. We have neither killed the river nor raped it, although people claim both are true. What has happened is closer to a failed marriage. Nature still exists on the Columbia. It is not dead, only altered by labor. It is the steam within Emerson's boiler. It lies hidden in aluminium factories and pulp mills, in electric lights and washing machines. The organic and mechanical have been merged, as Lewis Mumford hoped they would, but the results have not been what he intended."[53]

And White concludes that "there is no clear line between us and nature. The Columbia, an organic machine, as virtual river, is at once our creation and retains a life of its own beyond our control." It is an archive of choices made. White and other practitioners of environmental history direct our attention to the changing relationship between humanity and nature, in this case a river.[54] Similarly, William Cronon, in several works but particularly in his essay "The

Trouble with Wilderness," has emphasized the way in which natural systems have been socially constructed. What is natural, he also argues, is to some extent contingent.[55]

In a Canadian contribution to this discussion, Matthew Evenden has published *Fish versus Power: An Environmental History of the Fraser River*. The Fraser is in some respects the exact opposite of the Columbia. "On the Fraser, fish have triumphed, not dams," and Evenden explains why. But he also shows that human choices as well as natural forces have been at work on the Fraser. The complete history of its changing nature requires an understanding of human agency, social action, and politics. Like the Columbia, he argues, "The free flow of the Fraser River bears the consequences of history."[56] A coalition of conflicting interests formed around the exploitation of salmon and hydroelectric potential. The triumph of the salmon coalition prevented the construction of dams on the mainstem of the Fraser, displacing them to other rivers. Though the outcome is different, the Fraser is as much of a human construct as the Columbia.

In different ways literature, art, and history illuminate the reflective power of rivers. Thinking like a river forces us to think of ends. There is a reason rivers are central symbols in religion and literature. A great river moves both the great American novel and, arguably, its great Canadian counterpart. We discover an image of ourselves mirrored in the history of our rivers, and we confront the knowledge that who we are has in a large sense been shaped by rivers and nature. It is not surprising, then, that rivers form central symbols in our culture, that several generations of historians have devoted themselves to writing the history of rivers, and that we too should want to enter the conversation.

Back to the Bow

A venerable tradition in Western thought polarizes the concepts of nature and culture, separating humans from the physical world. During the early modern period, according to Clarence Glacken, three main problems focused discourse on this question: the extent to which divine design animates and is revealed in nature, the ways in which the environment influences humanity, and the role and responsibility of humans as modifiers of the environment.[57] In the modern period, as Keith Thomas and others have argued, the idea that humans had a responsibility to employ and improve nature triumphed and, in the process, justified exploitation of the environment for the public good.[58] But as Donald Worster has documented, this "imperial" tradition of domination had to contend throughout this era with an Arcadian tradition fostering a peaceful and less disturbing coexistence with nature.[59] Throughout, nature has been conceived as something apart from us, something for us to use for our benefit or, alternatively, to subordinate our desires to.

Following the lead of recent work in the field of environmental history, we would like to blur the boundaries between nature and culture. Leaving to one side the philosophical conundrum that nature can only be apprehended through culture, we will argue that nature and culture have become so intertwined in history as to make a sharp separation between them arbitrary.

The Bow River has a history that is very much bound up with human imagining, engineering, design, and profit. It must be one of the most thoroughly engineered and regulated rivers on the continent. It has been developed and used in a myriad of ways for more than a century. It is not an untouched wilderness river. Quite the contrary: it has been dammed for mountain storage, hydroelectric generation, and irrigation diversion. Its flow is monitored and managed like an assembly line. Its waters have been taken in vast quantities for cities, industries, and farms and in large measure returned in a fouled state. Oil refinery residues, fuel oil, creosote, and sewage have polluted its waters. It has provided a transportation corridor for railways, pipelines, transmission towers, and highways. For sixty years, logs from the mountains were driven on its spring floods to mills in Cochrane and Calgary. Mines washed their coal and flushed their boilers with its water. Its fish have been propagated in hatcheries along its banks. It has been bridged and dredged, its channel reamed and diked. It has been mapped, measured, mathematically modelled, painted, and imagined in law and literature. Invisible lines of jurisdiction have been drawn in its water. Its banks have been industrial sites, junkyards, and parks. The Bow is about as "human-made" as it is possible to be.

Yet at the same time it remains a "natural" phenomenon. Tourists come from all over the world to gaze at it, photograph its falls, roam its banks, canoe its rapids and reaches, and fish its waters. The river, enrobed with national, provincial, and urban parks, is much admired by locals and visitors. Its river is now very much a central aspect of Calgary's self-image, though it has not always been so. The Bow has been a site of conflict over uses and abuses, but it seems also to have become a place of resolution and accommodation. And what humans have seen reflected in it has changed dramatically over time. As a result, the Bow occupies a middle ground between the extremes of "natural wonder," on the one hand, and "a river no more," on the other. It may not be typical, but its experience in this middle ground is quite common, and some of the stories we tell will, with variation, seem quite familiar in other locations.

The Bow is a partially human construct yet one that is almost universally regarded as a natural phenomenon as well. Over time, the natural and the cultural have become intertwined in complex ways. The remarkable Bow River fishery, for example, is not just the result of human restocking programs with introduced species; as we shall see, the fish have thrived in a river enriched by nutrients from Calgary's sewers. Introduced species have hybridized with indigenous ones. Every tree in the much admired riverside parks is encased in a metal shield to defend against, not human vandals, but natural ones, nuisance beaver.[60]

The Bow is a joint project of nature and human culture. In this book we want to go beyond our political economy origins to delve more deeply into what is happening to the river itself as choices are made over it through conflicts, agreements, daily use, or inadvertence. Further, we want to place culturally induced change into the perspective of a river as a naturally evolving, constantly changing system. Many human-induced changes – wetlands preservation through irrigation, for example, or fisheries revitalization incidental to stream-flow regulation – have been deemed "beneficial." Change in and of itself is not the issue; it happens no matter what. We want to explore how different parts of the river become socially constructed and change over time. We want to examine more closely the ways in which humans and nature act together to create new, intermingled forms and processes.

This environmental history attempts to connect the popular political "discovery" of the Bow River in the 1990s with the preternatural river that evolved after the ice age. The history of a river cannot be reversed any more than the flow of the river might be permanently held back. Once it has been incorporated into culture, it can never be returned to its prelapsarian, untouched state. Nevertheless, the Bow reminds us that even after generations of use, the river still lives and inspires, even as it is put to work. And some of the more debilitating processes can be changed if not wholly reversed. Finding a balance between use in all its many manifestations and respect for natural processes, to the extent they can be apprehended, seems to be the central problem, and a history of the Bow might help us see more clearly how such balances can be sustained.

We will not go so far as to claim that our study has led us to think like a river, but it has led us to think hard about a river. We started our inquiry with a work plan divided into twenty-two separate chapter topics. Early on, we learned we could not write a comprehensive history covering all aspects of the river, even for a small river like the Bow. We had to be selective, as all historians ultimately must be. We have winnowed our thematic chapters to twelve, conscious that we have left out some important matters – or in some cases, dealt with them elsewhere. We have not done justice as we ought to have to mining, scientific activities, mapping, or art, though art forms the subject of another book, *The Painted Valley: Artists along Alberta's Bow River, 1845–2000.*[61] Throughout, we have attempted to wed political economy and its emphasis upon power and economic development with a greater sensitivity to the environmental impacts and imbrications of nature and culture. Nor do we see the river as simply enchained by human interventions. It has exerted its wildness in the past, and its protean powers still remain beyond human control.

A river is an archive; it records and retains what has been done to it and by it. The condition of rivers is in some sense the measure of the societies dependent upon them. Environmental history is, in Leopold's terms, "round." It deals not just with the power humanity exerts over nature but also with the consequences of those actions and reciprocally with the ways in which nature impinges upon human cultures. Environmental history seeks a fuller view of "the place of nature in human life,"[62] and that is our objective too as we set out anew to discover the Bow River.

2 · Homeland and Margin

In the one hundred years before Euro-Canadian settlement, the Bow River was at once a homeland and a margin. As a homeland, it lay at the centre of an indigenous world. Along the river, the Blackfoot Confederacy, comprising Piikani, Kainai, and Siksika groups, as well as the Tsuu T'ina and Nakoda, lived, hunted, and warred. They incorporated the river into their seasonal rounds, crossed and recrossed it, and gained sustenance from its waters. The river flowed through their homeland and gave it form and meaning.

As a margin, the Bow existed at the ragged southwestern edge of a continental fur trade, into which the goods of the trade slowly diffused and from which were drawn the skins and meat of bison and the furs of beaver, wolves, and foxes. Traders rarely ventured into the region because its resources were judged to be meagre; more commonly, indigenous peoples carried meat and furs to posts on the North Saskatchewan River or, in later years, south to American posts on the Missouri. From the perspective of the fur trade, the Bow was a distant upriver tributary of the South Saskatchewan which, upon investigation and after much effort, yielded little to the trade. Posts were built on or near the river, but none lasted more than two years. On some fur trade maps, the Bow was described as "the Bad River." This marginal position began to shift by the mid-nineteenth century, when the Palliser expedition, sent on behalf of the British Colonial Office with support from the Royal Geographical Society, traversed the plains and began to think about rivers and land from the perspective of inventorial science with a view to agricultural settlement and transcontinental communications. The expedition signalled a new interest in the area and anticipated a major shift in the status of the river, its region, and its peoples.[1]

Historians typically write of this period of flux and change, of new cultural interactions and mutual discoveries, under the sign of "encounter." The term is simply descriptive – people from different backgrounds came together – and avoids assigning the status of observer and observed, or discoverer and discovered, to participants. It also assumes that different ways of knowing the world came into contact with one another and transformed both. Richard White's related and evocative term, the middle ground, suggests a space of active exchange and mutual influence, outside a formal state sphere, where cultural groups had to develop new ways of understanding one another to cement bonds of kinship, community, and power. The Bow River from the 1750s to the 1850s can be thought about in these ways. No one group controlled the basin. Power was asserted, subverted, and challenged by all. Groups who entered the region from afar had to make sense of those they met, negotiate, and come to terms with them politically. Indigenous peoples on the ground faced similar challenges.[2]

Maps produced in this period point to some of the complexities of this process. They highlight profoundly different cultural understandings of land, rivers, and space, but also some surprising commonalities. They underline the importance of lived worlds and of social goals in shaping the observations of and values assigned to the Bow River across time and cultures. The Bow entered the life worlds of indigenous peoples, traders, and explorers as a series of places to travel along, hunt and gather around, and pass. When we consider this period, we need to think not only about peoples coming together but also about ways of knowing and experiencing a dynamic river that was many places simultaneously.

Shifting Territorialities

In the nineteenth century, the Bow River was not the preserve of one particular group of people: it was a space of shifting, overlapping, and contested territorialities. Looking at the plains region as a whole, anthropologist Alice Kehoe reminds us of the remarkable difficulty of stating precisely the relationships between people and places: "The Plains in the nineteenth century, already well populated, became a refugium for Midwestern Indigenous Nations forced out of their homelands by advancing invasions of Anglo-Euro-Americans. Add to this the refugees from Spanish invasions into the American southwest, and Numa migrations out of the Great Basin. Now pepper the mix with European epidemics, entrepreneurs trafficking in horses, guns, slaves, ornaments, amulets, and foodstuffs; toss in gamblers and adventuring youths, and the task of disentangling ethnic history is formidable." Ted Binnema, a historian of the indigenous northwest plains, argues for a regional rather than a tribal approach because ethnic boundaries were constantly shifting as band societies intermixed and came into conflict.[3]

Between 1750 and 1850 considerable fluidity and change marked the cultural geography of groups along the Bow River. In the upper basin the Nakoda, a

Siouan-speaking group, held sway, combining a mixed economy of resource gathering in the Rocky Mountains with bison hunting on the plains. Nakoda groups straddled the Rockies and plains and extended north and east towards present-day Edmonton. Parts of this territory had formerly been used by Kootenai bands, who relocated further west among the core settlements of the Kootenai, possibly in the early eighteenth century. They continued to travel into the region and hunt bison on the plains, though at considerable risk of harassment from plains groups. Further down river, a coalition of groups known as the Blackfoot predominated, occupying different sites over the year. In 1750 Kainai, Piikani, and Siksika bands lived seasonally on and around the Bow, though they moved much more widely on the open plains in pursuit of bison, as far south as the Yellowstone (according to Blackfoot oral history) and east into sections of present-day Saskatchewan. By the early nineteenth century, the spatial configuration had changed as the Piikani pulled further south into the foothills and towards the Missouri, the Kainai occupied the area between the Bow and Belly rivers, and the Siksika lived in and around the Red Deer River. These groups spoke Blackfoot, an Algonquian language, and intermingled frequently with one another. A smaller group of Athapaskan speakers, the Tsuu T'ina, had migrated into the region in the early eighteenth century from the boreal forest to the north. By the early nineteenth century they occupied sections of central Alberta, and by mid-century they had shifted still further south to the Bow. They adapted quickly to the plains environment, hunted bison, and maintained an alliance with the Blackfoot.[4]

Although these groups held precedence on the Bow, others occasionally passed through in the course of fur-trading activities, on raids, or because they lived among the dominant groups. In the mid to late eighteenth century, for example, a Hidatsa group travelling north from the middle Missouri region built a fortified encampment on the north bank of the Bow near present-day Cluny. Whether the people who abandoned the site merged with Blackfoot peoples or returned south to the Missouri is unknown, though either explanation is plausible. In one of the earliest written documents describing the cultural geography of the Bow River, fur trader David Thompson recounts his experiences overwintering with a band of Piikani near the Bow River in 1787–88. Much of his account reports the life story of Saukamappee, or Boy, a seventy-five-to-eighty-year-old elder who was himself a Cree, born far to the east. In telling stories of battle, Saukamappee explained to Thompson the recent attempt to capture women and children of Shoshoni bands to the south to help recover from the losses imposed by the devastation of smallpox, which had cut through plains groups in 1781–82. That a young man, born in England, should be listening to an old Cree living among Piikani and explaining the capture and integration of Shoshoni says something of the cultural complexity of the Bow and the surrounding country. Between 1750 and 1850 the northwest plains remained a fluid zone of movement and interaction in which dominant groups exploited resources over a wide range. River basins could be a principal homeland, but they were far from bounded spaces.[5]

The arrival of horses, guns, and epidemic disease reinforced and exaggerated existing patterns of mobility and cultural exchange. Horses arrived from the southwest as early as the 1730s and were acquired at different times by various groups. Peoples of the northwest plains already kept dogs and used them as beasts of burden, but horses introduced a new dynamic. Saukamappee told Thompson stories that referred to the arrival of horses, originally called "big dogs" (in Cree) or "big elk" (in Blackfoot), and recounted that the Shoshoni could ride "swift as the deer," but he did not explicitly mention the period of adoption among Blackfoot peoples. In time, the Shoshoni lost their monopoly, and other groups adopted an equestrian culture. When Anthony Henday visited the plains in 1754, it is likely that he witnessed Blackfoot peoples on horse, though his nomenclature for these people (Archithinue) is ambiguous. Patterns of hunting, gender relations, communication, trade, and warfare significantly shifted as pedestrian peoples variously integrated horses into their seasonal rounds. Movement over the plains could be accomplished more quickly and over greater distances. The bison hunt could be conducted on the chase. And a greater variety and volume of goods could be collected and carried using travois. Horses became a fundamental attribute of wealth and prestige and therefore the focus of recurring theft, feuds, and warfare. They also changed how indigenous peoples interacted with rivers, crossed them, and looked for shelter and pasture on their banks.[6]

Guns entered the region later than horses and from the opposite direction. As early as the 1730s, some guns were acquired through trade connections with Cree to the east and north, though they were not used decisively in battle. By mid-century the supply of guns from Hudson Bay increased. Whereas horse-rich groups amongst the Shoshoni and Crow had extended their dominance north in the previous decades, now the balance of power began to shift. Armed Blackfoot bands pushed south towards the Oldman and Missouri Rivers and "replaced Shoshonis and their allies as the dominant military force on the northwestern plains." Saukamappee made his mark as a young man in part by introducing guns into warfare on the side of Piikani in their fights against Shoshoni. Because southern groups lacked easy access to guns, Blackfoot and their allies successfully extended and maintained a new southern range, encompassing the Bow River but reaching well beyond it.[7]

A third shift occurred when epidemic diseases, formerly unknown on the plains, entered the region. In 1780–82 a devastating smallpox epidemic originating in the south worked its way north. By October 1781 it had reached Blackfoot bands along the South Saskatchewan River. Hunting bands, living closely together in teepees, spread the disease rapidly among one another. Since the virus could remain infectious even on corpses or clothing and the incubation period was relatively long (one to two weeks), it diffused unimpeded along lines of social interaction, warfare, and trade. Traditional means of coping with fevers may have helped in some cases, but a common practice of entering sweat lodges and then plunging into cold rivers or lakes probably made matters worse. What followed was a period of rapid and devastating population loss, remembered in narratives recorded in the early nineteenth century as a time of suffering, want, and turning inward. Saukamappee reportedly told Thompson, "Our hearts were low and dejected, and we shall never be again the same people. To hunt for our families was our sole occupation and [to] kill beavers, wolves, and foxes to trade [for] our necessaries; and we thought of war no more, and perhaps would have made peace with them [the enemy], for they had suffered dreadfully as well as us, and had left us all this fine country of the Bow River to us." Amid the misery, the smallpox assisted the Piikani and their allies in asserting and extending control over the Bow and regions to the south.[8]

Other epidemics followed. Measles and whooping cough struck the plains in 1819–20, and another devastating smallpox epidemic occurred in 1837–38. Because the size of the plains population in this period is difficult to know, the demographic effects of these events are hard to figure. Jody Decker estimates that Blackfoot groups experienced mortality of 30–50 per cent in the 1780–82 smallpox epidemic. Small bands may have folded into larger ones, and some groups may have ceased to exist. Over the longer term, Maureen Lux suggests, plains groups showed a remarkable resilience: "In the sixty years between 1780 and 1840, no Canadian plains group experienced a continual population decline, despite at least nine major epidemics." However one calculates consequences, these epidemics disrupted social relations and imposed a grim burden on survivors.[9]

Amidst these considerable changes, Euro-Canadian fur traders began to make tentative connections in the region. Although the trade had operated in eastern North America for centuries, the advance of trading posts into the western interior occurred more slowly and operated principally along the North Saskatchewan River axis, from which traders cut north into the Subarctic. Most of the effects of the trade were indirect. The closest points of contact were struck along the northern and western edges of Blackfoot dominance at Fort Edmonton and Rocky Mountain House. Following Thompson's visit to the Bow River in 1787–88 as a representative of the Hudson's Bay Company, other traders edged into the region. In 1800 Peter Fidler established a post at Chesterfield House, near the confluence of the Red Deer and South Saskatchewan Rivers, and was quickly joined by representatives of the North West Company and the XY Company. But within two years, the house closed. After the North West Company merged with the HBC in 1821, a Bow River expedition sought to survey the country and re-establish a fort near the confluence of the Red Deer and the South Saskatchewan, but poor fur returns and what was judged to be a meagre local resource base caused the company to close the new Chesterfield House within a year (1822–23). In a letter sent to Governor George Simpson of the HBC, Donald Mackenzie, the author of the expedition journal, reported, "I fully consider the undertaking [the expedition] to have received a fair trial and the result I look upon as a total failure." A decade later, another attempt was made farther west. Between 1832 and 1834, Peigan Post (sometimes called Old Bow Fort) sat on the north bank of the Bow near the confluence with Old Fort Creek and sought to draw Piikani trade at a time of looming American competition to the south. But once again, traders beat a hasty retreat when faced with poor returns and dangerous conditions.[10]

Within the geography of the fur trade, the Bow basin held a marginal place, useful as a provision source but markedly inferior to colder regions to the north that produced higher quality furs in greater abundance. Because it did not lead west to an easy pass through the mountains, the Bow after initial exploration was not used for east-west transportation in the trade.[11] The region fell in a broad zone of indirect contact in the fur trade, in which the major trade items were bison meat, pemmican, and robes, rather than furs. Contact occurred primarily when Blackfoot traders moved north or east to trade at forts at the edge of or outside their range. By the mid-nineteenth century, pressure from American traders to the south began to be felt at the southern edge of the Blackfoot coalition on the upper Missouri. The establishment of Fort Whoop-Up along the Oldman River spurred on the whisky trade with devastating social consequences for plains groups. By the mid-1850s the demand for bison robes in the American trade would pull an increasing proportion of Blackfoot trade south, weakening connections with the HBC forts to the north and ultimately contributing to the tragic destruction of the bison across the plains.

These regional changes unfolded within a broader debate about the future of the western sections of the continent within the British Empire. The HBC's monopoly over commercial trade in Rupert's Land was due to be renegotiated in 1859, and a variety of interests in Britain and the Canadas wondered if it was not time for the company to relinquish that exclusive control and open the region for settlement and expansion. In the mid-1850s two separate expeditions were formed to examine the western interior and offer scientific assessments of its resources and potential. One emerged from the heated settler politics of the United Canadas, where a perceived shortage of land in Canada

West fuelled a vigorous discussion about the desirability of extending authority across the continent to encompass many of the lands currently controlled by the HBC. Led by a chemist and naturalist, Henry Youle Hind, this expedition focused on the eastern prairie, in present-day Saskatchewan and Manitoba, and never ventured as far west as the Bow. However, the second, imperial expedition, led by Captain John Palliser, funded by the Colonial Office, and sponsored by the Royal Geographical Society, did travel further west, crossing and reconnoitring the Bow, and worked its way at several points into and over the Rocky Mountains.[12]

The members of the Palliser expedition brought the perceptions and concerns of nineteenth-century inventorial science to bear on the landscapes they encountered. They sought to accumulate facts about the natural world, conduct a vast survey, and justify their science as utilitarian. As a son of the Irish gentry, whose only relevant experience was a bison hunt on the Missouri and Yellowstone Rivers in 1847, Palliser made an improbable leader in this cause. But his social position and contacts in the Royal Geographical Society and the Colonial Office did help to make the expedition possible. Once the idea gained momentum, however, it attracted personnel who had considerable experience and talent. Eugène Bourgeau, the expedition's French botanist, came with the recommendation of William Hooker, scientific director of Kew Gardens, a major centre of botanical collection, who looked warmly upon the prospect of receiving specimens from such an unknown region. James Hector, the expedition surgeon and geologist, had the backing of Sir Richard Murchison of the Royal Geographical Society, as well as a solid training at the University of Edinburgh in medicine, zoology, botany, and some geology. In the end, the expedition would include members of the Royal Engineers, as well as Metis guides from Red River and a shifting cast of indigenous guides who, on the Bow, were generally Nakoda. The expedition worked its way west in 1857 and spent part of 1858 and 1859 along the Bow, having first travelled to Rocky Mountain House to seek the permission of Blackfoot chiefs to cross their territory. The expedition's final report covered a much wider area, of course, and contributed significantly to the reimagination of the western interior as a fissured land divided into a fertile belt in its northern portions and an arid waste in its southern centre. On the Bow the expedition focused in particular on the possible travel routes into the Rockies and conducted wide-ranging botanical and geological surveys.[13]

Mapping the River

Rivers flowed through this world. Perceptions of space, place, and environment were fundamentally structured in relation to them. To gain a sense of

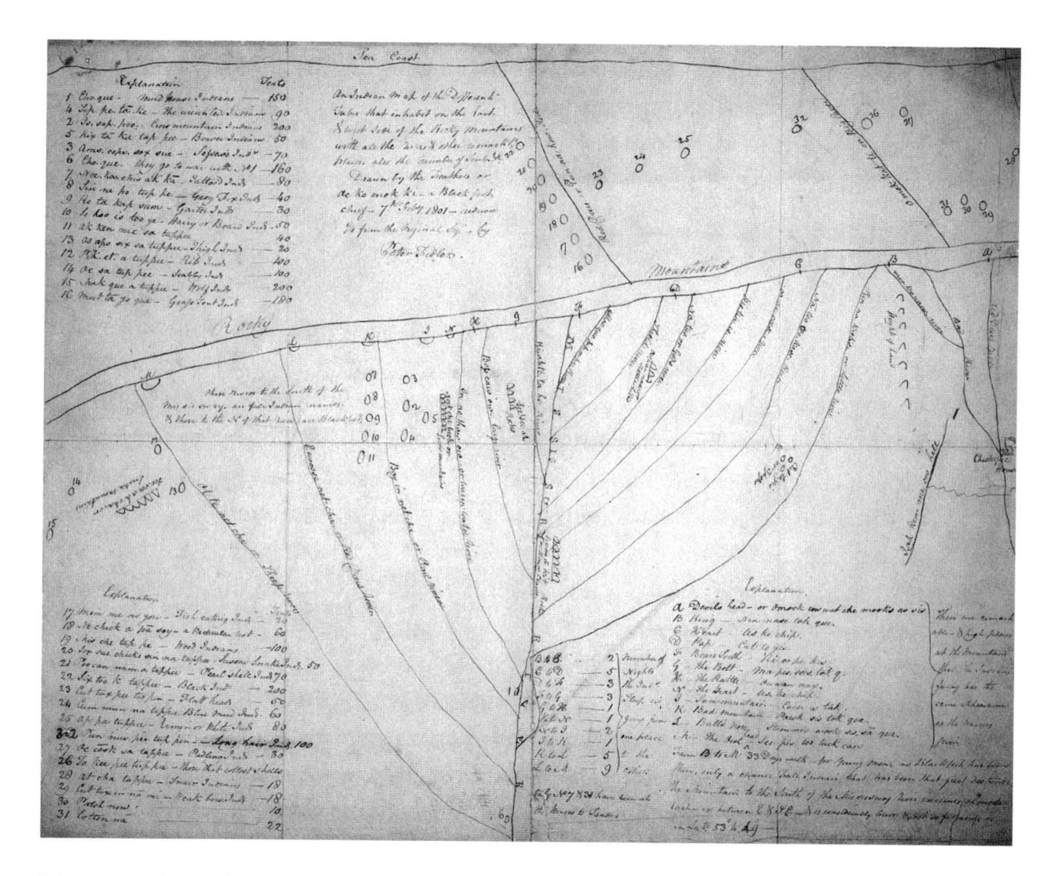

Map 2.1 The Ac ko mok ki map, redrawn by Peter Fidler, 1801. "An Indian Map of the Different Tribes that inhabit on the East & West Side of the Rocky Mountains with all the rivers & other remarkbl. places, also the number of Tents etc. Drawn by the Feathers or Ac ko mok ki – a Black foot chief – 7th Feby. 1801 – reduced ¼ from the Original Size – by Peter Fidler." (Hudson's Bay Company Archives, Archives of Manitoba, N4157)

how the Bow River figured in the geographical imaginations of indigenous peoples, fur traders, and inventorial scientists, it is useful to consult three early maps of the region: one based on a drawing in 1801 by Old Swan, a Siksika chief, reveals aspects of a known Blackfoot world in which the Bow figures as a northerly boundary or edge; another, composed in 1813 by David Thompson, summarizes the geography of western North America from the perspective of a surveyor engaged in the fur trade and places the Bow within a continental network of river systems and physical geographies; and a third, drawn in 1865 by James Hector of the Palliser expedition, inscribes the river basin with a new toponomy, resource assessments, human settlement sites, and expedition routes. All these maps reveal the significance of rivers as primary landscape signs, bounding space.[14]

The first map may be found in the Hudson's Bay Company Archives, one fragment of a collection made by HBC surveyor Peter Fidler, who ranged widely in the western plains and further northwest in the early nineteenth century. Old Swan drew the map at Fidler's request, to demonstrate a route taken by a war party travelling south. Subsequently, Fidler copied the map and circulated it within the HBC. The contents of the map were of interest to Fidler because they seemed to provide a first-cut representation of the region that offered insight and possibly advantage. According to Ted Binnema, much of this sampling and extraction misread aspects of the map because western cartographers had difficulty understanding the principles informing its representational strategies. Mountains were not plotted according to a bird's-eye view but as they might have appeared visually from the plains, as landmarks. Rivers were not traced to their sources, as in western maps, but to significant places. A convention of straight-line cartography represented the *mistakis* (or backbone), an important travel route, as a line running north-south, and rivers also as straight lines cutting east from the *mistakis* into the plains. This approach clarified the most important features of the landscape for way-finding and treated the twists and turns of the *mistakis* and rivers as irrelevant. Plains people did not use canoes or travel on rivers and therefore had little need for information about how river courses meandered. Straight lines made their point about spatial relationships more plainly than would have twisting lines. The map served its purpose, in the context of a story, according to a way of seeing.[15]

The second map now hangs in the lobby of the Archives of Ontario. Thompson drew a version of it for the North West Company at the end of his service, summarizing his geographical knowledge developed over the course of more than twenty years exploration, trading, and surveying in the western parts of the continent. Unlike Old Swan's map, however, it operates according to the geometrical principles of Western cartography, which seek to plot points according to astronomical observations and work out directions by compass. Thompson's rivers were drawn on the basis of field examination and the notes he kept during years of travel in the west, often led by indigenous guides. Although the ink on the original is badly faded, the Bow River appears, as well as several named upper-basin tributaries. Whereas Old Swan's map depicts rivers beginning from a certain point of cultural significance, Thompson's seeks to trace them to their physical source. While Old Swan's identifies important landmarks and the *mistakis*, Thompson's assumes an omniscient perspective and highlights mountains of significant elevation. Unlike Old Swan, who included elements of the cultural landscape, Thompson sketched the continent as physical space, without cultural elements, except names –

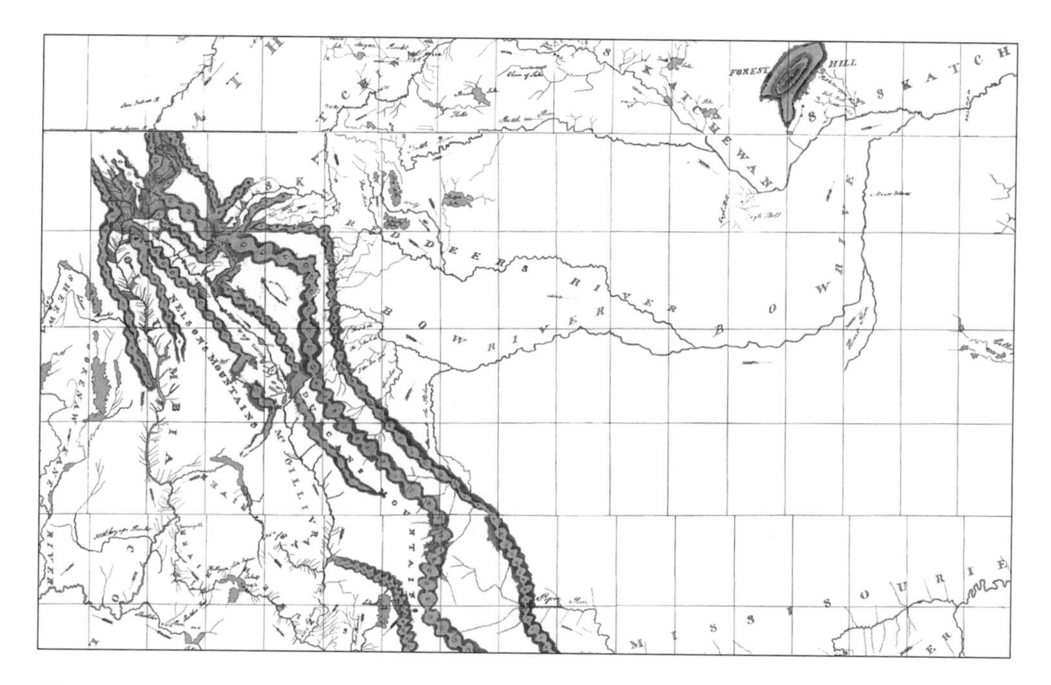

Map 2.2a A section of David Thompson's 1814 map of the North American west. "Map of the North-West Territory of the Province of Canada, from actual survey during the years 1792 to 1812. This map made for the North West Company in 1813 and 1814 ... by David Thompson, Astronomer and Surveyor, Sgd. David Thompson." (LAC/NMC 44302)

several rivers retain their indigenous nomenclature. Indigenous groups do not appear on the map, nor do trading posts.[16]

Hector's map seeks to narrate the Palliser expedition's progress and report its findings. Members of the expedition already had functional maps of the region to work with and so did not attempt to produce them anew. The expedition journals contain a host of astronomical readings as the members of the expedition sought to locate themselves within a previously defined, abstract space. Across the black and white print of the expedition report's map, Hector highlighted the routes with a bold red line, which along the Bow unfolded like a ribbon in various directions, north, south, and west into the Rockies. Along these routes, he named the leader and the date travelled, but gave no hint that indigenous guides had led or trail-blazed. However, Irene Spry, the editor of the Palliser papers, writes that "the Expedition travelled, to an extent that astonished its members ... over long-established trails." Alongside the etched lines which depict topographical features, Hector also inscribed names that the expedition had chosen en route, effacing indigenous nomenclature. And

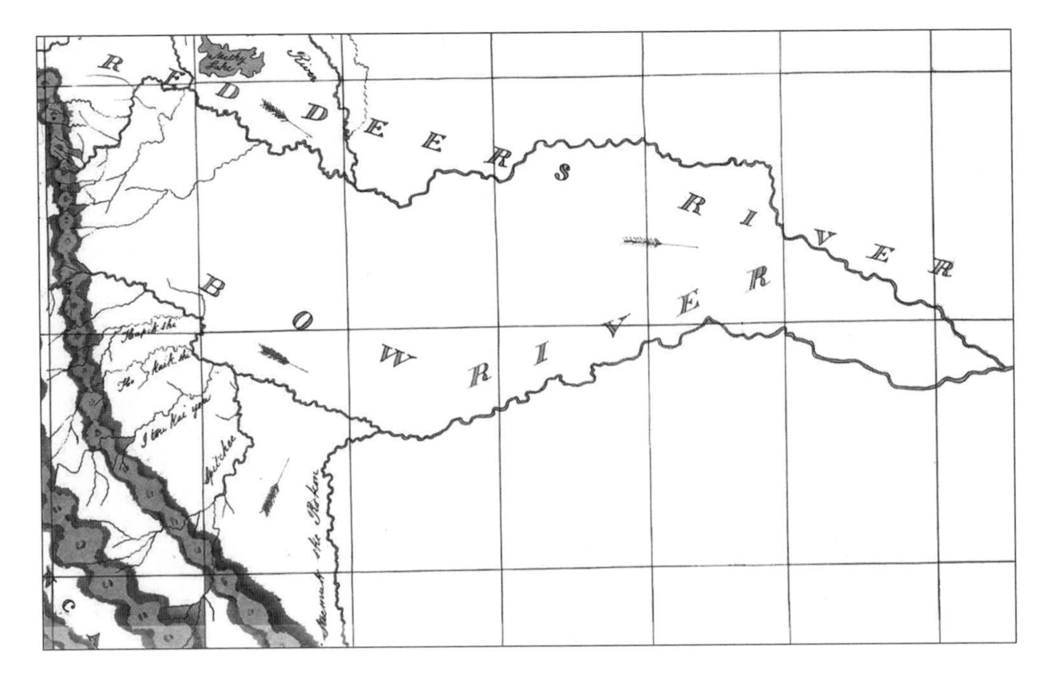

Map 2.2b A close view of the Bow River from Thompson's 1814 map.

so appeared such places as Mount Bourgeau, which Hector named in honour of the expedition botanist. Only a few indigenous names remained in the Bow Valley under his hand (Kananaskis and Ispasquehow, which was placed, parenthetically, after the Highwood River). Hector also inserted information about resource conditions, human settlements, and, in the Rockies, important passes. In the vicinity of the Highwood-Bow confluence, for example, he noted the presence of a Nakoda camp and described the local setting with the words "Great deal of Game & wild fruit; Val. Rocky & Wooded. Sandstones." The Bow crossed the expedition's division of fertile and arid land. South and west of the Highwood confluence, Hector located "good pastureland," but near the confluence with the Belly River, he wrote, "Arid Hills," and filled a blank space with "no water." The expedition's inventorial science produced a cartography that tended toward encyclopedic description, travel narrative, and imperialistic toponomy.[17]

All the maps are remarkable for their emphasis on rivers as fundamental features. Within this emphasis, the Bow appears as one among many. Old Swan's map sketches a route, but also a known area through which bands travelled and in which the Bow was an important element. Like Old Swan's map, Thompson's depicts a known area. Thompson had famously journeyed west down the Columbia and visited many of the rivers that he assiduously placed

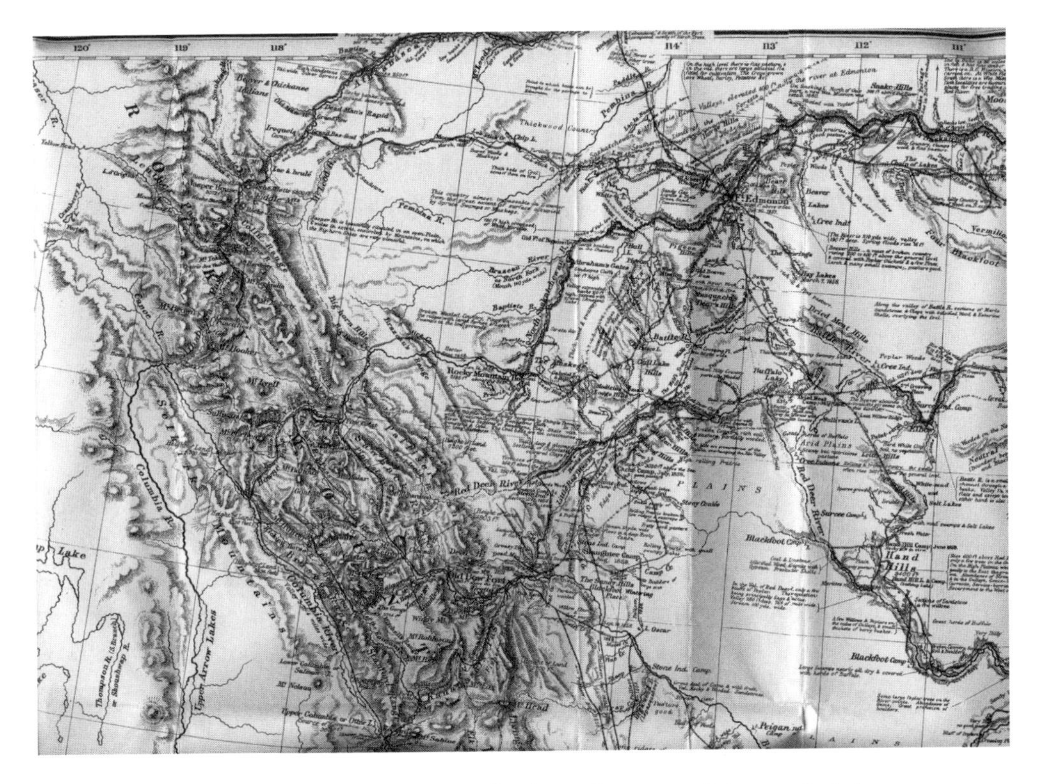

Map 2.3 A section of James Hector's map for the Palliser expedition, c. 1856–60. (Glenbow Archives, NA-789-46)

and drew. His personal travel narrative and therefore his map encompassed some of the outer limits of the fur trade's geography. The prominence of rivers in his map also speaks to the dominant concern of a fur trader to understand the interconnections among rivers for purposes of canoe travel. The bends in rivers that Thompson drew mattered as well for canoeists in a way they did not for plains groups travelling by horse. Rivers remain prominent in Hector's map, but the accent has shifted to emphasize land, which was carefully etched to reveal general topography and frequently described as to conditions and routes. Less frequently, Hector described changing river conditions on his map; "clear water" near the Highwood and "rocky current rapids" further north and west were the only two comments that ran along the Bow. Like the Blackfoot, the Palliser expedition did not primarily follow river courses or use canoes. It crossed and recrossed the open plains on horses, used carts to haul gear, and ascended difficult sections of the Rockies on foot. Hector admitted that he and Bourgeau followed the Bow River in their search for a pass through the Rockies in 1858 more out of curiosity than necessity: "We both chose this route as it allowed of our entering the mountains at once without travelling further in the open country, which yields little of interest either to the geologist or the botanist."[18]

These maps provide a first point of entry into the river of encounters, but behind them stood life worlds: of cultural interaction and change, of movement, warring, and trade, and ultimately of modes of interacting and living with rivers. To understand how a river mattered for indigenous groups, for a commercial fur trade, and in the vision of an expedition, one needs to ask how people encountered and made use of it.

Travelling with the River

To drink – this was the first necessity of peoples on the move in the northwest plains. Any indigenous groups venturing across the plains to hunt, trade, or conduct war had to know where water could be found or put lives at risk. Old Swan's map shows route ways and identifies the length of time between prominent points, several of which lie along rivers. "The Blackfoot," noted anthropologist John Ewers, "had knowledge of the locations of all running streams, clear lakes, and springs in and near their hunting grounds that afforded clean drinking water." Access to drinking water mattered for fur traders as much as for indigenous peoples. "In 1792," writes Ted Binnema, "[Peter] Fidler noted that the length of a day's journey was dependent on the distance between water sources." When the American fur trade began to reach north from the Missouri in the mid-nineteenth century, traders became intimately aware and concerned about the distances between rivers and the likely directions in which one might find water when crossing the open plains.

Alexander Cuthbertson's narrative of a trip north as far as the Bow in 1870 focuses primarily on the way in which he found his route and the struggle from river site to river site to find enough water to keep men and animals alive in the summer heat. Sometimes low-lying water would be found that was so brackish that his mules would not drink it. On approaching the Belly River, Cuthbertson reveals something of the desperation involved: "With anxiety and elated with the certainty of soon drinking and bathing in the cool water fresh from the snowcapped peaks of the Rocky Mountains, we hastened on our jaded and famished mules." Palliser also knew that worry and sense of relief. In August 1858 his party approached the Bow River after crossing the "arid plain." By nightfall they came upon a salty lake; its water was undrinkable. "When we awoke … we found ourselves about two miles distant from the river saddled up and hurried down there as fast as possible."[19]

Rivers provided way-finding points and sources of water, but they also had to be crossed. In the upper sections of the Bow, care and experience were required. The Wesleyan missionary Robert Rundle almost died in May 1847 on his first time across the river because, he confessed to his journal, he had been "too rash about it" and did not know how to lead his horse properly. He was also probably crossing a surging spring freshet. "After the Almighy preserved me," he wrote, he knelt and prayed on the river's bank. To judge by his journal, he made subsequent crossings by raft. Upper basin tributaries could pose difficulties too. When James Hector and his crew crossed the Ghost River in 1858, they "were obliged to seek a shallow rapid at where it joins Bow River, and it was only with the greatest care and trouble that we were able to get the carts down the steep bank and pass this point."[20]

These were difficult conditions, but they were more challenging downstream, where the river could flow high and broadly, particularly in seasonal floods. Trying to ford the river, carrying goods, leading children and the aged, all the while guiding horses, presented a potentially dangerous situation. One ethnographic account from the 1870s describes Blackfoot crossing a stream in Montana: "they made a kind of float from the skin covering of the lodges, upon or within which their effects were placed, men, women, and children swimming, the warriors towing the floats by a cord held in the mouth. Such horses as were fit for that service were also made to do duty in transporting their riders. By these means a village of 500 lodges would cross a considerable stream within an hour's time." One of anthropologist John Ewers's Kainai informants, Weasel Tail (b.1859; d.1950), told him in the late 1940s that "in his youth a man was careful to choose a horse known to behave well in water." Even as rivers guided travellers and provided them with essential fresh water, they also presented significant barriers and dangers.[21]

It is for this reason that the two principal fords on the lower Bow became some of the most significant Blackfoot sites on the river. At Blackfoot Crossing (or Soyo-powah'ko, which translates as "ridge under the water") a broad valley, well supplied with cottonwoods, was joined by an underwater bridge of boulders. On either side of the river, there were good sites to set up teepees and gather in large groups. In addition to its functional and strategic role as a point of safe passage, the crossing became an important social node and rendezvous point in the northwest plains: a place of contact, exchange, and festival. In the winter, Siksika used it as a camping place. American fur trader John Cuthbertson judged Blackfoot Crossing to be "the only good ford on the river" – he was apparently unaware of another ford further north and west. The Blackfoot also crossed the Bow near the confluence with the Elbow

River, located within the centre of present-day Calgary. At this point a series of Blackfoot trails converged. Both Blackfoot Crossing and the ford at Calgary proved to be enduring and important sites for Blackfoot groups and newcomers alike. Of all the possible points of meeting in southern Alberta, it was at Blackfoot Crossing in 1877 that Blackfoot, Tsuu T'ina, and Nakoda chiefs met Canadian officials to sign Treaty 7. Blackfoot Crossing would subsequently be placed within the boundaries of the Siksika reserve. The Blackfoot trails that converged at the ford near the Elbow provided an early focus for settlement at Calgary. River crossings also became sites of encounter.[22]

Fur traders brought horses with them to the Bow River, but their journals and written documents speak less about the river as a barrier or crossing than as a means of travel by canoe. David Thompson's journal of a trip along the Bow in 1800 contains numerous references to the conditions of the water, its speed, and its navigability. All these observations relate back to his calculations about the potential for trade in the area. On 29 November 1800, for example, he ends a long description of the upper river and its falls with the comment "the River every where navigable for large Canoes, loaded with 20 Pieces and so even at this time of Year, when the Water its lowest." Thompson's observation strikes an optimistic note that was not shared by Donald Mackenzie, who recorded the progress of the HBC Bow River expedition in 1822–23. Like Thompson, the expedition aimed "to ascertain how far it is practicable for Boats or Canoes to be navigated up those Rivers." It managed to move about freely on the Bow and Red Deer Rivers, but its members found the Belly River dry and reported that they had heard "on good authority" that it was navigable only in August. They dismissed it at once as a suitable means of transit and included the Bow in their negative depiction.[23]

Hunting Grounds

If rivers posed a barrier to passage because of their depth and width, then bison hunters found advantages in their steep cliffs and cutbanks incised by glacial meltwater. Rivers and associated landforms were woven into the bison hunt. Whereas in the final stages of the bison era hunts conducted on the open plains were made possible by swift movement on horses, in earlier times hunters had sought to organize the bison hunt in drives. These funnelled the charging bison into discrete channels, constructed with blinds and fences of various kinds. Hunters aimed to force bison to stampede into traps or pounds in which they could be easily picked off with bow and arrow. At other places, often along rivers, hunters drove bison off cliffs. The efficiency of this method sometimes produced a condescending reaction among Europeans who witnessed it. Alexander Henry the Younger, a fur trader who came into contact with Piikani during a westward journey in 1811, described hunting with bison jumps as evidence of indolence: "So much do these people abhor work that, to avoid the trouble of making proper pounds, they seek some precipice along the bank of the river, to which they extend their ranks and drive buffalo headlong over it." The most famous such site is Head-Smashed-In Buffalo Jump south of the Bow River, but several upper valley locations were used in this way, possibly more intensively in the pre-equestrian era. In Peter Fidler's translation of an 1802 Blackfoot map of the Oldman and Bow Rivers by Ak ko wee a ka, a buffalo jump is identified along the banks of the Oldman River as "Steep rocks river – where Buffalo fall before and *break their skulls in pieces*" (emphasis in original).[24]

In addition to the use of the riverbanks as an element in the hunt, rivers attracted game needing to drink. As Helen Meguinis, an elder of the Tsuu T'ina Nation, recently put it, her people followed the rivers "because the buffalo have to drink from the water." Fur traders made no significant use of the river's landforms or cliffs, but they shared an interest in the river as a hunting ground. David Thompson's Bow River field notes in 1800 may be read in part as a menu of animals hunted, slaughtered, and eaten along the way. On 29 November 1800 he and his crew crossed the upper Bow to the north bank. While Thompson set up his observational equipment, "Mr McGillivray, a Man & the Indian set off ahead to hunt the Mountain Goat, several Herds of them feeding a little Distance before us in our Road. Went on … having amused ourselves the whole After noon with running after Goats – 3 of them were killed by the Indian & a large old Buck by Mr McGillvray, which we skinned round – we found their meat to be exceeding sweet & tender and moderately fat." One gets the impression from Thompson's notes that the river acted as a magnet for game that could be easily killed. The record of the fur trade in the region, however, suggested that the commercial trade for furs could not be conducted so easily as Thompson's hunting day. Although both Chesterfield House and its successor, New Chesterfield House, took in a respectable number of furs, scouts who travelled up the Bow in 1822–23 found few beaver and judged indigenous reports of plenty to be exaggerated. Palliser's expedition, three decades later, never ran short of meat on the Bow, but there were lean days, waiting for Nimrod, the Nakoda guide, to bring back a moose captured at a distance.[25]

In search of shelter and seasonal foods, plains groups moved to river valleys as winter approached. Other plains hunters travelled north into the parkland and boreal forest, but for Siksika, Tsuu T'ina, and Nakoda bands, the migration progressed westward. Larger summer groups dispersed into smaller band

and family units. In winter, valleys provided some obvious advantages to the open plains: protection from the worst winds, local sources of wood for fuel, fresh water, and a range of local game also seeking shelter, ranging from antelope to deer to bison. Over a long winter, the availability of fresh food to supplement stored bison meat and berries was important. As horse-keepers, these groups also had to consider the availability of water and pasturage. With hobbles removed, horses could dig through snow to find grass. When no grass was available, they were fed the bark and small branches of cottonwoods that thrived on the river's banks. The search for pasturage might account for one or several relocations of camp during the winter. Although horses could eat snow, this consumed precious calories. As opposed to warmer seasons, when a considerable proportion of horses' water needs were met by grass consumption, in winter water availability was crucial. After the destruction of the bison and the signing of Treaty 7, some of these winter camp areas were set aside as reserves. The Nakoda reserve, for example, was centred on Morleyville, a former winter camp upon which Methodist missionary John MacDougall sited a church in 1874 and around which he encouraged Canadian government officials to place the reserve in the treaty negotiations for which he served as translator.[26]

The wood and bushes that could be found in river valleys such as the Bow provided more than fuel for fires in winter. Water birch trees were cut to make sweat lodges, and poplars were used for the centre poles of ceremonial lodges. Women cut lodgepole pines found in the upper Bow annually to renew teepee and travois poles and made bowls with them as well. Brush was often used in constructing markers to channel bison into pounds or over cliffs. Valleys were also the site of seasonal picking of berries, which would be eaten fresh, dried for winter consumption, sometimes mixed with bison grease and meat as pemmican, or stored in leather pouches with grease. Other berries and tree barks were important as medicines and teas. Ethnobotanist John Hellson noted in the mid-1970s that Siksika informants differentiated between three varieties of Saskatoon berries – river, coulee (cutbank), and mountain – preferring those found in coulees. Little Bear (Ki oo cus), a Siksika chief, drew a map of the northwest plains for Peter Fidler in 1802 which identified numerous sites, including two on the upper Bow, that were said to have "plenty of berries." Although bison hunting structured seasonal movements, the wealth of plants to be found in river valleys in all seasons contributed importantly to diets and material culture. Wood was a critical winter resource.[27]

Fur traders found less of use along the river. Because many of their materials, from nails to rope to medicine, were shipped in, they viewed the river and its valley with a narrower set of needs in mind. Principally, they needed wood for building the structures of the fort as well as defensive palisades. Near the lower river, where Chesterfield and New Chesterfield House were sited, the lack of good timber for building caused frustration and complaint. Fidler wrote on 25 September 1800 that "the woods here are few and bad for building with." The following day, however, he came upon a stand of poplars near the confluence with the Red Deer River that would suffice. When Donald Mackenzie set about building a fort near the same point twenty years later, he was less sanguine. The palisades, he wrote, "are far from being good[,] the wood in general being large and crooked tough Poplars." With perseverance, members of the Bow River expedition cut and squared timber for building and, by and by, raised a fort. They let it be known in their official record of the expedition, however, that the river was a poor place for such a critical resource. Peigan Post, established further west and north, could depend on a better timber supply, and Thompson early reported on the good stands of fir to be found in the middle basin stretching high up the shoulders of mountains into the Rockies. The Palliser expedition had little immediate need for wood, except for campfires, but members of the party proved more approving of the timber potential along the Bow, perhaps because they spent more time in the upper valley than the fur traders. While they similarly dismissed the lower river as part of an arid waste with little wood but poplars, the expedition journals contain a range of comments about the potential supplies of fir, pine, and even cedar. Mostly, the observations were of a general character, but sometimes Palliser would measure the circumference of a tree or muse on the history of forest fires along the upper Bow. Below the forest canopy, and less interested in utility, Eugène Bourgeau busily collected his floral specimens for eventual transshipment to Kew Gardens.[28]

Although the Bow River is today known for its trout fishery, of all the indigenous groups and traders living along the Bow River, only the Nakoda invested time and effort in fishing. Using spears, bone hooks, and lines of twisted willow, as well as fish traps, they captured a range of species. The fact that they spent more time in the mountains, near smaller streams and lakes teeming with fish, no doubt made the opportunities easier and more rewarding. Whereas bison could be stored, fish were generally eaten fresh. Although other groups might have resorted to fishing in lean times, generally they avoided it. During the summer, when fishing would have been most easily conducted, they were out on the plains following bison herds. This temporal obstacle coincided with cultural prejudices. "Fish is poor food" went one statement attributed to Arapooash ("sore belly"), a Crow, in 1830. The Bow River expedition journal never mentions fish. Thompson's notes of his 1800 trip along the river alludes to several types of game but no fish. Like bison hunters, fur traders had little time or energy to devote to fishing. By contrast, the Palliser expedition ate fish, caught by Nakoda guides, several times along the Bow. Hector was impressed when he wrote, "Our Stoney boy shot several trout as we came along [the river] with his arrows."[29]

Passages

The upper basin, in which the Nakoda predominated, was used differently in part because of its vastly different terrain. Against the gentle decline of the plains with its vast herds of bison, the Rockies stood out like a sharply pointed wall facing east. The farther up the river one went, the more vegetation, topography, and travel conditions changed. Grasslands and low-lying stands of poplars were replaced with firs, alpine meadows, and exposed sandstone faces. Although the height of the Rockies proved a barrier to traffic east and west, the Nakoda effectively transcended this obstacle through several passes, conducted hunts on both sides of the mountains, and traded with Kootenai groups farther west and south. Kootenai groups also passed through the region in the course of hunts and to engage in horse trading and theft on the plains. In the opposite direction, Piikani moved up the Bow and crossed the Rockies on warring expeditions against the Kootenai and with the intent of stealing horses. When David Thompson stayed with a Kootenai group on the Kootenay River in 1810, a group of Piikani threatened to attack. Traders rarely crossed the mountains via indigenous routes along the Bow, preferring to cut south from Rocky Mountain House, bypassing the Bow entirely and linking up the Columbia River trade with the northern axis on the North Saskatchewan River. For traders and for many surveyors who would follow them, the Rockies in the upper Bow did appear like a wall, even if it was one that Nakoda and other groups scaled frequently.[30]

Part of the Palliser expedition's aim was to discover usable routes through the Rockies to provide guidance for future land-based travel to the Pacific. It therefore surpassed the probes of earlier traders into the upper river and Rockies and over two years travelled through what came to be known as the Howse, Kicking Horse, Vermilion, and Kananskis Passes, reaching the headwaters of the Kicking Horse and Kootenay Rivers. Even with Nakoda guides, the going was tough. Apart from the grade, blown-down sections of forest impeded movement, unstable sections of shale proved dangerous, and the river itself threw up challenges, with impressive and dangerous falls in some sections and slack marshes in others slowing horses to a crawl. For Hector, the mountains provoked excitement and awe. "The snow of the mountains with the foreground sharply lined by projecting ledges of rock was quite exhilarating after the dreary monotony of the arid plains," he wrote on first sighting the Rockies up the Bow valley in 1858. Viewed from the Kananaskis Valley, the mountains were, in Hector's language, "bold, grotesque peaks"; after the party traversed Vermilion Pass, the "mountains form cubical blocks or ranges of battlement like precipices." For a geologist, these views were heaven-sent, but the travel was never pleasant. In the end, the forward-looking members of the Palliser expedition found it implausible to think that railways could ever cross the Rocky Mountains.[31]

In the period of encounter, mobility, not sedentariness, defined lives and ways of interacting with the environment. Rivers were so deeply embedded in cultural patterns and processes that they framed cartographic representations of the world, taking on far greater importance than any other landscape feature. People hunted along rivers, drank from them, and camped on their banks. Rivers were a barrier to travel for indigenous peoples, as well as a way-finding mechanism; for fur traders, they were the principal means of transit; for members of the Palliser expedition, they were both a route way and a store of resources. Take away rivers and the plants and animals that found shelter in their valleys, and one would lose basic elements of indigenous material culture, food, and habitat. Bison hunters of the plains are rarely described as river-dwelling people, but rivers were fundamental to their patterns of mobility and economy. Without them, traders would have had far greater difficulty penetrating the northwest plains and would have had no ability to transport furs, bison skins, or pemmican. Rivers were least important to members of the Palliser expedition, and yet their routes and observations reveal a keen interest in the Bow, its passes, and the surrounding valley. Despite the common importance of the Bow, it held particular values and meanings for different peoples. For indigenous peoples of Blackfoot, Tsuu T'ina, and Nakoda groups, the Bow was a principal territory and homeland, whereas traders viewed it circumspectly as a place of want and poor furs, worth defending and occasionally exploring but rarely with much intention of permanent engagement. For members of the Palliser expedition, the Bow provided a route to passes through the Rockies and carved a resource-rich valley inviting survey and settlement.

The Bow was marked by changes in the century between 1750 and 1850: horses, guns, and epidemic disease recalibrated political control of the region and introduced new ways of living. Newcomers came initially to trade but later to assess the country with plans to dispossess the original inhabitants. After 1850 the mobility of peoples in this landscape would become increasingly circumscribed. The remarkable and precipitous decline of bison, the occurrence of another and devastating smallpox epidemic in 1870, and the advance of settler colonialism, represented symbolically by the signing of Treaty 7 in 1877, all combined to reduce significantly the viability of mobile, indigenous economies along the Bow. Indigenous peoples would not disappear from the Bow – far from it – but their characteristic patterns of living would be radically changed and their ability to move about curtailed. Behind the traders and explorers came a new set of forces, including armed police, missionaries, ranchers, and eventually agriculturalists, who would have different ways of living on the river. In the next chapter, we begin to examine how settlement – sedentary living – remade the river and how the river shaped settlement.

3 · Home on the Range and River

Human activity had an impact on the Bow valley well before the arrival of Euro-Canadians, but the pace of human intervention quickened as a new population appeared to take part in what historians have traditionally described as "western settlement." Conversely, the natural environment continued to create opportunities and impose limits on human activity during the settlement era as it had done before.[1] Although historians have studied certain aspects of this process, writing the environmental history of the repopulation of western Canada has only begun. As a contribution to that unfolding story, this chapter examines the Bow River as a significant force in the settlement of southern Alberta between 1860 and 1930 and the ways that humans with differing needs, means, technologies, and economies adapted to the challenges and opportunities of living with a river.

The word "settlement" has misleading connotations of the continuous gathering, calming down, and gradual buildup of ordered society. We want to emphasize at the outset that settlement in southern Alberta was not an orderly, linear process of accumulation but, rather, a series of rising, falling, competing, enduring, and overlapping socio-economic regimes. Within this sequential process, different social configurations, economies, and agricultures placed different values upon the river, perceived its attributes in different ways, and oriented themselves towards it accordingly. A tight focus on the river as a factor in human occupancy allows us to see both the differences across social regimes and some of the continuities that persisted in the human-river relationship. We do not intend to retell the story of western settlement yet again, though we think the time for critical revision of the standard account is long overdue. Our focus will be upon the environmental history of the river in the complex settlement process, but to do so we must first establish some context.[2]

The New Order: Succession and Persistence

The history of settlement in the Bow valley begins with an environmental catastrophe. For generations, indigenous peoples had controlled access to the northwestern prairie and the resident herds of bison upon which they depended. The Blackfoot Confederacy reached the peak of its power in the mid-nineteenth century. That position began to weaken in the 1860s, when merchants and hunters from the southeast began to penetrate Blackfoot territory, trading whisky for bison hides newly valued as winter robes and for leather belting in the industrial United States. An international boundary, the Medicine Line, notionally divided the Blackfoot into two countries, but they continued to move about hunting bison, trading, and raiding, indifferent to its existence. The Blackfoot in Canada permitted a small detachment of mounted policemen to establish posts in their region in 1872 to control the often unscrupulous and socially disruptive whisky traders.[3]

Indigenous leaders grew apprehensive about the future as they observed a steady decline in the number of bison and an increase in Euro-Canadian migrants. In 1877 the Siksika, Kainai, Piikani, Tsuu T'ina, and Nakoda reluctantly sought protection from the government through a treaty which they believed gave them exclusive control over strategic territories, the right to continue hunting on their traditional lands, and guarantees of survival.[4] Government officials saw the treaty as a way of gaining title to the land and the settling of indigenous people in reserves to prepare them for the time when the bison would be gone, perhaps ten or fifteen years hence. But the bison numbers declined precipitously. In the 1880s the Blackfoot were forced to range much further south into their enemies' territory in search of food; then even the southernmost herds disappeared through over-hunting, environmental change, and other stressors.[5] An unthinkable catastrophe, the disappearance of the dominant mega-fauna on which their life depended, suddenly undermined established ways. The Canadian Blackfoot retreated to their reserve lands, where they became dependent upon the government for rations and sought to learn new agricultural methods. After the Riel Rebellion in 1885, the Canadian government introduced a pass system to limit and monitor the movements of indigenous peoples off the reserves. The system may not have been effective or consistently applied, but it signalled the concern of the surrounding settler society and of state authorities to impose a new form of control over indigenous lives.[6]

Gathered together for the first time in houses and teepees made of canvas and weakened by hunger, indigenous peoples suffered a further devastating blow as tuberculosis ran rampant through their communities. Within a generation a powerful group of hunter-gatherers who had dominated the prairie and intimidated the advance guard of Euro-Canadian settlement had to adjust to a new sedentary mode of living, overseen by Indian agents and missionaries and shaped significantly by fitful and inadequate state welfare.[7] The Siksika, whose reserve sat astride the Bow River in the heart of their traditional terri-

Illus. 3.1 The Bow River Horse Ranch near Cochrane about 1900. (Glenbow Archives, NA 2084-43)

tory, were thus involuntary settlers on their own land, their numbers reduced and their culture fundamentally disrupted by the double blow of an environmental disaster and a demographic crisis. In due course, the culturally distinct subgroups, the Tsuu T'ina and the Nakoda, acquired reserves of their own on the Bow further west.

The destruction of the bison and the marginalization of indigenous peoples opened a vast new territory, which was rapidly infiltrated by a new economic and social organization from the south. Open-range ranching had been developed in Texas and had spread northward to Montana on the lush grasslands on the plains and rolling foothills in the eastern lea of the Rockies. Senator Matthew Cochrane is credited with convincing the Canadian government that leasing land to large ranching companies promised the fastest and most profitable way of opening the far west.[8] Hardy cattle were set loose on the plains with a minimum of attention, and herds multiplied on the ample summer fodder. Cattle fended for themselves in the winter on the cured grasses kept clear of snow by the chinooks. In the spring the cowboys rounded up their cattle to identify the calves. In the fall they drove the fattened cattle to market. When all went well, it was easy money and not much work. The cattle and the form of organization originated in the United States. The capital, management, and the labour – the cowboys – came from eastern Canada, Great Britain, and the United States. Senator Cochrane located his first ranch – or ranche, as the Canadians and British preferred – in the well-grassed foothills west of Calgary. Huge leasehold ranches were also opened up to ranching companies in a line north from the border to the Bow River on the prairie and foothills and following the Bow River valley eastward below Calgary. Within a few years, hundreds of thousands of cattle replaced bison herds on the open range. A new economic organization, the corporate ranch, managed and operated by a small population of largely male immigrants oversaw production. The cattle were sold and butchered locally to supply rations to the indigenous peoples, provision the government police force, and feed the growing local population. After the coming of the railway, these cattle reached a much wider market farther east on the prairie, in eastern Canada, the American Midwest, and even transatlantic destinations.

The cattle kingdom, with its iconic proletariat of knights on horseback, flourished for a short time, but it too collapsed, undermined by challenging environmental conditions, the ambitions of small ranchers and homesteaders, and the behaviour of the leaseholders themselves. The land looked like ideal range country, and it was most of the time. But it was precariously situated on the northern margin of country suited by climate for open-range ranching. Every few years hard cold winters or heavy snows unrelieved by chinooks wiped out whole herds and pitched the ranching companies towards bankruptcy. Cattle froze to death or more often starved because they could not paw through even a thin layer of snow, like horses or bison, to reach the grass. Ranchers had to turn to the more labour-intensive process of putting up hay for winter, which added considerably to costs, as did the rounding up and herding in winter quarters. The economics of winter feeding favoured smaller ranches, which former ranch hands, ex-mounted policemen, and homesteaders seized every occasion to locate on particularly desirable land – around water holes on the prairie and valley bottoms along the southern prairie rivers, especially the Bow valley.[9]

With a strand of barbed wire strung across a bow in the river, the small rancher or homesteader could pre-empt hundreds of acres of prime summer and winter forage. But those strands of wire also reduced the value of the leased land on the high plains, which now lacked access to water and shelter. That is why the large leaseholders were sworn enemies of squatters, homesteaders, and settlement in general. The mounted police spent a great deal of their time in the unheroic task of dispossessing squatters on the ranchers' behalf.[10] Tensions flared between the two interest groups, a struggle eventually won by the

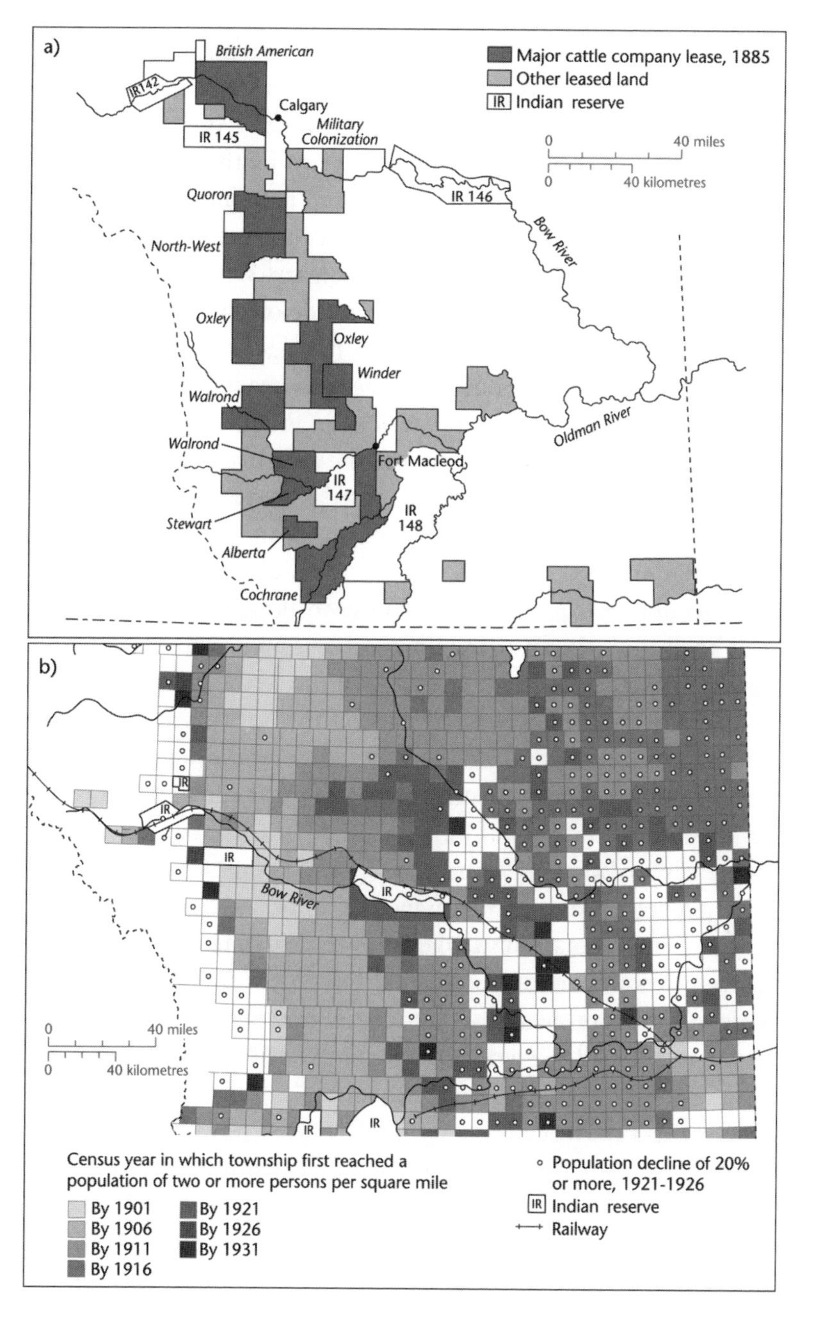

Map 3.1a Leases of major cattle companies, 1885. (Adapted from Simon Evans, *The Bar U: Canadian Ranching History* [Calgary: University of Calgary Press, 2004], based on Department of Interior orders-in-council)

Map 3.1b Population change in the southern prairies, 1901–31. Note the abandonment that occurred in the lower Bow Basin during the 1920s. (Adapted from William J. Carlyle, John C. Lehr, and G.E. Mills, "Peopling the Prairies," in *Historical Atlas of Canada*, vol. 3, plate 17) homesteaders and the small ranchers. After 1896 the new Liberal government, in response to pressure from aggrieved settlers, reduced or cancelled the leases and opened up most of the territory to homestead or freehold occupation. The large ranchers themselves contributed to their own downfall by the high-handed obduracy with which they fought off their rivals and by their own tendency to locate their freehold acreage on these very river bottoms. While squatters were having their houses and outbuildings torn down by the police, the ranchers and their friends were settling in unmolested. Slowly during the 1880s and more rapidly in the next decade, despite and in part because of the large ranchers, the most desirable lands along the Bow River began to be taken up by small ranchers and determined homesteaders as open-range ranching and leasehold tenure collapsed.[11] Even though leased acreage retreated, the total number of cattle tended by small ranchers continued to climb from 1901 to 1921.

The coming of the railway contributed first to the flourishing of the cattle kingdom and then to its eventual disintegration. In 1882 surveyors staked a beeline across the prairie, avoiding the deep coulees and steep grades close to the river and tactfully skirting the northern edge of the Siksika reserve. The next summer gangs of tracklayers hammered their way across the prairie, laying rails as fast as any track had ever been put down and then more slowly as they pushed the railway up into the mountains on the winding banks of the Bow.[12] Operation of the railway greatly expanded the market for western beef on the hoof as well as the local market for slaughtered cattle. But the Canadian Pacific Railway, a large, capital-intensive, technologically advanced corporation with its own needs, labour requirements, economic imperatives, and differential influence over decision-making in Ottawa, by degrees restructured life on the prairie away from the north-south axis of police posts and ranches to an east-west line of small towns strung along the railway and centred on Calgary.

The CPR exerted its influence on the struggle between ranchers and homesteaders in three ways. First, to increase its passenger and freight traffic, the railway preferred more intensive land settlement. This, along with steady pressure from newcomers seeking land, eventually broke the grip of the large ranchers and opened their leaseholds to occupation. Secondly, the railway connected the west to external markets more effectively. It transmitted market information, migrant farmers, equipment, and settlers' effects westward, gradually transforming western agriculture from ranching to wheat farming to serve international markets. Moving massive quantities of wheat and the inputs required to grow it created more profitable revenue opportunities for the railway than grazing. Finally, by building a north-south branch line to

Fort Macleod down the spine of the ranching leases, the CPR and its subsidiaries demanded and acquired land grants along their lines which had to be taken out of the leases. As the largest landowner in the region as well as a railway in need of traffic, the CPR determined that the future of the region lay with farming rather than ranching.[13]

As a corporation, a landholder, and a transportation system, the CPR restructured the economy, altered power relations in the region, and in the final analysis decisively tipped the balance away from large open-range leasehold ranches towards farming. In this newly emerging economy, farmers and ranchers moved up from the river bottoms to be closer to their fields on the high prairie and the grain transshipment facilities on the railway. After the turn of the century, thousands of hopeful farmers flooded into the newly available lands on the open prairie in southern Alberta. Land could be acquired at minimal expense to homesteaders or purchased from ranchers and the land companies which were disposing of CPR, Hudson's Bay Company, and school lands. Not only was land abundant; so too was moisture. This migration and wheat boom coincided with a period of greater than average rainfall. Farming, especially wheat farming, flourished on the rich soils of the broken prairie. In less than a decade a rectangular geometry of cultivated sections and long, straight range roads leading to the railway imposed a settlement grid on the formerly featureless prairie.[14] The accompanying graph, showing homestead entries in Alberta, suggests the breathtaking speed with which this land rush occurred. It also reminds us that this flood of humanity had both a flow and an ebb. While homesteaders staked claims to a great deal of land, a significant number of them failed for one reason or another and abandoned their land. Of the 34.5 million acres of Alberta homesteads claimed after 1905 more than 15.8 million acres or 46 per cent had been abandoned and cancelled by 1930.[15] Neither the rain nor the high prices for wheat lasted. During the drought-plagued 1920s, about as many homesteads were abandoned in the dust as were taken up by new entrants.

Some influential westerners had always argued that the semi-arid plains were no place for conventional farmers. The federal land agent in the west, William Pearce, had long championed open-range ranching as the best use. He insisted, for example, that only about 5 per cent of the stony river bottom lands were fit for cultivation. After the demise of the leasehold ranching system, which he fought tooth and nail to protect, Pearce would become an enthusiastic proponent of irrigated agriculture. Partly on the strength of his lobbying, the CPR acquired some 2.9 million contiguous acres east of Calgary in 1903, the CPR Block, as the last instalment of its federal land grant.[16] Amply supplied with capital and technical advice, the railway developed its property

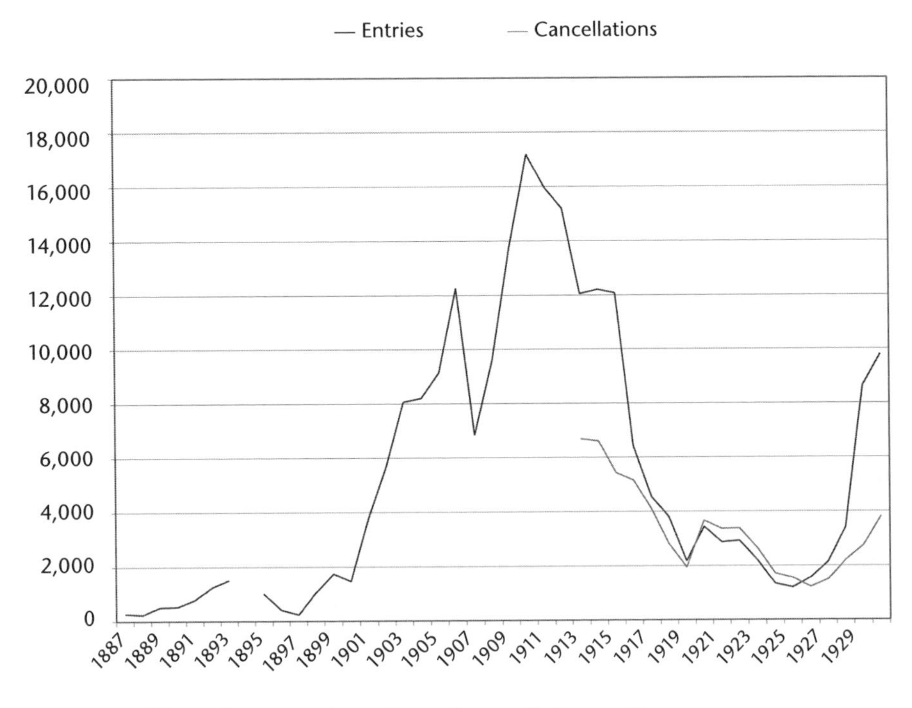

Figure 3.1 Alberta homestead entries and cancellations, 1887–1930.

SOURCE: M.C. Urquhart and K.A.H. Buckley, eds, *Historical Statistics of Canada* (Toronto: Macmillan, 1965), series K34-41, 320.

as irrigated farmland. The story of this attempt to change the environment by leading the river out of its bed through canals and ditches onto a fertile but semi-arid land will unfold in a separate chapter. Here we need to note that even as dry farmers were struggling in the drought and abandoning their properties in droves, other farmers were moving onto nearby irrigated land. Capital-intensive irrigation required close coordination among farmers, collective management of the resource, and a corporate form to build and operate the hydraulic system. Irrigated agriculture, too, would encounter its crisis in the 1930s. But that is a matter for later.

To sum up, western settlement was anything but settled. It was as tumultuous and tormented a process as it is possible to imagine. Settlement was not some steady accretion of culture but a constant struggle between regimes, and between people working within those regimes. The population, society, and economy of the region were built up in a complex and sometimes cruel calculus of pluses and minuses, of success and failure, of rising and falling fortunes. It was an unremitting struggle against government policy, the banks, the CPR, markets, elevator companies, middlemen, mange, insects, depression,

Table 3.1 Population of the Bow valley, 1901–31

	1901	1911	1921	1931
East of Calgary	1,412	9,306	12,766	14,589
Calgary	4,392	43,704	63,305	83,761
West of Calgary	740	2,392	3,497	4,516
Indian reserves	1,628	1,558	1,630	2,109
Total population	8,172	56,960	81,198	104,975
As a percentage				
East of Calgary	17.3	16.3	15.7	13.9
Calgary	53.7	76.7	78.0	79.8
West of Calgary	9.1	4.2	4.3	4.3
Indian reserves	19.9	2.7	2.0	2.0
	100.0	100.0	100.0	100.0

SOURCE: Census of Canada, 1931, vol. 2: 98–9, Population by Areas.

NOTE: Our statistical observations need to be understood as approximations. Census takers collected data from census sub-districts that bore little relationship to the shape or boundaries of the Bow River and its valley. We have selected those sub-districts that best encompass the valley and its communities while acknowledging that the fit is more or less precise, depending on the sub-district and its territorial extent. From the Calgary District (which covers much of southern Alberta) our sampling includes sub-districts 2, 6, 7, 8, 10, 11, 12, 13, 26, 28, 29, and 30. For a map of the sub-districts in 1906, see http://collectionscanada.ca/archivanet/02015304/020153040102_e.html.

rustlers, the weather, diseases, exhaustion, drought, wind, dust storms, prairie fire, and blizzards that left outcomes always uncertain. Some won; others lost and gave up. Settlement is like war in that the survivors get to write the history, and that imparts a misleading bias to the story. What would the process look like if the casualties also had a voice?

Table 3.1, compiled from the *Census of Canada*, captures the sum of this shuffling of humanity through the opportunities and challenges presented by the political economies of the region at four points in time. In 1891 census-takers counted 6,864 people in the Bow valley, most of them male, Canadian- or British-born. After 1901 the region grew quickly, especially during the period 1905 to 1921, the sex ratio became more balanced, and the proportion of children increased, as did the percentage of foreign-born. Growth proceeded but at a slower pace in the 1920s. The Bow valley covered a huge expanse (25,123 square kilometres), but the majority of the population settled in a comparatively small area, the city of Calgary. Settlement is predominantly a story of urban growth, notwithstanding the concentration of the literature (and this chapter) upon rural activities. Calgary comprised the majority of the population as early as 1901, and its dominance grew during the next thirty years. The valley downstream of Calgary, containing about 15 per cent of the region's population, grew rapidly to 1921 and then levelled off. Upstream from Calgary the population increased at a much slower rate and to much lower levels. Indigenous people on reserves declined in absolute and relative terms between 1901 and 1911 and then stabilized at about 2 per cent of the total population of the region.

But the census, with its rising columns, gives a misleading impression of stability. These totals were not just the same people, with new ones added. Rather, these snapshots taken a decade apart capture people in motion in response to their opportunities and their environment – some staying, some coming, and some going – making spaces others had filled. Births and deaths add and subtract as well. Migration to the city from the ranches, farms, and reserves subtracts in one column and adds in another. The population of five census subdivisions east of Calgary actually declined during the drought-stricken 1920s. Settlement involved a constant shuffling of new people in and old people out, with a few hardy or fortunate families persisting to become pioneers.

We have sketched a sequence of cultural regimes that followed one another: the indigenous prairie, the open range of the ranch companies, small ranches nestling in the valley and foothills, the railway establishing a metropolitan corridor through the valley, homesteaders and dry farmers breaking sod on the high prairie, and the irrigators spreading across the CPR Block. Each regime involved a different form of social organization, and each had to fight for survival. But the sequential metaphor should not be overstated. To some extent, parts of these regimes were simultaneous, or at least some elements of each regime persisted. The Siksika remained a presence throughout this period on their reserve, as cowboys, visitors, and casual labourers. Despite the remarkable sell-off of almost half the Siksika reserve south of the Bow River in 1911, along with other cut-offs in subsequent years, indigenous peoples managed to eke out an existence and develop new, if constrained, economies within the emerging settler society.[17] The large ranches, considerably reduced in scale though still quite large, nevertheless continued on, mainly in the foothills along the front range. The railway came, stayed, and grew into the most powerful economic organization in the region, exerting its influence as a transportation system, as a real estate enterprise, and as a corporation. Small ranches raising cattle, horses, and, in a few cases, sheep survived alongside the new mechanized wheat farms on the more expansive high prairie. As the ditches spread the hydraulic empire across the plains, farmers and small

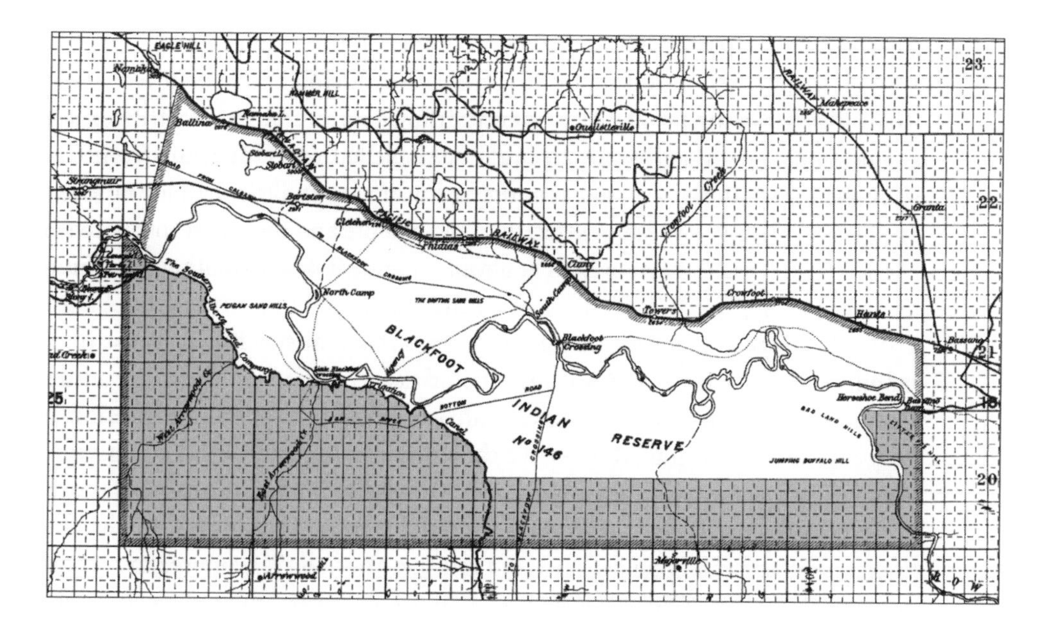

Map 3.2 The reduction of the Blackfoot reserve, 1911. A further section would be removed at the eastern end of the reserve to accommodate the Bassano dam on Horseshoe bend. (Source: Canada, Department of the Interior, "Blackfoot Sheet, Alberta, West of Fourth Meridian," *Sectional Maps of the West, 1:380,160/*, map 115, 7th ed. [Ottawa, 1 April 1915])

ranchers adjusted to the new possibilities. We have emphasized succession here to show more clearly the different regimes that together produced a composite society on the banks of the Bow River. But the notion of succession needs also to be balanced with the idea of persistence. Western settlement was the complex computation of change and continuity among different cultural regimes.

Each of these political economies of production had to struggle both with and against the Bow and formed its own particular relationship with the river. Some issues were common to several or all of them. The reciprocal relationship of the river and urban life will be dealt with in separate chapters, as will one of the most significant environmental transformations in the valley, irrigation. This chapter will focus on the river in indigenous reserves, open-range ranching, small ranching, railroading, and wheat farming, drawing its evidence from monographs, memoirs, novels, photographs, government records, and archival collections. The river could be a blessing and a curse. It could be alternately a docile and generous neighbour and a murderer.

A Well-Watered Domain

Water, a vital but notably scarce element on the western plains, was one of the reasons, but not the only one, that animals and humans especially favoured river valleys. Settlers and business organizations valued Bow River lands in proportion to their need for water.

Indigenous encounters with the river as mobile hunter-gatherers conditioned their approaches to the Bow in the settlement era. The Siksika, Tsuu T'ina, and Nakoda all located their reserve lands on the Bow and one of its tributaries. The Piikani and Kainai situated their reserves on the banks of the Oldman River to the south. The authorities, in laying out the reserves, were concerned, too, that indigenous people should have frontage on a river or a lake so that they would enjoy a good mixture of water, timber, and arable land as well as hunting lands.[18] Once settled on the reserve, the Siksika built winter houses for their elders on the river flats so that they could be closer to sources of water.[19] The Anglican mission was also sited near the river, though it had to be removed to higher ground after the turn of the century because of the pollution of its wells caused by frequent flooding.[20] Previously important river sites retained cultural values even as new forms of transportation and mobility appeared. The Sun Dance, a major summer rendezvous and ceremony, continued to be held at Blackfoot Crossing, for example, despite the discouragement of Indian agents and missionaries.

Early European travellers encountered the Bow River and its valley after enduring long, hard, disorienting rides across a treeless, seemingly endless prairie. The river and its valley came as welcome return to familiar terrain. Here was a land like home (the east) with trees, greenery, and cool, refreshing water. In high summer the river valley appeared a veritable garden oasis in the midst of a desert. Travellers' reports thus tended to extol the beauty and the agricultural possibilities of the garden lands spread out across the river bottoms.

In the summer of 1881 four different groups of visitors penned impressions of the region. The self-serving nature of some of these accounts, written as they were to encourage investors, curry favour with politicians, or describe breathless boys' outdoor adventures amidst exotic circumstances, presented a much more attractive representation of the country than the reconnaissance a little over two decades earlier by the Palliser expedition.[21] But in all four accounts the Bow valley became the main focus of attention. Duncan McEacheran, an eastern Canadian travelling north from Fort Benton in Montana to establish the Cochrane Ranch Company, felt immediately more at home as he crossed the invisible border. In his written account he imagined that the soil and pasture improved steadily as he rode north, reaching an ideal state when he

Illus. 3.2 Looking west up the Bow valley from the Rawlinson Brothers Ranch near Calgary circa 1905. (Glenbow Archives, NA 963-7)

arrived at his chosen destination on the Bow River.[22] Another ranch manager from the Bow district, Alexander Begg, composed a description of the river range for John Macoun's paean to the prairie, *Manitoba and the Great North-West*, published in 1882. Begg claimed that the grasslands along the Bow were much more suitable for ranching than the more famous ranges of Montana and would, by extension, support an even more prosperous ranching industry.[23]

The summer of 1881 also found the young George M. Dawson criss-crossing the southern prairie and probing the front range of the Rockies while engaged in compiling an inventory of the economic resources of the region for the Geological Survey of Canada. At the end of August he floated down the Bow River from Morleyville to the junction with the Oldman River sketching rock formations, the geological strata exposed by the cutbanks, and the meanders of the river.[24] Being a geologist, he was less concerned with agricultural possibilities of the scenery as he drifted downstream than with the faults, folds, fossils, and stratification of the earth. As a scout for industrial civilization, he kept a sharp lookout for outcrops of coal, signs of mineralization, and merchantable timber. Though he was disappointed that the Bow proved to be too shallow for steamboat navigation and that he found no traces of gold in the sandbars, he nevertheless became enthusiastic about an exposed coal seam and the heavily wooded banks around the confluence of the Highwood River. Farther downstream he admired the vegetation on the wide valley bottoms and sketch-mapped the many islands that dotted the winding river.[25] Dawson, an experienced, dispassionate scientist not given to the hyperbole of the promoters, generally subscribed to the view that the southern prairie was unfit for settlement. Yet his field notes do not give the impression he was passing through an arid landscape as did the accounts of his predecessors who had traversed the open prairie. The view of the southern prairie from the river was much more benign.

Dawson beached his rowboat at Blackfoot Crossing on 5 September to camp. He would then have been present on 7 and 8 September when Governor General Lord Lorne's entourage of twenty-one wagons, a fifty-man mounted police escort, journalists, and a professional artist, Sydney Hall from the *London Graphic*, arrived for a grand council with the assembled Tsuu T'ina and Siksika community, who had recently returned to their reserves from a failed hunt. After six wearying days in the saddle or lurching wagons in their journey from Battleford, crossing a featureless, dun-coloured, and unseasonably cold prairie unsettlingly devoid of even bird or insect life,[26] Lord Lorne's party finally arrived at what one of his companions described as "the most striking scene since we left the shores of the Great Lakes," the high banks overlooking the Bow River.

> Beneath lay a beautiful and well-wooded plain through which ran a broad river, dancing amid the long unfamiliar green which fringed its banks. Wild Indians on horseback were fording its waters and careening over the plain. Some 2,500 of them, Blackfeet and Sarcees, were encamped on the other side, and the smoke from their numerous tents added to the softness of the scene. But the charm of the picture was that long and magnificent line of gigantic peaks and mighty masses 120 miles away, on which the sun was going down in glory, throwing long bars of gold across the western sky.[27]

The next day the assembled indigenous groups put on a show of horsemanship and mock combat, after which Chiefs Crowfoot, Old Sun, Heavy Shield, and Bull Head pleaded with the governor general for help. Lord Lorne offered little by way of comfort to the petitioners except the advice they should forget the bison and take up agriculture.

After this gathering G.M. Dawson would have clambered back into his rowboat to continue his scientific reconnaissance downstream. The governor general and his entourage mounted up to continue their trek overland to Calgary, from where they made a brief foray to Morleyville to get a closer look at the mountains before heading south to Montana and home. As an artist and a Victorian traveller in search of the exotic, Lorne was primarily fascinated by indigenous peoples and was emotionally affected by their situation. Though he paused occasionally to sketch the scenery, his written account focuses almost entirely upon indigenous peoples.[28] Other members of the party expressed more interest in the countryside. The Reverend James MacGregor, reporting for the Edinburgh *Courant* and *Scotsman*, expressed astonishment at the scale of the newly established ranches, the fine condition of the cattle and horses recently arrived from Montana, and the fertility of the land. "I never saw finer vegetables than I saw here," he reported from Calgary. "Beets, parsnips, turnips, cabbages and potatoes had reached perfection." In the fields the oats stood four and a half feet tall. At the Indian Supply Farm at Fish Creek MacGregor marvelled at a pair of horses and a yoke of oxen ploughing a furrow more than a mile long in rich, black soil. The crops on the bench lands were in just as fine condition as those taken from the river bottoms. The land, he declared confidently, "was admirably adapted for farming."[29]

These observations, we now know, came during a period of cooler, wet weather in marked contrast to the drought conditions that had prevailed during the earlier surveys by the Palliser expedition.[30] To travellers in the early 1880s the Bow valley seemed a green and pleasant land, especially in contrast to the open prairie. The key difference, of course, was water. The abundant, crystal clear, fresh running water in the lush valley offered relief from the brackish sludge that passed for water on the prairie and made travelling thirsty as well as dangerous work for both people and beasts. Water made the valley hospitable, a bountiful land seemingly waiting for cattle, ranchers, and cultivation.

The managers of the ranching companies located their ranch houses, outbuildings, and corrals in the river valley. Senator Cochrane nested his first ranch above Calgary in a coulee on a broad, expansive flat. The government established the Indian Supply Farm to provide meat and produce to the nearby reserves on river bottom land where Fish Creek joined the Bow. Colonel Strange situated the headquarters of his ranch, Stangmuir, on a riverbank east of Calgary. Shareholders in the company had staked freehold land in the vicinity.[31] Lachlin McKinnon, a green hand just out from Ontario, worked on several of these properties in the mid-1880s. On the Goldfinch ranch part of his chores involved hauling water from the river for Mrs Goldfinch to do the laundry. McKinnon later leased the freehold land of the former 76 Ranch and named it Pleasant Valley. He later acquired a ranch of his own, but he built too close to the river. After several floods, he relocated on more protected land downstream, on flats that later acquired his name. As McKinnon shrewdly added to his property, his spread eventually covered over three miles of riverfront.[32]

A location close to the river had several advantages. First and most important was access to water for household use and for the animals in all seasons. The bottom lands were also the places where the native hay grew in the great-

Illus. 3.3 Lachlin and Sarah McKinnon on their wedding day in Calgary, 20 September 1893. (Glenbow Archives, NA 2198-1)

est profusion. Hay could be cut close to home and mowed adjacent to the animals' winter quarters. Hay and water kept the cattle from roaming too far in the winter. The rich alluvial soil on the bottom lands was also ideal for domestic gardens and the first field grains. Angus McKinnon and his brothers received 1 cent each for the tails of gophers killed on the prairie, but 10 cents each for the tails of the river-bottom gophers which damaged the crops and gardens.[33] William Pearce built his estate on low-lying land at a bend in the river where he could experiment with tree propagation and irrigation.[34] But some places in the flood plain could also be quite stony and gravelly and unsuited to cultivation. Water nearby was also insurance against the deadly prairie fires. As well, dipping tanks to protect the cattle against mange could be excavated in the coulees and fed by tributary streams.

Squatters or homesteaders – the term depends one's point of view – prized these bottom lands. In turn the ranchers and their allies raged against the interlopers. "Farmers coming here to establish themselves choose the lands on the rivers," a western correspondent explained to a federal cabinet minis-

Illus. 3.4 McKinnon's LK Ranch buildings on the Bow about 1914. (Glenbow Archives, NA 2511-7)

Illus. 3.5 The McKinnon family at their ranch near Dalemead east of Calgary about 1910. (Glenbow Archives, NA 3850-1)

ter in 1884. "It is a nuisance for us, because our cattle will soon be unable to reach water. Down with farming!"[35] But the small ranchers and homesteaders saw fertile land waiting to be cultivated and settled. Some of that hope is captured in Amelia Banister's autobiographical novel *Chinook Medley*, about homesteading in the Bow valley.[36] After a long journey by buckboard, the new schoolteacher, Fredina (or Freddy, as she prefers to be known), reaches the heights above her new home in the valley. Her driver, Brady, turns to her: "Take it in, Miss Macmillan … Let me introduce you to the Big Bend of the Bow." Fredina exclaims: "I had no idea the river bottom would be so extensive. It's a picture. I can't see the house though. How beautiful the river looks." Brady then points out her destination below in the lengthening shadows of a late fall afternoon:

> There is a grand view if you follow this bank down-stream until you're opposite John's house. Do you see it – that white house across the river with some smaller red buildings nearby and corrals built so close to that shimmering body of water, which John calls "Clear Spring"? It is fed, like the Banff pools, from numberless springs which bubble up through the gravel bottom. Such water is priceless to stockmen and is their reason for settling close to the river instead of near unsurveyed roads. But John will show you all the beauty spots. He's as keen about this place as I am about mine.

Then as Brady gingerly leads the team of horses down the long winding hill to the flats, he explains to Freddy the advantages of these locations on the river bottom. Soon they fetch up in front of the white house they had seen from the heights, right beside its fenced-in garden.

For the animals of the prairie, indigenous peoples, ranchers, and later homesteaders, water and the vegetation it nourished attracted them to the valley. Seen by eyes accustomed to moist, temperate climes, the Bow valley was a place of trees, gardens, hay, meadows, fields, and of course fresh water, a waiting Eden.

As Shelter from the Storm

A valley location also afforded protection from storms for both people and beasts, especially in winter. Bison sought out the valley in winter not only for the relatively abundant browse but also because of the protection it provided against the elements. Cattle, when they were introduced to the region in the 1870s and then in large numbers in the 1880s, behaved in the same way.

Illus. 3.6 Herd of cattle from the Burns Ranch on the banks of the Bow River southeast of Calgary early in the twentieth century. (Glenbow Archives, NB 16-300)

During the spring, summer, and fall bison and later cattle ranged freely across the short-grass prairie. In mild winters on the southern range they could survive out in the open as long as food remained accessible. It was not so much the cold temperatures of winter as the heavy snow that posed a danger. Bison could paw through a thin snow covering, but they could not handle a heavy snowfall. Cattle did not know enough to scrape away the snow and would starve to death next to a hay mow if it was covered in snow. In theory, the frequent chinook winds cleared the snow from the prairie and therefore spared the stock. However, a heavy snowfall followed by an ice storm or a melt and quick freeze doomed a herd of cattle on the open range. Horses were even more vulnerable.[37]

At the first sign of a winter blizzard, cattle headed for the protection of the valleys and coulees. Knowing this, ranchers always tried to locate their properties adjacent to rivers, not only for the water but also for the winter shelter. Angus McKinnon remembered that as a boy, growing up in the homestead down by the river, winter meant he could no longer play in the woods along the river. "Any sight other than an adult seemed to put these big steers into a panic." In May, after these animals had been moved up onto the prairie or taken to market, the kids would then reoccupy their beloved "bush."[38] Winter put humans as well as cattle at risk. Many cowboys perished in these winter storms trying to drive their cattle to safety. Ranchers and homesteaders built their dwellings and outbuildings in the lee of a hill, preferably on the south side of the valley, where they were protected against the winds and bitter cold of prairie storms.

The history of ranching in southern Alberta is punctuated with stories of horrendous winter stock losses. Many of the cattle Lord Lorne saw being driven onto the Cochrane ranch property in 1881 perished in their first winter. Senator Cochrane rebuilt his herd, only to have storms in 1882–83 kill off three-quarters of his stock. The carcasses of putrefying cattle poisoned the air and water above Calgary, as Charles Shaw learned when he surveyed a new main line for the CPR in the spring.[39]

Another hard winter in 1889–90 killed off most of the cattle on the Bow River range east of Calgary. Two years later, in April 1892, a snowstorm followed by sleet and fifteen days of cold weather killed more cattle on the range than these earlier storms, according to Lachlin McKinnon, who had just begun ranching on his own: "Cattle along the Bow fared pretty well as they got under the hills and were sheltered. Those drifting in from a distance hit the banks of the Bow the second day of the storm. The snow was drifted over the brow of the hills like a veranda roof, they wallowed blindly into it and perished."[40] He recalled, too, a September storm in 1900 that wiped out hundreds of cattle on the Burns Ranch. Once again McKinnon's stock weathered the storm in better shape: "We were always fortunate when these storms struck as our cattle were ranging not far from the Bow and always managed to take refuge below the hills along the river where they could graze until the storms were over." When L.R. Symons began working the range in 1914, the veteran cowboys were still telling stories about the big die-off in the winter of 1906–07, when half the cattle perished and the "big smell" hung over the valley for months in the spring. That year the losses amounted to upwards of 70 per cent of the entire herd.[41] The same storm tragically killed off the cattle that had been tended by the Siksika for their own consumption. Discouraged by the experience, many abandoned the cattle business. They were not alone in this reaction.[42] But if cattle could get to the valley bottoms and woodlands, near hay that the ranchers had put up for the winter, they could survive.

But getting to shelter proved to be an ever more difficult problem. William Pearce, the federal land agent, railed against the squatters who occupied bottom lands and then immediately fenced their properties, "preventing stock from reaching water and shelter during winter storms." The dogs of the homesteaders also, it was claimed, kept the cattle away. By interfering with the movement of cattle in winter, fencing led to the death of many cattle in storms. In February and March of 1904 storms on Lord Beresford's ranch drove about

12,000 head of cattle towards the protection of the river. But barbed wire held them on the high prairie, where they died by the hundreds, piled atop one another.[43] Those strands of barbed wire not only doomed the cattle; they also eventually strangled the business of the large ranchers. As fences carved up the open prairie and kept cattle out of the valley, open-range ranching was no longer viable.

The river afforded protection from another kind of storm, the firestorm. Prairie fires stampeded and subsequently starved cattle. They also killed the ranchers trying to cut fire breaks. Valley locations at least meant that buildings and people could be spared the worst of a prairie fire, and water was available to put out the most threatening flames. Upstream forest fires posed a hazard to prospectors and surveyors. Charles Shaw and his surveying crew sought shelter from a particularly vicious five-day forest fire by swimming to an island in the river near Lake Louise.[44]

The Bow valley remained a shelter from the storm, but as time passed, fewer people and cattle sought it out for that purpose. The shift to dry farming and wheat cultivation changed this orientation toward the river. Farmers now built their houses and their sheds on the open prairie, where it was easier for them to move their grain to the elevators. While he was a rancher and horse breeder, Lachlin McKinnon built two homesteads in the river valley. But when he became a wheat farmer on a large scale, he built a much larger house on the tablelands amidst his huge fields. In this new agricultural regime the danger and inconvenience of hauling heavy farm equipment and grain up and down the hills led to the abandonment of valley homes, notwithstanding the exposure to the elements involved with a prairie homestead. But by that time – in McKinnon's case, 1912 – the railway made better building materials and sounder dwellings more readily available.[45] The passing of ranching and the rise of dry farming drew farmers up out of the valley onto the plains.

A Transportation Corridor

Traditionally, the great rivers of Canada have been considered transportation corridors. To a certain extent the Bow River also served in that respect, but it was not ideally suited to the task. It worked for some purposes, but it possessed defects that could not be easily overcome.

As we have seen, G.M. Dawson did his geological reconnaissance scanning the banks over the gunwale of a rowboat as he floated down the Bow. When he got to the Oldman, he came across an "old flat bottomed boat stranded on a gravel bank," a sufficiently picturesque sight that he sketched it in his field book. Here was tangible evidence that others had used the river for transportation before him.[46] At Calgary a small fleet of boats had been built for Lord Lorne to take him back east down the Bow. He opted instead for an overland trip to Montana which brought him closer to a railhead.[47] Isaac Kerr, out from Wisconsin to scout timber limits in the Rockies in 1883, began his return journey heading eastward down the Bow in a rowboat to the end of steel at Medicine Hat.[48]

In the mountain reaches of the river, fur traders, prospector, miners, and railway construction crews all used the river from time to time to move equipment and supplies. As Charles Shaw and his crew were surveying for the CPR upriver from Kananaskis Falls early in 1883, they spied a raft with three men floating down the river. "We ran to the edge of the water and waved and shouted to them to come ashore as there were falls ahead. They must have heard us, but seemed to think we were joking, as they only laughed and waved their hats."[49]

The swift current of the upper river made it essentially a one-way street. Downstream travel was more likely than upstream. In the early days hay cut from the meadows around the Vermilion Lakes was rafted downriver to Banff for winter fodder for the trail horses, draught animals, and bison in the paddock. Eventually, lumbermen used the river as a transportation system to float logs downstream to the mills at Cochrane and then Calgary, as will be discussed at greater length in a subsequent chapter. Their workers also moved supplies from camp to camp by boat, pushing against the river with paddles and poles.

The river below Calgary could carry small boats, but its meanderings and frequent shallows meant that it was not entirely reliable even for small craft. It was wholly unsuited for steam navigation, as G.M. Dawson noted in his field notebook.[50] Moreover, since it was a mountain stream, the volume of flow varied considerably over the course of a year. On the prairie the river wandered, turned back on itself, carved new courses with each flood season, and in places was littered with snags, tree trunks, slides, and debris. Notwithstanding these deficiencies, both the Bow and the Oldman were classified as navigable rivers in Canadian law. Unlike the North Saskatchewan, the South Saskatchewan tributaries above Medicine Hat could not sustain commercial boat traffic without continuous dredging.

In different seasons and for a short time the river did provide an axis of transportation below Calgary. The line of homesteads and ranches along the river afforded a regular trail of de facto guest houses and bunkhouses for mounted police on patrols, surveyors, wolf hunters, hide and horse buyers, and indigenous people seeking work in order to eat. In winter, parts of the river became a highway. Indigenous people continued to use the river, espe-

Illus. 3.7 A "wannigan" delivering supplies to a survey party on the upper Bow circa 1910. (Glenbow Archives, NA 1432-13)

cially in winter, to move their camps. With some of the condescension of a prosperous settler, Lachlin McKinnon remembered the sights and the sounds of these regular migrations on the ice: "Their ky-yi-ing could be heard for a considerable time before the cavalcade of Indians of all ages riding single file would come into sight. There would be travois on both dogs and horses carrying the belongings of their masters. These processions were really interesting with the colours and accoutrement along with the singing and chattering as well as the yelping of many mongrel dogs." It was all very entertaining except, McKinnon observed, "when one of these parades would stampede our cattle off the feed grounds and when they would tear down the fences and hit for the hills."[51]

Until 1881 the transcontinental railway had been planned to cross the northern prairie, entering the mountains via the Yellowhead Pass. But strategic considerations induced the new Canadian Pacific syndicate to choose a more southerly route.[52] Major Rogers's legendary and highly contested exploits in 1882 and 1883 eventually located a way through the mountains, though it was far from ideal. The southerly route across the prairie had defects too; it discounted Palliser's evaluation of an arid wasteland, preferring instead James Macoun's more bountiful later assessment.[53]

When the CPR decided to run its main line across the southern prairie, Charles Shaw, the surveyor responsible for locating the route, kept as far away as possible from the rivers with their wide valleys, steep grades, and numerous ravines or coulees running in from the side. Once the route crossed the temporary trestle on the South Saskatchewan River at Medicine Hat, it headed straight northwest, angling away from the wandering deep valley and heavily incised flanks of the Bow River. Geometry ruled on the prairie, as surveyors sought the shortest distance between points with a minimum of bridging, cutting, and filling. Shaw avoided antagonizing the Siksika people by skirting the northern boundary of their reserve, which took him a little further north into steeper grades than he would have liked. The Siksika may not have approved of the railway, but they did not oppose it or intimidate the survey or construction crews. They displayed their displeasure in more subtle ways, according to Shaw. "Though they rarely tampered with survey markers or removed them, they would sometimes express their resentment by defecating upon the tops of every available stake, which added nothing to the amenities of the job."[54] In later years, when the CPR sought a right of way for an irrigation canal, the Siksika protested and demanded that the company make payments for horses that had been killed by the railway.[55]

Dodging northward around Carcass Hill, the surveyors then headed straight towards Fort Calgary. Trial surveys brought the railway along the north bank of the Bow to a point opposite the fort, but that left little room between the bluff and the river for yards and station. Subsequent surveyors improved on this location by bringing the railway onto the broad, flat expanse of the south bank. To get there, the railway finally had to come to terms with the Bow River. A long but relatively low bridge would be required to bring the rails onto the south bank of the Bow, where the track would pass by the existing townsite and continue across the Elbow River to a station more conveniently situated on the open spaces of Section 15, which the CPR owned and where the town site would relocate.[56]

Beyond Calgary the distant and tentative relationship between the river and the railway changed fundamentally. Where segmented straight lines defined the route on the prairie and avoidance of the river was the rule, from Calgary west the railway would bend and curve as it sought a gentle grade along the sinuous course of the river. The route of the river and the railway would be one. Major Rogers had followed the river quite closely in his trial surveys, snaking his line along its north bank up into the mountains to the height of land, too closely for some people's taste. Back in the Winnipeg construction headquarters, railway management despaired at the cost and the time required by his plan. Fifty years later Charles Shaw, who had had a great deal of CPR surveying experience in northern Ontario and the prairies, recalled being summoned to Winnipeg by James Ross in March 1883: "He unrolled about sixty miles of exceedingly heavy, difficult work, with a mud tunnel – that abomination to engineers – over a half a mile long … Examining the plan, I found that the line followed the North Side of the Bow River all the way. I at once said, 'I can get a far better line than that.' I told Ross, 'I will take my party out and locate the line from Calgary to the gap. If I don't save at least half a million dollars over the estimated cost of construction of this line, I won't ask any pay for my season's work.'"[57]

Shaw's self-assurance stemmed from a brief reconnaissance of the Bow valley the previous year. On a ride out from Calgary through the foothills to view the Rockies, he had taken note of the difficulties the valley presented to railway construction. In the spring of 1883 he returned with a survey party and mapped out a better line. The most obvious difference between the two surveys was Shaw's willingness to use bridges across the Bow. Major Rogers had located his line exclusively on the benches winding along the north bank.[58] To find a better grade involving less work, Shaw took his line back and forth across the river. When one bank was too rough, narrow, or otherwise inhospitable, he shifted to better terrain on the other side. Rogers's line contained no bridges across the Bow; Shaw's meandering line required three crossings. In contrast with the delicacy of locating the line in the vicinity of the Siksika reserve, the south bank route cut directly across the Nakoda reserve, though Shaw makes no mention of negotiations or compensation. His route also substituted an easy low-level crossing of the Kananaskis coming in from the south for a difficult high-level passage over the Ghost River where it joined the Bow from the north. It avoided some thirteen trestles across coulees in the foothills. In all, James Ross estimated the savings in construction alone at $1,350,000.[59] Shaw won his gamble and a further contract to survey the line beyond the gap to the Great Divide.

As the route approached the present location of Banff, the rails could not easily follow the river up through the winding, narrow defile at Bow Falls. Instead, Major Rogers had proposed to punch a tunnel through the mountain blocking the way, which inevitably came to be known as Tunnel Mountain. When he saw the plans, Van Horne raged at the expense and delay and ordered his staff to "take that damned tunnel out." Shaw, who was on the ground, was ordered to find another way. Some accounts have him searching fruitlessly until, at the point of despair, he discovered a way around Tunnel Mountain via Devil's Head Creek and received a handsome purse for his efforts. A more casual variant has him sunning himself on the side of a hill one Sunday afternoon, smoking his pipe, and taking in the vista, when he spied a small creek

Illus. 3.8 The newly laid CPR line along the Bow River near Banff in the late 1880s. (Whyte Museum of the Canadian Rockies, V10, NA 66-672)

running into the Bow that begged investigation.[60] Shaw's own account is more prosaic, and as is customary with engineer's tales, the main point is to deprecate the work of others: "On examination I found that Tunnel Mountain was an island with a valley to the west of it joining the valley of Devil's Head Creek. So I located the line up this creek and around Tunnel Mountain to the main Bow Valley, shortening the line by a mile, avoiding two long grades, and above all, eliminating the tunnel … Rogers' location here was the most extraordinary blunder I have ever known in the way of engineering."[61]

Shaw's revision of the routing around Tunnel Mountain is the only deviation away from the main Bow River valley above Calgary. Beyond what is now Banff, the bottom of the river valley provided a virtually ideal gravel-ballasted roadbed for the railway. The grade, though steady, was not steep, and there were few obstacles. Indeed, the rails at several points run virtually alongside the river. Where the river curves, the railway curves, though northwest of Banff both run fairly straight. At a small tributary running into the Bow from the west just beyond Lake Louise, the railway survey parted from the Bow River.

The approximately 305 miles of railway in the Bow River valley from Medicine Hat to the Great Divide entered operation in the spring of 1884. For two years most of the traffic involved shuffling people, machinery, and materials forward to construction sites in the mountain division. Work crews fleshed out the main line with sidings, stations, freight yards, telegraph lines, and maintenance, repair, and fuelling facilities. Gleichen and Canmore were the two designated divisional points along this section of the railway. Initially, Calgary was to be only a station and freight depot; much later (1911) repair shops and marshalling yards were added. Revenue traffic on this section commenced in 1885, and in early summer the following year the first through train climbed the Bow valley from the prairies into the mountains. Construction of the railway transformed the landscape of the upper Bow valley but left the lower Bow untouched. The riverside environment above Calgary and particularly the forested valleys of the eastern slope had been extensively burnt and cut over in the location phase. In later years the railway would have a less obvious environmental impact, but its capacity to reorganize space and move people, commodities, and information radically reoriented the river and surrounding region.

As a Barrier

The river was a trickster. It sometimes enhanced communication, but at other times it prevented it. Sometimes it was a highway; at others it was an impenetrable barrier. The river had seasons – settlers thought of them as moods – and a difficult geography that required adjustments in human behaviour.

The river effectively divided the range into northern and southern divisions. Cattle rarely crossed the river on their own account, except perhaps in winter when they wandered across the ice. They could with some difficulty be driven across the river after roundups or to be taken to market, but it was not something that they willingly did on their own en masse. Driving cattle across the Bow was ticklish, dangerous work. But in general the river acted as a de facto fence separating two ranges. In his reflections on the river, Lachlin McKinnon remembered: "It was a boundary used for many purposes and a dividing line between our range and that of the great ranges to the south and except for some straying across during winter time, it was only on rare occasions that stock crossed it of their own free will."[62]

The sloping, rounded hills of the valley could with some difficulty be negotiated by horse and wagon. But loaded wagons often careened out of control downhill or tipped on uneven ground. Getting threshing equipment up and down these hills took enormous effort. Angus McKinnon, who worked on his father's farm in the first two decades of the twentieth century, recalled how his family handled the problem:

> Climbing up and down the river hill did present some difficulties. There were really three grades as they were usually referred to. One

> was the saddle horse grade, almost straight up. Then there were two used for hauling vehicles up and down. One of these was known as the "old grade," and the other the "new grade." Both of these grades were upstream and required considerable detour and extra distance. The old grade was the closest and also the steepest and could be used for lighter traffic. The new grade was farther to the west but was not so steep and much safer, especially for hauling down the big loads of hay and the oats for winter feeding. Chaining one or both hind wheels was always advisable; and often during the winter snow, melting resulted in icy conditions which required both wheels and sleight runners to be rough-locked. This was wrapping a logging chain around several times to make the rough lock. It was not unusual even with all precautions for a big load of hay to slide off the grade and upset down into the bottom of the coulee. This was a real hazard and often resulted in injury to the teamster and busted equipment. These river hill grades were also the nightmare which so haunted the automobile salesmen when they would come to the ranch trying to make a sale.[63]

The steep-cut banks, where the river had eroded the land into cliffs hundreds of feet high, absolutely barred crossing.[64] Travellers had to journey miles upstream or down looking for negotiable grades.

Of course, the river could not be crossed safely when it was in full flood early in the summer. In spring or later in the fall it was possible, though never easy. The long period of freezing in the late fall and early winter, when the ice had not fully formed, absolutely cut off cross-river communication. Again in the spring the noisy breakup prevented movement north and south. But in winter, when the ice froze solid, regular intercourse across the river could flourish.

The many moods and challenges of the Bow figure prominently in Amelia Banister's unpublished novel of homestead life on its banks. The river forms part of the sometimes breathless plot. Men proved their manhood handling the river. Freddy admires the skill and strength of her man as he rows across the Bow: "It was only John who could tilt the boat at the right angle to send it up on the ice with the last stroke of the oars. Freddy held her breath as the strong arms did their work so accurately and her heart swelled with admiration. However fine he had appeared in uniform he looked finer in his ranch clothes."[65] Crossing on horseback and with a wagon and team required special skill. The perils attendant upon fording the river gave men the opportunity of saving women in trouble. After a harrowing rescue, Amelia's hero explains to Freddy: "The Bow takes some understanding. It is as capricious as a woman."[66]

Illus. 3.9 Eight thousand head of cattle crossing the Bow River near Carseland on their way to the CPR shipping point at Strathmere around 1900. (Glenbow Archives, NA 1888-3)

Crossing the river, especially at night, marks a poignant moment of separation between mother, father, and son:

> John pushed the boat to the water's edge, hopped in and took the oars. Man John drew his wife closer and they watched until they could no longer see the boat in the gathering darkness. The sound of dipping oars, usually music to her ears, was now a sigh and she drew closer into the circle of her husband's arm. When they heard the boat crunching on the beach on the other side and the sound of the oars being drawn through the rowlocks they knew that once again Boy John had arrived on his own soil. They heard him speak to his horses, no doubt as he mounted. Then his cheery whistle came across the water, growing fainter, as he neared home.[67]

In times of separation, lights burning in the night across the river provided reassurance that all was well. In Banister's plot the river symbolizes separation, reunion, danger, and a force to be mastered, usually by men.

The onset of winter imposed a season of enforced isolation between cross-river neighbours. Houses within sight of one another remained cut off for months. There was always anxiety along the river that the ice would not be solid enough to cross. Some of the tension in Amelia Banister's novel comes from the uncertainty as to whether or not the ice will set in time to allow a family reunion at Christmas. A timely forty-below-zero night closes the gap in the ice and brings the family together once again. The McKinnons had their own test as to whether the ice was solid enough to cross. A bachelor neighbour regularly crossed to the McKinnons' for supper, a game of cards, and some songs. "It was generally considered safe when our old neighbour Tucker Peach appeared on our side after testing every step of the way across with a heavy stick. Being an old Klondiker he never took any chances with the river."[68]

With the ice frozen solid, winter became a time of visiting. These were the joyous times of range hospitality and dances, of sleigh rides across the prairie and the ice, bundled up in the cutter.[69] In winter young men and women met, courted, danced, and made plans of their own. Amelia Banister claimed she could hear this season approaching:

> The silence smote his ears like a bombarding announcement – the silence of the frozen places of the world. When he came in for breakfast he put his fur cap over his wife's head, his coat around her shoulders and led her outside.
>
> "Hear anything?"
>
> "No. It seems quieter than usual."
>
> "It is. The Bow has lost its voice."
>
> "It seems strange not to hear it. How long will if be like that?"
>
> "Until the middle channel opens up in the spring, whenever that may be. It is late in freezing over this season. Often it freezes in November."
>
> "Can you skate on it?"
>
> "Sometimes. If the water rises and floods the ice and then freezes smooth it can be skated on. If the river rises and heaves the ice into huge boulders, it can't. It was a near thing this year. There remains only one clear day to make the ice safe for horses. We may have to walk. It always seems jollier when the Bow is frozen over. Father's daily walk is to visit us and Mother and Harriet drive across almost every day. We must make the most of them as neighbours – their last winter here."

On the way back the young couple sink through a hole in the ice. As they walk the rest of the way half-frozen, they can hear singing at the party that continues across the river.[70]

Equipment could be driven across the ice in winter to thresh stored sheaves. Cattle could also be herded across the ice with relative ease. Settlers on the north bank became accustomed to visits from the cowboys tending cattle on the huge Burns Ranch on the south range. They bought and hauled feed across the ice in winter. Some years the ice would form early; other years it would be late. Sometimes the surface would form smooth and hard; other times it would be ridged, rough, and broken. Each year the ice had its own particular character, which in turn determined human communication across it.[71]

Then in spring the ice became unreliable once again. The McKinnons on their ranch in the flats missed the regular visits of the Burns cowboys. Neighbours were once again cut off from one another. Everyone waited for the spectacle of the breakup to commence and for the ice to clear safely from the river. Angus McKinnon remembered breakup with delight: "Evidence of the break up would be extra rushing of water from melting snow and finally the loud cracking of the thick ice pushing and roaring in flood. My father always wore a beard during the winter, and he would shave it off on the day of the break up. School would be let out, and we would all hike up the hill to get a good view of this spectacular event." The eternally romantic Amelia Banister listened, too, for spring. The sound of water announced renewal: "The ice in the river was not long in melting and soon the music of flowing water mingled with the welcome sound of the spring breezes."[72]

As the population grew and economic activities intensified, movement became more important and river crossing more urgent. Rowboats would for a time suffice, but they only moved people. The men in Amelia Banister's fictional homestead regularly carried travellers across the river in their boat. This method, however, required the careful positioning of horses on both sides and close coordination among partners. The next step was a ferry. Where the river narrowed and the riverbanks flattened out as water deepened with not too much current, ferry boats began to appear up and down the river around the turn of the century. A photograph taken in the mid-1890s by A.B. Thom of Winnipeg shows another ferry on the Nakoda reserve near Morley, constructed of timber planks and rough logs and operated by a system of cables that used the river's power to pull the load across. Theoretically, anyone could operate a ferry, but in practice the operation of a ferry became a property right. On the Siksika reserve, for example, Eagle Ribs, a famous warrior, looked after a number of band members from the revenue gained from a ferry that "crossed the river at his flat."[73] Riparian owners, or people who controlled property, assumed this task and operated vessels that provided varying degrees of service. Upriver at Kananaskis a cable suspended across the gorge moved people and supplies across the river. Several cable ferries worked the river between Calgary and Carseland well into the twentieth century. And, of course, a car ferry operates to this day on the reserve near Cluny.

As the volume of traffic increased, bridges replaced fords, rowboats, and ferries as the safest and most reliable means of crossing the river. The CPR built the first bridges across the Bow and the Elbow in Calgary.[74] Upriver, as

Illus. 3.10 The Stoney Ferry on the upper Bow River, 1894–96. (Glenbow Archives, NA 2084-77)

we have seen, the route wound back and forth across the river seeking the best grades with the fewest obstructions. Bridges across pilings hurriedly built during the construction phase washed out quickly and had to be replaced with more substantial, longer-spanned steel beam or truss bridges. But these railway bridges did not officially carry horses, pedestrians, or wagons. The first road bridges appeared in Calgary, where communications north, east, and west all required river crossing. The road from the east was carried across the river at Calgary first in 1885 when Sir Hector Langevin visited the town and instructed the territorial government to undertake the necessary improvement. That bridge was replaced in 1910 with the colourful iron truss bridge which, much reconstructed, still serves as a secondary entrance to the city.

Upriver a flimsy structure on piles driven into a shallow gravelly bar first carried a path westward, but it did not survive the major flood of 1896. It was rebuilt several times until the concrete spans of the Louise Bridge, built in 1911, afforded secure communication in that direction. A real estate developer with property on the north bank of the river built the first wooden truss bridge across the Bow at Centre Street in 1906. The city purchased this bridge in 1911, but it too got swept away with fatal consequences for one of the observers in the flood of 1915. By that time the city had required a more durable structure,

Illus. 3.11 Calgary's Centre Street Bridge under construction, July 1915. (Glenbow Archives, NA 2752-1)

one that connected more directly to the residential heights north of the city. By this time the municipal authorities possessed the resources and the technical capability to build a structure that not only was functional but also symbolically and architecturally expressed the ambition of the city. Thus after the flood of 1915 the massive concrete Centre Street Bridge began to rise where it stands to this day, a municipal and engineering triumph over the river boldly proclaimed by the pride of British lions guarding its entrances.[75]

At Banff the spas and the hotels by the hot springs were located across the river from the train station. So a bridge appeared there very early. Though floods were not as frequent or as devastating in the upper reaches of the river, that bridge became vulnerable from time to time. Eventually, in the 1920s an elegant concrete bridge, its facade decorated with friezes of stylized Indians, established secure communication between the railway and the CPR hotel. Reconstructed several times since its opening, it continues to carry tourists across the Bow and attract photographers.

For much of the settlement era the Bow River acted as a barrier to movement. The river obstruction could be overcome with varying degrees of security and reliability, depending upon the volume of traffic, the local geography, and the institution involved. A railway had to solve this problem early with durable structures. Individuals and groups could more or less get by with makeshift measures for quite some time. Bridges were expensive and an insuperable burden for small communities. Nevertheless, the production of commodities for export, the construction of irrigation works, and the transportation of goods and materials for towns and cities required sturdy structures able to withstand heavy loads and, of course, high water. At Canmore bridges built by the railway, the mining companies, and the municipality connected the town with the coal mines on the southern benches. At Morley, on the Nakoda reserve, the federal government replaced a ferry with an iron bridge in the 1930s. At Cochrane a similar structure, built by the province, crossed the Bow at an arc in the river south of the town. Bridges slowly appeared in the lightly populated areas from Calgary eastward, at Carseland, and on the Siksika reserve on flats south of Gleichen, likely to facilitate the construction of the irrigation project to the south after 1913. The huge irrigation dam at Bassano, begun in 1912, also carried traffic across the river.

It took some time for different governments to summon up the resources and the technology to establish enduring links across the barrier. Indeed, providing bridges was one of the functions that communities organized governments to supply. Even bridges did not provide permanent solutions when a shallow, meandering stream could become a raging, boulder-hurling torrent.

As a Threat to Life and Property

The challenge of building and maintaining bridges reminds us that not only was the Bow River a difficult neighbour to live with, but it could also suddenly and quite unexpectedly become brutally dangerous. Travellers and mounted policemen on their circuit frequently drowned fording the river.[76] The spring roundup often coincided with a seasonal rising of the river waters coming down from the mountains. Horses, cattle, and cowboys could be and frequently were swept away and drowned. Lachlin McKinnon, bringing cattle home from the southern range roundup in 1892, had to push his reluctant herd through high water and almost lost one of his men in the process. Three years later, while he was taking his father across the river with a wagon and team, the horses balked in mid-stream. The wagon upset, throwing father, son, and their luggage into the river. The horses and passengers survived, but the wagon and its contents were lost. Crossing the Bow, Lachlin McKinnon ruefully recalled, was a "treacherous business" with sometimes fatal consequences.[77] The upper reaches of the river also harboured their own special dangers. Charles Shaw, fording the river on horseback near Lake Louise to survey the route of the CPR in 1883, almost drowned when his horse stumbled in midstream. He and his crew remained haunted, too, by the memory of those men on a raft blithely floating to a certain death over the falls.[78]

Drowning was the second most frequent cause of accidental death in Alberta, next to crushing in mining, transportation, construction, and agricultural accidents.[79] Each year the Bow claimed its victims, usually children playing unattended on the riverbank, swimmers, fishermen, and boaters who got into trouble, careless workmen, riders fording the river, and stockmen driving their cattle. Each spring and summer the local newspapers regularly reported these mournful misadventures: "Drowned in the Bow; Wm Murphy Finds a Watery Grave"; "Sad Drowning Accident; Three Lives Lost in the Bow"; "Little Georgie Tuttin of Hillhurst Lost His Life in the Turbulent Stream Yesterday." Attempts to rescue people in trouble in the water often ended in multiple tragedies. The river occasionally delivered unidentified bodies to downstream communities. For the desperate, the river afforded an exit from this world. Suicide by drowning was a gruesome choice, as these headlines attest: "A Nameless Woman Writes 'Good-bye. Am Tired of Life' on board and drowns herself in the Bow River"; "Tried to end life: Jumped into the Bow." Murderers dumped their victims into the river to get rid of the evidence. In 1911 fishermen stumbled upon the grisly scene of a man in workman's clothing with his throat slashed. A local businessmen was found in shallow water, his feet bound by wire to a timber. On at least one occasion a murderer escaped by swimming across the Bow.[80] The river provided a voluntary and an involuntary exit from life.

Two volumes of Amelia Banister's three-volume melodramatic novel, drawn from her experience living on the banks of the Bow, build towards climaxes centred on the river. In the plots of each the river actively plays the part of the villain and symbolically acts to mark a growing distance between the hero and the heroine. In the first volume the hero John has to navigate the Bow with a helpful neighbour to rescue his wife and ailing son on the other side. "My boat is safe and I know the Bow like a book," he reassures them as he manoeuvres the boat into position. "Both women shrank at the sight of the roaring river and it took all their courage to let John settle them in the boat before he jumped in and took the oars." They get safely across, but alas, the child, John and Freddy's son, died. The sad couple make the return trip across the river, now quite placid as the flood receded, with a tiny white coffin on the seat of the boat. On this river crossing the gradual estrangement between husband and wife commences.[81]

In the second volume Freddy loses her way riding home at night in a storm on her favourite horse, Mumps. In the darkness and driving rain, horse and rider plunge over a cutbank. In a flash of lightning Freddy tumbles towards the river roaring below. Mumps survives the fall and returns home riderless. The men, just back from fighting a prairie fire, follow the horse back to the scene of the accident, where, in the dark, they find Freddy miraculously alive, her body crumpled precariously at the foot of the cliff. She is rescued and rowed by strong men against the current, raging in the darkness, to a lantern guiding them to safety and the gravel beach on the far shore.[82]

The river was a force to reckon with. The ice could suddenly give way under humans and livestock. Besides the river itself, the banks could also be fatal. Animals stumbled to their death over the unfenced cliffs. The McKinnons lost their first Red Sussex cow, Rosy, when she fell over a cutbank into the river. Animals, farm equipment, and people could occasionally get trapped in quicksand. For the families living on the ranches and homesteads along the Bow, floods posed the most frightening menace, especially the flash floods that occurred often during the spring breakup when water dammed up behind ice jams.

Life on the flats had its advantages for small ranchers, but there were also liabilities. When the dominant form of agriculture shifted from ranching to wheat farming and it became more convenient to locate up on the prairie closer to the operations and the railway, the McKinnons and their neighbours left their old homes in the valley without much apparent regret.

As Provider

What the river took away, it also delivered – to someone downstream. It was, as well, a source of important materials upon which indigenous people, ranchers, railroaders, and homesteaders depended.

When Lachlin McKinnon lost his wagon crossing the Bow with his father in 1895, he salvaged only the front wheels. The next year one of his neighbours found the back wheels, and they were reunited with the rebuilt wagon. The flood of 1902, on the other hand, delivered a considerable quantity of drift material that McKinnon was able to salvage for use on his ranch. He and his neighbours depended upon the river for building materials. McKinnon gathered enough drift logs from the river to build his homestead. One year the Eau Claire Lumber Company booms broke, distributing the wood far downstream. Rather than haul it back to the mill, the company set up a small sawmill and sold the logs and lumber to the ranchers.

Wood, which was in very short supply on the southern prairie, could be found or cut on the riverbanks for building materials and heating. The river valleys contained the only significant stands of trees east of the foothills on the open plains. As we have seen in the previous chapter, indigenous peoples made a range of uses for the different tree species along the Bow – to supply fuel, building materials, and even medicines. Settled on the reserves, they gathered driftwood for burning on the river flats, often relocating to riverbank dwellings in winter to be close to shelter, fuel, and water. Both the Siksika and government officials appear to have settled on Blackfoot Crossing as a reserve site in part because it was heavily wooded.[83] Travellers, crossing the prairie, welcomed the sight of the river forest as a sign of life. Charles Shaw, the railway surveyor, remembered: "After a long, weary journey over a treeless and often waterless prairie, just before dark one evening we rejoiced to see a line of trees ahead." The railway surveyors, marking their line across the prairie, cut saplings and gathered driftwood along the river to serve as location stakes.[84] Homesteaders regarded trees as signs of fertility, biblical tokens of happiness and security. In Amelia Banister's fiction, "Great balm-of-Gileads and cottonwoods stood out majestically as if saying, 'the aborigines named us Ah-ka-ke-ke. We are not ordinary trees; we're the biggest trees on the Bow.'" Freddy, as a new bride, was overjoyed at the verdure of the location her husband John had chosen for their house: "This isn't bald-headed prairie. I'm surprised to see the trees. It will be easy to be happy there with you, John."[85]

Winter was a time for cutting wood. This was done on an industrial scale in the mountains, in the lumber camps, and in the railway tie-cutting operations. On the ranches the hands busied themselves in the winter cutting and hauling firewood in the river forest. In the early years, when coal was harder to come by and expensive, ranchers and homesteaders relied upon wood for heat and cooking. Besides tending to the livestock, Lachlin McKinnon and his sons passed the winter months gathering fuel: "There was plenty of driftwood along the river and we also cut down some big cottonwood trees. This was the best stove wood, even though it was heavy hauling it home and cutting it up." The wood had to be cut and split into proper lengths "to suit the women and their stove." As in any tree-felling operations, there were dangers. McKinnon's neighbour Jerry Newton was killed in 1914 cutting wood on his river flat.[86]

The river was the source of other essential materials and foodstuffs as well. The abundant wild berries growing on the river flats provided essential nutrients in the traditional indigenous diet. Ranchers and homesteaders, too, made an annual ritual of picking mainly the saskatoons and gooseberries but also wild raspberries, wild strawberries, and black currents. These would normally be made into jams and jellies. In season, the feasts of "wild strawberries, raspberries and saskatoons along with fresh unadulterated cream and sugar and of course fresh homemade bread and butter" lingered in Angus McKinnon's memory. "This wild fruit certainly broke the monotony from a constant diet of dried prunes, apples and so forth."[87]

Illus. 3.12 Hauling ice blocks from the Bow River at Bowness, 1911. (Glenbow Archives, NA 1604-2)

More prosaically, the riverbank and beds were a source of gravel for construction projects and ice for refrigeration. When the CPR built its right of way, the contractors were able to assemble some of the necessary construction materials locally. Railway ties, piling, and bridge timbers were cut in the mountain passes and floated down the river or moved by rail into position. The engineer P. Turner Bone, who was in charge of building some of the railway bridges over the Bow and the Elbow, excavated the boulders needed to rip-rap the piers from the river nearby. The McKinnons drew gravel from the river for the concrete foundations, lanes, and roadways of their new farmhouse on the prairie. Farmers excavated fertilizer from the foot of some of the old riverside bison jumps. All along the river settlers cut ice from the river and hauled it to their ice-house dugouts to preserve food in the summer. From 1900 into the 1950s the Alberta Ice Company maintained a very large commercial ice-production facility on the Bow River above Calgary to serve the town and the railway.[88]

The river also exposed and made accessible a promising alternative source of fuel: coal. Lord Lorne and his party publicized the existence of coal. G.M. Dawson and his colleagues at the Geological Survey identified promising seams. Charles Shaw even staked a claim in the mountains, only to lose it to bureaucratic chicanery. The ambition of one of the early ranches is spelled out in its formal title: the Military Colonization and Coal Company. On the lower portion of the river the boom town of Bow City briefly flourished.[89] Large, well-known commercial coal mines operated at Anthracite (1889–1902) and Bankhead (1902–22) in the national park at Banff and downstream at Canmore (1889–1950).[90] Less well-known is the existence of a coal mine on the Siksika reserve, where several seams were identified and coal extraction and sales occurred. An Indian agent with a mining background persuaded the Siksika to pool their efforts in one commercial shaft (1920s to mid-1930s). Each fall groups of men and their families would relocate to the mine to work the seam. In winter they would sell their produce in communities nearby. They became sufficiently competitive as coal merchants in the region that regulations were proclaimed by municipal government to discourage them. Coal companies based in Lethbridge, in Crow's Nest Pass, and on the upper Bow supplied coal in greater quantities and better quality more reliably and at a lower price. The Siksika mine on the Bow was thus restricted by the market to supplying local needs outside the cash economy and the requirements of the residential school.[91]

As a Meeting Place and Pleasure Ground

For all its menace, the river was an all-season source of pleasure and amusement. In the middle of his memoir, Lachlin McKinnon paused to correct a misleading impression he sensed he might be leaving: "The only reference to the Bow River thus far has been in connection with damages and tragedies. This is far from being fair to such a beautiful as well as useful river."[92] The river was a dream world of childhood. It nurtured the solitary pleasures of angling and boating, as well as the informal social intercourse of skating, swimming, courting, and communal picnics.

Fords and crossings were focal points in the indigenous plains and as a result also became the locations of a police post and trading posts. Out of some of these grew towns and of course the city of Calgary. Fort Brisebois, later named Fort Calgary, was situated at a traditional ford in the river. This site, though far from ideal, thus became a major rail point and later a junction and evolved into the market centre for southern Alberta. Canmore, Morley, and Cochrane were also situated at narrow river crossings.

Towns located their parks and pleasure grounds by the river. Calgary's amusement park and picnic ground, Bowness Park, evolved after 1911 eight miles above the town on flats on a bend in the river.[93] The federal government made a gift to the rising town of Calgary of islands in the river for park purposes.[94] At Banff the winding course, colourful lakes, and stupendous waterfalls of the Bow River became the most picturesque sites and amusement grounds for the majority of tourists in the park.

Fishing in the cold, swift-flowing stream afforded hours of pleasure and with luck some tasty meals for weary ranchers, farmers, construction workers, tourists, and town folk. Angus McKinnon reckoned that he enjoyed "better than average boyhood days along the Bow River, with plenty of big cotton wood trees to climb, clear pure water to fish and swim in, plenty of wild life and open spaces." With grasshoppers on his hook, McKinnon and his friends tried to catch the big fat trout they could see lurking in the pools.[95] The irrigation weir at Carseland, built in 1910, however, created slack water in which much different and less edible species of fish propagated. In a few places, under ideal circumstances, swimming was possible in the Bow, but generally the water was too cold even in midsummer. Side channels, old oxbows, sloughs, and streams joining the mainstem provided more endurable opportunities for bathing. The eccentric Colonel Strange, however, would hobble down to the river each day to have cold water poured over his badly injured leg in the hope of healing it. The notion of bathing and water as a cure lived on along the Bow.

If the flats provided meeting places for indigenous peoples, the ranchers, homesteaders, and CPR construction workers took advantage of the same geography for pleasure and recreation. The McKinnons never missed a Sunday picnic down by the river, with the men taking a day off from the irrigation works construction. McKinnon family picnics were held annually on their flats. In time, the United Farmers of Alberta organized the local farmers at picnic gatherings by the river.

Occasionally, the Bow arranged itself into an ideal configuration for that most Canadian of all pastimes, skating. When the good ice formed, the young ranchers and their boys played hockey on the river against a team from the Dunbrow Indian Industrial School. When the river flooded perfectly, the McKinnons and their neighbours could skate all the way upriver to Calgary.[96] Just such an event features in the second volume of Amelia Banister's novel. The temperature has plummeted during the night, and the river has frozen into a sheet of glass. The next day, in crystalline air "as intoxicating as champagne," the family either skate to town or are carried in their cutter behind horses dancing over the ice in spiked shoes.[97] Clear, smooth ice also drew out the town folk, the temperatures notwithstanding. Up in the mountains at Banff much lower temperatures made the thrilling sport of iceboating a regular winter activity on Lake Minnewanka. Spring breakup, announced by piercing rifle-shot ruptures as the ice cracked, great groaning plates of ice grinding at one another, and the sudden flood of water as the pressure was released, amused those on high ground, especially the children let out of school for the spectacle, untroubled by the damage being caused.

The Bow River had character or, rather, its neighbours anthropomorphized its changeability. It wound its way into the lives of many of the people who settled on its banks. Indigenous peoples, ranchers, townsfolk, and farmers had to adjust to what they called the moods of the river. They grew up with the river as a neighbour, sometimes generous, sometimes ornery, sometimes violent, but always memorable. The Bow was an independent, autonomous force. It gave the settlers great pleasure and sometimes great pain. It was a place of remembering and forgetting, of childhood dreams and adult nightmares. It formed a prominent marker in their lives, both a symbol of the things that divided them and a place of meeting where they formed a community. Respected and admired rather than loved, the Bow was best appreciated from a distance.

Settlers fundamentally altered the ecology of southern Alberta. Their cattle, sheep, and swine, ploughs and seeds, hunting and weeds, transformed the environment of the prairie in a myriad of ways.[98] But neither the settlers' economic needs nor their resources permitted them to think they could manage the river. In that respect they were more acted upon than acting. Though they would seize, possess, and more or less vanquish the land, they did not imagine they could tame or harness the river as the lumber, railway, hydroelectric, and irrigation companies might. Living with the river, they went with the flow.

4 · The Wooden River

People see different things in rivers. When Isaac Kerr, president of the Eau Claire Lumber Company, looked at the Bow River in 1883, he saw houses, schools, shops, and factories floating in it. As his gaze shifted from the forest-clad front range to the rolling foothills to the stream of water flowing out onto the broad prairie, he dreamed of choking that river from bank to bank with sawlogs jostling their way on the annual spring drive to his mills in Calgary. There, in two miles of booms, channels, and weirs, the logs would be stored, sorted, and steadily shuffled by the current to the screaming mill, where the river itself provided the power necessary to transform logs into timbers, lumber, shingles, and sawdust. The commercial and civilizing potential of the forest and river lacked only the capital, enterprise, organization, and labour that he could command. When the lumberman Isaac Kerr looked at the river, he saw a city.[1]

But when William Pearce, the redoubtable western agent of the federal Department of the Interior and irrigation promoter, looked at the Bow River at the beginning of the twentieth century, he saw fields and pastures. The river, when properly dammed, diverted, channelled, and ditched, could bring a thirsty land into cultivation. He worried about the long-term security of that water, however. What would happen to his dreams of irrigation if lumbermen such as Kerr had their way and stripped the protective forest cover from the foothills and eastern slopes of the Rocky Mountains, the source of the Bow River? For the prairie to flourish with irrigated farms and stable communities, the forest in the mountain headwaters would have to be protected. The Bow and its precious water depended on those trees. For Pearce, too, but in a different way from Kerr, the Bow was a wooden river.

Cruising

Soon after the ranchers and the railroaders, the lumbermen arrived. Indeed, a loose continental web of communications brought these groups together on the banks of the Bow. The story begins, as so many do in Canada, with government.

By the spring of 1883 the Pacific railway had been laid halfway across the prairie. The layout of the line had been settled as far as Calgary. Westward a route up the Bow River valley through the front range of the Rockies had been agreed upon and was being surveyed. Where the rails would cross over the Continental Divide into British Columbia remained to be decided. Anticipating the need for building materials, the Department of the Interior advertised for tenders for ten timber berths, or limits, covering about 500 square miles on the western slopes of the Rocky Mountains in the Bow valley. Government bureaucrats, without knowing much about the quantity or quality of the timber or the lands, simply plotted neat rectangles on the map along the river valley; surveyors planted stakes on trails at the most accessible boundaries; and the timber was offered for sale to the highest bidders. It was for the potential owners to estimate what these rights might be worth to them. Four of the berths straddled the Bow River from modern Canmore to the small mine site, Silver City, west of Banff; two covered the lower reaches of the Spray River; and four more were strung out along the course of the Kananaskis River, a tributary that joined the Bow at the eastern edge of the mountains.[2]

Timber sales constituted a significant source of government revenue before income, corporation, and sales taxes. This sale of federal timber licences in 1883 followed in a long and often tawdry succession of government auctions of public resources for private exploitation. A tried and true ministerial defence of opening the Crown lands treasure trove was fear that good timber would be lost to the fires that always accompanied railways. Sell for the public benefit and cut it down for private profit before it burns up went the logic. So the forest on the upper Bow would be sold for lumber.

Kutusoff N. MacFee, a Winnipeg lawyer, got early wind of the sale, tipped off perhaps by a relative, a clerk in the Privy Council office in Ottawa. Having heard glowing reports of the mountain forest and determined to benefit from his inside knowledge, he hastened to Eau Claire, Wisconsin, centre of the midwestern US lumber business, where he brought the impending auction of these limits to the attention of a group of established lumbermen. As these Wisconsin businessmen could see ahead to the not-so-distant future when all the good timber on the eastern edge of the Great Plains would be gone, they were predisposed to consider possibilities further west. Moreover, several of the people MacFee approached were Canadians who had followed the white pine lumber trade from Ontario across Michigan into Wisconsin. They had no hesitation about doing business north of the border. Thus near the end of June, fourteen lumbermen agreed to put up $100,000 between them to bid on these limits.[3] MacFee, who put nothing into this speculation but his wit and his connections, nevertheless demanded a substantial share of the benefit and, being a lawyer, drafted an ironclad agreement to protect himself. For a quarter share of the subsequent proceeds, MacFee undertook "to obtain the assistance of my friends with the Dominion Government to obtain all the privileges possible respecting the said timber limits."[4] While MacFee managed the tendering process in Ottawa, the group incorporated in Wisconsin under the name the Eau Claire and Bow River Lumber Company.

MacFee's informants on the ground – Tom Watts, agent of the CPR, David MacDougall, "resident for 10 years adjoining the limits," and Mr Alders of the Hudson Bay Survey – must have impressed him with the richness of the timber stands. Moreover, the Wisconsin group must have really wanted the timber because MacFee, on behalf of his associates, bid extremely high to be certain of getting it. When the results of the auction were announced on 3 July, the Wisconsin lumbermen had won provisional rights to nine of the ten berths with tenders on average two and a half times higher than their nearest competitors, at a total cost in bonus of almost $50,000.[5] To complete the deal the new owners had to file surveys of their limits, undertake to build a mill and commence operations by 1886, pay the specified annual ground rent per square mile, and submit monthly returns of timber taken to the Department of the Interior, upon which timber duties would be charged.[6] Upon receiving news of this success, Isaac K. Kerr, who had been born and raised in Russell near Ottawa, and two companions made preparations to go west to inspect the new properties. They would gather first-hand information to help the group decide whether to make the necessary investments to complete the terms and conditions of sale or flip the properties to other buyers as a speculation.

Late in July Kerr and his companions caught the train up to Winnipeg, where they took the CPR west to the end of steel. Throughout the punishing heat of July, track gangs, well supplied with carefully marshalled materials, were laying four, five, and, on the occasional big push such as 28 July, more than six miles of rails per day. During Kerr's trip the end of steel would have just reached a collection of tents, shacktown, and NWMP fort at the confluence of the Elbow and the Bow Rivers known as Fort Calgary.[7] From there the men rattled across the prairie in a democrat to David MacDougall's trading post at Morley. MacDougall, whose father, the Reverend George, had founded a mission to the Nakoda in 1873 and whose brother, the Reverend John, continued it, had operated a general store and trading post next door to the church for many years. He had been, it will be recalled, one of MacFee's informants on the quality of the timber limits. He and his brother had travelled up the Bow and the Kananaskis numerous times with the Nakoda and hunting, fish-

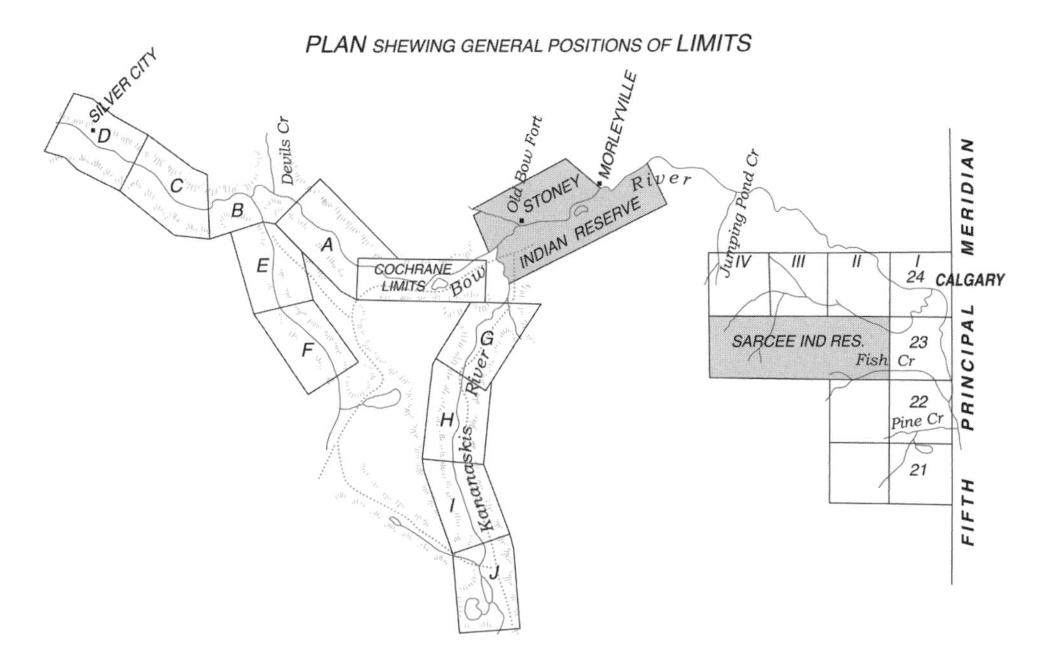

Map 4.1 Timber berths in the upper Bow basin. (Glenbow Archives, Eau Claire and Bow River Lumber Company Fonds, M1564, box 3, file 29, Timber Limits Maps)

ing, trapping, and trading with the Kootenai across the pass in the Columbia Valley at Windermere. According to a pioneer's recollections, MacDougall's store was something of a sight, a dim den filled with the hides of big game, especially grizzly, cinnamon, brown, and black bears that MacDougall and his indigenous companions had killed on their mountain adventures.[8] MacDougall arranged for "Indian George Kananaskis" to lead the three American visitors on a tour of their limits.

On Thursday, 10 August, Kerr's party left Morley on horseback. For the next six days they followed a path up the Kananaskis valley running more or less parallel with the river and traversing Berths G, H, I, and J. They crossed the summit into the upper reaches of the Spray watershed (where three inches of snow fell on them), intending to descend it to examine their other limits, but their guide refused to take them further on account of the numerous windfalls across the trail. He may also have been less than comfortable guiding them in territory frequented by traditional enemies, the Kootenai. They then retraced their steps back to the Bow, where "Indian George Kananaskis" parted company with them. After lightening their packs, they proceeded on their own upstream along the recently cleared railway roadbed to examine the other limits in the vicinity of present-day Banff. After some difficulty – their horses wandered off – and some chance meetings – with Sandford Fleming and Principal George M. Grant, who were on their way to examine Major Rogers's supposed route through Kicking Horse Pass – they traversed their three limits on the Bow: Berths A, C, and D. After several false starts they eventually located the mouth of the Spray River (locally called the Lorne) below Bow Falls and followed a rough trail up through Berths E and F until they found themselves about three miles from the point near the summit where they had previously turned back twelve days earlier. The second of September found them back down in the Bow Valley, then thick with tie-cutters and railway grading crews. A few days later they arrived back at MacDougall's store at Morley after just over three weeks in the saddle.[9]

When I.K. Kerr looked at a forest, his mind turned it into specific quantities of sawn lumber. His diary of that trip shows, as might be expected, a lively interest in the forest cover, especially the species of trees and their density, height, and diameter at the stump. The Bow valley itself and the lower reaches of the Kananaskis contained relatively little merchantable timber; however, as the inspection party moved higher up, the quality improved. On Berth I, for example, Kerr noted: "We came through some excellent groves of timber (spruce), some of it as large as 33 inches on the stump and a few trees [that] would cut four logs to the tree." Commenting on the forest above Bow Falls, Kerr observed: "The timber up towards the summit and for a considerable distance east of it is principally cypress and spruce pine, none of it growing larger than 18 inches on the stump but it grows to a good length, say from 90 to 100 ft high and very thick on the ground." On 1 September, tramping up through Limit E on the Spray, Kerr noted: "As we proceed farther up stream the timber grows larger and about half way on the limit we found some groves of very fine spruce."[10] The lumbermen passed through quite a lot of burnt timber in the main Bow Valley, as might be expected from the recent silver mining and railway activity. But Kerr seems to have been surprised by the amount of burnt timber they encountered in relatively remote regions high up in the passes: at the junction of the Kananaskis and Smith-Dorian Creek, for example, and on the upper reaches of the Spray. With his practised eye, Kerr estimated the quantity of timber in board feet that could be taken off each berth: 2 million board feet in A; 12 to 15 million in C; 15 million in D along the Bow; 2, 8, 40–50, and 25 million respectively in Berths G, H, I, and J along the Kananaskis; and 15 and 35 million in Berths E and F on the Spray. The limits might look much alike on the map, but topographical variation and fire made for an uneven distribution of commercial timber.

Kerr studied the rivers just as thoroughly as the forest. When he looked at a river, he instinctively imagined a transportation system for logs. Could it be driven? Could it be improved to make it drivable? Where could logs be stored in booms? Where were the possible snags and backwaters where timber could be lost? Where could dams be located to best advantage? For example, on this trip Kerr worried about the flat banks and shallows on the lower reaches of the Kananaskis. Where the river braided into several channels, he calculated which channels could be closed off with brush dams to prevent timber being lost. Two big waterfalls higher up on the same river set him to thinking about explosives:

> About halfway up this limit there are two falls about a half a mile apart. The first one has a short pitch of about 25 ft coming down between the rocks which rise to a height of 50 ft on either side, almost perpendicular. Situated near the foot of this fall is a large rock which would require dynamite to make it drivable. The upper falls are a success of pitches coming down between rocks 100 ft high in places for a distance of 40 rods. At two of the pitches the channel is not more than ten or twelve feet wide which I think can be enlarged sufficiently by the use of dynamite to make the driving of logs all OK. The water at these falls comes down with such force that nothing but the narrowness of the channel would stop the logs.

Kerr seemed pleased to note the heavy flow of water comparatively late in the season; indeed, he was prevented by the rush of water from crossing over to examine some promising parts of his limits more closely. He made detailed notes on the structure of Bow Falls, over which he would have to drive timber from some of the limits. In the upper reaches of the river system, however, the lumbermen would have to make their own current to move the logs. At lakes and sloughs Kerr had his eye out for places where dams could be built to create reservoirs, booming grounds, and flash dams to enhance the spring freshets. Illustrating his diary with sketches on the facing pages, Kerr scratched rough maps of the course of the rivers in relation to the stands of timber. But on the whole he liked what he saw. "The Spray River," he scribbled in his diary, "said by some to be so named by the Indians on account of the Spray Falls on the Bow at its junction, will make with very little improvement an excellent driving stream."[11] The transportation system was not perfect – it would have to be improved here and there with dynamite and some dams – but it was more than adequate.

After a brief rest at MacDougall's post, the American visitors traded saddles for oars, even though by this time the CPR rails had already reached Morley and were heading up the Bow valley into the Rockies. Clambering into a boat specially built for them by one of the teachers at the mission, Kerr and his party slipped downstream in the swift current. This way the Americans could survey every yard of the mainstream of the Bow from a log's-eye view. And they could keep their eyes peeled for the best mill site. They descended quickly down the Bow, past Calgary, and then on through slower water, where much rowing must have been required, to the big timber railway trestle at Medicine Hat. There they arranged for a young lawyer, James Lougheed, to sell their boat, while they hastened back to Wisconsin by rail to report on their findings to their partners.

Attracted by the timber and the prospects of the Canadian northwest, the Eau Claire lumbermen decided to invest. In the spring of 1884 Louis B. Stuart, a professional timber cruiser, was retained to conduct the detailed surveys of the timber required by the Department of the Interior before it issued a licence. Stuart essentially walked a line the length of each limit. In his report this line ran down the centre page of foolscap. In the left margin the cruiser noted the latitude and longitude. On either side of the centre line he sketched in important topographical features, groves of high-grade timber, burnt-over patches, rivers, ridges, and slopes. When the survey line ran up a river valley – the Kananaskis, for example – Stuart drew a map of the river, with its waterfalls, branches, and braids, winding back and forth around this centre line. At regular intervals running this line he noted the elevations of the valley, the species of trees (spruce, poplar, jack pine, cypress) and the condition of the forest (burnt, second-growth, small, medium, or large trees).

Stuart drafted a general descriptive report of each berth, which he attached as a cover memorandum to these reams of field notes. On the Bow valley limits (A, C, and D) he confirmed: "Most of the timber of any value is confined to the immediate banks of the river along which it forms a belt varying in width from one to ten chains, though clumps of good timber may occasionally be found back some distance from the river." Most of the trees were of small to medium size, though in a few places trees of 12 to 18 inches in diameter at the stump could be found. The best-timbered limit, in Stuart's view, was D, the furthest up the Bow. The lower reaches of limit E on the Spray River contained mainly small jack pine and were marked by "patches of prairie." Small streams fed the main river at about half-mile intervals, creating a deeply incised channel. Up into the Spray Valley, the timber increased in size, some trees reaching 2 feet in diameter. Stuart mapped several of what he called "belts" of heavy timber continuing into limit F. Moving to the other limits, he noted that fire had destroyed much of the forest on the lower Kananaskis, and he commented again on the increasing height, density, and diameter of the forest further upstream into limits H, I, and J. In a rare deviation from utility

Illus. 4.1 Eau Claire Lumber Company loggers in the Banff area about 1899. (Glenbow Archives, NA 265-19)

toward aesthetics, Stuart wondered at the different scenery he encountered on the upper stretches of the Kananaskis. On account of numerous fires, the shore of the lower lake on the river was "a scene of great desolation." By contrast, the upper lake – deep and clear, filled with fish, and fed constantly by torrents and springs, ringed as it was by a thick forest, steep rocky slopes, and jagged peaks – presented "a very picturesque appearance."[12]

In the summer of 1885 Kerr and Peter Prince, another Canadian immigrant to Wisconsin from Quebec, set out with their equipment to commence logging operations and establish their sawmill. Kerr became president of the Eau Claire and Bow River Lumber Company and Prince manager of the mill. Over that winter they had two crews in the woods, one based at Silver City above Banff, working limits A, C, and D, and another crew at the mouth of the Kananaskis on limit G.[13] It may seem incongruous now, a full-scale logging operation within the boundaries of the Rocky Mountains National Park and within sight of the town of Banff, but it was not viewed that way at the time. Indeed, the superintendent of the park noted that logging off the burnt timber in the region greatly improved the overall appearance of the park.[14]

Illus. 4.2 Log dump on the Eau Claire Kananskis River limits in the 1890s. (Glenbow Archives, NA 1360-7)

When the Eau Claire company and the railway contractor James Ross decided to give up their rights to the timber limits A to D after taking out the readily available merchantable timber, the Department of the Interior seized the opportunity to set aside a timber reserve for the Siksika on limit B. Other indigenous groups recently settled on reserves had tracts of timber allocated to them for fencing, farm buildings, and houses, but the Siksika had been overlooked. A berth was redrawn in the region of Castle Mountain, a cut-over and heavily burned area from the lumbermen's point of view but said by surveyors to contain enough timber for the indigenous people for a hundred years. The only problem, according to the surviving archival correspondence of the Department of Indian Affairs, was that the Siksika did not want timber that was so far away and difficult to get at. They preferred a location closer to their reserve in the vicinity of present-day Bragg Creek. Siksika woodsmen conscripted to mark out the Banff timber berth refused to work and defied the exasperated surveyor. After several years of discussing other possible locations for the limit, the issue appears to have been dropped without resolution.[15]

A century later, in the early 1980s, the Siksika made a specific claim to the government of Canada over the non-existent timber berth. In 1993 the federal government accepted that it "has a lawful obligation within the meaning of the Specific Claims Policy to set aside a timber limit as a reserve for the use

Illus. 4.3 Eau Claire crew driving logs on the upper Bow in the 1890s. (Glenbow Archives, NA 1432-11)

and benefit of the Siksika Nation." The two parties are currently in mediation on the issue.[16] In any event, commercial logging in the immediate vicinity of Banff ended in the 1890s, while logging continued on one Eau Claire limit up the Spray River, in more remote portions of the park, into the 1930s.[17]

On his voyage downstream in 1883, Kerr had spotted a near-perfect mill site on the Bow just north of the Calgary townsite. He and Prince acquired this relatively inexpensive riverside property, a half-mile or so north of the main real estate action along the railway line. Compared to the railway, the riverbank was the emerging city's back door. To confirm its riparian right, the timber company obtained a thin sliver of land running along the south bank of the river. By the fall of 1886 Prince had a small, 75-horsepower, steam-driven sawmill up and running. For the first season the company brought its logs in by rail from Banff. That fall Eau Claire crews drove piles in the riverbed and built a boom on the river ready to corral their logs from the drive next spring.[18]

Thus by the summer of 1887 Wisconsin capital and Canadian entrepreneurship and labour had turned the Bow into a classic Canadian log-driving river, a forest connected to a mill by a swift flowing stream. The Eau Claire people were not the first, nor were they alone, but eventually they would have the river to themselves. In the winter months loggers felled the trees on the moun-

Illus. 4.4 Log drive at Bow Falls in the 1890s. (Whyte Museum of the Canadian Rockies, NA 66-1348)

tain slopes and piled them in dumps on the banks of the upper branches of the river. Come spring and early summer, intrepid woodsmen with pikes and peavies would coax hundreds of thousands of logs downstream to the mill on the flood each year. Early tourists to Banff would have watched Eau Claire logs shooting over Bow Falls.[19] In the pool below the falls, at the confluence with the Spray, logs from the upper Bow and the Spray would be boomed and then sent on the next stage downriver together, over the shallow flats at Canmore, through the rapids at The Gap, and out of the mountains into the foothills. At the mouth of the Kananaskis another boom of logs joined the pack for the swift journey over Kananaskis Falls, then Horeshoe Falls, and down the long reaches through the Nakoda reserve, past Cochrane, and around the grand sweeping bows in the river above Calgary. The first drive in 1887 was a nightmare, according to Theodore Strom, who worked in the mill. Six of the river drivers were killed when an unexpected freshet swept their boat over Kananaskis Falls; three men miraculously survived. On account of the many windfalls and rocks obstructing the flow and the unseasonably cold weather, it took most of the summer to bring the logs down to Calgary.[20]

Nevertheless, the drive of 1887 established a pattern that became a seasonal ritual. For almost two months beginning in late May, gangs of thirty or forty men, working dangerously long hours, would coax a ten-mile flotilla of logs downstream three or four miles a day. With tremendous physical labour and great dexterity, the drivers would keep the wooden river flowing, forcing logs into the main channels, prying and winching them off snags, preventing jams at narrows and in rough water. Then, once they had reached Calgary, the logs had to be boomed, sorted, and shuffled up to the mill. In the off-season, river crews would improve the river by constructing brush dams at troublesome spots, hauling out snags, dynamiting open tight channels, and building flash dams higher up to augment the flow. Here and there a lake had to be raised with strategically placed dams to store water for the drive.

Isaac Kerr realized his ambition; for the rest of his life he filled the Bow River with logs and drove them to his mill. In a matter of years, between 1886 and the height of production in the 1920s, the Bow from a commercial perspective became primarily a wooden river.

Millstream

If the Bow and its tributaries had to be improved to make them better driving rivers, so too the Bow at Calgary had to be modified to make it a better millstream. In short order, Kerr and Prince had plans drawn up for a dam to be built in the Bow which would raise the water level by about twelve feet. This structure, which connected with gravel bars and an island, would divert water into a narrowing channel leading to the company's mills. By cutting a channel through the gravelly isthmus that connected this island with the shore and building a power station at that point, the company would be able to float its logs right into its mill and, with twelve or more feet of head, generate the hydroelectric power necessary to saw the logs into lumber. The Bow would thus be engineered to both transport the logs and power the sawmill.

Kerr and Prince organized the Calgary Water Power Company as a subsidiary of the Northwestern Lumber Company of Wisconsin to hold the necessary water licences from the Department of the Interior, to supply electricity to the mill, and after negotiating a franchise with the city, to sell surplus alternating current to the budding municipality and its merchants. In a race to meet the stiff terms of the franchise, the company attached a generator to

Illus. 4.5 View of the Bow River at Calgary looking west towards the lumber mill on Prince's Island, 1893. (Glenbow Archives, NA 1172-2)

its steam engine, threw up a two-storey shed across its millrace, installed a turbine and dynamo, and commenced business as a hydroelectric company in competition with an existing and highly unpopular direct-current street-lighting monopoly, the Calgary Electric Company. With the help of Senator James Lougheed, the Calgary Water Power Company negotiated the rights to the entire flow of the Bow River at Calgary. It then built the L-shaped dam, which raised water levels, forced the current past its mill, and increased the efficiency of it power turbines.[21] The Eau Claire group thus possessed two extremely valuable rights early on: a virtual perpetual franchise from the city to operate as an electric utility and what amounted to sole rights to extract power from the river at Calgary. Soon its rival, the Calgary Electric Company, collapsed. Thus for a time, from 1894 to 1906, the Calgary Water Power Company also enjoyed a monopoly in supplying electric light and power to the city.[22]

But the river could be capricious. It could give and take away, and at times it could be too much to handle. "The Bow River was tricky then as it is now," Theodore Strom recalled in the 1930s. "Sometimes there was hardly enough water for our boilers, and then there would be big floods that would drive us off the river." The river showed its unpredictable nature most vividly during

Illus. 4.6 The Eau Claire log boom during the flood of 1915. (Glenbow Archives, NA 4355-14)

low water in the autumn of 1893 when the company was building the Hillhurst dam. As Strom tells it,

> We had all our scaffolding built across the river to carry our pile driver and engine and were just starting to drive our solid work. The weather had been fine but it had been colder up west than it had been at Calgary. There had been a big ice jam up the river somewhere that had held the water back. As I remember it, it was on Thanksgiving Day at about three o'clock in the afternoon and we were all working on the river, when a wall of water and ice five feet high came sweeping down on us.
>
> We all ran for shore as fast as we were able to. The ice jam took all the false work out except the span where the pile driver and engine were standing. The flood handled the big 34-foot square timbers as though they were matches. As nothing more could be done that day, the boss said we could go and have our Thanksgiving dinner.[23]

Several more flash floods complicated construction before the dam could be built. That year the water ran the two turbines satisfactorily all winter. But most years, even with the dam diverting all the flow into the power canal,

Illus. 4.7 The Eau Claire Lumber Company mill on the far left and the Calgary Power Company turbine house and spillway, centre, around 1911. (Glenbow Archives, NA 1044-6)

there was not enough water to drive the generators. Lights flickered, dimmed, or went out completely until the steam plant kicked in.

Growth in the demand for power both from the sawmill and the city led to constant expansion and reconstruction. By 1900 the ramshackle power shed had been replaced by a much neater, white-painted, black-trimmed clapboard structure, matching the design of the rest of the mills. More turbines were added until the plant reached its maximum output of 600 horsepower about 1904. The Calgary Water Power Company also expanded the capacity of its steam plant, fuelled by mill waste not only as a standby but also to carry the base load during the winter periods of low water. Needless to say, these frequent interruptions in service, combined with a somewhat crusty "take it or leave it" monopolistic attitude on the part of Peter Prince, led to tremendous hostility between the company, the city, and its customers.[24]

By 1894 this man-made waterpower site, which took advantage of the local topography of the river, was in full operation. The Eau Claire company built a much larger mill, storage yards, and outbuildings. Gentrified remnants of this industrial site remain to this day as theme buildings in the Eau Claire Market urban renewal project. But from the 1890s through to the 1940s this was a gritty industrial area, in milling time ringing with the screech of the saws, the air redolent with the sweet smell of sawn lumber, everything covered in

Illus. 4.8 Colonel Walker's first sawmill at Kananaskis, 1886. (Glenbow Archives, NA 387-24)

a fine dust, and underfoot the spongy give of sawdust. From midsummer to fall logs filled the sorting jacks in the river and the channel between the south bank and what was now called Prince's Island. Lumber was piled in every conceivable space, drying, being sorted, awaiting delivery. The workmen, some of them Scandinavian woodworkers, lived in small houses built along the streets immediately around the mill, Their handsome Lutheran church and a few of these wooden houses still stand at present beyond the reach of modern office and condominium development. Sawdust from the mill, rather than being thrown into the river as at other mill sites across the country, was used to stoke the steam engine or sold to Calgarians as fuel. In winter, activity in the mill quieted down, in part because of a shortage of electricity. Then some of the workmen would migrate to the logging camps up in the mountains, returning with the logs on the drive in June. In their absence the Eau Claire lumberyards would be busy with deliveries to retailers and hundreds of construction sites across the city.[25]

But it would not be easy building this business. If there were obstacles in the river, there were also organizational difficulties to overcome. In the first place there was competition. Eau Claire was not the first lumber company. When the Cochrane Ranch was established, Colonel James Walker had arranged for a timber limit, the Cochrane Limit along the Bow River through to Canmore, to supply wood for ranch operations. He built a small sawmill near the mouth of the Kananaskis, where from the early 1880s he began selling timber in the region. When he left the ranch in 1884, he moved his small steam-driven mill to the banks of the Bow downstream from the Eau Claire plant by the Elbow River. In the mid-1880s Walker drove about eight thousand logs down to his mill from limits around present-day Canmore. The river was so swift, he reported to the newspaper, that a raft of logs could be brought from Morley to Calgary, a distance of about fifty miles, in less than ten hours. But thereafter he shipped his logs by rail, likely because the Eau Claire company, having acquired all of the available timber berths and having made improvements to the river, thus had the right to demand payment for their use, which in effect gave the company a de facto exclusive right to drive the river in perpetuity.[26] Another small firm, the Calgary Lumber Company, operated a small sawmilling operation at Cochrane as well. Taken together, these two operations already sold about 200,000 board feet of lumber annually at the time the Eau Claire company entered the market.

The Wisconsin newcomers basically doubled the supply of lumber for the growing town. All three lumber companies faced competition from British Columbia once the railway was completed through the mountains. This supply tended, the Crown timber agent in Calgary noted, to moderate prices in the Calgary region. Kerr and Prince had to operate in a growing market, yet in frequent oversupply, with a price ceiling set essentially by the cost of rail imported lumber from British Columbia. But the Eau Claire lumbermen, using the river to transport their timber, enjoyed a significant cost advantage over their smaller competitors, who moved their raw material by rail. Colonel Walker complained constantly to the CPR that its freight rates were driving him out of business. In 1887 he claimed that he paid $3.00 to ship 1,000 board feet of timber to Calgary, whereas it cost the Eau Claire Lumber Company a mere 50 cents for the same quantity. When the railway lowered his rates to 7 cents per 100 pounds, he was by his own admission able to struggle along. Walker still grumbled that Eau Claire's sawdust and slab sales effectively covered their log drive costs.[27]

From a supply perspective, some of the Eau Claire limits on the upper Bow were terribly exposed to trespass and burning. The CPR, for example, considered it a right to simply take ties, cordwood, and building timbers off the company's limits along the Bow. When the company could not get any legal redress against this trespass, and in view of the fact this timber was not of the highest quality in any event, it surrendered limits A, C, and D running west and north of the Banff town site.[28] Eau Claire seems also at this time to have acquired an additional limit on the upper reaches of the Kananaskis, perhaps by way of tacit compensation. A little later, around 1906, the company obtained berth 468, another large timber tract on the upper reaches of the Ghost River, to feed its Calgary mill.

Internally, the company faced problems of its own making. The original agreement with Kutusoff MacFee, who had retired from the hurly-burly of frontier speculation to England to live off his Canadian rents, burdened the operation with onerous obligations. Without putting a penny into the capital of the company, MacFee was owed one-quarter of the profits. Kerr and Prince yearned to be free of this requirement. Lawyer James Lougheed advised them against any sharp practices, such as winding up and reincorporating. Even according to the Wisconsin lawyers, there was simply no way around the provision. It necessarily reduced the profitability of the investment for those who had put up money. Eventually, in 1903, after exploring every other possible avenue, the Wisconsin partners held their noses, negotiated a buyout of their silent partner's one-quarter share, and thus freed themselves of the incubus of the free rider MacFee.[29]

Despite these internal problems, the Eau Claire company was much better capitalized and managed and controlled more timber than its two direct competitors. Although the Crown timber returns are not complete for this period, the picture that emerges is fairly clear. For a time there was room for all three companies, but Eau Claire quickly established its dominance. By 1888 the company was sawing approximately 78 per cent of the lumber in the local market. The Calgary company eventually disappeared, leaving only the Eau Claire and James Walker sharing the much smaller market in the recession of the 1890s. By 1894 the Eau Claire Lumber Company was turning out about twice the volume of the Walker company, which also folded soon after in any event. From then on the Eau Claire company was the sole producer sawing logs in Calgary. Other lumberyards, of course, brought in carload lots from British Columbia and, after the construction of the Calgary and Edmonton and the Fort MacLeod railway branch lines, from mills in the northern and southern foothills of Alberta. But Eau Claire and the Calgary Water Power Company ruled the Bow River.

Urban rhythms regulated the flow of the wooden river. Logs were stripped off the slopes and drawn down the river by Calgary's demand for lumber, and that demand was by no means constant. Though it is not so obvious now, Calgary was once a city built almost entirely from wood. Houses were not only framed with wood; they were roofed with shingles and clad in clapboard. Only a few examples of these buildings remain, having survived fire, decay, and demolition on the avenues south of the railway tracks, and on 1st, 2nd, and 3rd Avenues on the western edge of the downtown core. Urbanization imparted a powerful demand for sawn lumber. But the demand was erratic, rising and

Illus. 4.9 A block of frame houses in the Eau Claire mill district awaiting demolition in 1973. (Glenbow Archives, NA 2864-23803)

falling with the boom-bust cycle characteristic of western towns. Through the 1880s, for example, the demand for lumber grew steadily to plant row upon row of stores, hotels, and residences along the streets and avenues laid out neatly on the prairie. And of course, Calgary supplied the small towns and ranches over a wide area with lumber. Then, with the recession of the 1890s, construction ground to a halt, and the demand for lumber collapsed. Operations in the woods, on the river, and in the mill were cut back. Those were lean years. Logs tumbled down the river each year, but not in such great volume as before. Competitors fell by the wayside. But in the late 1890s business began to pick up once again, migrants started to flow into Alberta, houses sprouted in suburbs, and multi-storey businesses in the downtown core were rebuilt, some with stately sandstone facings to protect against fire. Soon the Eau Claire Lumber Company could barely keep pace with demand, routinely cutting, driving, and sawing five times as much wood as in the 1890s and about twice as much as the peak years of the 1880s.

Data taken from the company records and from the returns of the Calgary Crown timber agent published in the Department of the Interior's *Annual Report* record these waves of timber flowing down the Bow River.[30] Eau Claire lumber output rose sharply during the initial construction boom of the late 1880s and then dropped off substantially for most of the 1890s. At the end of

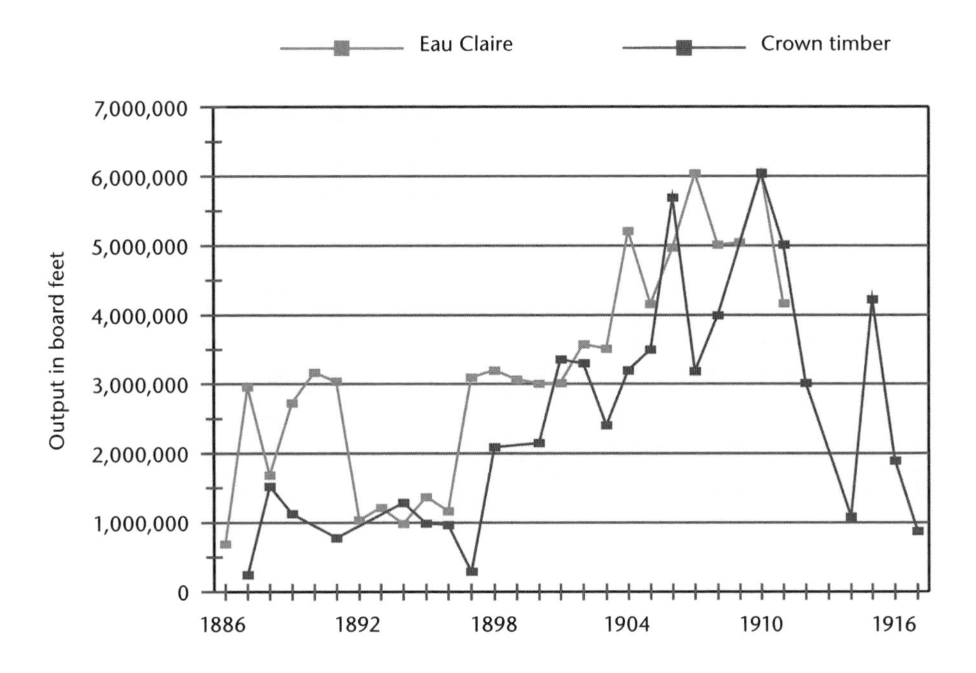

Figure 4.1 Output of the Eau Claire Lumber Company, 1886–1916

SOURCE: GA, Eau Claire and Bow River Lumber Company Fonds, box 2, file 15.

the nineties, production levels had been restored to previous levels. During the first six years of the new century, output doubled from around 3 million board feet annually to just over 6 million. The log drives from 1900 to 1907 were the largest ever organized. The recession of 1908 briefly halted growth, but the construction boom returned with renewed vigour for the next four or five years. At first the company responded by increasing output, and then, for some reason, during a period of extraordinary growth, output levelled off and began to decline. The company may have faced more aggressive competition from sawn lumber imported by rail, it may have experienced difficulty in its mills, or the company may have been carrying uncomfortably high inventories of sawn lumber. The decline continued into the recession of 1913, when the company operated at 1890s levels of output. There was a brief revival in 1915, but the following year brought a return to minimum operations. In the years of peak production before World War I the company typically cut about 100,000 logs on its limits and drove them down the Bow to its Calgary mill, where it maintained an inventory of about 125,000 logs from year to year. The volume of the annual cut and drive rose and fell depending upon production and sales the previous year and the number of logs on hand in the inventory.[31]

Looking back in 1913, the company calculated that it had sawn more than 84 million board feet of lumber in its first twenty-seven years of operation, more than half of it in the last decade. It had paid the government of Canada $50,000 for its original limits and $40,000 for its Ghost properties. Over the years it had expended a total of $24,870 in ground rent and an accumulated royalty to the government on its sawn lumber of $40,000. All told, it had put out a total of $154,882 for the rights to its lumber supply.[32] Not included in the calculation were revenues, costs of production, and profits. At $16 per thousand board feet, the Eau Claire and Bow River Lumber Company probably grossed something close to $1.34 million over this period, not a princely sum but a substantial amount. Costs and profits cannot easily be reckoned. The firm provided reasonably steady employment to a labour force of about fifty skilled and semi-skilled workers. But at the end of the day there were respectable profits to be distributed to the shareholders. Both Prince and Kerr became local notables, investing in several other domestic businesses, including Robin Hood Mills. Prince built a substantial pile to advertise his wealth, a mansion now relocated to Heritage Park which stands as a monument to Calgary's gilded age.

The Calgary Water Power Company also faced competition and self-imposed entrepreneurial limitations. Poor service, high prices, and monopolistic arrogance led the city to build its own municipal steam-powered electric plant in 1905. Rates fell significantly. In 1911 the city entered an arrangement with a new regional hydroelectric utility, the Calgary Power Company, to receive and distribute much cheaper power generated at Horseshoe Falls on the Bow; we will learn more about this enterprise in the next chapter. Rates again fell quite dramatically. This competition confined the Water Power Company to the city centre, while the municipal system captured the growing suburbs. Capacity at the hydroelectric station, with its seven turbines and generators, never exceeded 600 horsepower, while the steam plant had an eventual capacity of 4,000. The name of the company was, as the historian of electricity in Calgary observes, "a misnomer."[33] The Calgary Water Power Company depended overwhelmingly upon thermal generation. The Bow never effectively lit or powered Calgary or even, for that matter, the Eau Claire Lumber Company.

Boxed in by municipal competition and a much larger hydroelectric producer in the region, Calgary Power, the Calgary Water Power Company nevertheless seemed content with its situation. It had more or less reached the capacity of its low-head hydroelectric site at Calgary. Growth could only come from increasing high-cost thermal capacity or finding new sources of hydroelectric power elsewhere. Having experienced the fickle nature of the river flow for many years, the company was disinclined to risk further investment in water power. Thus as the city grew, the municipal system took on the marginal load. By World War I the Calgary Water Power Company had been reduced to a niche market locked in the city centre.

Both the water power company and the lumber company matured in the years before the war. As their founders grew older, more risk-averse, and more embattled in their Calgary market, their companies seemed more prepared to stand pat and less inclined to meet the opportunities of growth. Kerr and Prince had perhaps come to the conclusion that they had got the most they could get from the river.

Tailrace

The resumption of Calgary's growth at more modest levels in the 1920s coincided with a passing of the generations at the Eau Claire and Bow River Lumber Company and the Calgary Water Power Company. Kerr and Prince both lived on into their nineties, as stubborn and ornery as ever. Prince died in 1925, Kerr in 1929. As the founders transferred their properties to the next generation, they and their successors discovered there were new forces to contend with on the river.

Years of quarrelsome relations between the city and Calgary Water Power came to a head in the mid-twenties. The company refused to sell its plant and constantly violated the various "gentlemen's agreements" struck to regulate the business. The city decided to eliminate the anachronism and the nuisance of competition once and for all. In 1926, with the co-operation of the provincial government, it obtained an amendment to its charter that effectively eliminated the company's claim to a perpetual franchise. The bill also gave the company ten years to wind up its affairs, during which time it could not expand its business. In 1938 its municipal franchises would be declared "forever terminated, extinguished, cancelled and annulled." The end was then but a matter of time. However, to avoid such an agonizing death, the company agreed to sell out to Calgary Power for $600,000 in 1928. Calgary Power clearly had no interest in the existing thermal and hydroelectric production facilities in Calgary or the distribution system; however, for future purposes it probably did covet the Calgary Water Power Company's perpetual licence to divert the entire flow of the Bow. Under new management the small hydroelectric station was scrapped and the steam plant was mothballed in 1928. Calgary Power ran the old Water Power distribution system until 1938, when, under the terms of the charter amendment, it and the two thousand customers it served passed into the hands of the municipal electric utility.[34]

By that time the life of the Eau Claire Lumber Company, after a brief revival in the 1920s, was also nearing its end. Charles E. Carr, the mill manager, had taken control of the company from the estates of its original Wisconsin shareholders in 1928. He reincorporated and recapitalized the firm as a Canadian company, Eau Claire Sawmills Limited. The timing was right. Production in the 1920s had returned to levels not seen since the first decade of the century. The prospectus of the new company sent to prospective preferred shareholders noted: "The business is one of the oldest in the City of Calgary and has been very successful and profitable ever since established." By way of reassurance, investors were told that over $600,000 in cash dividends had been paid out over the previous twenty-five years. The prospectus implied that the company had something akin to a monopoly in its market:

> The bulk of the lumber sold by the Company is cut from the Company's own timber limits in Alberta, west of the City of Calgary. The logs are floated down the Rivers to the Mill at Calgary, where they are manufactured into lumber in the Company's Mills which are situate almost in the centre of the City. Most of the output of the Mills is sold retail directly to the consumer in the City and surrounding district, by the Company through its yards. By doing this, the Company receives the entire profits from its sales. This gives the Company a decided advantage over its competitors who have to buy their lumber from other mills and ship it in from Northern Alberta or British Columbia by a long rail haul.

Carr told one interested party: "There are no other saw mills operating on the Bow River or its tributaries and the company has special rights on the river which would make competition very difficult." Demand was therefore strong for the 7 million board feet per year the company produced, and, Carr claimed, "the lumber business is improving." Moreover, it possessed a secure supply of timber as well, he asserted: "There are large quantities of timber on the Bow River and its tributaries which, with growing trees, will enable the company to continue its operations on the present basis for an indefinite period."[35] The company listed assets of $450,000, including almost $250,000 of logs and lumber in inventory. In this rosy view the lumber company looked more like a utility – well established, a monopoly, and secure. The forest and the river could be transmogrified into a steady flow of dividends on preferred and common shares. On the strength of this information, 38 Calgary doctors, lawyers, widows, brokers, and businessmen subscribed to 11,000 shares of the issue. Two years later the number had swelled to 131, of whom 7 were doctors and 37 women.[36] The company was now knit into the local fabric in another way, as a place where savings could be pooled, invested, and multiplied as capital employed in a business.

Illus. 4.10 Employees of the Eau Claire Lumber Company posed in front of their sawmill about 1895. (Glenbow Archives, NA 353-4)

The company was also a place of employment, no longer one of the largest work sites in Calgary but still significant. According to income tax returns filed for 1929, it employed thirty-seven men. Almost all of them lived on 1st Avenue West, within walking distance of the mill. Most had been employed for between ten and twelve months, though a few either casuals or newly employed had been on the payroll less than half a year. The total wage bill in 1929 amounted to $114,054, including Charles Carr's $5,000 salary as manager. Excluding him, the wage for twelve months of work in 1919 ranged from $2,400 for some middle managers to $1,050 for some single labourers and averaged around $1,400.[37]

All of this occurred, as we can see from hindsight, at the peak. The Depression descended upon the west with a special fury. The expansion of the city came to an abrupt halt; the demand for lumber dried up.[38] Once again the lumber company had to scale back its operations to survival mode.

There were other forces at work placing new obstacles in the way. For the first two decades the Eau Claire Lumber Company had ruled the Bow River. In 1911 the Calgary Power Company had thrown a dam across Horseshoe Falls and soon after another just upstream at Kananaskis Falls. Later a third major dam obstructed the path of the company's logs at the Ghost River. In all three cases arrangements had to be made after the fact for logs to pass around these dams on flumes, but this process caused delay and expense as logs had to be boomed and coaxed through a narrow aperture. In the 1920s, when the power company applied to establish upstream storage dams on the Kananaskis, the lumber company was again not consulted, nor was provision made for the passage of logs in the plans.[39] After 1911 the lumber company had to work around the structures and the altered river flow of the higher priority hydroelectric operations on the river. Eau Claire might have exercised an effective monopoly over logging and river driving on the watershed, but it had lost its supremacy on the river.

Moreover, the economics of the business had fundamentally changed. Eau Claire had now become a high-cost producer. It had to drive its logs over long distances to its mill. The small quantities of merchantable timber remaining on its limits were high up in the mountains and difficult to get at.[40] The company competed for labour in an expensive urban market. Relatively low volumes of output reduced efficiencies. It was cheaper to deliver BC lumber by rail to Calgary than to produce it locally. The last river drive occurred in 1944. After surviving ten years of depression and four years of war, Eau Claire Sawmills Limited eventually gave up the ghost and wound up its business.[41] As an indication of how anachronistic logging and river driving had become in modern Canada, the company's remote woods operations were commandeered during the war as prisoner of war camps. Something that had once been a staple of Canadian life had, with changing technology and lifestyles, been turned into a prison.

The regime of the wooden river lasted only about fifty-eight years, from 1886 to 1944. A good portion of the city of Calgary had, as I.K. Kerr imagined, been floated down the river. Thereafter it would come from other sources off the road and railway. The river had many other capabilities, but for some reason the Eau Claire managers chose not to attempt to take advantage of them. Their hearts and minds stayed firmly fixed at their mill site in Calgary; they did not adjust to new hydroelectric technologies or venture into the forest products industry using other methods. They continued to operate a classic, centrally situated, river-drive operation whose costs eventually overwhelmed it. The company's activity on the Bow was limited by the vision of its founders, that of a wooden river.

Thus in the late 1940s the Eau Claire site on the southern bank of the Bow River in Calgary became a spectacle of deindustrialization. Machinery was removed, most of it sold for scrap. Buildings fell into disrepair and were pulled down. Weedy vacant lots replaced the scented piles of lumber. Logs jammed the river no more. The abandoned dam remained astride the river, but it grew more jagged and derelict with each spring breakup. Prince's Island elided on some maps and in more minds into Princess Island, and the channel separating it from the mainland silted up. The Eau Claire district acquired an unsavoury reputation to go with its neglected appearance. Eau Claire would bring other associations more quickly to mind than wood or Wisconsin. Constrained by their imaginations, the managers of the Eau Claire company could never get enough power out of the river to move to higher stages of production. Others would.

Conservation

If Kerr and his lumbermen colleagues regarded the front range of the Rockies as a warehouse of building materials and the river a transportation system and power supply, men such as William Pearce, whose main concern was agriculture, tended to look on the front range as a giant sponge, soaking up precipitation in the winter and spring and then slowly releasing it in the summer to nourish a thirsty prairie. For Pearce and others like him, forest conservation ought to have been the highest priority of government to ensure reliable river flows. For that reason Pearce advocated vastly extending the boundaries of the national park to include much of the headwaters of the Saskatchewan, thus bringing the forest under government protection. He had no problem with parks serving several purposes. As we will see, his preoccupation with using the front range as an irrigation reservoir led him to break with his colleagues in the national park movement in the 1920s over the issue of hydroelectric storage. He actually favoured storage reservoirs in the park because they would enhance the retention capacity upstream, and he lamented the purist parks doctrine that led inevitably to the redrafting of park boundaries and reduction of park area to exclude non-conforming uses such as hydroelectric storage. Eventually, at about the time the Bow River forest industry went into decline, this concern for protecting vital water resources did become the basis of public policy, though not in the way Pearce envisioned – through a latitudinarian park policy.

The federal government transferred Alberta's natural resources to provincial control in 1930. The formal transfer of forest reserves and fire-ranging districts followed two years later.[42] This left the province responsible for the management of the area outside the national park system, including 6,300,000 acres of woodlands in the Rocky Mountains Forest Reserve, which constituted the watershed of the north and south branches of the Saskatchewan River. Though covering just 12 per cent of the total watershed, the eastern slopes of the Rocky Mountains supplied about 87 per cent of the river's total flow, water used by most of the people of Alberta for sanitary, irrigation, industrial, and power purposes. "Water," as Alberta's chief forester put it in 1972, "is the most valuable produce of the mountain slopes in Alberta." Moreover, both branches of the river ran eastward across the adjoining province of Saskatchewan to their junction near Prince Albert, and thence the water moved across Manitoba to empty into Hudson Bay. As a result, protection of this water supply was of great significance to many people in all three Prairie provinces.[43] Whatever happened upstream not only had major downstream impact as in all rivers but also, in a federal system, carried with it significant interprovincial, even national, consequences.

The idea that the federal government should take some responsibility in the management of forest and water resources on the front range of the Rockies emerged, ironically, from the province of Alberta in the 1930s. In the years immediately following the transfer of resources, Alberta's principal concern with its woodlands was forest fire damage. From 1925 to 1930 fire losses on the Saskatchewan headwaters had averaged 11,750 acres per year, but in the drier years between 1931 and 1936 fire consumed an average of 47,000 acres annually. In December 1937 the provincial government, its finances hard hit by the Depression, proposed that in view of the importance of forest cover in maintaining stream flow for the rehabilitation of drought-struck farmland in both Alberta and Saskatchewan, the federal government should put up the money to pay for firefighting. Alternatively, if Ottawa would take over fire protection, the province of Alberta offered to hand over administrative control of the watershed lands and grant the federal government the surface rights to the lands, while retaining all subsurface entitlements to future oil and mineral developments. The federal minister of the Interior, T.A. Crerar, replied that neither proposition appealed to him or his cabinet colleagues. If Ottawa began to provide Alberta with grants for forest fire protection, every other province would be certain to demand the same treatment. The revenues from timber licences alone were likely to be insufficient to cover firefighting costs for many years to come. So Ottawa would only consider resuming this responsibility if Alberta agreed to hand back both the underground and surface rights to the entire Saskatchewan River watershed. The provincial government showed no enthusiasm for surrendering jurisdiction so recently acquired after long-drawn-out negotiations, and there the matter rested for the next decade.[44]

The proposal languished during the war, but it arose again afterwards at Alberta's insistence. That before oil Alberta was a have-not province can be too easily forgotten. By 1947 it had spent $4,169,000 on the woodlands of the eastern slopes of the Rockies, organizing them into provincial forest areas and planning reforestation, but it still found the costs burdensome. As federal Mines and Resources minister J.A. Glen reported on discussions with Alberta

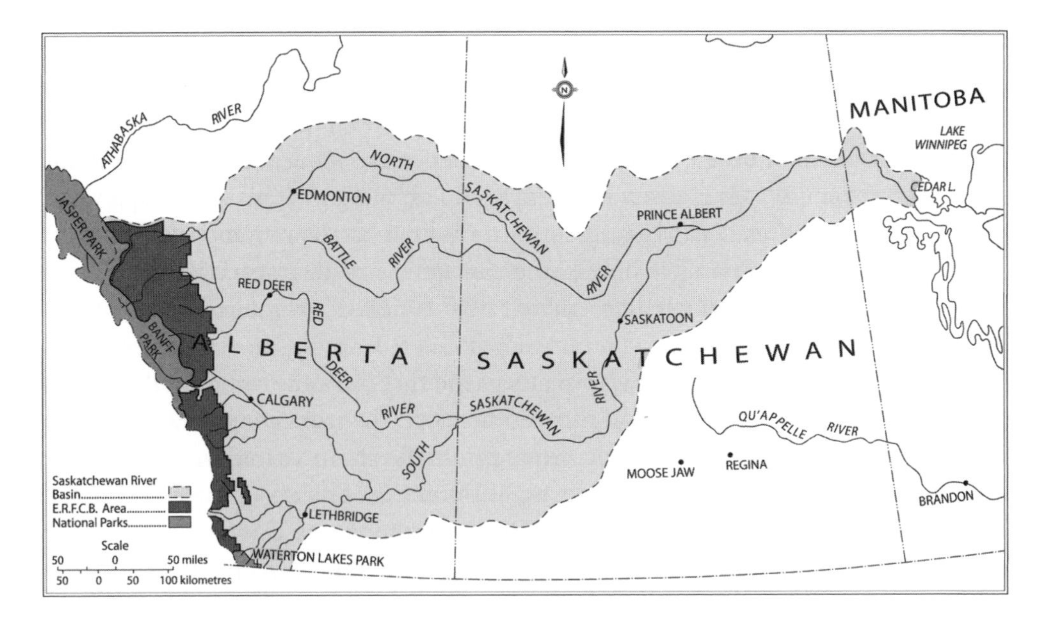

Map 4.2 The Eastern Rockies Forest Conservation Board's mandate area in the context of the South Saskatchewan basin. (Adapted from *History of the Eastern Rockies Forest Conservation Board, 1947–1973*, a report by W.R. Hanson in collaboration [Calgary: Eastern Rockies Forest Conservation Board, 1973?])

premier Ernest Manning and his Lands and Mines minister, N.E. Tanner, "The government of that province has frankly admitted its inability to adequately protect that area and took the view that the expenditures it was making were all that could be justified in the interest of the province alone, and that the problem was of vital concern not only to the provinces of Saskatchewan and Manitoba but to the whole of Canada."[45] Though Alberta remained adamant that it would not give up the administration of the subsurface resources, both governments agreed that protection of this watershed was a "national problem." The ministers therefore proposed that concurrent legislation should be passed to establish a joint federal-provincial body to protect the headwaters of the Saskatchewan River.

The province and the federal government established the Eastern Rockies Forest Conservation Board (ERFCB) in 1947. The agreement had a life of twenty-five years, and for the first six years of its life, during which the board would administer a capital budget of $6.3 million, it would consist of two federal representatives and one Alberta delegate.[46] After that the proportion of federal and provincial board members would be reversed. Up to $300,000 annually was to be provided jointly by the two levels of government to support fire protection, commence a forest inventory, and start reforestation in the Rocky Mountains Forest Reserve. The provincial government would retain the revenues from forestry, though with provision for its share of the costs to rise if returns exceeded its contributions. Operations would commence on 1 April 1948, with the Alberta Forest Service carrying out the actual work.[47]

The goal of the board in regulating forestry on the front range was to ensure "the maximum possible flow in the Saskatchewan and its tributaries." Solon Low, Alberta Social Credit member of the House of Commons, in commenting on this legislation in committee, pointed out: "From the beginning of this discussion the province has taken the stand that protection of this east slope is of national importance, and should, therefore, by financed largely by the federal government." And he went on to praise the province for its extraordinary efforts to promote "the sacred trust" of water conservation. James Gardiner, speaking along with C.D. Howe on behalf of the Liberal government, pointed out that this was but one of two government initiatives to conserve water at its source so that it did not just "rush down to the sea in a short time without doing much good."[48]

During the first six years the ERFCB spent its federal grant of $6,300,000 on the construction of 274 miles of trunk roads and 176 miles of secondary branches across the watershed of the Saskatchewan River. As an unexpected consequence, these roads soon facilitated a marked increase in tourist travel.[49] Crossing and bridging passes not only opened the country for better fire suppression but also made the backcountry more accessible for hikers, hunters, skiers, and of course loggers. Forestry operations continued in the area under provincial and ERFCB regulation, and these roads replaced the old log drive as trucks now hauled the wood out of the forest to mills in the foothills towns.

Meanwhile, the Alberta Forest Service began studies to discover how much water actually ran down the Saskatchewan. So little expertise with hydrology existed in Canada that the assistance of the United States Forest Service was sought for the first general survey. With the help of the federal Water Resources Branch (later renamed the Water Survey of Canada), gauging stations were put in place by 1953 on all the major tributary streams where they drained out of the forest reserve. At the outset, maximizing the flow of water was the "over-riding consideration in all the Board's activities," and since both the ERFCB's members and its staff were mainly engineers, their attention naturally turned to the possibilities of damming the flow of the Saskatchewan's tributaries to control floods, produce power, and even out flow. Most of the best sites were outside the Rocky Mountains Forest Reserve, but an aerial survey in 1950 revealed that there remained more than twenty-five possible sites for construction projects. These locations, however, proved to be too small or geographically isolated, so that the ERFCB did not become a dam-building agency as might have been expected.[50]

Ottawa's role in the board's affairs reduced dramatically when all the initial capital had been spent by the mid-1950s. At that point and henceforth the provincial government appointed two of the three board members. The basic hydrological studies were now complete, and the staff was scaled back. By 1956 Alberta had agreed to bear the entire cost of the board's operations. Two years later the federal government suggested that while recognizing a continuing responsibility to protect and preserve the flow of the Saskatchewan River, the ERFCB should be expanded to include representatives from the provinces of Saskatchewan and Manitoba, but Alberta rejected this idea. The following year the province assumed full control over management of the Rocky Mountains Forest Reserve.[51]

The ERFCB engineers realized that simply maximizing the flow of water in the Saskatchewan system might easily worsen problems of flooding and erosion downstream. As a result, their objective shifted to ensuring an "optimum" flow with due regard to quality and timing. In 1959 the first watershed management research project undertaken anywhere in Canada started, when thirteen agencies from both levels of government agreed to establish the East Slopes (Alberta) Watershed Research Programme. By the early 1960s the ERFCB was seeking to discover how timber harvesting affected water runoff and to manipulate high-altitude spruce-fir forests to protect or improve a watershed.[52] Research led to the apparently paradoxical conclusion that "good" forestry practices did not necessarily produce optimum flows. If 100 per cent of a woodland was growing vigorously, the loss of moisture through interception and transpiration might actually increase, so that watersheds in "pristine" condition would produce lower snow accumulations (and hence less runoff) than ones with patchy forest cover.[53] Removing rather than protecting trees improved the flow of water in the river system.

After a decade of study, the board concluded that it was possible to increase the flow of certain streams through the removal of forest cover to promote snow accumulation. But as ERFCB forester W.R. Hanson reported, no such projects were actually started because there was "not a general demand for more water in the Saskatchewan River. Regulation of flow for flood control and to supply water in low-flow season was more important, and large-scale removal of forest might contribute to floods and seasonal scarcity."[54] On that rather melancholy note the swan song of the ERFCB sounded. By then the province of Alberta had grown impatient with its narrow pursuit of fire suppression and forest management at the expense of outdoor recreation and tourist development. The original intergovernmental agreement, which had come into effect in 1948, had run its term of twenty-five years, and on 31 March 1973 the board handed over its responsibilities to the Alberta Forest Service and ceased to exist.[55]

During its quarter-century of existence the Eastern Rockies Forest Conservation Board did not significantly affect the environmental conditions in the valley of the Bow, one of the principal tributaries of the South Saskatchewan River. Topography decreed that the board would not become a major dam-building agency even if it had possessed the necessary authority. The ERFCB's responsibilities lay mainly in the area of data collection and research, and it did establish gauging stations that could measure the flow of the streams in the watershed as they left the Rocky Mountains Forest Reserve. These data provided hydrologists with the information needed to estimate "optimum" flows in the Saskatchewan basin.

What effect did a century of logging, park development, and forest fire prevention have upon the Bow valley forest? And in turn, to what extent, if any, did forest change affect the river? The answers vary depending upon which portion of the river is under examination. For the Kananaskis, for example, the subalpine forest of the mid-1970s closely resembled the forest almost a century earlier. But the much more intrusive human presence in the montane environment of the Bow valley above Canmore changed the forest significantly.

Logging, driving, and the associated modifications of flow necessarily altered the nature of the Bow River, but by how much is difficult to judge. The removal of the forest cover through logging in and of itself would not have altered the river. The amount of forest removed each year relative to the total forested area was quite small. Compared to fire, logging played a very minor role in denuding slopes resulting in more rapid runoff. Blasting to open up primary channels and the removal of obstacles certainly increased the velocity of the river in some places. This intervention probably locally exaggerated peak flows in subsequent flood events. But the crude dams built to hold back water to create artificial freshets would have had the opposite effect. It is unlikely that the limited flow-improvement efforts of the log drivers permanently rearranged the relationship of riffles and pools on the mainstem. Scouring and downstream siltation would have had an adverse impact upon fish-spawning habitat. But overall, the river's fluvial processes and the construction and operation of hydroelectric stations were likely more effective than the lumber company in rearranging its channels and altering its flow. Dislodged bark from the logs settled in pools and lined the bottom in places. But in the upper reaches the swift flow would have swept much of this organic litter downstream. The relatively small volume of timber transported in the short annual driving season would not have overwhelmed the river's capacity to digest this bio-loading of bark fragments slowly over time.[56]

Despite a century of commercial logging and vigilant state-sponsored fire suppression, the Kananaskis valley subalpine forest in the late twentieth century had not changed fundamentally since the lumberman I.K. Kerr and the timber cruiser Louis B. Stuart surveyed its forest products potential in 1883–84. A comparative analysis of modern aerial photographs with the timber cruiser's systematic notation suggests substantial similarity in species quantity and distribution, the age and extent of the forest, and the dynamics of forest growth. The reason for this stability over a long period would seem to be that the forest fire frequency of the past century closely resembles the fire history of the front range forest as it can be determined by scientific analysis for the preceding century or more. In the absence of significant and sustained human-induced transformation, fire remains the leading disturbance factor determining forest dynamics in mountain valleys. Tree ring analysis, examination of fire scars, and other tests suggest that roughly the same areas of the forest were burned over during the twentieth century as in the nineteenth century and at about the same frequency – every ninety years. Modern research suggests that fire suppression may not have become even partially effective until the 1980s.

The most notable discontinuity in fire patterns and thus, by extension, forest dynamics seems to have occurred not in the twentieth century but, rather, early in the eighteenth century. Fire patterns changed in approximately 1730. Before that date fires burned at roughly fifty-year intervals; after that date, at ninety-year intervals. This change predates European contact. It does not appear to have any relationship to changing indigenous hunting, occupation, or burning behaviour. Rather, the shift is most closely correlated with climate change, a relatively sudden shift from warm, dry summers to cooler, wetter ones, a regime that lasted up into the 1940s. Thus the forest that I.K. Kerr examined in 1883 is largely similar to the one a tourist sees driving into Kananaskis country today. Forestry, hydroelectric storage reservoirs, and, more recently, recreational uses have not had a measurable impact upon the forest cover.[57]

This conclusion would probably hold for other valleys on the southern front range similar to the Kananaskis. The same could not be said, however, for the mainstem of the Bow River above Canmore, where an evolving transportation corridor, extensive human habitation, and a more vigorous fire suppression program would have been more active factors in shaping forest dynamics. Before penetration by the railway in the early 1880s, this region was a Douglas fir savannah. On the valley floor fires – lightning-sparked and indigenous hunting fires – regularly burned in the open areas between the trees, clearing brush but leaving the tall fir trees relatively unscathed. Close study of photographs, fire scars, and carbon deposits in pond sediments suggest a sharp spike in severe fire activity during and immediately after construction of the railway.[58] Exploration, prospecting, and railway surveying were unintentionally incendiary operations. After about 1890 these surface fires diminished in number and frequency, probably as a result of vigorous suppression efforts by park rangers. After 1940, fire was almost totally eradicated from the park. This outcome in turn has allowed a succession forest to infiltrate the open spaces, colonize them, and create an aging, dense, closed canopy forest.[59]

As a result, by the mid to late twentieth century, the forest cover of the upper Bow valley was probably as thick as it had ever been. But the species composition had changed over time and the forest had aged. Trees that needed fire for reproduction, such as lodgepole pine, or trees of the savannah gave way to spruce varieties and poplars along the riverbank and creeks. The changing forest had implications for wildlife, the most significant of which for our purpose was the gradual colonization of the valley by beaver, drawn up from the foothills by the more abundant food sources in the riverside deciduous growth. The aging of the forest, with its attendant buildup of flammable material, and its increasing density have in recent years led foresters to reintroduce fire to the valley to mimic earlier fire regimes.

The ironic conclusion is that after a century of logging, the forest of the upper Bow valley is in some places virtually unchanged from its historical characteristics, or it has grown thicker and more continuous as a result of effective fire protection. The forest continues to be logged selectively outside the park boundaries. A huge sawmill in Cochrane draws its supplies by logging trucks off the slopes of the front range. A heavily mechanized woods operation runs twelve months a year, with loggers commuting from their homes.

The Bow is as much a wooded river as ever, but wooden thinking no longer rules the river. The mountain forest continues to be protected in public policy as a water reservoir for the entire prairie by the government of Alberta. But that fixation with protecting water supplies and forestry has given way in the last generation to multi-use resource thinking, as we will see in subsequent chapters. Outdoor recreation and tourism are now more important to the regional economy and the self-image of the province than forestry and hydroelectricity. The shape and flow of the river have changed considerably, but changes in the forest cover and wildlife have likely had a negligible impact on that aspect of the river. Instead, human beaver have been much more influential, throwing dams across the branches and mainstem of the river and building storage reservoirs, a subject to which we now turn.

5 · Power and Flow

A river reveals its power most dramatically when it falls. Once upon a time the Bow River made an open display of its energy in a series of spectacular waterfalls. These magnificent cascades, through their force and their location, exercised a seductive influence. During the first half of the twentieth century, however, all but one of them disappeared.

Waterfalls have always stimulated contradictory human responses: dread and hope, the religious and the carnivalesque. The roaring maelstrom, thought to be the emanation of resident spirits, has made waterfalls sites of pilgrimage, prayer, sacrifice, propitiation, and for a time (at Niagara Falls) consummation. In the nineteenth century the sensation of being simultaneously close to the hand of creation and sudden death inspired an aesthetic of melancholy contemplation and shuddering awe known as the sublime.

Yet the close study of falling water also gave rise to another line of thinking: calculation as to how to put that energy to use in replacing animal and human labour. At waterfalls a river does its hardest work, sculpting a new bed for itself against stiff rocky resistance. For several millennia humans have been redirecting river flow near falls to use that energy to do other kinds of work, turning wheels attached to ever more ingenious machinery. Cataracts historically became hives of industry as mills of various sorts sprang up to exploit the power. Then at the end of the nineteenth century advances in hydroelectric technology and especially long-distance transmission made it possible to separate the source of the energy from its place of consumption. Industry no longer had to go to power; power came to it over wires. Coincidentally, electricity was made to perform many new functions: transportation, ventilation, cooking, driving machinery, and brightly lighting cities. In the process falling water acquired great value.

To be valorized, waterfalls had first to be controlled with words. This process gave rise to a revealing language that echoed the imagery of the dominant agricultural economy. Waterfalls had to be tamed, their power harnessed. Like wild stallions, rivers had to be broken, domesticated, made docile and useful. Thrifty people deplored profligate nature; all that energy being dissipated, wasted in spectacular display, when it could be more usefully and gainfully employed. More fabulously, the genie of the falls had to be released from its demonic possession by superior intellect and set to more benign purposes. The roar of wilderness had to be disciplined into the hum of controlled power (always measured as horsepower). Thus the vocabulary of subjugation surrounding development legitimated the gradual suppression of the waterfalls. Linguistically, what they might be made to do had more value than their sublime presence.

On the Bow River, however, the waterfalls were not always what they seemed. In some seasons these falls thundered, but at other times they merely trickled. Early photographs, normally taken in summer, invariably show the water in full spate. Snapshots of the river, frozen motionless in January, would give a much different impression. Nature was an imperfect machine in other ways; engineers often calculated that the river, diverted into a new course, might be made to deliver more power. Thus for hydroelectric purposes it was not sufficient simply to dam up the waterfalls of the Bow; the river itself must be redesigned.

A Mountain River

Because the Bow upstream from Calgary is a mountain river in a northern latitude, its flow varies not only annually with the amount of precipitation but also seasonally with changes in the ambient temperature. The snow and rain falling upon the Bow valley derive from the Pacific Ocean, borne aloft by the prevailing westerlies and deposited on the slopes and valleys of the Rocky Mountains. As a broad generalization, it could be said that the melting of the snowpack, along with the spring rainfall on the foothills, jointly govern the variation of flow. The Bow, over the course of its 375 miles (600 kilometres) drains about 9,810 square miles (25,000 square kilometres). The mountains and foothills along the upstream reaches of the river receive a good deal more precipitation than the prairie downstream. Lake Louise, for example, receives 30 inches (72 centimetres) of rain and snow annually; where the Bow joins the Oldman to form the South Saskatchewan, the average annual precipitation is less than half that, only 14 inches (35.6 centimetres). Moreover, precipitation runs off more readily into the river in the mountainous upstream reaches than on the dry prairie, where evaporation and absorption dissipate the precipitation before it reaches the river. Thus about 60 per cent of the total flow of the river comes from the mountains, which comprise approximately 20 per cent of the drainage basin.[1]

Twentieth-century science has, of course, gauged this flow. Accordingly, it is now possible to attribute with some precision the proportions contributed by the various tributaries and other sources. Flowing at its long-term mean levels, the Bow passing Banff comprises about 31 per cent of the flow of the river measured at the Bassano Weir. The Spray adds a further 11 per cent, and the Cascade River, coming down from Lake Minnewanka, 6 per cent. Farther downstream the Kananaskis flowing in from the south contributes 12 per cent, and from the north the Ghost, 6 per cent of the water of the Bow. At Calgary the Elbow adds a further 8 per cent. The residual, 26 per cent of the flow at Bassano, drains directly into the Bow itself through many smaller creeks and streams, including the Highwood system.

In years of low volume the mountain tributaries are more important components of the flow, supplying 88 per cent as opposed to 74 per cent during mean periods. By contrast, in years of higher flow more water – 38 per cent – comes directly into the river from the Bow's downstream drainage basin and only about 62 per cent from the mountain tributaries. The mountains thus constitute a gigantic reservoir, releasing water on a seasonal cycle to the foothills and prairie. For much of its length and most of its volume, the Bow is predominantly a mountain stream, as its cold temperature, clarity, and swift flow would suggest to the less technically inclined.[2]

Snow does not fall evenly upon the mountains. Nor does the same quantity of snow or rain fall each year. And on the dry southern Alberta prairie the rain is famously unreliable. The flow of the Bow – and all other mountain rivers, for that matter – fluctuates with these climatic variations. In retrospect we can observe long-term cycles in the patterns of precipitation and flow: long wet periods, long dry periods, intervals of stability, and periods of erratic weather and corresponding flow behaviour. As a result, the flow of the Bow as measured at Calgary, for instance, varies enormously. In some years the river can be carrying 20, 30, and even on one occasion 60 per cent more water than mean volumes. In other years water levels can drop just as much. And mean flow is not something that can be relied upon; it is, rather, a statistical construct. The mean is a product of memory and measurement over time, not immediate observation. People who lived near the river learned to expect change from year to year, rather than sameness and constancy.

The annual cycle of the seasons in the years 1912–78 also brought changes in river levels from month to month (see fig. 5.1). In the dead of winter, moisture froze, solidified in snowdrifts and glaciers, and clung to the slopes. As the snow was piling up in the mountain watershed, the flow of the river, again measured at Calgary, diminished to an average of 1,559 cubic feet per second (CFS) in January. It continued to flow at this rate in February and March,

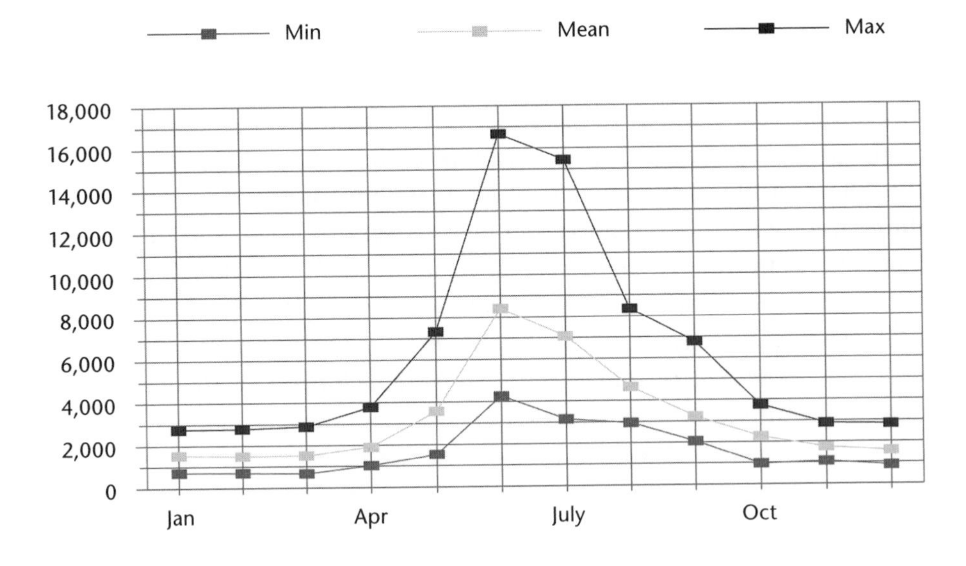

Figure 5.1 Bow River flow: monthly variation at Calgary in CFS.

SOURCE: J.R. Card, F.D. Davies, and R.A. Bothe, *South Saskachewan River Basin Historical Natural Flows, 1912 to 1978* (Hydrological Branch, Technical Services Division, Alberta Environment, 1982).

increasing slightly as temperatures began to warm up in April. During the month of May river volumes doubled to 3,592 CFS, and they doubled again to 8,405 in June. The flow began to subside somewhat in July to 7,075 CFS. Water levels began to drop significantly in August (to 4,640 CFS), a trend that continued through September and October until by freeze-up in November and December river flows were back in the 1,500 CFS range. During six months of the year, October to April, only 25 per cent of the annual volume of the river passed Calgary. Then in the next three months almost 50 per cent of the annual flow swept through the city.

This, of course, is the flow of a fictitious river, a statistical artifact. As it happened, the Bow River varied quite dramatically around these means. A July flood could be twice as great as the July mean and five times more than the recorded minimum for the month. Sometimes, depending upon the temperatures at higher altitudes and the vagaries of summer weather, the high water came in May or August rather than June and July. In some winters the river flow under the ice fell to under 1,000 CFS; other years the river ran at May volumes all winter, spilling out and creating huge new sheets of plank ice on the flood plain. There could be high and low water levels in the spring, the same in the autumn. The accompanying figure 5.1 marks these minimums, means, and maximums at Calgary. But it should not be imagined that the river actually flowed in this manner. These monthly variations could be experienced across the year in a bewildering array of possible combinations. These are but three flow possibilities among many, but these lines on paper do establish the numerical margins, the highest and lowest recorded levels, and the fleeting mathematical mean. At any given time actual readings could fall anywhere within these boundaries, though obviously closer to the mean than the margins. The Bow shared this pattern of long-term cyclical and seasonal variation with most other rivers in a northern temperate climate. Rarely does a river flow uniformly season after season or year after year, but the mountain origins of the Bow greatly exaggerated these characteristics.

The rough-and-tumble nature of the river was, of course, what attracted hydroelectric developers. Between Banff and Calgary, a distance of 93 miles, the Bow River descends 1,100 feet from an altitude of 4,500 feet to 3,400 feet, most notably in three spectacular cataracts. As every tourist knows, soon after the river flows under the bridge on the main street of Banff, it makes a sharp turn as it is deflected between Tunnel Mountain, Sulphur Mountain, and Mount Rundle. As it is forced into a narrow defile, it plunges 63 feet over one of the most picturesque and most photographed sites in the Rockies, Bow Falls. In 1883 I.K. Kerr, a lumberman concerned about a river drive, made a careful study of this falls, identifying and sketching thirty-seven distinct steps in the descent.[3] Because this waterfall is located in a national park, it is still there to enjoy from the riverside viewing platform or the golf course. As we shall see, location in a national park may be a necessary condition of survival, but it is not sufficient.

Just as the Bow leaves the mountains proper, it tumbles – or rather, it used to fall – about 120 feet at two sites only a couple of miles apart. At Kananaskis Falls rocks pinched the river into a narrow channel, where it dropped over four huge steps. Just downstream the Bow made an S turn above Horseshoe Falls, where, as the name implies, it plunged over a U-shaped ledge into a seething caldron 70 feet below. At these two places the narrowing of the river and exposed bedrock for anchoring structures made for ideal hydroelectric sites; deep gorges and steep riverbanks also provided natural containment for the forebays. Both of these sites were within easy reach of a likely market for the power generated, Calgary.

Similarly, the tributaries of the Bow fell from great heights, but they contained considerably less water. From the lakes at its head the Spray dropped over 1,400 feet before it joined the Bow; the Kananaskis, 1,700 feet.[4] Several

Illus. 5.1 Horseshoe Falls on the Bow River, 1879. (Glenbow Archives, 87 5501-12)

excellent hydroelectric sites could be identified along the turbulent courses of these rivers as well as the Cascade further upstream and the Ghost further downstream.

Thus there seemed to be thousands of horsepower ready for the taking from the rapid descent of the Bow. The flow was not constant, of course, but it seemed more than adequate to meet local demand. On the strength of this perception, millions of dollars were invested in building hydroelectric facilities to exploit its power potential. Had it been known at the time how thoroughly the Bow was a mountain stream and how capricious its flow, the outcome might have been quite different. But the capital was sunk before the nature of the river was properly known. As a result, nature had to be changed to preserve an investment.

Calgary Power

Gaining control over the energy of the river required other forms of power, finance capital, political influence, and technical knowledge. Thus the men who were first able to impose their will upon the river and profit from its control lived far away from it. The Nakoda, who were the riparian owners of the best sites, were at first unaware of their utility. Other local people might sense an opportunity and stake a claim but lacked the resources to capitalize upon it. They could raise thousands of dollars, but hydroelectric works on this scale required hundreds of thousands, sometimes millions. Moreover, the corridors to this power lay far off in Ottawa, where distant bureaucrats operating under obscure rules and in the sway of meddlesome politicians held the keys. Finally, extracting the power of the river turned out to be a more complicated engineering problem than at first it seemed.

Hydroelectric promoters approached the Bow, not as a system to be developed, but rather as a series of individual choices, each historically conditioned, which then had to be made to work after the fact. The convoluted story of the hydroelectric development of the Bow River, necessarily imbricated with the history of Canada's premier national park, is sufficiently interesting to warrant a separate volume all on its own.[5] Here we will compress the narrative fairly ruthlessly, our principal concern being to explain what happened to the river as a result of business and political decisions.

Something, however, needs to be said about the choice of technology in this situation. There were other ways of making electricity. The choice of hydroelectricity was not simply economic. Small-scale thermal power was the old way; it was being replaced. The insistence upon hydroelectricity, even in circumstances in which it was not ideally situated, was as much a cultural statement about modernity as an economic choice. It was also a manifestation of the reach and impetuosity of metropolitan finance capitalism.

In the spring of 1909 the mercurial young Montreal financier William Maxwell Aitken (later Lord Beaverbrook) spied Kananaskis Falls from the window of a CPR train while en route to examine his British Columbia hydroelectric properties; soon afterwards things suddenly began to happen along the Bow.[6] As Calgary's population increased rapidly after the turn of the century, pressure rose to secure electricity to light its streets, homes, and offices and run the street railway with hydroelectricity more cheaply than the power supplied from the local steam plant. Rights to Horseshoe Falls had already been acquired three years earlier by two local businessmen, W.J. Budd and W.M. Alexander, who had an interest in a nearby cement works at Exshaw. They bought the necessary property from the Nakoda on very reasonable terms, acquired a licence to develop the power from the Department of the Interior, and negotiated contracts with their own cement plant and with Calgary. Still, Budd and Alexander could not arrange financing for their deal. Their engineer, C.B. Smith, who took over the project, interested Aitken in the promotion.

Aitken already knew something about Calgary, where in 1897 he had cut his teeth as a capitalist and where he had sent his brother to scout street railway prospects. Through Royal Securities, his investment bank, Aitken aimed to promote a series of hydroelectric utilities across the country and abroad upon which a flood of stocks and bonds could be released at great profit. He also knew a good deal about cement-making and its power requirements; he was at that moment in the midst of floating Canada's biggest and most controversial merger in that industry, the Canada Cement Company. Lighting Calgary, a small-scale proposition with lots of growth potential, fit his plans nicely. Within the year he had organized the Calgary Power Company, folded into it the hydro properties of one of the Alberta cement companies, brought his old chum, Calgary lawyer R.B. Bennett, into the company to renegotiate the power contract with the city, and in 1910 commenced active construction at the Horseshoe site.[7]

Calgary Power would not be one of Aitken's more successful promotions. Expensive problems reared their heads right from the start. No systematic stream-flow measurements had ever been taken of the river. The engineer C.B. Smith had simply contented himself with observing, "A valuable feature of the water supply in this river is the fact that all the head waters above the power site are situated within the Rocky Mountain[s National] Park, ... which will be very slowly deforested if at all. The future constant flow of the river is thus ensured."[8] In February and March of 1910 construction proceeded on a practically dry river bottom. Then in the June flood the Bow burst through the coffer dam, swept away a great deal of scaffolding, and undermined the footings for the main dam. As a result of this accident, the spillways had to be redesigned to take into account a larger flow. At the same time the winter horsepower rating of the plant had to be drastically revised downward. Smith had estimated that the absolute minimum flow in the most extreme winter conditions would be 960 cubic feet per second with an average flow of about 1,600 CFS, permitting annual average production of 10,000 horsepower. Federal government engineers later discovered that the Bow at the mouth of the Kananaskis River just above the dam carried an average of only 725 CFS in March and 880 in April.[9] Smith, Kerry, and Chace, the engineers, were summarily fired for incompetence.[10] But by this time capital had already been committed.

When the generators were started in May 1911, it was clear that the company could not supply even the 3,000 horsepower (at a cost of $30 each per annum)

Illus. 5.2 Calgary Power Company spillway, penstocks, and powerhouse nearing completion at Horseshoe Falls, 1912. (Glenbow Archives, NA 3544-28)

contracted for by Calgary with any degree of reliability during the winter months. This was a serious problem, for a run-of-the-river hydroelectric facility could only be rated at its minimum annual production capacity. Moreover, it left no window for the growth upon which a successful financial promotion depended. Long winter nights, of course, were the very times electricity was in the greatest demand; that was when the company was least capable of meeting it. The city naturally began to look elsewhere for its power. Thus, even before the completion of the Horseshoe plant, the company had to find ways of producing more power in the winter to meet its contractual obligations to the municipality and the cement companies and to accommodate the booming domestic and commercial growth of the city. Two strategies, neither entirely satisfactory, recommended themselves. First, it could build more dams and power stations, but they would have the same problems. Secondly, the company could try to arrange to store water upstream, releasing it when it was needed. But upstream all the lands lay within a national park. Nevertheless, survival depended upon pursuing this two-pronged strategy, however fraught it might be with problems.

With R.B. Bennett at the helm and Max Aitken having moved on to bigger and better things while coaching from a safe distance, the company managed to persuade the Department of the Interior, with some difficulty and in the face of competition, to grant it hydraulic rights to Kananaskis Falls.[11] More land was carved out of the Nakoda reserve to accommodate the plant, the reservoir, and the transmission line, although this time the Nakoda protested menacingly that collusion between the company and government officials and long delays by the company had deprived them of a fair return, and they were right. In 1912 a short dam across the first narrows raised water levels, flooding hundreds of acres of land; the water of the river was channelled through a large penstock carved through the rock down 72 feet to a powerhouse at the bottom of the falls, which now fell silent. The Horseshoe dam backed up water right to the Kananaskis powerhouse. The most turbulent section of the Bow had been turned into two long narrow lakes. The deep gorges in which the two plants were situated lowered the costs of dam construction but limited the amount of water that could be impounded. Power from Kananskis began flowing into Calgary in 1913, and the project was completed the next year. The Kananaskis run-of-the-river plant effectively doubled the capacity of the company, but from a hydroelectric point of view it was still terribly inefficient, its winter capacity, like the Horseshoe plant below it, being a fraction of its full potential.

Upstream storage aimed at making the existing plants more efficient by supplying them with more water when they most needed it. Electricity could not be stored; the surpluses of June could not be sold in January, but water could be held back in the spring and summer to be released in the fall and winter. Lake Minnewanka, the prehistoric riverbed of the Bow now blocked off by glaciers at its northeasterly end, could provide an excellent storage site. And while objections might be reasonably expected regarding turning a lake in a national park into storage reservoir, the park authorities themselves had already created something of a precedent in 1908 by building a small dam at the outlet leading into the Cascade River, in order to raise water levels four feet to improve navigation for small excursion steamboats. Calgary Power's request to store water inside the national park was received sympathetically in Ottawa. The Parks Branch, under the guidance of J.B. Harkin, operated under a policy that some have characterized as a doctrine of "usefulness."[12] To attract the resources necessary for stewardship, the parks must demonstrate their attractiveness to society; they must be accessible, and when accessed, they must welcome and serve users. Preservation of the natural environment was not an absolute goal: nature needed to be enhanced here and there in the interests of tourism, and tastefully constructed amenities provided to make the stay enjoyable.

Raising the level of Lake Minewanka further could be accommodated within this commodious policy, even though legislation had to be passed to permit parks land being used for these purposes. The engineers who staffed

Illus. 5.3 This dam, built in 1912, raised the level of Lake Minnewanka for hydroelectric storage purposes twelve feet. A small power station, built by the national park below this dam in the 1920s, supplied electricity to Banff. These works were subsequently flooded by a much higher dam built during World War II. (Whyte Museum of the Canadian Rockies, V108-885)

the Water Power Branch, which became a separate entity within the Department of the Interior in 1911, could be expected to support the interests of one of their clients in moderate distress.[13] The Forestry Branch took the view that protecting a small amount of scenery should not stand in the way of "the development of the whole country."[14] The Laurier government was already in the process of drafting a new Forest Reserves and Parks Act, which gave the cabinet the power to make regulations covering the exploitation of natural resources, including waterpower developments and transmission lines. This legislation passed through parliament in the spring of 1911, attracting little attention.[15]

Early in 1912, therefore, Calgary Power received permission to build a 16-foot-high dam at the outlet of Lake Minnewanka, raising the water 12 feet above its current level. Anticipating an increase in its capacity, the company signed a revised contract with the city of Calgary in 1913 to supply up to 5,000 continuous horsepower of electricity (for $26 per annum); in addition, bulk power was sold to the Canada Cement Company. Meanwhile, park officials had begun to think about going into the hydroelectric business themselves. As one of the terms of the lease, they required the company to install a "thimble" in its dam capable of releasing 150 cubic feet per second of water for their use should they decide to construct their own generating station to supply Banff. The townsite at that point received its electricity from a thermal plant operated by the CPR at its nearby Bankhead coal mine, also located within the park.[16] Thus with this 1912 dam at Lake Minnewanka, Calgary Power began the long struggle to redesign the upper reaches of the river to increase hydroelectric production.

The new storage dam at Lake Minnewanka helped a little, but not as much as had been hoped. Only about 20 per cent of the water released during the winter months actually reached the headponds of the power dams a few miles downstream; the rest was trapped in ice-choked channels or sank into the deep gravel bed of the Bow.[17] Calgary Power found that even with the additional water, it sometimes remained unable during the winter months to produce at the rate of 5,000 continuous horsepower at its Horseshoe and Kananaskis plants (which had a combined nominal capacity of 31,000 horsepower). The company thus applied in 1914 to increase the capacity of Lake Minnewanka by dredging out the old dam and drawing the water level down further. The Water Power Branch was quite amenable to this proposal, and the Parks Branch does not seem to have raised any objection. However, as a sign of things to come, the proposal met vigorous opposition from federal fisheries experts, who predicted that such low water would prevent the trout in the lake from spawning. The application was therefore turned down, and Calgary Power had to manage without further storage capacity.[18]

World War I offered the company some momentary relief by dampening the rapid growth of the city. Nonetheless, Calgary Power continued to face difficulties every winter in fulfilling its contracts with the city and large consumers like the cement plants, particularly in years such as 1920, when the flow of the Bow fell to just 60 per cent of the normal April mean. This situation required the Calgary municipal utility to retain a thermal generating plant with a capacity of 14,000 horsepower, though the current that it produced was much more expensive than hydro (3.80 cents versus 0.44 per kilowatt hours in 1919). When the war ended, the city briefly explored the possibility of buying the company, but even it was reluctant to take over the private utility without some guarantee of additional water storage to operate the company's plants more efficiently.[19]

When urban growth resumed in the 1920s, the company renewed its quest for more upstream storage and laid plans for new hydroelectric projects. The management of the company now openly admitted that if full information about stream flow had been available in 1910, the hydro plants would never have been built.[20] But since they had been, every effort had to be spent making them more efficient and the company more profitable. In 1921 the company renewed its application to expand the storage capabilities of Lake Minnewanka by removing the old government dam and drawing lake levels down six feet further. The Water Power Branch endorsed the idea, but Parks Commissioner Harkin and his staff strongly opposed the idea, fearing that the mud flats exposed in early summer would drive away tourists from one of the most popular locations in the national park.[21]

Undeterred, Calgary Power proposed a much bigger plan to raise the level of Lake Minnewanka 30 feet to feed a new power plant, to be located on the Bow at Anthracite. This project would not only store water for the system (supplemented by the diversion of the headwaters of the Ghost River to the east into Minnewanka) but add a further 18,000 horsepower of capacity to the grid. The new station could supply Calgary's base load in winter, while the run-of-the-river plants along the Bow would operate in summer when flow was high, at a time when Lake Minnewanka filled up for the tourists. The result would not only rectify the existing design flaws but provide lots of surplus power, which might permit the extension of the company transmission lines north to Edmonton and south to Lethbridge.[22]

Calgary Power did its very best to turn Lake Minnewanka into a reservoir. Company president V.M. Drury declared that future increases in power demand in southern Alberta rendered some such development inevitable. In fact, he argued that a higher dam would drown the scrubby jack pine which grew along the low-lying shoreline at the lower end of the lake; the result would actually improve the scenery. Even with wide mud flats along the shores exposed to the early summer visitors after the winter drawdown, the estimated 145,000 tourists who visited the lake each year were unlikely to complain. Putting the best possible face upon the plan, Drury said the view for them would simply "resemble a bold seacoast at low tide."[23]

The Parks Branch had gone along happily enough with the pre-war scheme to raise the Minnewanka dam by 12 feet (though not to lower the water levels even further thereafter), but it was not prepared to endorse the construction of a huge earthen dam to push the high-water mark up another 30 feet. Not only would there be a large generating plant beside the Bow at Anthracite visible from road and railway but high-tension power lines running right alongside the river to the eastern edge of the national park en route to Calgary, which would be, in the words of Commissioner Harkin, "a very great eyesore."[24] As one bureaucrat wrote in the autumn of 1922, "power engineers seem incapable of recognizing the filthy mudflats and bare shores without a vestige of timber or flower growth destroy scenery. But the average person who has seen what the small dam at Lake Minnewanka has done will … hold a different view." A visit to the lake during the summer by Interior minister Charles Stewart (a former Alberta premier), accompanied by Harkin and the park superintendent, helped to convince the minister to announce that for the time being there would be no new concessions granted for power development within the national park system. Despite grumbling from Drury about the minister's "complete and almost dictatorial power," there was little that the company could do.[25] In a sublime act of bureaucratic hypocrisy, the park officials meanwhile pressed ahead with plans to build their own power station in the park, taking water from the dam at Lake Minnewanka.

Though prevented by concerns for tourism from expanding the amount of storage at Lake Minnewanka, Calgary Power remained determined to rectify the defects of the original run-of-the-river plants and to provide for future expansion of low-cost generating capacity in the Bow valley. The company looked elsewhere for water storage in more remote sectors of the park. The Spray River, which I.K. Kerr had explored on his 1883 timber cruise, drained northward to join the Bow between Sulphur Mountain and Mount Rundle just below Bow Falls. Even before World War I the company had examined this stream, which fell some 800 feet from the mountain valley of the Spray Lakes at a steady rate of about 30 feet per mile, but in the absence of any abrupt falls, it had concluded that Lake Minnewanka would provide lower-cost water storage.[26] Now the company's chief engineer, G.A. Gaherty, proposed an ingenious and ambitious plan to dam up the outflow of the Spray Lakes to create a huge reservoir from which a tunnel could carry water directly to a pair of new power plants in the Bow valley above Canmore, where a working head of 1,100 feet would create much larger amounts of power without problems caused by winter freeze-up. The "scientific combination" of low- and high-level plants would provide maximum power from the water available, an initial investment of $6 million permitting the company to produce an additional 16,000 horsepower, with the existing generators on the Bow putting out 20,000 horsepower more than at the time. Though the Water Power Branch favoured the proposal, the deputy minister of the Interior told Gaherty from the outset that "in his opinion there was not the slightest chance of an application for power in the Spray basin being accepted." Nevertheless, the engineer commenced a campaign for the project which lasted for the next thirty years.[27]

The Spray Lakes plan proved important in another way, because it led to a debate about the purposes and value of a national park system. Federal officials had always held an ambiguous position on development inside parks. Commissioner Harkin argued, "The Parks Service strongly opposes any form of commercial invasion of the National Parks. It feels that if a precedent like the Spray scheme is ever established it would be impossible to prevent the disintegration of the Parks." Yet he based his case on economic grounds, contending that the benefits accruing from "selling scenery" to tourists, particularly Americans, far outweighed the benefits to electricity consumers. In a nice piece of irony he contended that keeping the landscape around the Spray Lakes unspoiled would lead such people to prolong their stays, especially once a planned "scenic highway" to the foot of Mount Assiniboine was opened up.[28]

Harkin's opposition found a ready audience. The chair of the Calgary branch of the Alpine Club of Canada immediately wrote to the Interior minister opposing the conversion of the Spray Lakes into a power reservoir: "'[H]ands off our national parks,' we say." In the summer of 1923 Harkin put his case to a national meeting of the Alpine Club, resulting in a unanimous resolution to form a new Canadian National Parks Association and to condemn any further alienation of resources lying inside the park system without a vote in parliament. The executive of the new CNPA commenced a campaign, and letters from across the country poured in upon the minister of the Interior over the next couple of years.[29] Not just predictable allies such as the Alpine Club joined in, but some unlikely bedfellows such as the Western Canada Coal Operators Association were recruited to the cause by the CNPA on the grounds that, unlike hydroelectric stations, thermal plants would consume thousands of tons of coal.[30]

Yet the Spray Lakes project attracted some equally fervent supporters. Most vocal of those was William Pearce, who tried to exert his influence by harking back to his role in the creation of Banff National Park (discussed in a later chapter). In mid-1923 Pearce unleashed a blizzard of letters to the minister of the Interior, local MP R.B. Bennett, the local Board of Trade, and the editor of the *Calgary Herald*. A strong proponent of irrigation from his early days in western Canada, Pearce was convinced that by storing more water it would be possible to more than double the amount of irrigable land downstream. At the same time such a reservoir would, he insisted, provide valuable protection against downstream flooding. He pooh-poohed the idea that the fluctuating water levels in the Spray Lakes would detract from the scenery; in typical Pearce fashion he proposed a detailed plan to clear the banks of the reservoir of all trees and brush between the high- and low-water marks and cover them with gravel, coarse enough that it would not be washed away but fine enough that those few tourists who did penetrate to the area could "prowl around barefoot."[31]

Questioned in parliament about whether he would grant permission to develop the Spray system, Interior minister Charles Stewart temporized.

While he conceded that more electrical energy would eventually be required in southern Alberta, he observed that that his officials in the Parks Branch strongly opposed Calgary Power's plan, and he promised only to keep "a perfectly open mind about it." Doubtless he was mindful of the fact that the Liberal minority government depended upon the votes of Progressive MPs from western Canada and that opinion in Calgary seemed generally favourable to the power company's application. Moreover, the government of Alberta now entered the picture with its own application to develop the Spray Lakes for hydroelectricity through a provincially owned utility or as a means to secure more control over Calgary Power regarding rates and service.[32]

At the end of 1923 Alberta premier Herbert Greenfield sought advice from the engineers at Ontario Hydro. This report, received in early 1925, was daunting since it concluded that to take over the existing Calgary Power plants and complete the first phase of a new development at the Spray Lakes would cost between $12 and $14 million.[33] Such an investment to serve Calgary would tax the province's financial capacity severely; nor was it likely to be popular with the voters who formed the backbone of the ruling United Farmers of Alberta since rural electrification was then almost non-existent. Premier Greenfield contented himself simply with reiterating his government's claim to priority in developing the Spray Lakes. When the matter came up in the House of Commons, Alberta MPs argued that if Calgary really needed the power, the development should be permitted to go ahead since there was already plenty of parkland in the province set aside. Stewart sought to escape responsibility by proposing legislation that would only permit the alienation of resources located in national parks upon passage of a private member's bill by parliament. There matters stood during the next eighteen months, since the provincial government was not prepared to make a commitment to develop the Spray itself.[34]

Yet as a result of the deficiencies of the private company's system, the city of Calgary continued to face winter power shortages which forced it to operate its expensive thermal power station. In January 1927 a meeting was finally convened in Ottawa between Interior minister Stewart, the new premier, John Brownlee, and civic and company officials. The latter declared themselves ready to start work at the Spray Lakes as soon as the necessary licence was issued. To the dismay of opponents of the scheme, Stewart said he had no objection to the creation of a reservoir there, but insisted that enabling legislation authorizing the passage of a private member's bill to remove the area from the national park must come first.[35]

Behind the scenes, however, Brownlee and Stewart seemed to have concluded that the best means to settle the dispute was to redraw the boundaries of the Rocky Mountains National Park, a task that surveyor R.W. Cautley was quietly despatched to study. But they failed to reach an agreement based upon Cautley's report because the province refused to renounce all future claims to minerals or other resources found inside the park system in Alberta.[36] Yet in the fall of 1928 Prime Minister Mackenzie King travelled to western Canada to meet with the prairie premiers and, in the hope of strengthening the Liberal party in a subsequent general election, offered to return their resources to the provincial governments. Following further discussion, the resource agreement was formally announced in the spring of 1929.[37] The next year Stewart introduced legislation "to protect the parks from private exploitation in future" by cutting 630 square miles of the Spray watershed out of the national park. Sensing a lost cause, the Canadian National Parks Association gave reluctant support since future excisions from the park system would now require the passage of a private member's bill.[38]

The way for the Spray project lay open, but another obstacle now surfaced. G.A. Gaherty had based his plan upon the assumption that Calgary Power would be permitted to cut off the Spray River and reverse its entire flow through the power canal above Canmore. But the river flowed through the golf course at the Banff Springs Hotel just before it entered the Bow, and the Canadian Pacific Railway was not prepared to see the riverbed almost dried up. The railway reminded Charles Stewart, "The company has spent many millions of dollars on our hotel, and your department has gone to great lengths to retain the natural beauty features in the district." In April 1928 he advised the power company that in summer 500 CFS of water, about equal to the normal natural flow, would have to be released from the dam to maintain the scenic appearance of the golf course. Gaherty replied that such a release was "utterly out of the question," as it would reduce the storage capacity of the Spray reservoir so much as to render the entire development uneconomical. He offered to release no more than 200 CFS, which with careful engineering would maintain the scenery. With the CPR's backing, Stewart, though prepared to haggle, would settle for no less than 350 CFS, and Gaherty refused to go ahead.[39]

Thus Calgary Power's effort to re-engineer its Bow River plants with the Spray development collapsed in 1930, just when it finally appeared set to go. Facing a looming power shortage in Calgary and with its power supply contract with the city due to expire in 1928, the company had already cast about for another solution. Wary of a political outcry, the Interior department quickly agreed to grant the rights to the Radnor site at the confluence of the Bow and the Ghost River just below the Horseshoe plant. There a head of some 75 feet could be obtained, though the valley was much wider at Radnor, requiring a substantial dam, and any water stored there could not be used at

Illus. 5.4 Ghost dam on the main stem of the Bow River. (Authors' photo)

the two upstream plants. Nonetheless, during the late twenties the company went ahead with a 37,000 horsepower installation, which entered service in 1929.[40] (See the appendix.)

The company now had three hydroelectric stations strung out on the mainstem of the Bow, all subject to the same defect: low winter stream flow. It had only been able to install limited upstream storage at Lake Minnewanka, but more reservoir capacity could be made to work three times over as the water passed in succession through the Kananaskis, Horseshoe, and Ghost power plants. The election of 1930, which brought long-time Calgary Power insider R.B. Bennett to power as prime minister, seemed to guarantee a favourable hearing for its proposals in Ottawa. In addition, the company had begun to consider increasing its storage in the Kananaskis watershed, which lay outside the national park boundaries. Yet the steadily worsening economic depression soon flattened the rising demand for power in southern Alberta and ended any prospect of large new developments. Having paid its first dividends only in the late 1920s, Calgary Power now had to struggle to survive. Accordingly, the company's only construction project during the following decade was a small dam on Upper Kananaskis Lake in 1932, which increased its storage capacity and made the three plants on the Bow slightly more efficient; but that hardly mattered when Calgary Power had excess capacity.

War changed everything. The demand for power, especially industrial power, shot up. In the summer of 1940 inquiries were made of Calgary Power about supplying 26,000 horsepower annually to the new Calgary plant of the Alberta Nitrogen Company to produce anhydrous ammonia from natural gas for the munitions industry. Immediately, the proposal to build a 60-foot dam at the outlet of Lake Minnewanka was revived, but now the approach to the Department of the Interior for permission came not from the company but from the power controller in C.D. Howe's influential Department of Munitions and Supply. Charged with allocating energy supplies among strategically important consumers, the controller, H.J. Symington, was actually a member of Calgary Power's board of directors.

The company aimed to flood 1,900 additional acres around the shores of Lake Minnewanka and use the 150,000 acre-feet in this vastly expanded reservoir not only in its plants lower down on the Bow but at a new 23,000-horsepower station employing the 300-foot drop from the dam to the banks of the river just east of Banff. Asked by a federal bureaucrat "whether or not the company were using the war as a lever to obtain increased pondage at Lake Minnewanka, the rights for which they were not able to obtain on a previous application, Mr. Gaherty acknowledged that, in a sense, they were using such leverage."[41] The moral imperative of doing everything possible for the war effort trumped the high cards previously held by the Parks Branch and other interests groups, which had effectively stymied hydroelectric development in Banff National Park during the 1920s.

Why, asked the Parks Service people, did the company, rather than drastically altering an area highly popular with boating and fishing enthusiasts, whose use of the lake had risen by 40 per cent during the past decade, not build new storage elsewhere, such as the Spray Lakes, since 1930 outside the national park and still largely unfrequented by tourists? That, said the company, would cost too much and take too long: up to three years to drill the tunnel to the rim of the Bow valley above Canmore. Why not store more water in the upper Kananaskis valley on lands under provincial control, which would feed the three run-of-the-river plants? Twenty-five miles of roads would be needed, and the site for a big dam might prove unsuitable. Why not a thermal plant near the coal mines at Drumheller, which could probably be completed sooner than a dam? Perhaps so in peacetime, but now steam equipment would be difficult to procure in Britain, and the company preferred not to use American machinery.[42]

Park officials put up steady resistance, though with a glum sense that this national emergency could not be ignored. There were, after all, admitted Banff park superintendent P.J. Jennings, examples in other countries such as Switzerland of mixed commercial and recreational use of reservoirs. A new reservoir and plant at Minnewanka would be a "gilt-edged bargain" for the company once the war ended, said J.H. Byrne. "Under ordinary circumstances and normal conditions the above drastic action could hardly be accepted by the Parks Bureau without vigorous protests and opposition, but as matters stand at present this might be futile and unavailing." The Engineering Branch noted that it would take a couple of years for the reservoir to fill, and branch chief J.M. Wardle admitted that "once the Lake Minnewanka project is well under way the scenic value of that area is irretrievably lost." Yet when the acting deputy minister and his aides met with Gaherty and other company officials, though they promised to consider alternatives, discussion turned almost immediately to practical requirements such as workers' housing and road realignment, as though the decision was a foregone conclusion. Company president Gaherty, who was, after all, a hydroelectric engineer, was convinced that if the dam was finished in time to catch the spring runoff in 1941, the new generating station could be in operation by the autumn, when the nitrogen company required the power.[43]

When the bureaucrats continued to press the alternative of a thermal station, the company persuaded C.D. Howe to write to the minister of Mines and Resources, T.A. Crerar, stating that in view of the "urgency of the situation I would appreciate the favour if you would expedite the granting of the licence to Calgary Power to the fullest possible extent." Crerar's officials persuaded him to resist and trotted out all the old arguments against a power reservoir in a national park. The company was using the war emergency to make another effort to remedy the original design defects in its system. But the Canadian National Parks Association did not weigh in so strongly as it had in the 1920s, though Selby Walker did repeat his long-standing opposition. Only the Alberta Fish and Game Association charged that the Calgary Power had deliberately sought a contract with the munitions producers as "a cleverly contrived scheme to wangle something that the Power Company knows that they would never otherwise get." The town of Banff Advisory Council, however, gave support to the plan on the grounds that cheap power would attract industry to locate in Alberta once the war was over.[44]

By the end of November 1940, Calgary Power was acting as though a decision in its favour was a foregone conclusion, simply adding to its application a proposal to divert Carrot Creek into Lake Minnewanka without even bothering to consult the parks people. When Alberta MPs complained to Howe about the lack of new war industries being located in their province, he blamed the current power shortage in Calgary. On 30 November Crerar gave way and promised to approve the company's application.[45] But was it legal to undertake such works inside a national park? The initial response from the Justice Department was that federal powers under the War Measures Act were not sufficient to authorize activities specifically forbidden by other legislation, and that the National Parks Act must be revised to permit the dam. Howe wanted immediate action so that the reservoir would begin to fill during the spring runoff in 1941. Eventually the lawyers were persuaded to agree that an order-in-council might be issued to grant temporary approval, provided that it was specified that the legislatures in both Ottawa and Edmonton would pass enabling laws later on. Alberta premier William Aberhart signified his enthusiasm: "The province is anxious for development of power to aid industrializing our province and to do our utmost in contributing to the war effort." The necessary order was passed in mid-December 1940.[46]

Still the Parks Service continued a small rearguard action, insisting that before an interim licence was issued to the company, an independent landscape architect must certify that the dam, power station, transmission lines, and any other installations inside the national park were designed to be as unobtrusive as possible. With great reluctance, Calgary Power eventually agreed that landscape architect (and designer of the Banff Springs golf course) Stanley Thompson should be retained and that any design changes he considered necessary would be carried out. That settled, the clearing of bush and timber from the shores of Lake Minnewanka could get under way in February 1941, the contract being let to a contractor who hired members of the Nakoda band, who lived just outside the eastern boundary of the national park near Morley.[47] In raising the level of Lake Minnewanka, the new dam would also submerge the park's own small hydroelectric station. Henceforth power would be delivered to the park over Calgary Power transmission lines, some of it generated by the new much larger power station prominently located along the main route into the park.

When Crerar got around to announcing the amendment of the National Parks Act in the spring of 1941, he admitted "a drastic departure from the policy established for many years." His claims that the needs of war, combined with the fact that Calgary Power had already been using Lake Minnewanka as a reservoir for years, left Selby Walker of the Canadian National Parks Association unpersuaded. After listening to a description of the project presented to the Calgary Canadian Club by an engineer, Walker complained that the company seemed intent upon actually drawing attention to the new works: "Their power plant will not only be plainly visible from the road, but the motoring public will be encouraged to take a good look at the project because a road is to be constructed, leaving the highway near the power plant and running beside their hydro pipe and across the top of the dam to connect with the one-time Minnewanka highway." But Walker was a lonely voice. Before the House of Commons, Crerar insisted that the government would never have agreed to the application except for the wartime emergency. "We are in a desperate war ... [M]unitions must be supplied. The government felt that under the circumstances this plant should be brought into operation at the earliest possible day, and the importance ... is the justification for a departure from what has been the policy of all governments in this country for the last twenty years. It does not establish any precedent." The bill speedily passed into law.[48]

Despite continuing pressure from the Parks Service, the power company dragged its feet about the costly landscaping. After a couple of years G.A. Gagherty offered to pay $35,000 but failed to produce a final agreement, and the issue remained unsettled for years afterwards. Not until 1947 was the money finally paid over. Meanwhile, a powerful pump was quickly installed behind the new dam, so that even while Lake Minnewanka was filling up, Calgary Power was drawing water levels down below those previously permitted to supply water for its plants downstream on the Bow. The new Cascade plant started to operate in the summer of 1943, leaving mud flats twenty feet wide around the shore, even though it had always been assumed that the reservoir would be allowed to fill up during the tourist season. The following spring the company claimed there existed an unprecedented demand for power as a result of a coal shortage in Edmonton that had forced it to ship current northward. When government officials investigated these claims, however, they discovered that Calgary Power had been given a special allotment of coal to produce power in its thermal plant near Edmonton. At the same time the company sought permission to draw down the lake level because the ammonia plant was operating full blast and a late spring breakup had cut down on the flow of the Bow. A special drawdown was therefore approved, though the onset of the spring thaw rendered it unnecessary in the end.[49]

All that the Parks Service could do was to try to reduce the visual impact of the new Cascade power plant, particularly its tall, water-tower-like surge tank standing on the rim of the Bow valley where the feeder pipes plunged downhill. This effort led to a mildly comic exchange of correspondence between Canadian bureaucrats and the British High Commission in the spring of 1942 concerning a visiting British camouflage specialist. It turned out that the expert was a sometime landscape artist named Ironside, who looked at some photographs and suggested that it would be a mistake to paint the tank grey. England, he said, was full of gasometers and water towers painted grey, which only rendered them ugly as well as eye-catching. His advice was to paint the surge tank white, which would not make it inconspicuous but might make it more pleasant to look at. Dissatisfied with this advice, the Parks Service won-

Illus. 5.5 Cascade powerhouse beside the highway in Banff National Park showing the surge tank required to absorb the shock to the penstocks of closing off the flow of water to the turbines. (Authors' photo)

dered if it would be a good idea to coat the tower in "haze" or "mist," paint, which the National Research Council had developed for camouflaging ships at sea, on the grounds that from the highway the tank silhouetted against the sky somewhat resembled a ship's upper works. When the NRC was not able to supply the new paint, the decision was taken to use "invisible grey" paint, perhaps with a little blue added to help blend with the mountain sky.[50]

Even after the war ended, power shortages led Calgary Power to run down lake levels so far that extensive mud flats were exposed at Lake Minnewanka well into the summer tourist season. Since the company's interim operating licence contained no precise requirements about levels, the Parks Service found it impossible to exert control even when the water was fifteen feet below the promised level on 1 July 1946. The only lever the government possessed was that the terms of the final licence had not been settled, and by 1947 the company was eager to secure this before it issued some new debentures. In the end the minister of Mines and Resources was persuaded by the company to intervene and order the bureaucrats to issue the licence, but the company still refused to commit to keeping the water in Lake Minnewanka at any specified level in the summer, pleading continuing power shortages. Instead the minister was granted authority to regulate the depth, though Calgary Power was given an incentive to keep the water up in summer through a sliding scale of rentals. With this undertaking, by May 1947 the lake had been formally and finally converted into a power reservoir in the midst of a national park.[51]

Alberta's vigorous economic growth in the postwar era meant continued high demand for electricity. Anticipating rising use, Calgary Power was already investigating new developments, but the only new construction undertaken by the company was the Barrier dam, built in 1947 on the upper Kananskis with a small generating station that could produce 16,000 horsepower using a head of 155 feet, as well as storing water for the downstream plants on the Bow.[52] For any major expansion of capacity, company president G.A. Gaherty still favoured his original plan to dam the Spray River to reverse the flow out of the Spray Lakes to high-head plants on the Bow near Canmore. Each cubic foot of water could thus be made to yield more energy, not only from the great leap of over one thousand feet into the valley but also downstream to help smooth out the seasonal flow variations which plagued the run-of-the-river plants. When it became clear that the provincial government of Ernest C. Manning did not intend to create a publicly owned electrical utility, the company faced even stronger pressure to expand its generating capacity.[53]

A formal application in 1948 proposed spending of $18 million over five years to add about 100,000 horsepower to the company's generating capacity. Since the Spray Lakes had been removed from the national park in 1930, the provincial government bore the principal responsibility for approving the plan. However, the resource transfer agreement still left Ottawa with the power to fix the flow of all watercourses entering the national parks in order to preserve their scenic beauty. As in the late 1920s, any proposal to cut off the Spray altogether was bound to arouse the opposition of not only the Parks Service but also the Canadian Pacific Railway, whose golf course lay at the junction of the Spray with the Bow. To avert this resistance, Gaherty offered to spill enough water from the reservoir to keep 180 cubic feet per second flowing out of the mouth of the Spray during the tourist season. To release more would "seriously detract from the economic value of the whole Bow River development including the Spray … If there is an imaginative [*sic*] difference of opinion about this, the company claims that any possible adverse effect upon the scenery in a spot in this tremendous park is infinitesimal compared to the importance … of the power which can be developed."[54]

At a meeting with Gaherty, Parks Branch chief Roy Gibson bluntly told his minister and the deputy that any more power developments would "be resented by a great many people who have the interests of the park at heart." Yet Premier Manning continued to press the federal government for action to deal with an increasingly serious power shortage in southern Alberta. Surely, contended the powerful federal minister of Trade and Commerce, C.D. Howe, some scheme could be worked out to release just enough water to make the Spray still look attractive. After all, "the industrial growth of the province must be a first consideration. The province of Alberta is having a spectacular industrial expansion and I would be sorry to see anything happen that would interfere with this very desirable development." Mines and Resources minister J.A. Mackinnon, himself an Albertan, agreed and promised to do all he could to get the matter approved as soon as possible.[55]

Calgary Power's plans for the Spray development were modified somewhat from the pre-war era. A dam 740 feet long and 192 feet high would be built at the head of the Spray River to raise the levels in the Spray Lakes in order to create a huge storage reservoir to feed three new power plants near Canmore. Instead of a tunnel, the company now proposed to create a diversion canal along the Goat Creek valley to the headpond between Mount Rundle and Ka Ling Peak. This plan also required a control dam 2,100 feet long and 48 feet high and an earth dike running through the Goat valley. Since this area lay inside the 1930 boundaries of the national park, an act of parliament would be required to remove another twenty square miles of territory to be used for commercial purposes.[56]

By 1948 the CPR had indicated that it would not object, provided that 200 cubic feet per second flowed out of the Spray during the summer months. Alberta officials agreed that this volume would be sufficient to maintain the appearance of the golf course, which would require the use of 2,500 acre-feet of storage capacity. Increasing the flow by just 50 additional CFS would require four times as much stored water, serving "to show the extravagance of endeavouring to improve the scenery at the expense of the power development."[57] Premier Manning was keen to see the new legislation passed immediately so that additional power could be available as early as 1950, but despite Mackinnon's efforts to rush amendments to the National Parks Act through parliament in 1948, his colleagues refused to alter the parliamentary timetable.[58]

With a federal election in the offing, Mackinnon was eager to show his fellow Albertans of his awareness of the need for additional electrical energy. When he went home during the summer of 1948, he was persuaded to fix the summertime flow at the mouth of the Spray River at 200 cubic feet per second, the order having been drafted by Calgary Power's solicitor. The federal cabinet approved the plans that September, and the company set about ordering the generating equipment and planning construction with the aim of starting work before the autumn freeze-up. The bureaucrats were completely excluded from the final negotiations, glumly reporting that "the minister was determined to grant the request and that he had received authority from Council [i.e., the cabinet] to do so." Events moved so swiftly that they had no opportunity to whip up any effective opposition from amongst wilderness preservationists. A protest from Selby Walker was blandly turned aside by the minister with the comment that the Spray development was the "most expeditious" way of coping with a serious power shortage.[59]

The legislation cutting the Goat Creek lands out of Banff National Park made its way through parliament in the spring of 1949, the second-reading debate in the House of Commons lasting no more than ten minutes. But even with construction on the Spray project under way, the severity of the power shortage was made clear. Unusually low precipitation led to Lake Minnewanka being drawn down to unprecedented levels that summer, so that by autumn the lake contained only about one-third of the usual amount of water. After visiting the area in the late summer, Parks Branch chief Roy Gibson observed that the shoreline reminded him "strongly of the back view of the head[s] of some of the boys who used to come down from the Gatineau after a winter in the woods; … the barber used to shave their neck[s] halfway up the back of their heads. Long [mud] flats are noticeable, and while these are reasonably tidy they are certainly not attractive, and it is altogether a most unnatural layout … [I]t would seem that the action that was taken by parliament has convinced the company they can do about as they like in the National Park."

However, the 1947 Minnewanka licence specified only maximum and minimum levels, and the lake was currently 11 feet higher than the minimum. When the autumn rains did not prove heavy enough, Calgary Power requested emergency permission to draw levels down 4 feet below the previous minimum in the spring of 1950. All the company's steam plants were running flat out, and with the hope that the Spray dams would be ready by the fall of 1950, no water at all would be released into the Bow from that watershed during the subsequent winter. The Banff park superintendent advised making concessions to avoid a great public outcry over power shortages.[60]

The Spray project also created continuing problems. The canal dike to carry the water from the lakes to the head of Whiteman's Pass above Canmore was constructed along the side of the valley of Goat Creek on slopes of between 50 and 60 degrees using frozen material. With power short, the company increased the flow in the canal in November 1950 to feed its Bow River plants, whereupon 800 feet of the dike bank melted and slid into the Goat valley. In order to keep the precious water flowing, the dike was then breached in several other places, so that the entire flow was carried down Goat Creek, creating "terrific havoc" and doing $100,000 worth of flood damage. Meanwhile, Calgary Power set about building a wooden flume to carry 700 cubic feet per second, but another breach in the dike in January 1951 led to the abandonment of the canal plan. The engineers now reverted to the original idea and started drilling a 1,400-foot tunnel through Ka Ling Peak. Until that was completed, the flow out of the Spray reservoir was cut off altogether, and as the snow melted in the spring, water began to pour over the control dam at the head of the Spray River at the rate of 1,630 cubic feet per second. A valve in the diversion tunnel could have been opened to cope with the problem, but it was feared that it might be impossible to close it again because of the water pressure.

Since the company had never expected to release more than a few hundred feet of flow down the Spray River, no proper channel had been built from the spillway to the riverbed capable of carrying such large volumes of water, and an estimated $350,000 worth of damage was done to lands and buildings in Banff National Park and considerable fish-breeding habitat destroyed. Downstream, with the Spray in full spate and without its lakes to act as a natural reservoir, the CPR's golf course had to be closed for a time in the spring of 1951. Who would pay for all the damage? A formal claim for $244,000 was countered by an offer from Calgary Power for only $103,000, eventually raised to $131,000, though G.A. Gaherty insisted the liability was really less than $100,000 but that he would be magnanimous in light of a desire for friendly relations with the bureaucrats.[61]

As a result of these engineering problems, the Spray project was not completed until late 1951, when the new reserves finally ended the company's power shortage. But when the next summer rolled around, the CPR was soon complaining that the 200 cubic feet of flow in the Spray had left only a small stream trickling along a wide rocky riverbed through the golf course. The Parks bureaucrats pointed out that they had long predicted this outcome, but that the railway had agreed to the deal in 1948. Narrowing the stream bed by landscaping seemed unwise because of the volume of the spring runoff, and the only solution was to build a series of weirs that would distribute the flow more evenly. Calgary Power eventually agreed to bear the cost of these works.[62]

The Spray reservoir added 210,000 acre-feet of storage capacity, bringing the company's inventory up to 585,000 acre-feet. The 900-foot drop from the headpond to the Spray plant created 49,900 kilowatts, another fall of 320 feet to the Rundle station produced 17,000 kilowatts, and the final 65 feet to the Three Sisters added a further 3,000 kilowatts. The total of 69,900 kilowatts compared with 82,800 kilowatts available from Calgary Power's existing plants along the Bow. Moreover, with additional water the Kananaskis plant would be able to turn out a further 8,900 kilowatts, so that in a single year, 1951, the company's generating capacity was nearly doubled. After nearly forty

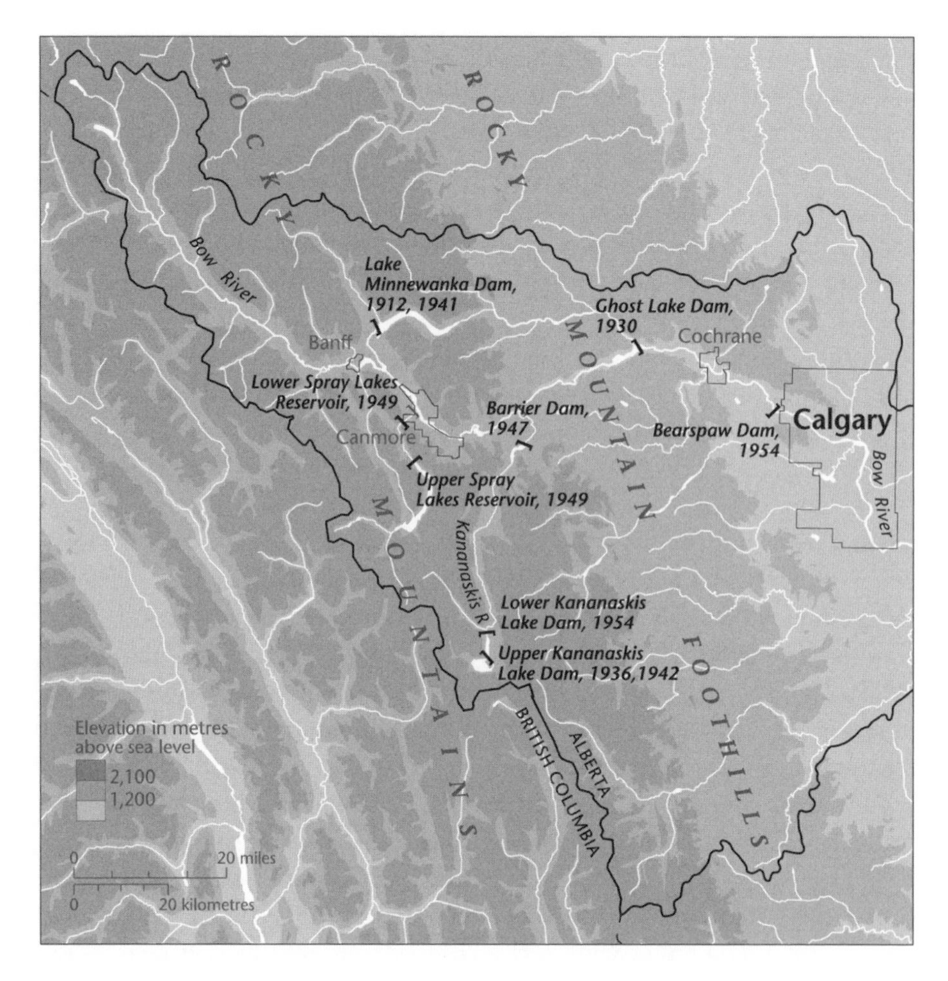

Map 5.1 The course of dam development in the upper Bow basin.

years, the use of the watersheds that lay in or around Banff National Park as power reservoirs was conceded.

Throughout the 1950s the company did its best to squeeze as much hydraulic energy as possible from the Bow watershed. To improve the efficiency of the plants at Horseshoe (opened 1911), Kananaskis (1914), and Ghost (1929), which were already using water released from the Minnewanka and Spray reservoirs, Calgary Power created additional storage on the Kananaskis. In 1955 another storage dam was added on its headwaters at Pocaterra, which permitted the development of 14,900 kilowatts with a head of 207 feet, and at the same time the small Interlakes station was added between the Upper and Lower Kananaskis Lakes, which could turn out 5,000 kilowatts from a 127-foot head. On the mainstream new generators at the Ghost plant added 22,900 kilowatts in 1954, bringing total capacity to 50,900 kilowatts. That same year the company was pressured by the provincial government into building the 50-foot Bearspaw dam just outside the western limits of Calgary for flood control purposes, though it managed to recoup some of the costs by creating a 16,900-kilowatt generating station there. In 1955 the company applied for the right to add new generators at the Cascade and Spray plants which effectively doubled their capacity. The Parks Service did not even contest these requests. By 1957 new equipment was in place at Cascade which created 17,900 kilowatts of new capacity, and in 1960 the Spray got 52,900 kilowatts of additional power, while a second unit at Rundle raised its potential to 49,900 kilowatts.[63]

Though a hydroelectric engineer at heart, Calgary Power president G.A. Gaherty came to understand that the relentless rise in demand for electricity in the booming southern Alberta economy would require a change of course. In 1955 he told company shareholders that thermal generating stations must be built to supply the ever-growing baseload.[64] As new thermal plants came on stream at Lake Wabamun west of Edmonton, where there was a huge surface deposit of low-sulphur, sub-bituminous coal, the hydroelectric stations along the Bow could increasingly be relegated to meeting peak requirements.

By 1960 the Bow River had been hydraulically redesigned to meet the needs of the hydroelectric utility. Diversion projects on the Spray and in the upper reaches of the Ghost changed the direction of flow to extract more energy. Storage reservoirs held back water in the spring and summer, delaying its downstream destiny until the fall and winter to keep the turbines turning. The most turbulent reaches of the river had been blocked by dams and flattened into a series of long, narrow lakes. In the process, lands in and around Banff National Park had been turned into a massive hydroelectric storage reservoir. All this river engineering was driven by the desire to salvage and improve upon the ill-considered investment decisions to construct run-of-the-river generating stations taken a half-century earlier. But increasingly these massive installations sat silent, guarded only by barbed wire and signs warning swimmers and boaters away from their forebays and outflows in case they should suddenly surge into life.

A River by Design

Some things, of course, about the environment could not be changed. Neither government edict nor corporate investment could change the total volume of water flowing down the river. This was governed by environmental factors. The pattern of precipitation changed from year to year; fluctuations in temperature held or released different quantities of the accumulated inventory of

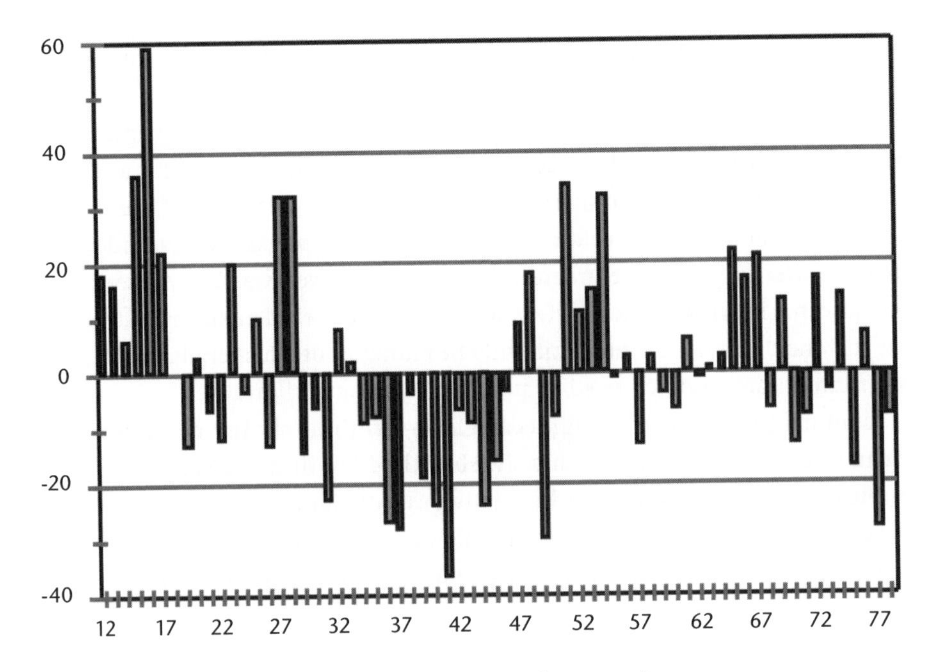

Figure 5.2 Flow variation, 1912–78: percentage deviation from mean.
SOURCE: See figure 5.1.

ice and snow each year. Rain and snow continued to fall with varying degrees of intensity and consistency over the watershed. And thus while the flow of the river might be altered temporarily by storage reservoirs before being fed through the turbines, the total volume of water in the river was governed by natural forces and continued therefore to rise and fall. Except now these fluctuations were closely observed and recorded by scientists trying to understand the behaviour of the system.

Figure 5.2 shows the recorded annual flow of the Bow in acre-feet measured at Calgary. Over the first half of the twentieth century the total annual flow of the river gradually decreased. These data indicate that during the construction of the Horseshoe and Kananaskis Falls plants the Bow was flowing at historically high levels. After 1916 the total flow of the river dropped back to mean levels and fluctuated within in a range of plus or minus 20 per cent. During the power shortages of the late twenties, flow increased for two years to 30 per cent above normal. However, the dryness of the thirties can be seen in these data as well. From 1919 until 1946, river flow remained well below the norm with the exception of two years. For much of this period Calgary Power had electricity surpluses. However, during the war years, when the company did need the power, total flow remained low. These were some of the driest years on record. Figure 5.2 shows a slight long-run tendency of diminishing natural flow to mid-century.

But as we have seen, the total flow of the river was not nearly as important as its distribution throughout the year. During the high-volume months of April, May, and June the company's power dams had to spill water. They could operate at capacity – if there was demand for all that power – and there would still be water to spare. The greatest challenge, however, was the winter, when the flow diminished. Then all the water of the river could be diverted through the turbines, and there would still only be enough flow to operate at a fraction of installed capacity.

A return to figure 5.1 highlights the long-term minimum, maximum, and mean river flows recorded at Calgary. Here the May-through-September peak can be clearly seen. So too can the relatively low volumes in the other months. This, however, was a pattern that could be changed or at least shifted. And it was to this end that from 1912 to 1951 the Calgary Power Company invested heavily, whenever it could, in building huge reservoirs upstream to hold back some of this peak flow for use later in the cycle. Beginning in 1912 with 44,000 acre-feet of storage at the first Lake Minnewanka project, the company pursued a long-term strategy of increasing storage to improve the efficiency of its run of the river plants. The Ghost plant added 75,000 acre-feet in 1929, and the Kananaskis Lake dam a further 36,000 in 1932. However, the major expansion in storage capacity occurred during the war, when the Minnewanka site was reconstructed to add 180,000 acre-feet of storage and when Kananaskis Lake was raised to accommodate 100,000 acre-feet. The massive Spray Lakes project, completed in 1951, built another 210,000 acre-feet of storage into the system. By 1952, then, the company had the capability of retaining over half a million acre-feet of water.[65] Its ability to exercise some measure of control over the seasonal flow of the river thus began in a significant way in the 1930s but increased dramatically in the late 1940s and 1950s.

What effect did this have upon the flow of the river? Figure 5.3 compares the much elevated January stream flow measurements from the 1950s, when most of the storage reservoirs were in operation, with the much lower recordings from the 1920s, when only the first expansion of Lake Minnewanka functioned in that capacity. Figure 5.4 compares two years of almost identical annual volume of flow: 1912, before any storage had been built, and 1966, when all the storage reservoirs were in full operation. This figure shows the cycle from May through to the next April and thus a full cycle of storage and release. These data demonstrate how these dams were able, to a certain extent, to shave off some of the summer peak. In the chart the area between the 1912 line and the lower 1966 line from June through September represents the volume of

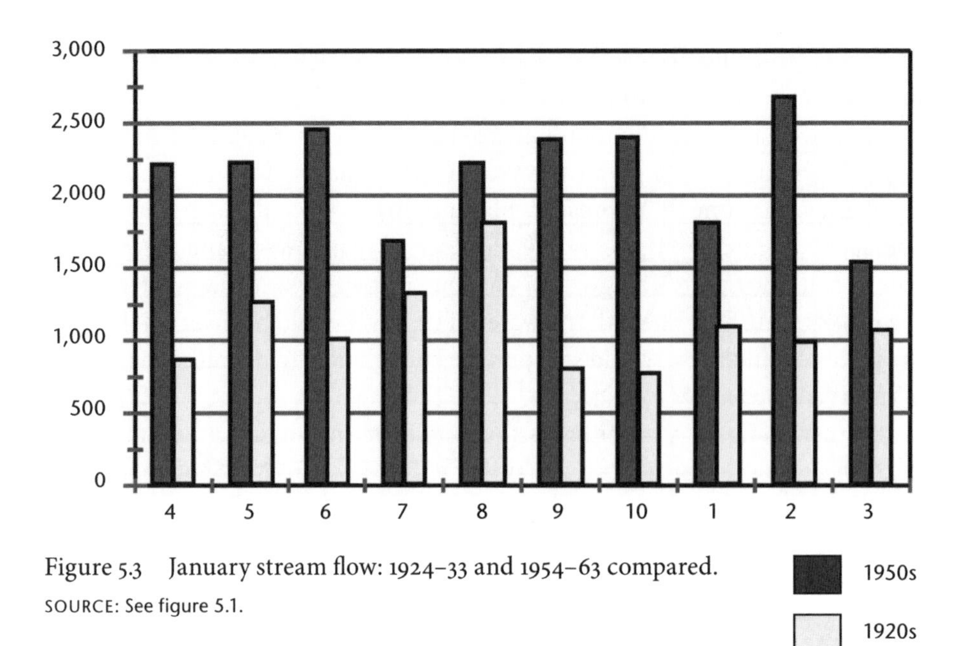

Figure 5.3 January stream flow: 1924–33 and 1954–63 compared.
SOURCE: See figure 5.1.

water being held back in the mountains for release in November, December, January, February, and March.[66] Here it is important to note that through the use of storage reservoirs the company could effectively more than double the winter flow of the river from about 1,000 cubic feet per second to about 2,500 CFS. This 150 per cent increase in flow could be utilized by all the company's Bow River plants, greatly improving the overall efficiency and profitability of the company. For this reason in the 1950s the company was keen to add new turbines and generators to the three existing plants at Kananaskis, Horseshoe Falls, and Ghost.

A river, of course, is more than its flow. Beyond the physical changes noticeable from year to year in the transformed flow regime, alterations also occurred in the riverine ecosystem. As we will note in a later chapter, the presence of dams had both deleterious and surprisingly productive effects on fish populations: destroying habitat in many upstream locations and improving spawning grounds for trout in the gravel reaches near Calgary. As dams progressively reduced the upward extremes of summer floods, the river morphology began to bear the signs of a more regulated flow. Channel movement, river braiding, and island formation all decreased in the Bow's mainstem, particularly after mid-century as major floods diminished, though it is arguable if this change had much to do with upstream storage. This new pattern of flow and sediment movement had its incremental effects, in turn, on shoreline vegetation; riparian woodlands that depended in part on the clearing effect of floods to

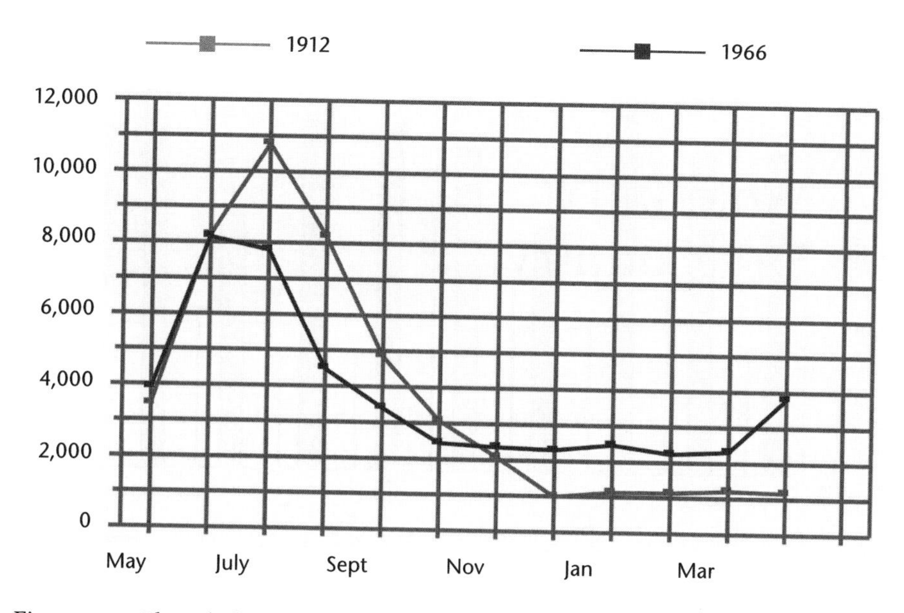

Figure 5.4 Flow shift with upstream storage: monthly mean flow in CFS, 1912 and 1966 compared.

SOURCE: See figure 5.1.

colonize new areas became attenuated. As on other dammed prairie rivers, the loss of cottonwoods and willows pointed to the reverberating effects of upriver dams on the biota downstream. The relegation of the Bow River facilities to peak power production in turn created pronounced daily fluctuations in stream flow that affected ice stability in winter and recreational uses of the river close to the dams in other seasons.[67]

The Bow in the 1950s was quite a different river from what it had been a decade earlier, and it had certainly changed markedly from four decades previously. Only Bow Falls remained as a reminder of what the river once looked like before it was tamed. Upstream its tributaries had been turned into new courses, dammed up, held back, and then released out of season. The Spray and the Cascade River could be turned on or shut off as required. Summer peak flows had been significantly reduced, and winter flow more than doubled. In the city summer floods became a thing of memory. The Bow, for all practical purposes, had been brought under control for the sake of light and power for the city of Calgary. Except, for some unexplained reason, the river had begun to flood inconveniently in the dead of winter. It remained a river full of surprises.

6 · Watering a Dry Country

As the Bow River flows out of the foothills and onto the prairies, it enters dry country. East of Calgary, the surrounding land and climate is semi-arid, receiving only about 35 centimetres of precipitation a year. The farther east the river travels, the drier conditions become. In the late nineteenth century, after the destruction of the bison, much of the lower Bow was cut into large ranching leases as well as the Siksika reserve. At the margins, squatters and settlers experimented with dryland farming, but without much success. Only at the turn of the twentieth century did the Canadian state seek to organize the area for large-scale irrigation development. First the CPR and then a group of ranchers and English investors obtained massive land grants, invested heavily in major water control structures, and sought to attract settlers and build communities. Although irrigation came to be practised in several areas along the lower river, the CPR's Eastern Section, located between the Bow and Red Deer Rivers, grew to be the largest and most important project. By looking at the particular experiences of this section, it will be possible to understand the general patterns and processes of change, while also analyzing how one area and set of institutions led the way.

Irrigation began as a hopeful idea. When we see the mundane canals and sprinklers in place today, it is difficult to recapture the sense of promise with which promoters extolled irrigation's virtues. Irrigated lands were sold as ideal homesteads, able to withstand a capricious climate and well-suited to market-oriented production. They were said to anchor rural life and community. Promoters claimed that irrigation was a practical science which magically brought life to a dry and dusty landscape. By the 1910s, as intending farmers streamed into the area, the shortgrass prairie had yielded to a landscape grid of settlement and grain and livestock production. But just as irrigation became established, World War I undercut the flow of settlers and placed expansion plans on hold.

The onset of war was not the only harbinger of difficulties. While developers struggled to attract farmers, the controlled environmental conditions they promised frustratingly came out of control. Canals transported weeds as well as water; muskrats colonized ditches; and seepage from canals turned some lands into marshes, contaminated by salts that rose to the soil surface. Adding to these environmental surprises, the economic challenges produced by the collapse of agricultural commodity prices in the early 1920s made the higher production costs of irrigation farming less attractive. Farmers contested their land contracts and argued that degraded land should be removed from lease and sale agreements. Environmental change became a focus for conflict. Many settlers simply left. The lower Bow had been imagined as a landscape of hope; its transformation had seemingly delivered despair.

By the early 1930s, after several decades of financial loss, the CPR had begun to divest itself of its holdings in irrigation projects. Over the next decade other corporations also exited, leaving a series of farmer-run irrigation districts to assume infrastructure and operations. These districts drove down administrative costs and pursued new government sources of funding to expand projects. Whereas in the late nineteenth century the state had acted to establish law and order in irrigation development in favour of large corporations, by the 1930s a substantial state financial commitment to irrigation projects had developed as government agencies replaced private corporations as developers. This new organizational structure gained momentum as a result of the rise in agricultural prices during and after World War II, which improved the conditions of individual farmers and placed irrigation districts on a more sustainable financial footing. As state funding increased in the late 1940s, the potential irrigable area expanded. This process intensified land use and provided the impetus for increased water withdrawals from the Bow. By the postwar era, irrigators had become the major users of the Bow's water and a powerful lobby in provincial politics. A long period of state-supported growth prompted prominent irrigators and institutions to ask just how much the river could sustain.

Origins and Structures

In 1894 the United States government published an analysis of irrigation agriculture in the western states. For the Canadian officials who had just finished framing irrigation legislation for the Canadian prairies, its findings might have caused envy. Neighbouring states such as Montana already had over 350,000 irrigated acres. Overall, the western states contained a combined total of 3.6 million irrigated acres. In Canada at the same date there was talk of irrigation and several ambitious plans but almost none in practice.[1]

There were a number of reasons why irrigation had not yet stirred great interest in the southern prairies. Semi-arid lands could scarcely be settled with conventional farming methods, but there was still plenty of land in more hospitable climatic circumstances that could be claimed under the Dominion Lands Act. The Department of the Interior had actively discouraged homesteading in some of the driest sections of the southern prairies, judging it unfeasible, and placed control over huge sections of land in ranching leases.[2] This policy meant that a considerable portion of the future irrigation region was practically removed from settlement. In any event, no large developers seemed interested in assuming the costs of building irrigation infrastructure. Although the CPR's main line cut through some of the driest portions of the southern prairies and ran parallel to the Bow River, company officials judged the lands of the lower Bow to be unattractive. The railway had been promised land "fairly fit for settlement" in its land grant from the federal government, and the lower Bow seemed unfit.[3] Finally, the federal government had attempted to squelch talk of irrigation because it challenged the image of the Northwest as a farmers' paradise that had been cultivated in so many posters and pamphlets distributed across North America and Europe.[4] To move irrigation development forward, a number of basic conditions would have to change.

Despite the cautious approach of the Department of the Interior towards irrigation, prominent members of its staff had begun to explore and promote the concept. Chief among them was William Pearce, whom we have met before, ostensibly the superintendent of mines who also held considerable influence over western settlement policy. From his base in Calgary he approved and denied settlement applications, set aspects of ranching policy, and shaped the system of settlement. Pearce had become convinced some years earlier that the southern prairies could not be settled without irrigation. He had observed irrigation projects in Utah and Colorado, had a clear sense of the challenges posed to farmers by semi-arid conditions, and looked with excitement upon the prospect of damming eastward-flowing rivers. In the late 1880s he made these views known in a range of public documents and reports, speeches, and letters to persons of influence.[5]

Irrigation held an element of personal mission for Pearce. At his home in Calgary, on the south bank of the Bow River, which he affectionately dubbed "Bow Bend shack," he hired men to dig canals and water fields so that he could instruct his neighbours in the benefits of irrigation.[6] Just west of Calgary, he collaborated with local entrepreneurs to establish one of the first irrigation projects in the region, drawing water off the Elbow, a Bow tributary. The plan never amounted to much.[7] Its primary success was to bring the matter of irrigation to a public audience. Before long, Pearce found himself corresponding with the president of the CPR about his venture and the importance of irrigation to opening up the west.[8]

Pearce's vision sprang from his desire to mimic and improve upon American irrigation development. He was impressed by the rapid expansion of irrigation projects across the American West in the late nineteenth century and foresaw the potential of this settlement strategy in parts of the Canadian prairies. He studied the irrigation laws of different states, sought to attend international irrigation congresses, and corresponded with prominent American irrigation experts and surveyors. "We can gather a great many hints from the experience of the United States," he wrote to one of his superiors in 1891. One of Pearce's correspondents, Andrew Schumacher, of the Canadian Immigration Bureau in Chicago, tried to convince Pearce to attend the World's Fair in 1893, so that he might see the wonderful displays on American irrigation – "the idol of your heart."[9]

Although Pearce admired the promise of American irrigation, one aspect of its development concerned him. He judged American irrigation law to be confused and ill designed. The legal regimes that had developed in different American states variously blended riparian use rights drawn from the English common-law tradition and prior appropriation doctrines, which vested priority in use to those who had developed water first ("first in time, first in right").[10] This mixture did not impress Pearce (he referred to it as "peculiar laws"); nor did the range and volume of litigation in American states over water rights.[11] He subscribed to the view current with some American irrigation experts that irrigation law needed to be overhauled to place greater authority in the hands of governments, supported by experts, who would be able to devise principles and practices for dispensing and cancelling water rights to various interests.[12] This approach could remove the ambiguity of state legal traditions, decrease litigation, and make irrigation development planned and rational. It was a perspective that called for a strong state and rule by experts, one of whom would be Pearce himself.

By the early 1890s, the prospects for developing irrigation in the southern prairies had improved. Dry years in the late 1880s had suggested the need for diverted water. Pressure from squatters on ranching leases had increased, as had calls for settlement in dry sections from Liberal Party supporters suspicious of the influence of ranchers on dominion lands policy.[13] In a series of promotional articles published in 1892, the *Calgary Tribune* cited evidence of ranchers constructing ditches to illustrate the promise of irrigation or referred to Pearce's company operating off the Elbow.[14] At the national level, Department of the Interior officials finally gained approval to move ahead and draft legislation that would provide a basis for irrigation in the Northwest. Pearce moved to Ottawa to help with the legislation, and J.S. Dennis, chief dominion surveyor, toured western US states to consult experts and observe irrigation in practice.

The statute that emerged from this process both created the possibility of irrigation development in the Northwest and controlled how and where it might be accomplished. The North-West Irrigation Act (1894) aimed to establish basic rules for obtaining and maintaining rights to water. The Crown held powers to grant water rights and to cancel the same if beneficial use was not realized. Water users had to apply for rights at particular diversion points, file surveys, and maintain infrastructure and beneficial use. If a holder of a water right allowed canals to deteriorate or did not use water in a beneficial manner, which meant invariably for agricultural purposes, then the right could be revoked. The basic aim was to cancel riparian rights before claims had been made in law and to establish the primary power of the state to plan, grant, and control water rights. With the American experience in mind, Pearce insisted upon the importance of cancelling riparian rights, but also bolstered the authority of the British Crown in water law, drawing on the example of a statute from Victoria, Australia. Although Department of Justice lawyers framed some of its language and various interests and American experts responded to drafts of the legislation, the law and its basic tenets were Pearce's.[15]

Before the act, irrigation projects along the lower Bow had been few in number and small in scale. When, in the mid-1890s, Department of the Interior officials sought to catalogue and convert the pre-existing water rights on the Bow, they developed a rather short list of users. William Maloney, a farmer from Calgary, had constructed a ditch for irrigation in 1893. Further west, a rancher and two other farmers had surveyed and constructed ditches for irrigation purposes. Possibly there were other small projects, diverting water into bottom lands for stock-watering on ranching leases or on to small patches of land used by squatters. One irrigation project existed on the Siksika reserve east of Calgary, a pet scheme of the local Indian agent, who had organized indigenous labour to dig the ditches and who hoped to convert his subjects into hearty yeomen farmers. There is no indication that the Siksika shared his goals but some indication that they resented the forced servitude involved in construction. Locals later referred to the project as "Ponton's folly," after the chief surveyor. During a series of wet years, the irrigation canal went virtually unused. Another attempt by an Indian agent in the early 1920s to resuscitate the scheme similarly led nowhere. Otherwise the Bow was virtually untapped for irrigation purposes. Pearce had asserted that no vested interests yet existed in the Northwest to complicate water planning and policy, and he had been substantially right.[16]

Canadian irrigation policy had never been aimed primarily at smaller users such as these. Rather, part of the rationale for creating a legal framework for development was to attract individuals or companies with access to capital and organizational expertise. In the United States, large colonization companies had developed projects in California and other parts of the southwest. Some of these were subsidiaries of railway companies seeking to promote settlement along their lines. Pearce and others now hoped to interest the CPR. In the years before 1894 Pearce had sought to engage CPR president William Van Horne in discussions of the potential of irrigation and had received courteous attention in return but no commitments.[17] This position shifted in the late 1890s as the federal government placed pressure on the company to assume its land grant so as not to delay settlement in the region. At the same time, Department of the Interior officials made arguments in favour of a flexible land grant policy that would allow the CPR to assume contiguous sections of land to facilitate irrigation projects. With no other prospects for these areas in sight and wishing to dispense with the problems of the CPR land grant, the minister of the Interior, Clifford Sifton, consented to these calls and passed legislation to put them into effect.[18] The way was cleared for the CPR to assume a large territory along the lower Bow and to develop it for irrigation. For the company this prospect held out the possibility of profits in land sales and water charges as well as increased traffic along the line. When the specific plans for construction developed in the eastern portions of the CPR land grant, it would also become necessary to cut off sections of the neighbouring Siksika reserve to facilitate the construction of a dam and reservoir.[19] Like waterpower projects further up the river, irrigation development impinged on reserve lands and required a further act of dispossession to be completed.

Systems and Settlement

To envision irrigation as Pearce had done was one thing; to plan projects, build them, and people them was quite another. The irrigation projects envisioned by the CPR on the lower Bow assumed a massive scale, in millions of acres. Land would have to be surveyed; water conveyance systems, including dams, aqueducts, and canals, planned and constructed. A management regime to operate the irrigation projects would need to be devised. And last but not least, settlers would have to be convinced that irrigation farming was superior to dry farming and worth the additional investment.

When the CPR first considered the Bow River irrigation scheme, very little was known about the land base or the topography of the region adjacent to the lower river. To facilitate development, the Department of the Interior in 1894 conducted irrigation surveys to map possibilities for investors.[20] The following year J.S. Dennis, the chief inspector of surveys, reported 223 miles of main canals surveyed; levels, elevations, and contours had been determined over 1,000 miles, and close surveys made of over forty reservoir sites.[21] A year later Dennis felt confident that a large project could be realized comprising 6,000 square miles, or 3,840,000 acres, in which possibly 60 per cent of the land could be irrigated. He argued that the gradual decline in prairie levels from west to east allowed for the possibility of a gravity-fed system, supplied by a diversion near Calgary in which water could be conveyed by a network of canals and reservoirs over hundreds of miles.

Dennis brimmed with optimism at the prospect: "The settlement of the district capable of being served by the Bow River Canal on this system is sure to be followed by success and by the rapid springing up of numerous small prosperous villages with contented settlers. The soil, climate, fuel supply and possibilities of the district for the outdoor grazing of cattle and sheep are now thoroughly understood and appreciated, and it only remains to provide, by irrigation the moisture which nature fails to supply."[22]

With this kind of boosterism from the chief surveyor, it is little wonder that the CPR expressed some reservations about the reliability of the surveys before moving forward with the project. Nor did the federal government's reservation in 1897 of a water right for an envisioned Bow River project prompt the company into action.[23] At this date, the company was still engaged with land sales in Manitoba and Saskatchewan. After the passage of several years and under a new president, Thomas Shaughnessy, the company commissioned another survey in 1901 by leading American irrigation engineer George Anderson at the expense of the Department of the Interior.[24] Anderson largely confirmed the findings of the previous surveys and judged an irrigation project on the Bow to be feasible and promising. None of these preliminary surveys had taken into account detailed considerations of site and planning, water conveyance, and storage. Nevertheless, by 1903 the CPR had moved to obtain title to the irrigation lands from the federal government and proceed with development. Later, more detailed surveys conducted by the company had to consider which coulees and valleys would pose a problem to a gravity-fed system and which lands and topographical regions could be irrigated. On the basis of these surveys, the large initial tract was cut into three: a western section that would be supplied by the Bow from a diversion at the eastern limits of Calgary; a central section that was deemed too uneven in topography and therefore unfit for irrigation; and an eastern section, located between the Bow and Red Deer Rivers, that would draw water from the Bow at Horseshoe Bend, four miles southwest of Bassano (see map 6.1).[25]

For just over a decade, from 1903 to 1914, land along the lower Bow was transformed into one vast construction site. Engineers and construction firms on contract from Chicago, Boston, and Ogden, Utah, crossed and recrossed

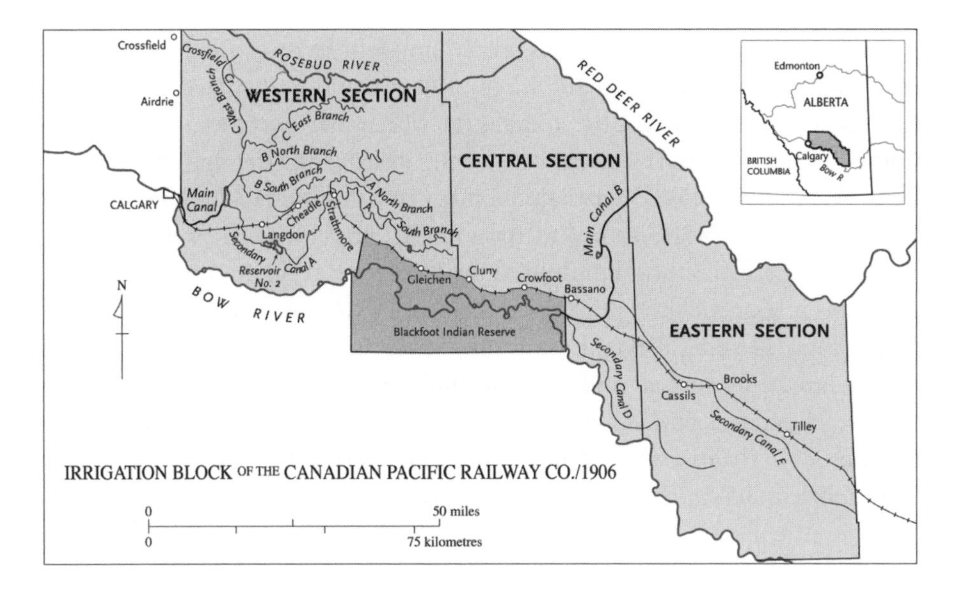

Map 6.1 CPR Bow River scheme, 1906. Note: The Blackfoot reserve as it existed in 1906 has been added to data from the original. (Adapted from A. Mitchner, "The Bow River Scheme: The CPR's Irrigation Block," in *The CPR West: The Iron Road and the Making of the Nation*, ed. Hugh A. Dempsey [Vancouver: Douglas and McIntyre, 1984])

the territory, drawing up plans and putting them into effect.[26] Crews of engineers camped out in tents and ate by fires fuelled by buffalo chips. By day they drove about in horse-drawn democrats, set up their survey devices, and fought off curious cattle that wandered over to take a look. Fred Cross, the engineer in charge of structures, spent his weekends sketching and painting the emerging irrigation works against a brown prairie landscape. In the winter, engineers skated on the Bow or rode the line to Vancouver for a break. Labourers were also attracted from near and wide, including a considerable number of sojourning Italians.[27] They settled in ephemeral work camps along the railway's main line and near to major work sites. Workers dug canals using teams of horses pulling large metal shovels. Against the skyline, wooden structures emerged to convey water. Later, large hulking reinforced concrete works – canals, weirs, aqueducts, and dams – were built. By 1909 the CPR had completed the first phase of the irrigation block, and the Western Section began to place settlers; a year later work began on the Eastern Section. By 1914 the main structures were complete.

The project attracted a good deal of attention. In 1911 *New York Times* published a full-page spread on the building of the Eastern Section and described the irrigation block as the "biggest irrigation plant in America."[28] In 1915 *Scientific American* called the CPR scheme "America's Greatest Irrigation Project."[29] In 1914, the year of the completion of the Bassano dam, the International Irrigation Congress met for the first time outside the United States, in Calgary; delegates travelled by special train to view the Bassano dam and the canal head gates. Word spread in the immigrant press. The same year forty Swedish-American journalists toured the CPR project. Behind them followed groups of engineers from the United States, as well as municipal and government officials from Alberta. From overseas came experts in search of inspiration. Messrs Wilanksy and Gutman of Palestine and California, Zionists seeking ideas for irrigation ventures in the Holy Land, visited in 1920, as did Mr A.N. Robertson, chief engineer of the Punjab, two years later.[30]

The scale and scope of the projects impressed observers. The Western Section comprised over 1 million acres, the Eastern Section even more (1,156,224 acres). Both held massive irrigation potential, with an anticipated 400,000 acres susceptible of irrigation in the Western Section and 440,000 in the Eastern.[31] The Western Section's major works included a weir at the eastern limits of Calgary that diverted water off the mainstem into a large eastward-flowing canal, as wide as 120 feet in places. Secondary canals diverted water from the main trunk to different sections of the project. Combined, these secondary canals measured about 250 miles in length. A reservoir, formed in a low valley, provided storage. A third order of ditches cut off the canals and reservoir and delivered water to individual farms. Throughout the system, water passed through structures encased in wood and concrete and flowed through unlined canals and ditches. In the Eastern Section the Bassano dam spanned the Bow River and rose to a height of sixty-five feet (see illus. 6.1).

The main canal ran east into the section, colonizing former creek beds, and crossed a low valley by means of a two-mile long concrete aqueduct. Lake Newell, the project reservoir, named after a famous US irrigation proponent, captured the flow, stored it, and allowed for diversions to various secondary canals throughout the section. The reservoir took years to fill. By 1923 officials reported that its surface area extended roughly four by eight miles; nineteen earth dams impounded the flow. These secondary canals had a combined length twice those in the Western Section and conveyed water to ditches and eventually farmers' fields. Some of the water diverted into the system returned to the Bow as runoff, and a few canals at the base of the system produced return flows to the river; but much was also diverted out of the watershed and ultimately into the Red Deer River to the north. All of this involved massive quantities of water. By 1919 the Eastern Section alone was diverting almost 200,000 acre-feet to the section during the irrigation season, while supplying

Illus. 6.1 The Bassano dam coming to completion in about 1914. (Glenbow Archives, NA 3641-1)

some areas with water stored in its previously filled reservoir. Two years later, over 360,000 acre-feet were being diverted from the Bow.[32]

Diversions and canals channelled water, but the operation of these systems depended upon continuous management. The CPR kept a hydrographer in Brooks to monitor the flows; gauges were distributed throughout the system. An electrician operated the Bassano dam as well as the canal head gates. In each of the two districts of the Eastern Section, a foreman oversaw canals and water deliveries. In the irrigation season a small army of temporary employees called ditchriders provided support. They rode horses up and down the canals, opening ditch gates for farmers. In 1915, eighty temporary workers were on the payroll of the Eastern Section. All these employees brought the concrete and flow of the system to life and made decisions about when in the spring to divert water into the system and when in the fall to shut it off. They had to assess the efficiencies of the network, contend with farmers who wanted increased or decreased water deliveries, and ensure system-wide coordination.[33] The irrigation network was as much a human organization as a collection of canals.

With infrastructure in place, the CPR entered the settlement phase. Agents representing the company purchased advertisements across Canada and the United States and maintained contact with others in Europe; they led tours of the project and arranged for special train cars to gather prospective settlers from various points in the American West. The company published and distributed a range of promotional pamphlets, altering them according to the regional market. The pamphlets offered practical suggestions to intending farmers, spoke of the glorious opportunities of the Bow valley, and outlined policies and costs.[34]

One pamphlet opened with a sumptuous image of an ideal farm and promised "Facts Concerning the Bow River Valley."[35] A handsome farmstead, surrounded by a fence, drew the eye. To one side lay a field containing a bounteous crop; on the other, a herd of grazing cattle. Through the middle of the image ran a prominent irrigation ditch. Opening this pamphlet, intending settlers learned that they were about to farm soils "almost beyond belief," a vague phrase that covered the lack of detailed soil surveys. The pamphlet continued with superlatives: the irrigation project contained "the finest winter wheat lands in America" and the best pasture "in the world"; the CPR had constructed "the largest irrigation system in the Western Hemisphere" and had attracted farmers "from all over the world." Combination farming, the pamphlet instructed, provided the primary opportunity. Settlers could raise wheat on non-irrigable land, alfalfa and other fodder crops on irrigable land, and livestock on the irrigated feed. The terms of settlement were presented plainly: "Non-irrigable lands are sold at prices ranging from up to $15.00 per acre and irrigable lands up to $25.00 per acre … The uniform terms upon which the Company disposes of its lands are: One-tenth of the purchase price in cash and the balance in nine equal annual installments with interest at 6 per cent on the unpaid balance." Prices for irrigation water would be assessed at 50 cents per acre per year. Although the pamphlet acknowledged that these costs would be greater than for a homestead, it stressed the importance of the connections to the railway system that the project provided and highlighted the reliability of irrigation against drought years.

However, the idealized portrait that the CPR produced could not overcome the practical problems. Although the land and settlement costs in the project did remain relatively low before the mid-1930s in comparison to equivalent American projects, and the company managed to attract a range of experienced American irrigators as a result, the irrigation block faced stiff competition from the land market in other prairie sections and the United States.[36] Agents representing the CPR had difficulty convincing potential land buyers that the promise of irrigation far outstripped other farming practices and deserved a higher price tag. This was particularly so in the first decade and a half of the century, when weather conditions were generally favourable to dryland farming techniques in most areas of the prairies. Furthermore, the

Illus. 6.2 The Royer homestead in the Gem colony of the CPR's Eastern Section in 1914. (Glenbow Archives, NA 500-5)

timing of these project starts was not ideal. The Western Section opened in a period of considerable activity in land markets and general optimism about farming opportunities in the region, but the larger Eastern Section became available in 1914, months before the opening shots of World War I. Thus the CPR had difficulty staffing its operations, let alone attracting settlers. For four long war years, the overhead costs of these projects weighed on the company. Although the end of the war offered a renewed opportunity, a brief phase of increased land sales from 1918 to 1920 was followed by a decade of static or declining settlement, driven in part by the plummeting price of grain. Since the opening of the project, the price of land had also increased, and despite a wide range of payment options, the high price of $50 an acre for irrigable land must have driven away all but the most determined would-be irrigators.

Even if the irrigation block did not live up to the idealized portrait sketched by the CPR, settlements did begin to take form, farms were developed, and a new cultural landscape shaped and bounded the land. The first phase of settlement in the Eastern Section occurred during the difficult years of World War I. In 1915 the CPR listed 60 irrigable farms occupied and noted that 16 new farms had been purchased during the year.[37] Although settlement expanded somewhat over the war years, it was not until 1918 that prospects improved. In order to capture an anticipated settlement surge, the company reserved lands for returning veterans, offered improvement loans, and placed ready-made farms on the market, with a house, fencing, and a planted crop.[38] The number of farm units in 1919 totalled 488; a year later they reached over 900. Settlers

Illus. 6.3 Irrigation pump, 1918. (Glenbow Archives, NA 4389-1)

arrived from various locations. The first group, experienced irrigators from Colorado, settled near Bassano in 1914. Others arrived from various parts of the United States and from surrounding prairie districts. By 1919 several townships along the Bow Slope had been settled by Swedish Americans, immigrating from the Dakotas, Nebraska, and Washington State. French Canadians moved in around Rosemary. Mormon settlers from other parts of southern Alberta and Utah took up sections around Cassils and Rosemary. Belgians purchased land around Patricia. In the 1920s several Mennonite communities took up land in the section, immigrating from Pennsylvania and Russia.[39] Riding a horse for a day across the section in the late 1910s, one encountered a diversity of building styles, churches, peoples and languages. (See illus. 6.2 and 6.3.)

The settlement pattern emerged in relation to the railway, the canal network, and the survey system. At first, the CPR sought to open lands within a limited area: near the rail lines and adjacent to one of the main canals. Proximity to the railway allowed settlers to ship in supplies and effects with ease and take products to market efficiently at a time of poor road services. Grain elevators were built along the line in conjunction with the settlement of surrounding lands. Placing settlers on the canals also eased problems of water distribution for the company and ensured that these early settlers would have consistent deliveries when they needed them. On farms without wells, the canals provided settlers with water for domestic use, stored in large cisterns. In the Eastern Section four settlement clusters were made available within the first years of operation. Two were located on the north branch canal, near Bas-

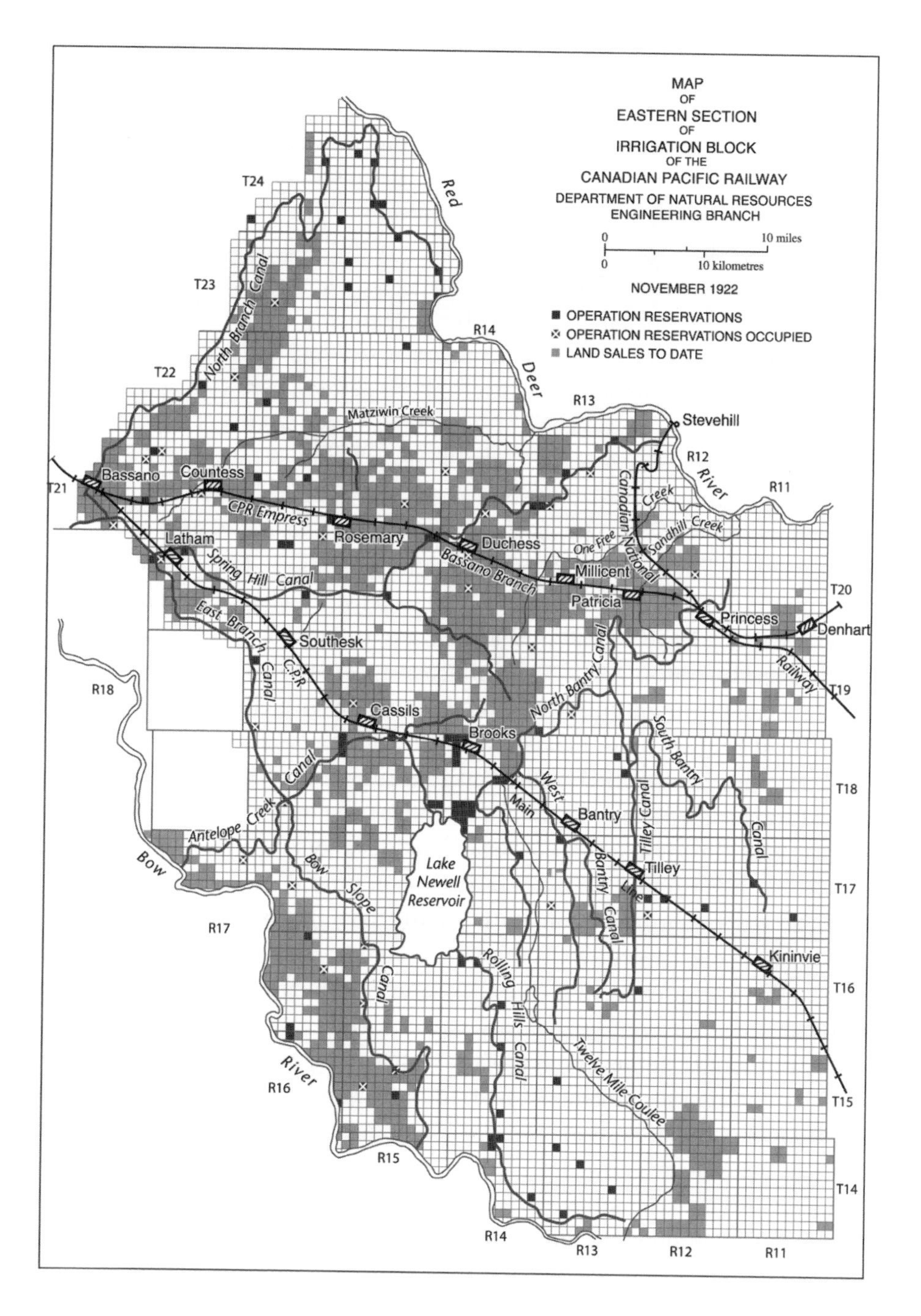

Map 6.2 Map of land sales in the CPR's Eastern Section, 1922. Note the tight relationship between settlement, railways, and canals. (Glenbow Archives, EID Papers, M6388, box 3, Annual Report, 1923)

sano, another on the North Bantry Canal, closer to Brooks, and a fourth on the West Bantry Canal near Tilley. Most of these clusters of settlement were a short distance from the CPR main line; others were serviced by a cut-off called the Bassano-Empress line, completed in 1914, which reached settlements around Duchess, Patricia, and Millicent. Within the Eastern Section a few large landholders established themselves in the first years of operation. These included the Sutherland estate, near Brooks, a sprawling beef-producing operation owned by a distinguished British family. For the vast majority of settlers, however, landholdings encompassed between 70 and 320 acres; the average holding was about 100 acres.[40] These parcels were sold according to standard property definitions within the survey grid (see maps 6.2 and 6.3).

Within a short period of time, the landscape of the irrigation block was converted from shortgrass prairie to a narrow range of cultivated crops and pasturelands devoted to extensive grazing. Annual farm censuses conducted by the CPR on the Eastern Section suggest emerging patterns of production and land use. In the earliest stages of development and statistical collection, company staff identified cattle as a mainstay of farm units. Hog and sheep production were less central to farm economies but also important. Chickens were kept in high numbers, and some farms also raised a few turkeys. Most farms kept horses or mules as beasts of burden. The arable land base was devoted primarily to wheat, as it was in most prairie districts around World War I. Fodder crops such as alfalfa, timothy, and green feed were also raised for livestock. Small quantities of vegetables and root crops, probably for local markets or domestic consumption, were also produced. Farmers planted trees, provided free by the CPR. Caragana, Russian poplar, laurel willow, green ash, Manitoba maple, spruce, golden willow, and other species were placed near homes, along fence lines, and across the formerly treeless land. They provided shade and wind protection, but also the comforting landscape effects of other places and former homes. Although the relative proportion of land devoted to any given crop or livestock use changed over time, from the beginning most farms in the Eastern Section operated on a mixed-use model of livestock production and wheat-raising.

The Muskrat Cometh

In the 1890s, when William Pearce had first promoted irrigation agriculture as a new model of settlement, friends, associates, and superiors had asked him questions about the ultimate effects of delivering water to a dry land. Would the climate change? Would malaria gain hold? Would irrigators get rheumatism from long periods spent in the water? Would alkali salt turn up on the land, as it had done elsewhere? Pearce had always responded to correspondents with an optimistic appraisal of environmental effects. Climate might well change, he judged, but only advantageously. Summer temperatures would be moderated and cooled, and winter temperatures would rise. Possibly, summer frosts would disappear.[41] The incidence of rheumatism would probably decrease, not increase. "What rheumatism we have here is attributable to the dryness of the climate and irrigation may mitigate it or remove it altogether."[42] Alkali would not be an issue because "the waters to be applied are derived from the mountains and traverse no alkaline beds and will not contain those deleterious salts in quantities to prove injurious."[43] The river's water was like a great elixir, responding to the many practical needs of farmers and healing the ills of the land and society. Pearce imagined water as a force for good, but he did not pause often to consider what might be the various effects of pouring vast quantities of water onto dry prairie.

Perhaps the least anticipated consequence of building irrigation projects on the lower Bow was the expansion of habitat for wildlife from the margins of the river into the hundreds of miles of canals, ditches, and reservoirs. These developments opened new channels for animals to colonize and exploit, ecological niches that had not formerly existed because of the prevailing semi-arid conditions. Within a few years muskrats and beaver had moved in from the Bow River and staked their claim as vigorously as any human settlers. Early CPR reports contain photographs of canal walls penetrated by muskrats, full of holes. A manager on the project suggested in 1918 that "unless action is taken at once to destroy the animals, possibly before we know it they have a hole dug through the bank which may cause a very serious break."[44] Beaver also found the main canals to be passable habitat and promptly set about damming them. In the 1930s bounties were set on muskrats, and local farmers-turned-trappers were encouraged to hunt and be merry. In the early 1940s, irrigation officials sought special privileges to hunt muskrats in the area, because of a decline in trapping during the war. Without such measures, they feared that the area would be turned into a "muskrat farm."[45]

Fish also entered the irrigation systems by migrating through weirs and diversion channels into complex webs of infrastructure, where they met an uncertain environment. In summer months, aquatic weed growth could be intense and water slack and warm. Fish could become trapped in relatively shallow channels and die because of a lack of oxygen or high temperatures. Worse yet, in the fall, at the end of the irrigation season, when irrigation workers shut down the system, fish that had not already made their way into reservoirs were trapped. As water drained from canals, fish would die, slapping about in the warm fall breezes. Some settlers would attend the annual closure of the Brooks aqueduct in expectation of easy pickings. One CPR engineer, Floyd Yeats, recalled that locals "would just get sacks of fish, whitefish mostly, that were trapped when the water was shut off."[46] Irrigation systems posed risks for fish that entered them, but also created some productive habitats. Those fish that did swim into major reservoirs survived in good numbers. From early in the century self-sustaining populations of pike were found in Lake Newell. By the mid-1930s introduced whitefish flourished and supported commercial fishers.[47]

From above a range of new bird species appeared. Particularly around large reservoirs, new habitat for birds had been incidentally created. Marshes, riparian zones, and trees all provided migrating birds with choice places to land, feed, and rest. Between 1943 and 1945 an early bird census in the area noted 213 species, many of which were uncommon on the prairies.[48] Although the irrigation landscape no doubt removed habitat for dozens of bird species that used the shortgrass prairie, the heavy heads of wheat on project lands near lake and wetland areas proved to be an important draw for a range of migrants. Compared to other regions of the prairies where moist sections called potholes were actively being drained to make way for more farmland, the irrigation project provided new habitat.[49] With time, the presence of birds and waterfowl around reservoirs led to attempts to set aside recreational space for bird use and human play. A park would be created to preserve Lake Newell's margins and wetland areas, and Ducks Unlimited and the irrigation district would establish a partnership to protect wetland habitat primarily to enhance duck-hunting opportunities.[50]

Between the river, reservoirs, and fields, the canals carried water but also an unwanted cargo: seeds of plant species deemed to be weeds that would compete with cultivated plants for scarce resources of land and water. Early on in the project, CPR officials connected the dispersal of weed seeds to water deliveries in the irrigation network.[51] Farmers inadvertently turned weed seeds onto their land along with water from the ditches and provided superb, moist growing conditions for the unwanted plants.

Weeds that fell into in canals produced a problem of a different order. They affected system functions. Weeds such as Russian thistle and tumbling mustard gathered in canals, slowed the flow, and raised water levels. Photographs contained in internal reports show employees standing on top of these jams trying to hack them apart. "When the small ditches are carrying water," irrigation officials reported in 1917, "they have to be constantly patrolled as one large Russian Thistle might easily cause a break … In the larger Canals the weeds drift into piles on the curves and at structures across the ditch section, at times as high as the Canal banks." On the main canals, teams of men led horses dragging chains to release weeds and pull them to gathering points where they could be hauled out and burned. More than one prairie fire resulted from these weed eradication exercises. Annually, irrigators had to devote time, labour, and expense to this problem.[52]

The canals were also unstable. Only major system structures were built with concrete or wood. Most canals were unlined or were formed in coulees, valleys, and depressions, all susceptible to erosion. Flowing water scoured steep sections, undermined overhangs, and built sandbars. Irrigation officials estimated that water levels dropped by about a foot in sections of serious scour, and they had to devote time to clearing the sediment deposits downstream or risk more serious blockages and breaks. In 1920 a landslide deposited an estimated three thousand cubic yards of material into the main canal of the Eastern Section, a quarter of a mile from the intake. Company officials noted soberly in the annual report that this slide "seriously interfered with operation." Little could be done about it, except to monitor conditions and clear sediment.[53]

In the background lay a more continuous and serious problem: seepage. Unlined canals could direct water but not contain it. In 1921, internal studies by the CPR suggested that approximately 14 per cent of the water diverted into the system was lost through seepage.[54] This was probably a modest estimate. Five years later company officials provided a more far-ranging assessment: "[B]etween 50% and 60% of the total water diverted into the system is lost in the form of seepage from canals, deep percolation after the water has been applied to the land and careless waste … A large part of the loss finds its way into the numerous and widely scattered basin-like depressions and is thus often the first stage in the creation of a non-productive area. An undetermined amount sinks into the deep sub-soil to join the underground reservoir and may or may not re-appear to cause further trouble at lower elevations."[55]

Lands adjacent to canals could be inundated by breaks in the system, but they could also be turned into marsh through persistent seepage. In areas with highly porous and sandy soils, problems occurred frequently and forced irrigation officials to build drainage structures alongside canals. Even when water was delivered to fields, seepage problems did not stop. Inexperienced irrigators flooded fields far more than was necessary, causing them to be waterlogged.[56] This practice contributed to the seepage problem locally and on downslope properties.

Irrigation water delivered an unexpected product from beneath the surface: salt compounds, commonly referred to as alkali. When water percolated deep below the soil surface, it absorbed soluble salts. As surface water evaporated, subsurface waters were drawn upwards, bringing these salts to the surface.

After several years of alkali emergence on water-logged land, farming conditions declined. Land became boggy and difficult to work. Salt grass moved in and crops died. Crusts of white alkali formed.[57] References to seepage and alkali emergence studded reports from the early years of settlement. By 1923 irrigation officials were writing in frustration, "The seepage must be stopped, or controlled, and the accumulated alkali must be removed and to do all, or any of these things may be, and usually is, complicated, difficult and expensive."[58]

Irrigation Blues

The CPR spent the first decade of development on the Eastern Section finishing the infrastructure and placing settlers on the land. The assumption was that, with time, settlements would grow and prosper. Farmers would get established and pay down their land contracts. The company's heavy investments would begin to bear fruit as revenues and freight charges increased. In the 1920s, however, the prairie agricultural economy entered a period of uncertainty as grain prices plummeted and immigration slowed.[59] In areas of southeastern Alberta encompassing the Eastern Section, episodes of drought produced dust bowl conditions. Officials complained of ditches filling with topsoil.[60] Rather than reaping returns on investments, the company found itself arguing with settlers over land contracts and the deteriorating quality of soils, repairing system infrastructure, building drainage ditches, and eradicating weeds. The range of difficulties befalling the project made circumstances difficult for settlers and fostered resentment against the CPR. In the 1920s, hundreds abandoned the Eastern Section, quit land contracts, left without notice, or leased land to others. Annual reports produced by company officials in the mid-1920s read like tales of woe. Officials worked hard to place new settlers on the land, only to discover at the end of the year that they had a negative balance because of the number of abandoned farms.

Settlers left the project for a range of reasons. Some found it difficult to adapt to irrigation methods. Others gambled on crop and livestock mixes that did not pay off. Financial distress commonly loomed in the background. Some were fed up with the CPR and wished to extricate themselves before matters became worse. The combination of plummeting grain prices and high settlement costs forced many to reconsider their options. Company officials looked on the number of settlers who had opted to leave with barely suppressed alarm. They wrote to superiors that settlers had different reasons to leave and dismissed some as mere "rovers." Nevertheless, they acknowledged a "complex combination of factors" at play. Chief among them was "anxiety as to the future and as to their obligations to the Company."[61] David C. Jones

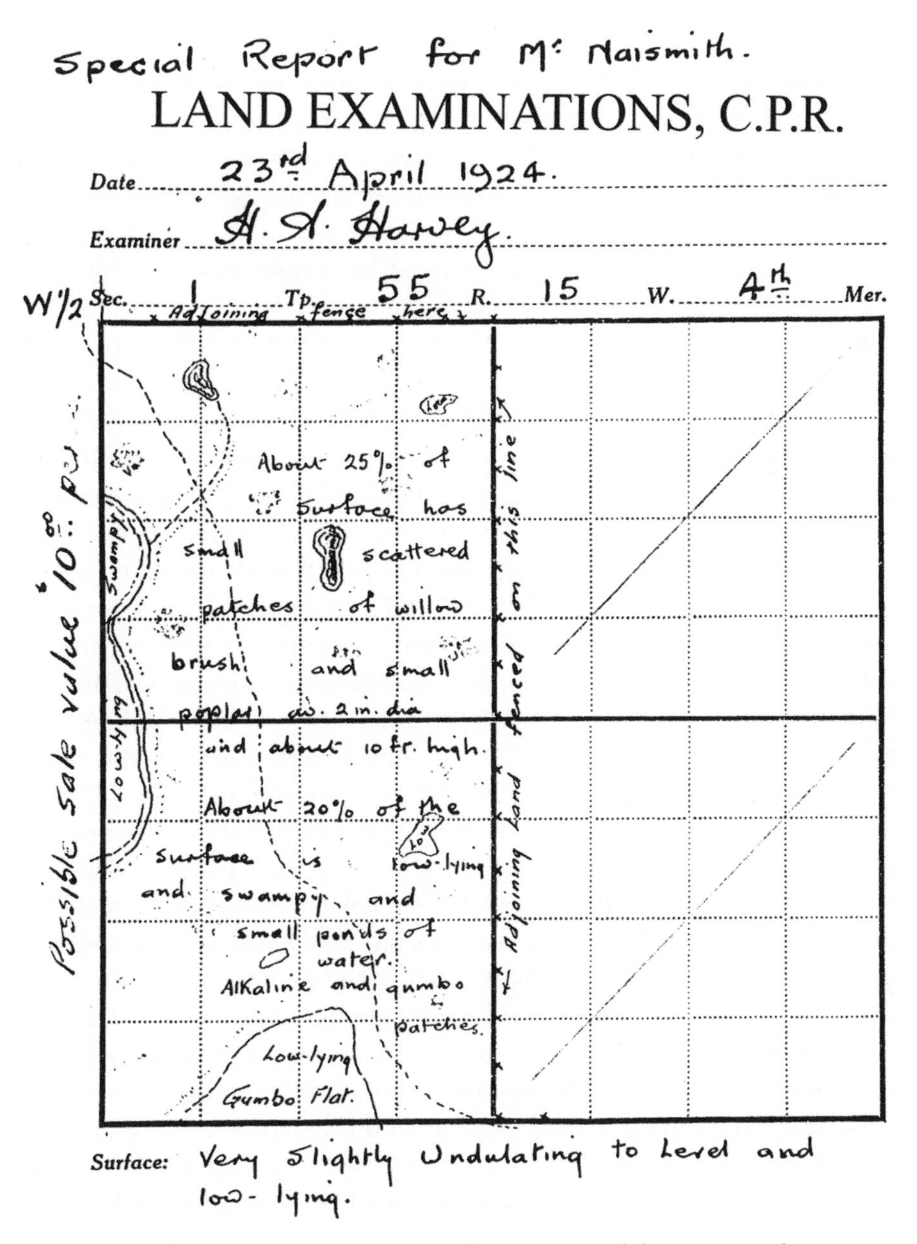

Special Report for Mr Naismith.

LAND EXAMINATIONS, C.P.R.

Date 23rd April 1924.

Examiner H. A. Harvey.

W ½ Sec. 1 Tp. 55 R. 15 W. 4th Mer.

Adjoining fence here

About 25% of surface has small scattered patches of willow brush and small poplar av. 2 in. dia and about 10 ft. high. About 20% of the surface is low-lying and swampy, and small ponds of water. Alkaline and gumbo patches.

Swampy

Low-lying

Low-lying Gumbo Flat.

Adjoining land fenced on this line

Possible sale value 10.00 pa

Surface: Very slightly Undulating to Level and low-lying.

Map 6.3 Land examination survey map. This survey card demonstrates how the alkali controversy drove a detailed resurvey of the land base to locate land degradation and alkali distribution. (Glenbow Archives, CPR Land Settlement and Development Papers, M2269, box 20, file 243)

analyzes the effects of dropping prices and drought conditions in eastern Alberta in the 1920s and identifies a wide-ranging crisis in dryland farming areas.[62] Matters were different but not necessarily better in irrigation areas. For while irrigation farmers had the capacity to produce even under drought conditions, they had to maintain payments not only on equipment and other farm costs but also on land contracts and annual water fees. Limited evidence suggests that farmers on the CPR project experienced an unusually high debt load.[63] In addition, the problem of alkali had focused attention on the cost of land and the need to adjust contracts to reflect the loss of farmable area within property units.

Alkali problems were sources of conflict between settlers and the CPR. Because the ultimate causes of seepage were complex and could potentially implicate both parties, neither group wished to assume responsibility. Since the productive land base had been reduced, settlers argued that land and water costs ought to be lowered. In 1925 alone, settlers lodged 238 complaints about alkali damage in the section.[64] Company officials, on the other hand, viewed many of these claims as trumped-up charges. "The poor class of farmer," wrote one official in 1926, "may not be greatly concerned over the ultimate outcome of his alkali affected areas, but sees in his alkali troubles an excuse for continual complaint, hoping to receive thereby from the Company, contract terms more favorable to himself."[65] Through the 1920s, nevertheless, the company employed surveyors to visit farms across the section and conduct alkali surveys. Lands that were classified as damaged were removed from calculations of the water fees, and adjustments were made to land contracts. The total irrigable acreage had stood at 400,000 acres when the Eastern Section had been planned. After the reclassification surveys, it fell to 250,000 acres, a massive drop with consequences for the CPR's revenues from water fees.[66] Following the surveys, drains were surveyed and constructed to reduce potential seepage from canals. Farmers were also instructed in land-levelling techniques to work out depressions on fields. By the end of the decade, officials judged that conflicts with farmers over alkali damage were decreasing.[67] Nevertheless, the alkali controversy had discouraged farmers and fomented disagreement and distrust between the company and settlers.

In the first half of the 1920s, the settlement statistics of the Eastern Section stood, as one CPR official put it, "at a standstill" (see table 6.1).[68] As the market conditions for agriculture continued to deteriorate, hundreds of settlers moved on. In 1921 the project held 1,062 occupied farm units. By 1925 the figure had dropped to 908. No marked improvement would be noted until the end of the decade. The number of contracted water users experienced a similar drop, from a high of 924 in 1921 to a low of 578 in 1927. All of this

Table 6.1 Farm units occupied and number of water users in Eastern Section, 1919–34

Year	Farm units occupied	Water users
1919	488	454
1920	910	760
1921	1,062	924
1922	996	893
1923	1,021	809
1924	968	733
1925	908	699
1926	929	638
1927	910	578
1928	922	587
1929	955	635
1930	1,003	761

SOURCE: GA, M6388, Eastern Irrigation District Papers, box 4, CPR Eastern Section Annual Report, 1935, 15.

NOTE: We have not been able to determine the precise definition of a farm unit. For the purposes of record-keeping and general analysis, CPR officials used this term and these data to make broad judgments about the success of the project. The figures give a basic sense of farms occupied but do not suggest, necessarily, whether owners or renters occupied farms. The water users figure provides a different measure, which suggests that not all settlers in the section took irrigation water and that some cancelled or ordered water on a shifting annual basis.

amounted to less land under crop. Abandoned properties cost the CPR in terms of maintenance costs and lost revenue. Alkali surveys reduced the costs of land contracts and therefore depleted company revenues. To accommodate settlers and under considerable public pressure, the company revised mortgage rates downwards, provided flexible debt repayment schedules, and waved various annual fees. By 1930 these efforts had borne some fruit, as the result of an assisted scheme to place about a hundred families on the project in cooperation with Mennonite settlement boards.[69] Without the addition of these settlers, settlement figures would have remained static.

Although many settlers moved on, a core group stayed. They became increasingly organized politically to challenge the dominance of the CPR and to lobby the company for contract revisions to decrease debt loads and cut costs. A few charismatic leaders such as W.D. Trego, a farmer from the CPR's Western Section and a persistent critic since the 1910s, led the movement.[70] Farmer protest also built around community institutions, particularly United Farmers of Alberta (UFA) locals. Carl Anderson, a Swedish-American immigrant who had settled on the Bow Slope near Scandia in 1918, was one such farmer who kept a minute book for the UFA local in Scandia. His notes suggest the prevalence of the contract issue in settler politics. In the midst of other community projects, such as lobbying for better postal service, holding community picnics, and leaning on the Swedish consul to advocate a rail link to Scandia, the local returned time and again to the contract issue. In November 1923 it hosted the local member of parliament. The record of the meeting expresses the charged atmosphere:

> This special meeting was … formally opened by the audience singing "O Canada." Mr Bergtson spoke on why we came here, our present situation and our prospects. Mr Garland, MP for Bow Valley, was then introduced. Mr Garland spoke on immigration, emigration, advantages of Canada, especially Alberta, banks, railroads, national debt, operations of government at Ottawa, and the future rosy prospects of this country. Owing to the general depression in farming, and to the nice promises given at the time of purchasing land here, several men were called on to give their opinion of this district, which did not seem rosy. After a short political speech, Mr Garland informed us that in his opinion the solution of the wheat problem was in sticking by the wheat pool, – that the CPR cannot sustain the present attitude in regard to our contracts, and that it's up to us with patience and perseverance to make this country better.

Underneath this description, Anderson scrawled "some meeting."[71]

Eastern Section farmers not only appealed to the CPR for better terms, but also sought to apply various forms of political pressure. On the advice of W.D. Trego, some people posted "for sale" signs in visible locations and threatened to move en masse to Bolivia if the situation did not improve.[72] By the early 1930s a contract holders' committee was representing the interests of settlers to the CPR and lobbying for various financial adjustments. Farmers wished the company to recognize the challenging economic environment and recalculate charges. But this primarily economic concern drew also on the conflicts over alkali, which extended back over a decade. The economic issues were experienced within the local context of environmental change. Because so many settlers did move on during the 1920s, the CPR had to take the threats and negotiating positions of farmer groups seriously. Settlers successfully exploited the situation to wrest significant changes to the financial arrangements and to place the burden of responsibility for maintaining infrastructure and repairing alkali lands on the company.[73]

Table 6.2 Livestock and animal holdings on Eastern Section farms, 1928

Animal	Quantity
Cattle	8,320
Hogs	10,117
Horses and mules	7,056
Sheep	30,363
Poultry	33,002

Apart from campaigning for better terms with the CPR, those who persisted increasingly reoriented their production towards livestock raising and finishing. By 1928, for example, the number of sheep raised in the section had increased almost thirty times from levels recorded in 1919 and had become the principal animal produced in the region (see table 6.2).[74] On abandoned lands and unsold sections of CPR property, farmers utilized the land as range and built up herds. Before 1931 the company allowed farmers to graze open land for free. In the late 1920s it and settlers organized a livestock feeders' association. Lambs were imported from various parts of Alberta and Saskatchewan in the fall for fattening.[75] When they were fit for slaughter, at about 100 to 110 pounds, they were sold and the profits divided between the rancher and the farm feeder. "Almost every farmer in Tilley took about 300 or more lambs each winter," Carl Anderson recalled, "and in this way kept busy, and was able to displace surplus alfalfa, hay and grain ... Even though in the depression years of drouth and low prices, when hardly anyone had any money, we all had plenty to eat, and with our large gardens and hogs and lambs we were well fed."[76] Although the irrigated farming of the Eastern Section was too distant from urban markets to allow for specialized crop and truck farming, the area's farmers increasingly took advantage of opportunities relative to dryland farming, particularly in terms of free, open range and abundant fodder crops, provided in part by irrigation.

In the early 1930s the irrigation areas of southern Alberta entered a period of difficulty and crisis. On the CPR's Eastern Section project, costs did not decline, but revenues generated from settlers did. In 1931 an internal report prepared by company staff provided a grim accounting for the Board of Directors. It found that since initiating operations in the Western and Eastern Sections, the company had sunk almost $41 million into development and operations, whereas only about $26 million had been realized on the projects as income. The company was owed over $15 million in back payments and debt deferrals, little of which was likely to be recovered.[77] Annual upkeep costs outstripped income by about half a million dollars. Henceforth, the CPR quietly set about to develop a strategy to unburden the company of the irrigation sections, while still collecting the lucrative freight charges on the shipments into and out of the region. At the same time, farmers in the Eastern Section became increasingly vocal and organized in opposition to the CPR. Carl Anderson and several others formed a committee on behalf of contract holders in 1934 to negotiate better terms and possibly to assume the entire operations of the section as a cooperative venture. The following year the CPR and the contract holders' committee struck an agreement to place the Eastern Section in the control of a farmers' co-operative. Outstanding debt would be forgiven, and the company would help to fund the organization in its first year of operation, up to $300,000.[78] The corporate era in irrigation had closed; the co-operative era had opened.

Irrigators, the State, and Expansion

Following the transfer, a co-operative institution formed under the provincial Irrigation Districts Act (1915) and the CPR's Eastern Section became the Eastern Irrigation District (EID). Many of the same farmers who had served on the negotiating committee with the CPR became members of the EID's initial board of trustees; this group hired E.L. Gray, formerly the deputy minister of municipal affairs in the United Farmers of Alberta government, to act as the district's manager. In a short time, the EID forgave five-sixths of the personal land debt of farmers and implemented a drastic reduction in operating costs. Farmers gained the satisfaction of landownership and had new funds available to invest in machinery and buildings.[79] The EID also began to seek out new opportunities. During the late 1930s the district entered the feedlot and livestock trade by building auction yards in Bassano and Brooks and combining several livestock shipping associations into a more integrated venture.[80] Livestock finishing, increasingly involving cattle, became the pillar of local farm production. In a very short period of time, the circumstances of EID farmers began to change for the better. The district that had been formed to deliver water became a more formidable institution with a range of roles and interests. In the press, local boosters described the EID as a "bright spot on the bald prairie."[81]

The EID had come into being in the same year as the federal government began a support program for prairie agriculture under the Prairie Farm Rehabilitation Act. The act was one piece in a flurry of legislation introduced by the Bennett Conservative government in 1935 to respond to the depression crisis shortly before a federal election.[82] Although it mimicked US New Deal legislation, the Prairie Farm Rehabilitation Administration that the legislation created was a relatively modest initiative. It drew on staff from agricultural experiment stations in the prairies region to deliver programs and services to farmers in distress. With a five-year mandate and a budget of $1 million, PFRA staff were called upon to instruct farmers in land husbandry, buy up and conserve drought lands, and build small water projects. However limited, this program signalled an important shift in the role of the state in irrigation development. Whereas earlier in the century the federal government had served as a promoter of irrigation and sought to organize the legal and institutional framework so that large corporations could enter the field, now the state would act to support farmers and invest directly in a host of rehabilitation projects. Within a few years the PFRA's budget was doubled; within five years its mandate was extended indefinitely.

Although much of the agency's early work in water development focused on small projects in which PFRA staff would assist a single farmer to improve water storage, by 1937 its mandate had been extended to community rehabilitation as well.[83] The PFRA's community agenda had two aspects: to establish community pastures on the re-grassed lands of drought-stricken regions and to transfer the displaced farming population to irrigated lands that could sustain them. Although many irrigation projects had experienced difficulties in the previous decades, the PFRA's irrigation plans were embellished with optimism. In agricultural literature the model irrigated homestead was replaced by photographs of agricultural experts, dams, and symbols of order and reconstruction. The agency's name emphasized rehabilitation: the sick land would be healed, partly with water. Most of the new community projects would be located in Saskatchewan, where the federal and provincial governments were in broad agreement about the direction of rehabilitation policy. In Alberta, however, the provincial government suspected an invasion of jurisdiction and opted initially not to participate in the community pastures program.[84]

The EID's new success and entrepreneurialism and the PFRA's expanded mandate and need for resettlement areas combined to produce a model experiment in community development in the late 1930s: the Rolling Hills resettlement project. Rolling Hills was little more than a name on a map in 1937, a section of undeveloped prairie used for grazing in the southern portion of the EID. Although the CPR had built a canal to the area, it had never been used because of low settler demand. Following invitations to consider an investment in the project by the EID manager, Jimmy Gardiner, the Liberal minister of agriculture, toured the district that year.[85] Shortly afterward, the EID and the PFRA entered into a partnership. The EID had already received $15,000 in PFRA funds to improve infrastructure.[86] The new project would provide a further $50,000 of PFRA aid as well as support from staff for development and settlement. Ultimately, the project would account for a full tenth of all

PFRA spending in Alberta before 1942.[87] The district's employees would level the land, renew canals and ditches, and in some cases plant a first crop. For its part, the PFRA would organize the settlement process, which would draw primarily on Saskatchewan settlers from drought areas, as well as provide transportation and supplies for the first year. Land would be charged at the modest price of $8 an acre.[88]

The agreement between the EID and the PFRA held out different benefits for the two parties. In the case of the EID, the money received for the project helped to cover the costs of renewing infrastructure in Rolling Hills but also other parts of the district. The potential increase in the number of farmers also held out the promise of future returns. The district board explained to water users that the new settlement would "materially aid in enlarging our capital reserves" by increasing the number of water contracts and land sales.[89] For the PFRA, the Rolling Hills project offered an almost immediate resettlement site in an established irrigation region of increasing prosperity. Other planned irrigation projects at Val Marie and Eastend, Saskatchewan, would take years to develop and were novel irrigation experiments. The virtues of immediacy and experience found in the Rolling Hills project were important for the PFRA because they allowed the administration to complete its tasks promptly and with some assurance of success.

One of the most difficult aspects of the Rolling Hills project involved the selection of settlers. This was formally under the control of the PFRA, but practically the EID had a strong role in shaping the process. The problem was the opposite of what the CPR had traditionally faced. Rather than a shortage of settlers, the project faced too many. Although 180 quarter section parcels were planned for the project, by 1938 the EID office had received over 600 applications. "I heard that you had some land for sale that you opened up last fall," Mike Heisler of Mendham, Saskatchewan, wrote, in a typical letter of inquiry. "How are you selling it? I'd like to take some up too if I can get some. Is there buildings on it too? Let me know as soon as possible."[90] Privately, the EID manager worried that they had received applications from "every 'deadbeat' in the two provinces."[91] To head off potential difficulties, the EID board established a stringent selection process. Settlers would have to be experienced farmers, "free of debt; secure absolute release from any responsibility for seed grain advances, [and] hospital accounts." They would need to own "sufficient livestock and equipment in their own names … free of encumbrances." They were expected to have sufficient capital for a year and would need to pay the water rate in advance for one hundred acres. In addition, they would be subject to a three-year probationary period during which the district claimed "full control." The application fee of $175 further limited the field.[92]

Although the EID and the PFRA could agree on the terms of settlement, the project gained unwanted provincial attention in 1939 which threatened to disrupt this first, important federal experiment in irrigation development.[93] Publicly, the EID board of trustees faced criticism that the project would cater mostly to Saskatchewan farmers.[94] Federal civil servants had considered a more general settlement policy, drawing on a range of possible source areas, but agriculture minister Jimmy Gardiner, a prominent Saskatchewan MP, insisted that the farmers who lost land to community pastures in that province had to take priority.[95] Before long, local grumbling reached the Alberta government. A self-identified Social Credit supporter from the area wrote to the provincial minister of agriculture that resentment had built up over the matter.[96] Another complained that "we Social Credit farmers have no chance to get any of that land."[97] Wilson E. Cain, a Social Credit MLA, wrote to Premier William Aberhart that the province needed to take a careful look at EID affairs so that its assets would not be "dissipated to build politicians in Saskatchewan at the expense of our Alberta farmers."[98]

The province had earlier made its suspicions of the PFRA public when agriculture minister D.B. Mullen had characterized the community pastures program as a brazen "land grab."[99] In January 1939 N.E. Tanner, the minister of Lands and Mines, wrote to the EID asking for information on how many Albertans had applied to the project. Gray, the EID manager, replied somewhat disingenuously that "the number of good farmers who applied from the Alberta drought areas, or special areas, was to me unexpectedly small. A large number of the applicants seemed to me more or less drifters."[100] Provincial interest in the project coincided with a new level of concern about the EID in general and particularly its manager. Gray had recently become the new leader of the provincial Liberal Party. In response to complaints about EID affairs, the provincial government moved to dismantle its management structure.[101] First, the government dismissed L.C. Charlesworth, the one-person provincial irrigation council who oversaw irrigation districts for the province, and replaced him with a three-person board ready to do its bidding. The government then removed the EID manager and board of trustees and launched a judicial investigation into EID management.[102] The local response was immediate and intense. Eight hundred people gathered in Brooks to protest the action.[103] The *Calgary Herald* published thirty articles on the affair in just over two months.[104] In the end, the judicial inquiry cleared the manager and the board of all allegations.[105]

In the background to the storm and stress of the provincial inquiry, settlers started to arrive at Rolling Hills. Railcars from Saskatchewan disgorged families carrying boards for temporary shacks, farm equipment, and the domestic

Illus. 6.4 Making a homestead at Rolling Hills, May 1939. (Glenbow Archives, NA 3092-75)

remains of former homes. Around them, experienced Alberta irrigators took up land and helped the newcomers adjust to a new mode of farming. This was the start of some lean years for the new arrivals, who had to sink savings into new homesteads, cover the water rates, and raise crops. Disagreements inevitably came up. Was it the PFRA or the district which had offered free alfalfa seed? Nobody could remember.[106] Was it fair that the district took a portion of crops in lieu of cash payments for water? A petition from water users in Rolling Hills in 1940 stated that it was not.[107] Although the literature surrounding the Rolling Hills project emphasized an orderly transition to a new tomorrow, the experience of the new irrigators was bumpier. A local history published in 1965 contained this comment from Jack Holt, one of the original settlers: "We were more or less dumped here by the powers that be – PFRA and the CPR – and left to find the way out of a dilemma."[108] Holt was one of those who had complained about the water rates in that first year. An EID investigator stated at the time that Holt had "not helped himself to any extent."[109]

And yet he persisted, as did most settlers to Rolling Hills. In the first three years of settlement, only 3 of the 180 parcels settled in Rolling Hills were abandoned.[110] Those who stayed did reasonably well. Average capital per farm quintupled from 1939 to 1946.[111] The war, and the rise in commodity prices which came with it, certainly helped. But so too did the stringent settlement process. The depression and drought context had allowed the EID and the PFRA to hand-pick their settlers and put in place one of the most important aspects of settlement success: competent and equipped farmers, free of debt, surrounded by experienced irrigators. The early CPR project, which had lost settlers faster than it could attract new ones, had none of these advantages. Despite the difficulties of the transition, the Rolling Hills experiment came to be touted as a model for a new era of irrigation development. As late as 1952, it gained appreciative commentary in the royal commission report considering the South Saskatchewan River Project.[112]

Part of the reason why Rolling Hills remained a relevant case in discussions after World War II is that it initiated and modelled so many of the organizing principles of postwar irrigation development. It built an alliance between farmer-led institutions and the state, it depended on state funding, it proceeded according to a planning process, and it expanded irrigable acreage dramatically. Some of the particularly charged aspects of the case, on the other hand, did not set precedents. After the jurisdictional frustrations and suspicions of the late 1930s, the federal and provincial governments sought to complement each other's policies and sometimes collaborate in new projects. Relations between irrigation districts and the provincial irrigation council also settled into a regular pattern of discreet oversight rather than muscular intervention.

The EID charted the way for district development along the Bow (see map 6.4). Although irrigation district legislation had been on the books since 1915 in the province, only a few districts had formed in the provincial south. The conversion of the CPR project into a farmer-led institution marked a dramatic transfer of authority and wealth. After the EID proved to be a functional institution that indeed thrived in comparison to earlier years, other farmers' groups began to emerge to press for the transfer of corporate projects. In 1946 the Western Section of the CPR followed the EID model and became the Western Irrigation District. The CPR had hoped to unload the project earlier, but the transfer had to await a farmer-initiated conversion.[113] Further east and south, the Canada Land and Irrigation project, which had gone into receivership in the early 1920s and had rarely offered farmers a stable infrastructure, was converted into a state-led irrigation operation: the Bow River project. Although the project engaged farmers in its operations to a greater extent

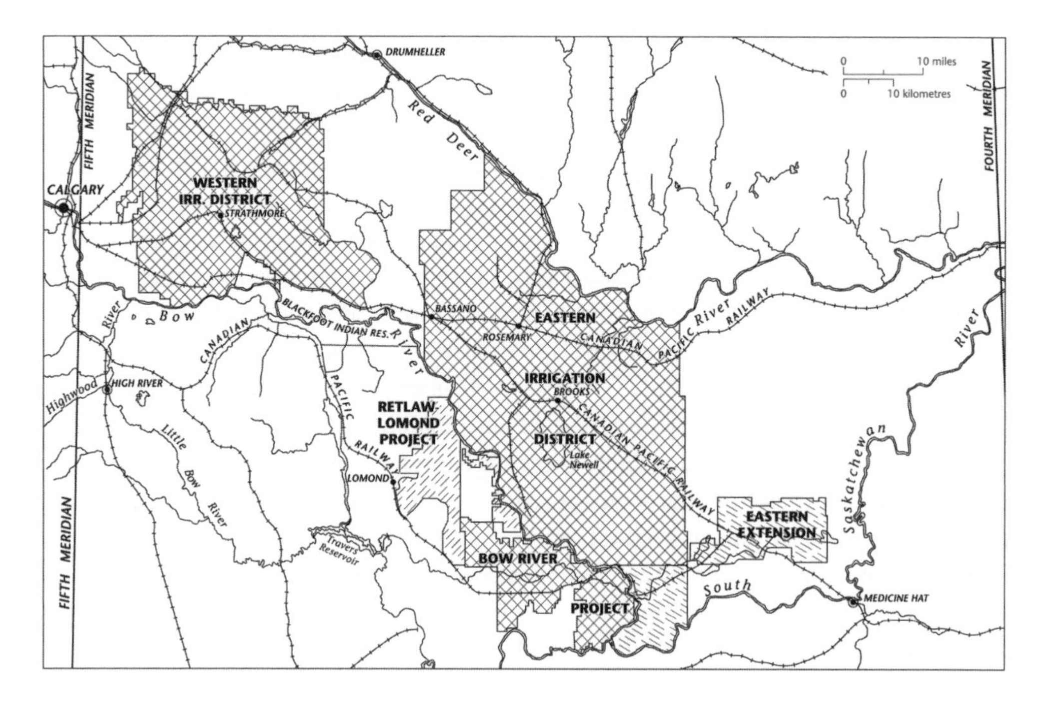

Map 6.4 Bow River irrigation districts, 1958. (Province of Alberta Water Resources Office, "Map Showing the Irrigation Districts Operative & Possible Extensions in Southern Alberta" [1958]; held at AA, Ref: 75.305/180)

than previously, it placed control under a provincial Crown corporation and the PFRA.[114]

By the early 1950s no major corporate irrigation projects were left on the river. Together, the irrigation projects formed a powerful coalition. In 1947 an Alberta Irrigation Projects Association came into being, uniting the irrigation projects on the Bow with those further south on the Oldman in order to press provincial and federal governments on irrigation legislation, funding opportunities, and agricultural policy. One of its first acts was to lobby the provincial government to decrease the summer storage of the Calgary Power Company on the upper Bow so that irrigation districts would not run short.[115] The formerly disgruntled and disenfranchised farmers of the CPR project rose in the postwar period to assume a major voice in provincial politics.

Irrigation districts were independent institutions only to a point. They relied, of course, on provincial oversight and authority, but they also depended on both levels of government for grants to support infrastructure development. Ever since the Ewing Commission of 1936, which had investigated failing irrigation projects in southern Alberta, it had become a common refrain in irrigation circles that settlers could not be expected to pay for all the development costs involved in establishing irrigation.[116] Such sentiments increasingly became policy in the late 1930s as the PFRA was turned into a permanent, rather than crisis-oriented, institution and the province began to work cooperatively to rehabilitate projects. With federal and provincial dollars came a measure of state control that varied over time and from project to project. In the so-called older districts, such as the EID, state intervention generally came in the form of infrastructure grants. When the Bassano dam was rehabilitated in the 1980s, for example, the two levels of government, not the district, covered the majority of the capital costs. Both levels of government supported infrastructure redevelopment, with the PFRA engaging in major structural renewals, such as the Bassano dam, while the province began an irrigation infrastructure cost-sharing agreement in 1969 which placed 86 per cent of the costs on the province and 14 per cent on the districts.

In the years that followed, the infusion of provincial dollars into the irrigation districts became institutionalized. The major jump occurred in the mid-1970s, when annual allocations for the program rose from under $5 million annually to over $30 million by the early 1980s and never less than $20 million thereafter.[117] This funding was justified as a general benefit to economic development, but from a more jaundiced perspective, Alberta taxpayers had become irrigation benefactors. Carl Anderson, never shy about stating his views, railed against the new state of affairs in a series of letters to the provincial government in the mid-1980s stating that, since the launch of the 86-14 formula, consulting engineers and members of the EID board had substituted "Greed for Need."[118] "We have a new generation of welfare bums," he stated with his usual bluntness. "The old people that in 1935 took over this project from the CPR wanted to stand on their own two feet. And Did!"[119]

State intervention delivered a planning process as well as dollars. At Rolling Hills this meant that the PFRA took an active role in recruiting settlers, determining policy, and organizing resettlement. In the Hays District in the Bow River project the agency created a new model community in the early 1950s in which all aspects of the settlement were devised by planners, including the town layout.[120] As in Rolling Hills, drought farmers were the major clientele, but now they came from both Alberta and Saskatchewan. Although in the mid-1950s the PFRA would back away from direct project management (at which point the provincial government assumed authority), the administration had demonstrated a wide-ranging capacity to develop and plan new projects. This capacity would prove important as it moved towards megaproject development on the South Saskatchewan River in the mid-1950s. Planning also extended to engineering matters. New federal and provincial infrastructure grants delivered engineering expertise to ensure that new canals, for example, did not seep as formerly. New techniques were devised to line canals with non-porous membranes, and gravel banks were built into old canals to reduce erosion.[121]

Despite the different administrative arrangements that had emerged by the early 1950s along the lower Bow, all districts held growth ambitions. District administrators and PFRA officials wished to build up irrigable acreage, improve infrastructure, and promote new sprinkler systems that could water uneven topography and distribute water more efficiently. What all these growth agendas shared was an optimistic sense that water would continue to flow; the Bow would deliver what was needed. This perception set off alarm bells in the EID office in the early 1950s, especially as the PFRA-led Bow River Irrigation District announced plans to water 240,000 acres, a sizeable increase from its historic scale of about 40,000 acres. At the same time, the EID hoped to expand and reach 280,000 acres. Add to these figures the 50,000 acres irrigated in the WID and the problem became apparent. As the EID board of trustees tried to explain to the provincial Water Resources Branch in 1952, the river could not sustain such growth. Over the previous decade and half, particularly in dry years, the EID had diverted the entire river's flow during irrigation season. Apart from irrigation withdrawals, the storage needs of power projects upriver created additional complications. The only way that bold expansion plans could be met would be with a massive increase in irrigation storage on the lower river.[122] As figure 6.1 shows, the fears of the EID board of trustees did not materialize – acreage *actually* irrigated grew much more slowly and modestly than growth projections of the early 1950s had envisioned. Nevertheless, a new period had emerged of perceived and potential *water scarcity* in which individual projects began to view themselves in a competitive position and in which Alberta would have to regulate the river not only to please a diverse provincial constituency but also to meet downstream needs under the interprovincial agreement with Saskatchewan.

The notion of scarcity that began to emerge in the postwar years contrasted strongly with the early visions of prosperity and plenty which promoters had drawn for expectant immigrant farmers. In so many ways, the period from the late thirties to the end of World War II acted like a breaking point in the political economy of irrigation development on the lower Bow which had various consequences for river use and river politics. Before the mid-1930s, irrigation development had been a corporate enterprise, organized and facilitated by the state, in which farmers acted as independent commodity producers. As a result of the perturbations of war and the sharp drop in grain prices in the

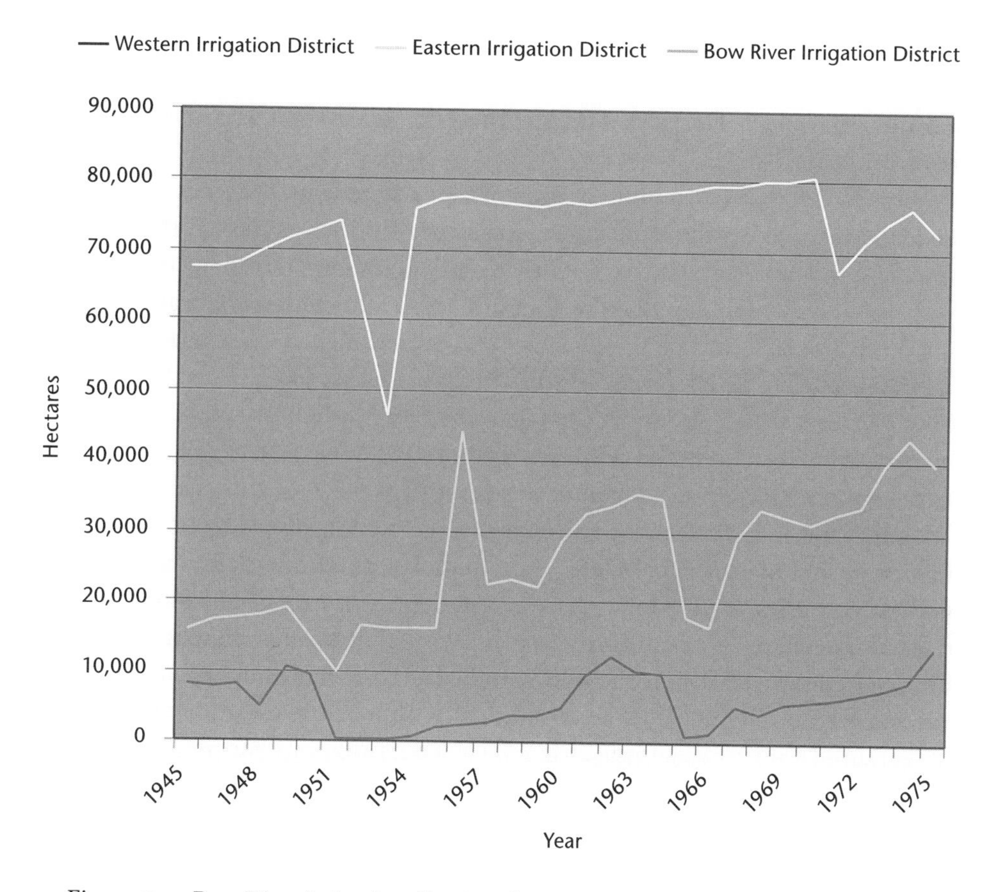

Figure 6.1 Bow River irrigation districts: hectares actually irrigated, 1945–75

early 1920s, the irrigation venture began amidst considerable difficulties. Early irrigation also threw up environmental surprises: not only the curiosities of fish and fauna in the canals but also profit-killing weeds and alkali salts on the land. Economic hardship and environmental change combined to focus a polarized debate between the CPR and farmers over who should bear the costs of development after the predictions of progress and success had worn thin. By the mid-1930s the corporate irrigation program had reached its breaking point, and a coalition of farmers banded together to see if they could maintain irrigation infrastructure through a co-operative plan. The CPR's exit from the Eastern Section in 1935 and the creation of a farmer-led irrigation district presaged other corporate retreats. At the same time as farmers sought to restructure their institutions and integrate functions, new sources of outside assistance began to appear. The creation of the PFRA in the mid-1930s produced a new form of state intervention: delivering dollars for infrastructure and expertise for development as well as management in new projects. The role formerly assumed by corporate developers was now being jointly assumed by a series of farmer-led irrigation districts and the federal and, later, provincial governments.

What had been created in the process was a vast network of canals, ditches, and, increasingly in the early 1950s, sprinklers drawing water off the Bow, storing it in reservoirs, and dispersing it across a massive territory. Irrigation had a profound effect in tying the river and the land together and transforming both. The river was dammed across its mainstem; weirs diverted its flow. At the height of the irrigation season, the river ceased to exist on the lower Bow – at least within its banks. As water coursed through dry prairie, developers and farmers wrestled with beaver and muskrats for control of canals and watched moving sediment and weeds plug up the works. The promise of flowing water in a dry land was also considerably undermined by the belated discovery that irrigation could do harm as well as good. Seepage from canals and over-watered fields drew up salts to the soil surface and made farming difficult and more expensive. In the postwar period these problems did not disappear. Historical air photograph studies suggest that they grew steadily worse after 1950.[123]

As dreams of ever more irrigated acres passed through irrigation districts, the PFRA, and the planning process, the flow of the river did not increase. Rather, new legal constraints began to be put in place to ensure that upstream users respected downstream needs – on the Bow and on the South Saskatchewan system more broadly. As interprovincial agreements set targets for water levels passing provincial borders, an extra burden of control fell upon irrigators to manage the river and their diversions. We discuss this new era of water powers in chapter 12.

7 · The Sanitary Imperative

Sanitation is a word with opposite meanings. The dictionary definition equates it with healthy. The synonyms of "sanitary" – antiseptic, disinfected, germ-free, hygienic, immaculate, spotless, sterile – point mainly towards ways of maintaining a healthy condition through cleanliness. Sanitation is associated with bright, sparkling surfaces, and latterly porcelain. The advertising phrase "Sanitation for the Nation" comes unbidden to mind in this connection. As this phrase is usually encountered in public toilets, one is also reminded of the other meaning of the word, which links it to garbage and sewage, the necessary by-products of cleanliness. There was once a time when garbage trucks from the sanitation department carted rubbish to sanitary landfills. Sewers are still divided between storm sewers for runoff and sanitary sewers for the outflow of our drains. Sanitary thus means both clean and dirty.

Humans have historically favoured riparian locations for their communities because of the sanitation – in both senses of the word – rivers provide. That, of course, is not the only reason. Rivers have also been the source of power, commerce, communication, protection and safety, food, pleasure, and beauty. While human communities, like human beings, are made from water and thus frequently choose to locate on rivers, inevitably they foul this necessity of life with their presence. And rivers were not always clean in their natural state. Once humans simply hurled their waste water and excrement into the streets. It was a great leap forward for civilization when these substances were confined to drains, directed to underground chambers, and eventually poured through sewers into the nearest streams. Then the river ran through, entering more or less clean and exiting soiled. It carried off refuse, offal, debris from the streets, broken and discarded property, the occasional body, and of course sewage. A river thus has the great virtue of removing the filth of civilization, the discarded residues of a city, downstream, out of mind, sight, and smell. A river is a site of forgetting until, *in extremis*, it throws up awkward reminders of the past.[1]

The sanitary imperative means, then, that rivers must be simultaneously clean and dirty. But a river is sanitary in another sense; under certain circumstances it cleans itself. Organic matter added to the flow gets broken down by bacteria, turned into gasses, absorbed into plants and water creatures, and taken up higher in the food chain by fish, birds, animals, and humans. Thus a river does not simply relocate the waste of a city; it eats it, transforms it, turns it into other things, making some of it magically disappear. Depending what and how much is put in relative to the flow, distance, and downstream conditions, a river will clean up after a city. However, it cannot completely purge disease-causing bacteria, inorganic materials such as heavy metals, and toxic chemicals. These it can only transmit downstream in diminishing proportions.

Cleanliness is not a universal or timeless virtue. It is an ideal that, to be realized, must be taken up in some form by social groups, directed towards specific objectives, and given political expression. Sanitarians, or public health advocates, appeared in Calgary as they did elsewhere in the cities of the continent around the turn of the century. Abundant clean water, free of disease and freely available, was one of their main goals. Sanitarian logic insisted that purging the environment of filth and disease was the first step towards freeing the community from poverty and ignorance. Removing disease-causing sewage from the urban environment was as important as ensuring the supply of pure water, milk, and food. There was, however, a curious blind spot in sanitarian ideology. Sanitarians cleaned the city but fouled the river. Early public health advocates worried more about the intake than the outlet; they treated water but not sewage. In Calgary as elsewhere these forces were typically not as concerned about the state of the river as they were about the health of the population because they laboured under the conviction that the river, unlike the city, could cleanse itself. Thus as towns grew up along the Bow in the twentieth century, the river became both a fountain and a sewer. There was no necessary contradiction here. With good luck and moderation, it could be both.

Fountain

Calgary grew at first with its back to its river, facing the railway from either side of the tracks. In the 1880s and early 1890s urban development clustered along the CPR. It took a decade or more before town lots filled in, with buildings north to the banks of the Bow River and south to the Elbow. This turning away from the river and alignment with a more powerful communications force was also symbolic. At the outset, water was a secondary consideration in Calgary, perhaps because it was so readily available. The water table lay close to the surface in a town situated on a broad gravelly stream bed. Thus for domestic purposes abundant water of good quality could initially be readily obtained with a shallow well or, failing that, from water carts. Individual homeowners could attend to their own needs. What prompted community thinking about waterworks was not the domestic needs of householders but, rather, the continuous presence of fire in a town made of wood. Fires, civil engineers never tired of saying, were great promoters of waterworks. Fire could not be left to individuals to manage as it necessarily threatened the property of innocent neighbours. Thus fire prevention required collective action. There was a further cause for concern. Owners of buildings in communities with fire hydrants paid much lower fire insurance premiums than those without. In this sense, it was argued, through the lowering of insurance costs, waterworks paid for themselves.[2]

Soon after its incorporation, the fledgling municipality of Calgary had to address this issue. In March of 1885 the committee struck to consider the question of fire protection concluded that, in view of the numerous demands upon the municipal finances, "a suitable system of Water Works for this Municipality is without our reach at the present time." Instead the committee recommended the placement of public wells at the corners of the most thickly settled blocks. Wells could be dug for $1.50 a foot and need only be sunk 25 feet, which, with cribbing and pump equipment, would make the cost $82 per well. Three strokes of a pump handle would fill a pail, and in this manner a bucket brigade could be well supplied. By October the same year nine wells had been dug and were in operation. But this was clearly a stopgap, born of penury. Nor was it effective as a fire prevention measure. Within two years the inadequacy of this approach forced the city council to reconsider the question.[3]

The town needed a proper system of high-pressure mains capable of delivering a continuous stream of water at a fire. Such a system required a reliable source of water, such as a river, lake, or underground source; a means of pressurizing the system, usually a pumping station; reservoirs for storage; and a network of underground frost-free water mains which could be tapped at intervals by hydrants. Individual property owners could also connect with the system to obtain domestic water supplies, as the many cesspools in close proximity to the shallow wells had begun to foul the water supply. The technology was already well established and in operation in many towns and cities, but it was expensive. There were essentially two ways of doing the job: the municipality could take out a loan to construct the system itself and then operate it as a public work; alternatively, it could franchise a private company to build and run the system subject to terms and conditions. Seeking guidance in 1887, the council turned for advice to T.C. Keefer, Canada's foremost civil engineer with much experience in waterworks. Keefer's report expressed no preferences with respect to the public versus private options; rather, he focused upon the quantity of water required for a population of up to 6,000, the number of fire hydrants needed, and the potential sources – he preferred the cleaner-flowing and cheaper Elbow to the Bow. If the private franchise option was chosen, he recommended that the city negotiate a price per hydrant, a schedule of rates for private users to be regulated by the contract, the employment of a competent engineer to run the waterworks, and detailed provisions for the subsequent purchase of the system by the city should it so wish.[4] Following receipt of this report, Calgary city council, after discussion at a public meeting, a canvass of the experience of other municipalities, and fearing that a waterworks debenture issue and borrowing for sewers would not both be simultaneously approved by the ratepayers, decided in the spring of 1889 to enter into a ten-year contract with the newly formed Calgary Gas and Water Company for the provision of water.[5]

The private company proved to be a flimsy affair financially. Organized by a Regina entrepreneur named Lucas and immediately sold to George Alexander, a Calgary rancher and businessman, the Calgary Gas and Water Company tried to raise money by marketing heavily discounted bonds. These, it turned out, attracted no buyers and were ultimately taken by the company providing water pipe as payment.[6] The company dug a well beside the Bow River just west of the town (near present-day 10th Street), installed a small steam engine, laid shallow mains in the central business district, and rented hydrants to the city for the specified rates. Everything was done on the cheap, and the system performed accordingly. Almost as soon as the waterworks went into operation in 1891, city officials complained that the water pressure was not sufficient to throw a stream of water 60 feet into the air as required by the contract. Thus began a decade of wrangling over the quality of service.[7] The surviving diary of the waterworks manager, James Russell, shows in painful detail his day-to-day struggle to maintain adequate pressure when the fire alarm rang, keep the steam engine running, repair broken mains and split pipes, thaw out frozen hydrants, and calm unsatisfied customers – usually the principal hotels.[8] It was a losing battle. In 1893 council passed a resolution expressing its desire to take over the waterworks, and a special committee was struck two years later to open negotiations with Alexander, who showed no inclination to co-operate even though he claimed to be losing money.[9] After five fruitless years of negotiations, council broke the deadlock in 1900 with a credible threat to build a competing system. This led the nervous bondholders, represented by R.B. Bennett, to revolt. In April of 1900, after the city charter had been amended to permit borrowing to operate water and gas works, the company was sold to the city for $85,000.[10]

A mere change in ownership did not, of course, solve the problems. The system needed major reconstruction and expansion. Rapid urban growth had removed the municipal fiscal constraint somewhat. Thus after purchasing the old company, Calgary went to the limit of its borrowing power in redesigning and rebuilding the water system. Fire requirements remained the primary concern, but water quality had become a much more significant issue to homeowners as well as the city. The city engineer noted that the Bow River as a source met most requirements: it did not contain many bacteriological contaminants, and it provided an ample supply. But in other ways it was defective: "as the river when in flood is exceedingly muddy and as a consequence deposits a large quantity of MUD, SLIME & ETC in the mains, we consider it rather unsatisfactory in this respect."[11] Domestic users periodically complained of taste, smell, and turbidity. In the long-drawn-out negotiations with the private company, the city had already begun to agonize over its options. Should it continue to pump water from the Bow? Should it install expensive filtration and treatment equipment? Alternatively, should it take cleaner water from higher up the Elbow River, a tributary of the Bow that joined the mainstem in Calgary, and deliver it to the city by gravity, a more reliable force than a sputtering steam engine? In the end the city opted to do both.

First, Alderman Watson led the campaign for a pipeline from the upper reaches of the Elbow, thus earning himself the endearing nickname "Gravity" Watson. In 1907 city ratepayers passed a "Gravity Bylaw" by the necessary two-thirds majority. Fire protection remained a primary concern, but on this occasion municipal politicians appealed for support from the women of the city, who, it was said, knew "What a good supply of pure water means to the household." In 1908 Calgary built a wooden stave pipeline to bring water down into the city from the Elbow.[12] Six years later, partly in response to periodic problems with service from the Elbow pipeline and partly to serve the growing northern districts, the city built a new steam-powered pumphouse upstream on the Bow that added water to the system from that source. This pumphouse, much modified over the years, still stands in splendid isolation on the banks of the river, an object now of urban renewal quaintly doing service as a theatre. Through the teens and twenties, therefore, gravity and steam power brought water from two rivers, the Elbow and the Bow, to fill reservoirs and water towers in the city, which pressurized a reconstructed network of water mains and fire hydrants. This revitalized system not only served the downtown but also expanded each year with the rapidly growing subdivisions.[13] When the system became fully operational, approximately 75 per cent of Calgary's water came from the Elbow and 25 per cent was pumped out of the Bow.

The modernized system provided a more reliable supply of water in much larger quantities than before, but water quality did not improve. The Elbow intake regularly clogged up with ice, cutting off the flow. The pipeline occa-

Illus. 7.1 Calgary Fire Department pumping water from the Bow River, 1914. (Glenbow Archives, NA 1918-11)

sionally burst, flooding nearby neighbourhoods and shutting the system down temporarily. In all seasons Bow water contained more silt than that drawn from the Elbow. In flood season (May, June, and July) water from both the Bow and the Elbow became clouded and seemed muddy to the taste. Every summer and whenever the Elbow system went out of commission, Calgary residents had to get used to sediment in their drinking water, murky rinse water for washing, and grit in their baths. This was on the whole a minor aggravation compared with the second problem with the water – it had become contaminated with disease-producing organisms. Upstream pollution from towns, farms, and train traffic had begun to contaminate the Bow, but even the Elbow was not immune. The frequent outbreaks of typhoid fever in the summer, according to Dr Mahood, who had revitalized the Calgary Public Health Board, could be directly attributed to contaminated water, both from wells and pumps in the city tainted by septic tanks and sewers and from the polluted Bow. His extensive reports set the agenda for a campaign to sanitize the city in which purification of the water supply was an important step. By 1914, after several years of gentle criticism, Dr Mahood began to refer to the municipal water as "dangerous" at certain times of the year.[14] In 1916 the city chemist, F.C. Field, reported that from 22 May to 23 September the city water supply had been continuously polluted with typhoid and coliform bacteria. Water drawn from the

Elbow was free of typhoid, but it suffered from other bacterial contamination. Field estimated an upstream population of only about 500 people lived in the Elbow watershed. But on the Bow in summer the contaminants of upwards of 6,000 people went into the river. Moreover, waste water from trains deposited on riverbanks and bridges amounted to a population of about 2,000 more.[15] The message was fairly clear: water from both sources needed to be treated to remove suspended solids, but especially to eliminate harmful bacteria. For both clarity and purity reasons, water from the Elbow was preferable to that from the Bow. Sanitarian ideology, a vital part of progressive-era urban reform thinking, had by the second decade of the century displaced fire protection as the primary requirement of a municipal water system.[16] Water must be pure for public consumption as well as properly pressurized for fire protection purposes; it should taste good and be free of disease-bearing organisms.

Thus in the winter of 1916 and the spring of 1917 Calgary plunged into a debate about what to do about the unsatisfactory water. The People's Pure Water Committee, a citizens' group led by Mrs C.R. Edwards, and public health officials mounted a spirited campaign to improve water quality in the city. The city engineer defended drawing water from the Elbow, which descended from "a wild and mountainous country devoid of any settled population." However, to address the problem of water-borne disease, he recommended moving the pipeline intake two miles farther up the river, constructing new head works to reduce the ice-clogging problems, and installing a filtration system to remove solids.[17] The cost of these works, estimated at $600,000, turned every municipal politician into an instant waterworks expert. Alderman A.W.E. Fawkes, who made much of the fact he was an engineer, championed a modified filtration system. Alderman Graves insisted that a rival sedimentation reservoir system could be built for about half the cost of filtration and would provide comparable cleansing. Water at certain times of the year might have an "opalescence," he admitted, and the city engineer agreed it would lack the "polish" of filtered water, but it would do. While citizens and politicians debated the relative merits of the two systems and juggled cost estimates, fiscally conservative ratepayers and aldermen pointed out that in the midst of a world war, with many taxpayers in arrears and municipal credit already badly strained, Calgary could not afford either scheme. In the end the matter was referred back to committee, where it quietly died.[18] As a quarter-measure, it was agreed simply to patch up the existing system and relocate the Elbow intake farther upstream.

Problems with the water supply thus continued into the 1920s, with the added complication that the capacity of the system could no longer meet the needs of the growing population. A devastating influenza epidemic in 1919 and periodic outbreaks of typhoid in the summer heightened awareness of public health issues in the city.[19] Beginning in the 1920s, water drawn from the Bow had to be heavily chlorinated to kill off the bacteria, often making it unpleasant to drink. "Black water" was a regular summer occurrence as silt levels rose with the spring thaw and summer floods. As the Elbow pipeline aged under the increased pressure of an intake at higher levels, it burst more frequently. With the system reaching its limits, cost-conscious aldermen considered metering water supplies, as opposed to flat-rate charges, to reduce consumption. Calgary had become profligate with water, it was argued. In Edmonton, where most water service was metered, consumption averaged 85 gallons per capita per day, whereas Calgarians consumed more than 200 gallons per day. In cities throughout North America, limiting consumption through measurement and payment per unit of use was a common stopgap response to looming water shortages.[20] Metering, however practical, was nevertheless an extremely unpopular measure. Citizens had come to expect more or less unlimited quantities of pure water as a right. Public health advocates deplored a measure which would force the poor to skimp on an item important to health and sanitation. Civil engineers preferred the equity and economy of meters; politicians feared the wrath of ratepayers determined to retain their birthright of flat-rate water. The campaign of public health officials for clean water, aided and abetted by local medical doctors and organized women interested in social issues, eventually triumphed when water quality and quantity concerns reached crisis proportions in the late 1920s.

Public exasperation with frequent system breakdowns and murky and suspicious-tasting tap water finally broke through the inertia in 1929. City council, realizing it would have to make a major investment both to reconstruct the failing system and to provide more abundant quantities of pure water, commissioned the leading waterworks specialists in the country to redesign the system. Gore, Naismith and Storrie responded in a mere four months with a proposal for an elegantly arched dam on the Elbow just beyond the city limits which would create a 900-acre reservoir in a park setting. This reservoir would provide enough water to meet the needs of a city of 200,000 and could be expanded to serve half a million. As to quality, beside the dam the engineers proposed a laboratory, a filtration and treatment plant to purify the water, and new mains and storage tanks to ensure secure delivery to the city's distribution network. This massive structure, of modernist design and estimated to cost $3,777,000, would deliver to Calgary for the first time water meeting suitable standards of purity and clarity.[21]

Predictable opposition arose on account of the expense, but the most surprising critique of the plan emerged from the Calgary Medical Association, which objected to the location of a reservoir so close to the city and associated dangers of contamination. Public health advocates were subsequently drawn into the planning process to ensure the effectiveness of the compensating treatment facilities connected to the dam and reservoir. With this scheme the Bow River would be completely abandoned as a source, in favour of its cleaner tributary. Yet the Elbow was not pure enough or clear enough to drink without proper treatment. By 1930 Calgary had given up on the hope of its two rivers providing pure water in the natural course of things. Henceforth purity and clarity would have to come from filtration and treatment with germicidal chemicals under the watchful eye of laboratory technicians and medical officers of health. Nature supplied enough water for fire protection purposes, but under the influence of sanitarian ideas, expectations about purity and clarity exceeded nature's unaided capacity to deliver.

The Glenmore project, as it came to be known, was the least expensive of the several options canvassed by the engineers, as well as the one they recommended. Council drew its collective breath and approved the scheme, a decision ratified by the ratepayers by a considerable majority though in a meagre turnout.[22] A project conceived in the relatively affluent late twenties had to be built, however, during the onset of the Great Depression. It thus became more of a fiscal burden to the city and at the same time an opportunity to create employment. Building Glenmore as a Depression relief work, pressed by zealous labour advocates on the city council, not surprisingly inflated the eventual cost of construction almost $300,000 over estimates, further distorting municipal credit and unleashing a flood of post hoc recriminations fully ventilated before a judicial commission of inquiry. Unseemly scandals about inflated costs of land acquisition plagued the project, though in retrospect the most unsavoury aspect of the land assembly was the scandalously low price paid for property obtained from the Tsuu T'ina. Water rates had to rise 25 per cent to pay for the facility, which unleashed a flood of ratepayer resentment against the Tyndal stone and French marble which adorned the filtration plant. The Glenmore plant therefore opened without a formal ceremony under something of a political cloud. Nonetheless, thousands of curious citizens flocked to the site to stand in awe at the scale and, indeed, the striking industrial beauty of the largest built structure in their community and, it was hoped, the solution to their water supply problems for years to come.[23] Even while under construction, it proved another aspect of its utility by holding back a raging flood on the Elbow in 1932, therefore sparing residents of the Mission district yet another destructive inundation. By the mid-1930s, therefore, the city of Calgary had achieved the sanitarian ideal of abundant, clean, pure municipal drinking water from this massive project on the Elbow, but in the depths of a Depression it was hardly in a mood to celebrate.

Illus. 7.2 The Glenmore dam soon after its completion in 1933. (Glenbow Archives, NA 2159-13)

The town of Canmore, also situated on a gravelly flood plain 20 kilometres downstream from Banff, drew its water supply indirectly from the river. The town system relied upon a series of wells tapping groundwater sources. In effect, the town's water supply was naturally filtered through sand and gravel. Still, from time to time Canmore experienced typhoid scares.[24] The coal mines near the town, mainly on the south slopes of the valley, obtained all the water they needed for their steam engines and washing operations from local creeks. The situation farther upstream at Banff will be discussed later in chapter 10. Most small towns along the Bow River valley got their water either from tributaries running into the Bow or from wells. The main exception was Bassano far downstream on the lower reach of the river. Situated on the open prairie in a semi-arid region, the town had to pump its water up from the Bow River, several kilometres distant. For this reason, and particularly before the town installed clarification and purification facilities, Bassano residents experienced a heightened level of concern about the quality of the water coming from the river, as we shall see.

Calgary was by far the largest municipal water-taker throughout the twentieth century. At first it had relied directly upon the Bow as its source. Then it turned to a Bow tributary, the Elbow, for a "gravity" supply. That erratic system had to be supplemented with water pumped from the Bow until the massive Glenmore project made it possible for the Elbow reservoir to supply a city many times its size. The Glenmore treatment plant, expanded in 1957 and again in 1965, served the entire needs of the city of Calgary from 1933 until 1972. It was not until the late 1950s that the city had to revisit the question where it should obtain additional water when the Glenmore plant reached the limit of the permitted flow of the Elbow River.

Filtration and chemical treatment had established themselves as effective means of counteracting naturally occurring and human-induced contaminants in surface water. The liabilities of the Bow River as a source, which had led previously to the retreat to the Elbow, could now be readily counteracted. Moreover, as the city grew southward, the Glenmore reservoir, almost completely surrounded with subdivisions, was no longer as free of contaminants as it once had been. The Bow was once again the obvious source of water to accommodate the rapid urban growth experienced in a booming oil city.

How the water might be taken from the river had to be carefully studied, in view of past difficulties. The idea of taking water from the Bow and diverting it into the upper reaches of the Elbow, thus taking advantage of the treatment facilities already in place at Glenmore, proved upon close examination to be too costly. Water taken from the reservoir impounded by the Bearspaw dam, built just upstream from Calgary on the Bow River in the 1950s, offered a simpler and cheaper solution.[25] The dam was already in existence for flood control and hydroelectric purposes. The city need only tap into the reservoir on terms agreeable to the power company which operated the dam. A second treatment plant would, however, be required, as well as a major feeder pipeline to the city. Accordingly, in 1972, after an interval of forty years, Calgary went back to the Bow for its water. With Glenmore operating at full capacity, all growth in the system had to be served from the Bow at Bearspaw. After a major expansion, the Bearspaw plant has supplied slightly more than 50 per cent of the water for the city of Calgary.[26]

The city drank, washed, flushed, and watered its lawns from the river, but not without the erection of enormously elaborate filtration and purification fortresses standing between the river and its consumers. These were all the more necessary because the river had also become a sewer.

Sewer

Taking water from the river for public consumption is only one-half of the sanitary story. A good deal of the water removed by towns and cities – on average about 80 per cent – returns to the river, but in a much altered state.[27] Moreover, the drainage system of the city gathers up stormwater running off the streets, roofs, pavements, and yards, soiled with urban contaminants, and deposits it in the river too. One way or another, the city of Calgary, for example, puts back about as much water as it takes out of the river. These large-volume return flows of sewage and stormwater in turn affect the character of the river.

Historically, cities have been more mindful of the quality of the water they withdraw from rivers than the flow they return. It is not just that sanitarians have been more concerned with purity and water-borne diseases than waste disposal. Indeed, public health advocates throughout North America also crusaded vigorously for improved sewers to remove disease-causing cesspools from crowded downtown districts. By removing pathogens from the environment, sanitarians proposed to advance physical and moral improvement. Medical doctors, public health authorities, and urban reformers were also great promoters of sewers amongst other instruments of civic prophylaxis. But in most cases sanitarian interest ended with building sewers and the removal of refuse from the city. What happened at the end of the pipe was beyond contemplation as long as it was downstream and out of sight. Even cities that drained their sewage into the source of their drinking water, as did Toronto, solved the problem by treating their drinking water rather than their sewage. During the course of the early twentieth century, cities built expensive engineering works at the intakes to protect the population from water-borne contaminants but showed much less interest in similarly expensive treatment facilities at the outfalls to return the water in comparable condition. The cleansing of the city had tremendous consequences for rivers all over the continent, the Bow being no exception.

Civil engineers charged with dealing with sewage contributed to this myopia. Conventional wisdom in the engineering profession during the early part of the twentieth century held that dilution was the best means of dealing with sewage. Essentially, that meant adding the sewage mixture to a flowing stream or body of water. The degree of dilution of the downstream plume depended upon temperature, the amount of dissolved oxygen in the water, and distance. As the leading textbook on the subject argued in 1912, "The disposal of sewage by dilution is a proper method when by dispersion in water the impurities are consumed by bacteria and other larger forms of plant and animal life or otherwise disposed of so that no nuisance results." In Calgary city engineer G.W. Craig subscribed to this view in his estimates for a major intercepting sewer in 1914: "In a community with a stream into which sewers may empty, if the flow exceeds 5 cubic ft per second for 1,000 persons, crude sewage may be discharged directly into it without nuisance, if the outfall is properly designed to ensure a good mixture of sewage with the water in the stream."[28] Below that population/flow threshold some form of screening, filtration, or septic tank processing would, however, be required. Craig and engineers like him knew that there were limits to the amount of sewage that could be dumped into a river without causing harm, but they were equally certain that moderate discharges into a steady flow could be absorbed and purged as they were borne downstream. In this case, with a population of around 56,000 in 1914 and an average mean flow of approximately 4,149 CFS in the Bow during these years, Calgary exceeded the threshold by a comfortable 74 CFS of river capacity per 1,000 people. Theoretically, the Bow could safely accommodate sewage from a city fifteen times bigger without the need for treatment.[29] It was this reassuring message from the end of the pipe that sustained the construction of an urban sewer system and delayed the construction of expensive sewage-treatment facilities.

Soon after the new town council of Calgary had made its decision on water supply in the late 1880s, it resolved also to borrow money for one of its first public works projects, the installation of sanitary sewers. Night soil collection was merely a stopgap measure, and the cesspits were becoming both a nuisance and a danger to public health. In 1890, with strong encouragement from the press, the ratepayers voted 146 to 37 to build a network of sewers along the major city streets and some back lanes. Construction of the system was attended by the usual scandals, difficulties with contractors, and inevitable recriminations, but by 1891 the sewers had commenced operation, about the same time as the private waterworks started up. Lateral sewers along 4th to 8th Avenues drained eastward towards an intersecting trunk sewer which ran north along 1st Street SE (then called Osler Street), where an outfall deposited the city's waste water in the river. Gradually hotels, businesses, and private dwellings were connected with the city sewers. Among the major buildings, only the federal courthouse was notably lax in this regard. Storm sewers and sanitary sewers were combined in one system which, in theory, ought to have been self-flushing with the spring rains. In practice, low flow in the sewers required constant manual cleansing. Inconveniently, from time to time unusually heavy downpours overloaded the sewers, backing raw sewage up into basements. Nonetheless, the advantages of the system outweighed its occasional disadvantages. Racist paranoia provoked an unedifying debate as to whether the sewage of different races should be allowed to mix when Chinese laundries sought access to the system. But by the turn of the century, operation of an effective system of sewage removal from all buildings had become a firmly established municipal responsibility.[30]

When this first system was designed, there was still a good deal of open space between the town, clustered along the railway and the river. At that point the Osler Street outfall was located at the "back door" of the town. But over the next decade, as that space filled in with houses and shops, urban growth

Illus. 7.3 Construction of Calgary's first large sanitary sewer in the first decade of the twentieth century. (Glenbow Archives, NA 2902-15)

put the sewage outfall in the very centre of the city. Moreover, with the growth of the sewer system – there were 26 kilometres in operation by 1906 – evidence of the outfall became more noticeable. Periodic contributions from Pat Burns's slaughterhouse also drew unwelcome attention to the sewers. While there were no strong objections to pouring sewage into the river, concerns did arise as to the best place to do it. In 1902 a prominent citizen advocated relocation of the outfall farther downstream, to a point well south of the city. Still, he had to defend the practice by relying on the dilution argument. "I do not believe that enough sewage would be dumped into the Bow River to be of serious consequence," J.T. Child told the *Albertan*. "The amount of water in the stream as compared with the sewage that would be dumped into it is so large as to make the sewage barely noticeable. Besides, the Bow River is a swift stream, leaping and gushing over a shallow bed of rocks and giving the best opportunity possible for the oxidization of the water." Plans for just such a project were prepared by the city engineer in 1907. It took a little time, but by 1914 the city had built a large trunk sewer to carry the waste of the city farther downstream to a district called, without apparent irony, Bonnybrook.

Even as Calgarians expanded their sewage system and relocated its outfall, reassured by professional engineers of the efficacy of the dilution principle, doubts as to the wisdom of pouring raw sewage into the Bow River had begun to creep in. The city engineer, who recommended the new trunk sewer in 1907, chose a location for the outfall where space on the river flats would eventually allow for construction of filter beds, septic tanks, or other treatment facilities when the provincial government called upon the city to provide a sewage disposal system. In 1912 the city struck a committee to visit several towns and large institutions in the province to report on the most advanced methods of sewage treatment on the grounds that, in the words of one Calgary newspaper, "The old plan to dump sewage into the river is obsolete." According to the letter of the Public Health Act, passed by the province in 1907, the Provincial Board of Health could order the construction of municipal sewage-treatment facilities. In practice, it moved slowly, in consultation with the municipalities. In 1916 the board, one of whose members, Dr Lefferty, had been an early mayor of Calgary, initiated talks with the city over sewage matters. These discussions were closely followed by vigorous protests from the town of Bassano downstream, which had been complaining that Calgary's sewage had contaminated its municipal water supply since 1912.[31]

Caught between the Board of Health, downstream users, and its own conscience, the Calgary city council capitulated, at least rhetorically. At the height of the war effort the council passed a resolution assuring Bassano that while it would be impossible to act under the present circumstances, "the first money by-law to be submitted to the ratepayers of the city covering capital expenditure will be to provide for the installation of a sewage disposal plant." Plans were in fact drawn up in 1919 for the construction of a sewage-treatment plant, but the $350,000 price tag apparently prompted a re-evaluation of municipal priorities, the bold promises notwithstanding. Calgary simply reneged on its promise to Bassano. Throughout the 1920s the city merely expanded the mileage of the collection system with public borrowing rather than investing in even elementary sewage treatment.[32]

It was only a matter of time, however, before some kind of sewage treatment would be forced upon the city by higher authorities, in this case the Provincial Board of Health, as complaints from Bassano and other downstream users intensified.[33] During the late twenties the board, once again responding to downstream complaints, began to exert pressure on the city of Calgary, perhaps emboldened by the knowledge that the pending Natural Resources Transfer Act would, among other things, give the province unimpaired responsibility for the Bow River. Nor did the city resist very forcibly. It knew that the era of dumping ever-increasing quantities of raw sewage into the river must eventually come to an end. In January of 1929 the Board of Health unanimously ordered that a sewage- treatment plant in Calgary "must be started this year and carried through over a period of years to completion."

Within a month city council undertook to comply by building a $250,000 plant in $50,000 increments, thus keeping the borrowing below the threshold that would necessitate approval through a bylaw vote. Thus in the early 1930s Calgary embarked upon two major public works: a reconstruction of its water supply at the Glenmore site and its first gesture towards sewage treatment at Bonnybrook.[34] It voluntarily entered into one responsibility; it had to be compelled by external agencies to undertake responsibility for its waste water. And to meet its obligations, the city had to proceed in such a way as not to request formally authority from the municipal ratepayers, whose sympathies, in the darkening economic environment, could not be relied upon.

In the mid-1930s the Bonnybrook Wastewater Treatment Plant went into operation, providing primary treatment for up to 16 million gallons of sewage per day. Primary treatment involved first screening the incoming flow to remove rags and large objects. These were scraped off the screens by hand. The sewage then passed into a grit chamber, where suspended solids settled out, to be carted away to landfills later. The fluids passed on into a large digester tank, where bacteria began the process of breaking down the organic materials. Residual sludge from the digester also had to be disposed of, though it could be used as fertilizer. Effluent from the plant then flowed into the Bow River, semi-cleansed. Not all of Calgary's sewage passed through the Bonnybrook plant. Sewers from some parts of Calgary continued to empty directly into the river through five outfalls. Two oil refineries took cooling water from the river and returned it after skimming off residual oil; the CPR Ogden shops also drained their waste water from holding ponds into the river.

Primary treatment typically removed about 50 per cent of the suspended solids and reduced the oxygen demand placed upon the river to complete the digestion of the organic waste by about 25 per cent. But there were exceptions. Heavy storm runoff, racing through the combined sewer systems, usually overloaded the treatment facility. On those occasions operators simply bypassed the plant, allowing the excess to flow untreated into the river. This also happened without human intervention as about thirty stormwater outfalls along the Bow River in Calgary poured backed-up sewage into the river. On normal occasions the plant had no difficulty dealing with the domestic waste water of the city. Industrial effluents, however, posed a greater challenge. Three packing houses emptied their sewage into the municipal system. In theory, the beef and pork packers used everything but the moo and the oink in their operations. Important meat by-products were rendered from the blood, bones, fat, intestines, organs, cartilage, and skin. Only "paunch wastes," the contents of the alimentary system of animals when they were slaughtered, entered the sewers.

Occasionally, however, the grates at the sewage-treatment plant would clog up with animal residues, guts, and bristles. This ought not to have happened, and the city maintained a constant watch on the packers to ensure compliance, but to avoid embarrassments a telephone call would alert the plant to occasional negligence, and these wastes would be bypassed into the river before they clogged up the sewage-treatment works. But the worst offenders were not the industries – the packers, breweries, and distilleries of the city – but rather the city stockyards. When maintenance men at the stockyards cleaned out the pens and hosed down the gangways, the manure flowed through the city sewers to the Bonnybrook plant, where the straw and organic material matted on the intake grates, effectively shutting down the plant. On those occasions, too, the municipal sewage-treatment plant workers, resenting having to clean up after the stockyard workers, simply let the manure flow past the plant into the river.

Calgary's sewage treatment, begun in the mid-thirties, nevertheless reduced the urban impact upon the river substantially under normal circumstances. Storms, industrial accidents, and negligence greatly decreased the effectiveness of the plant and increased the burden upon the river. Nor was organic waste the only problem to be coped with. Often, grease, oil, gasoline, dyes, and cleaning fluids from the city's automobile garages, factories, railways, and refiners found their way into the river. The Calgary urban area was, of course, the most important "point source" of pollution along the Bow River system, but it was not by any means alone. The first reasonably comprehensive "Provincial Survey of the Bow River and Its Tributaries," conducted in 1951–52, identified the major pollution sources, forms of waste-water treatment if any, and approximate volumes of discharge in millions of gallons per day.[35] By far the greatest volume of waste water (82 per cent) flowing into the Bow during 1951 came from the municipalities, primarily Calgary (73 per cent). In round numbers a population of about 140–150,000 emptied its wastes either directly into the river or indirectly after some form of primary treatment. Industries, primarily oil refineries and gas-scrubbing plants, contributed the equivalent to the load of a population of 66,000 people, or 18 per cent of the total effluent discharged into the river. The human population and industry in the watershed at mid-century poured approximately 55 million gallons of waste water per day into the Bow River or its tributaries. This figure may overestimate the quantity of municipal sewage, but it is certainly an underestimate of the total volume of waste water flowing into the river. The thirty or more storm sewers in Calgary alone poured contaminated stormwater directly into the river in unknown volumes. The total pollution load on the river system could have been as much as 20 per cent higher than the estimated 54 million gallons per day (250,034 cubic metres). That seems like a lot of waste water, but it constituted only about 2 per cent of the total river flow at the Carseland Weir below Calgary in 1951.

What effect did this charge of organic material, suspended solids, chemicals, and bacteria have upon the river? The Provincial Board of Health scientists performed numerous tests of samples taken at several points along the river to determine the acidity, the carbon dioxide content, the quantity of dissolved oxygen present, the biochemical oxygen demand (BOD) created by organic materials present, and coliform bacteria. No problems were observed with acidity, alkalinity, and carbon dioxide content. Similarly, in all the samples dissolved oxygen measured at or near capacity. The scientists concluded that "the Bow River above Calgary is polluted very little … With the exception of Banff, wastes flowing into the Bow River above Calgary are negligible." Below Calgary it was a different story. Despite a heavy burden of waste water to digest, the turbulent flow of the river kept dissolved oxygen at capacity, even in winter with a thick ice cover. On the other hand, "the effect of by-passing raw sewage into the river is in evidence all along the banks." In several places the scientists observed large beds of slime, "black-gray flocculent masses" lining the bottom of the river, and considerable brownish grey flocculent suspended in the flow. Coliform counts did constitute a health hazard immediately below Calgary, and the smell of oil from the water could be detected as far downstream as Carseland. However, bacteria could not long survive in the cold water, twenty-four hours at most. Rapid die-off reduced coliform numbers as they were borne downstream.

The Board of Health analysts drew a surprisingly upbeat conclusion from these observations after warning that the Bassano water system required more than chlorination to protect consumers from water-borne diseases. Given the remarkable capacity of the river to maintain the dissolved oxygen necessary for the digestion of organic material, the scientists calculated that at current loads the river could support a population of 950,000 or its equivalent. In winter, when ice cover reduced the oxygenating capabilities of the river, the maximum population load fell to 230,000. Nonetheless, they concluded rather optimistically: "There is therefore very little possibility in the near future of overtaxing the Bow River as far as the dissolved oxygen balance is concerned." As a result, the scientists saw no compelling need for secondary treatment of Calgary sewage to reduce the biochemical oxygen demand imposed upon the river by municipal waste water. The bypassing practices should be stopped and maximum effort directed towards removing "all settleable solids to avoid sludge banks in the river and remove all visible floating material in order to keep the river clean."[36]

The resilience of the Bow River in taking on this urban burden of nutrients impressed the provincial government scientists. While the board did not give the Bow a clean bill of health, it did not find it sufficiently contaminated to order more extensive sewage treatment in Calgary. Rather, it compelled Bassano downstream to purify dangerously polluted water taken from the Bow with a more sophisticated treatment system. Calgary was ordered to separate its combined sewers, close off some outfalls, and modify equipment and procedures at the Bonnybrook plant to avoid sewage bypass.[37] On the strength of this report and under orders from the Provincial Board of Health, the city of Calgary more than doubled the capacity of the Bonnybrook primary treatment facility in the late fifties and built a completely new treatment plant on the flats where Fish Creek joined the Bow to service the burgeoning southern suburbs. Both of these plants, and later a small plant at Ogden, provided primary treatment that removed about half the suspended solids and 25 per cent of the organic content of domestic and industrial sewage.

It had not escaped the attention of the scientists and engineers conducting the study that within recent years a major recreational fishery had developed in the river reaches downstream from Calgary, a matter discussed in more detail in the next chapter. Anglers from the province had long known of this. However, the fame of the Bow as a trout stream had begun to attract avid fishermen from all over the continent. In this respect there was a silver lining to the cloud of treated and untreated sewage poured into the Bow at Calgary. The lower Bow had become a fly fishermen's paradise.[38]

The only puzzle with this pollution-enhanced fishery was a persistent oily taste in the fish caught immediately below Calgary that made them inedible. As the city expanded in the oil-induced boom of the fifties and sixties, the output of refineries and oil-processing plants pouring effluent and returning cooling water into the Bow River increased accordingly. The industries of Calgary were not particularly water-intensive, but they did have their needs and in turn placed their mark on the flow of effluent downstream. New technologies and changing industrial structures altered the impact on the river. The diesel engine, for example, greatly reduced the CPR's need for water. On the other hand, the oil and gas industry used water as a coolant in its refineries and in some of its gas-scrubbing processes. Untreated waste water returned to the river laced with oil residues and phenols.

During the sixties, however, Bow River water quality downstream from Calgary changed dramatically. As a downside to improvements in personal hygiene as standards of living rose, widespread use of laundry detergents added much larger quantities of phosphates to the sewage effluent. Population growth also raised the volume of waste water. With this increased load of nutrient-laden waste water, the carrying capacity of the Bow River quickly reached its limits. Thick growths of pondweed of several varieties appeared in the river and in the irrigation canals. In the mid-sixties fishermen complained that "the city of Calgary had turned the Bow River into an open cesspool." As for the fishing experience, a local clergyman replied to a puff piece in a US fishing magazine that fish were abundant but inedible; moreover, "the stench of human excreta and other filth requires a strong stomach." Sewage in large quantities could be too much of a good thing.[39]

In response to public complains and the evident change in river water quality, the provincial Department of Health resumed systematic study of the Bow River and its fish in the early sixties.[40] Multiple water samples were taken from various points along the river in all seasons. Wild fish were caught up and down the river and tested for oily contaminants. Captive fish were immersed in the waters around the refinery effluents and the survivors tasted. Reports by R.H. Ferguson and P.H. Bouthillier in July and August 1965 painted a picture of a river in serious trouble. Ferguson's tests showed that dissolved oxygen levels dropped to dangerous levels under the ice in the reaches below Calgary. A daily BOD load from Calgary sewage of 70,000 pounds overtaxed the river's digestive capability, reducing oxygen levels to the point where fish might not survive. "Given the proper conditions of complete ice cover, low river flows, and a heavy oxygen demanding load," Ferguson warned, "the result would most certainly be a fish kill." Hyperactive weed growth, stimulated by the phosphate-rich water, further exacerbated the oxygen deficiencies in the summer. In warmer temperatures dissolved oxygen levels fluctuated widely between daytime and nighttime as weeds alternately contributed and consumed oxygen. Before dawn, dissolved oxygen fell to levels where fish could not sustain themselves. The thick weed growth in shallows slowed the waters, which had the effect of raising water temperatures, further endangering fish habitat. Unless this weed growth could be curtailed, summer fish kills remained a distinct possibility.

Bouthillier's study of the petroleum taste in fish showed that the odour in refinery effluent was not proportional to the amount of oil in the water and, secondly, that *any* presence of the odour causing phenols in water influenced the taste of fish. Even small quantities could taint the fish. He concluded further that the musty fragrance of the river, which also affected the flavour of the fish, could be attributed to sewage effluent. Both reports concluded that Calgary must begin secondary treatment of its waste water to reduce the load on the river; Ferguson recommended further phosphate removal treatment to curb weed growth.[41]

The scientists, with arcane language and bewildering statistics, documented what everyone knew to be the case: the Bow River below Calgary had become dreadfully polluted.[42] Like a canary down a mine, the condition of the fish signalled impending danger. Neither the Provincial Board of Health nor the city could pretend otherwise. Armed with these reports, the provincial health administrators summoned officials from the city of Calgary and the two oil refineries to a meeting in October 1965. L.H. Hogge, chairman of the Board of Health, calling the current state of the river "unacceptable," put the refineries and the city on the defensive, demanding to know what they intended to do about it. The refiners quibbled that seepage from a creosote factory, railway oil storage, and garage wastes in the municipal sewage might also be implicated, but the Board of Health's studies pointing in their direction were irrefutable. As a result of this discussion, the refineries were ordered to eliminate oily odours from their effluents. Whereas the refineries caused the smell and taste, the oxygen demand of the effluent from the city was suffocating the river. The city was similarly ordered to stop using the river for the disposal of sludge and to begin to investigate methods of secondary treatment to reduce the biochemical oxygen demand upon the river. In the future some kind of phosphorus removal treatment might also be required to control weed growth. The board ultimately imposed a 40,000-pound (18,143-kilo) per day limit on the BOD the city could place upon the river from its waste-water outfalls and ordered these improvements to be in place by 1968.[43]

Shamed by its own practices and fearful of a public relations nightmare associated with a major fish kill and the impact on tourism, Calgary readily complied with the Provincial Board of Health rulings. The city quickly built the required secondary treatment facility for the Bonnybrook, a 236,300-cubic-metre (52-million-gallon) per day plant that became fully operational in 1971. This significantly reduced nutrient loading of the river.[44] In 1980 the city expanded and upgraded the Fish Creek plant to provide secondary treatment for 72,600 cubic metres (16 million gallons) per day.

The end of the 1960s witnessed the conclusion to the policy of doing the bare minimum to treat urban and industrial effluent based upon the assumption that the river would carry away refuse and cleanse itself. That might have been true up to certain limits, but those volumes had long since been surpassed. For most of this period stubborn municipalities, anchored in their belief by frugal ratepayers, had preferred to do the bare minimum while brushing off downstream complaints. The provincial government, with its higher authority for public health, chose to use that power with discretion verging upon laxness. But in the end, when the problem became unavoidable, the province intervened to force the city to clean up its act, responding to the needs of a wider geographical base and to scientific advice.

Calgary, along with other cities, belatedly began to treat its sewage more effectively in the seventies under external pressure from the province, but also driven by what might be thought of as an internal municipal environmental politics. The late sixties marked the end of the era of thinking of the river as a sewer. The Board of Health's pollution summit in 1965 seemed to represent a real turning point in Calgary's thinking about the river. Up to then the city had done only what the province ordered it to do. From the seventies onwards it showed leadership. The health of the river had become inextricably connected to the self-image of the city. It would not do that a major recreational resource, a world-famous fishery, and a major tourist attraction might perish through municipal neglect. Already, isolated fish kills were being reported.[45] Moreover, the oil boom, and economic expansion generally, had increased the wealth of the city and its ratepayers. What the city could afford to do had thus changed. Frugality was not longer so much of a virtue in a self-consciously modernizing city. As we will see in subsequent chapters, Engineering Department officials could execute capital projects supported by a popular environmental consciousness that had recently focused on the river, without fear of having plans shot down by cheese-paring city councillors beholden to indignant ratepayers. After the mid-sixties, municipal spending on pollution abatement measures, especially sewage treatment and sanitary sewer construction, rose absolutely but also in relation to other civic spending.

Compliance with provincial regulation explains only part of Calgary's response. By the 1980s the city had moved well beyond the requirements of meeting provincial standards. It went from being a laggard to a continental leader in sewage-treatment policy as well as a leader in technological innovation. During this period a new environmental outlook had begun to drive policy at the provincial and municipal levels of government, within both bureaucracies as well as the elected bodies, and amongst the electorate more broadly. This change in attitude could be illustrated by the new Lougheed government's passage of a Clean Air and Water Act in 1971 and the creation of the Alberta Environment Ministry to administer it and thirty-one other pieces of provincial legislation.[46] Engineers within the municipal public service began to think of efficiency in broader terms, to include the environmental impact of projects.[47] Citizens' groups protested tree removal, even in the name of flood protection. Crusading journalists, such as the young Ralph Klein, made their mark drawing to public attention to the public shame of the "Filthy Bow."[48] This dramatic shift in public opinion and policy within Calgary and Alberta was, of course, part of a much wider cultural transformation in the early 1970s across North America and Europe as the environmental movement developed a much broader public constituency.[49]

By the late 1980s Calgary's primary and secondary treatment capacity exceeded a half a million megalitres. Secondary treatment in tank farms of huge digestors reduced the suspended solid levels in effluents and biochemical oxygen demand by more than 90 per cent. The residual sludge, carefully monitored as to its chemical and biological content, began to be applied to acreage south of the city with a specially designed injector-cultivator. In the early eighties, on its own initiative, the city installed tertiary treatment for chemical phosphorus removal at both plants.[50] This partnership with a private company exploring tertiary treatment at scale led to the development in Calgary of a technology that was subsequently marketed across the continent. Chlorination, introduced to kill off residual bacteria in the effluents, which inadvertently also killed off fish and plants in the river, was replaced by ultraviolet disinfection. Indeed, by the 1990s Calgary's efforts to provide state-of-the-art treatment of its waste water had begun to receive continent-wide attention.[51] At the end of the twentieth century the city was returning its effluent to the river in near state-of-the-art cleanliness.

Calgary's aggressive steps to reduce the impact of its sewage upon the river could be seen and smelled. Petroleum vapours diminished; the fish no longer tasted of oil. The clarity of the water improved markedly, and after the introduction of phosphorus removal, weed growth in the river became less of a problem. Secondary treatment dramatically reduced the city's per capita biochemical oxygen demand upon the river. In June of 1968 sewage effluent from the Bonnybrook plant imposed a BOD upon the river equivalent to 70,953 pounds (32,183 kilos) per day and poured in 42,109 pounds (19,100 kilos) of suspended solids per day. By December, as the secondary digestors came on line, that figure had been reduced 6,200 pounds per day for Bonnybrook and 11,950 for the entire system (or 5,420 kilos per day). With the growth of the population and effluent volume, this loading number gradually crept upwards to over 8,000 kilos per day in 1977 and above 9,000 in 1981. Then, when the Fish Creek plant received its full secondary treatment array, the total system biochemical oxygen demand on the river fell to an average of 1,811 kilos per day, 5 per cent of the levels in the late sixties for a population 65 per cent larger. In the same period daily deposits of suspended solids in the late eighties were ten times lower than the sixties.[52] On those vital measures, Calgary had dramatically diminished its oxygen demands upon the river.

Changing Outlook

Throughout the twentieth century a steadily increasing proportion of the Bow River's flow had been taken from its course, diverted through a ramified social and industrial system, before being returned – highly transformed in the process – to its bed. What had started as a trickle in the 1890s, running from George Alexander's pumphouse on the banks of the Bow and ending at the sewer outfall on Osler Avenue, had become an invisible torrent coursing through a vast arterial and venous network nourishing and cleansing a social and economic system. The city of Calgary dominated this socio-economic stream diversion, accounting for more than 80 per cent of the water-taking and effluent return on the Bow River, but similar smaller cycles occurred in other towns up and down the river and on some of its tributaries.

By the late 1980s the flow of the subterranean sanitary river in Calgary alone averaged 630,000 cubic metres per day. That represented a river about 10 per cent of volume of the average flow of the Bow through Calgary or, for purposes of comparison, a river about the same size as the Kananaskis. As the sanitary river wound its way through this underground labyrinth, it received a changing charge of organic material and chemicals from the society it passed through. Following the logic of the dilution principle and the inertia of an upstream user, the city of Calgary continuously increased the total burden on the river to a peak in the late 1960s, when, as we have seen, there were serious indications that the Bow below Calgary had approached the upper limit of saturation.

During the 1970s and 1980s the Alberta Ministry of the Environment, in co-operation with other similar departments across the country, established detailed guidelines governing water quality. Specific permissible limits for biological matter, bacteria, heavy metals, and other contaminants were established for drinking water, surface water, irrigation, and agriculture. The province of Alberta then established precise quantitative limits on the effluent flowing into rivers from sewage-treatment facilities. As a first step, for example, the province put a cap on daily BOD loading of rivers by sewage-treatment works. Towns and cities had to meet these standards as a condition of holding their licences. Other loading limits followed.

Initially, water quality was the primary goal. In time, however, the concept of environmental protection was broadened to ensure the stream flow needed to maintain a healthy aquatic ecosystem.[53] Cities such as Calgary could not longer expect to continuously increase their water-taking from river sources to meet domestic demand. The province thus effectively capped the quantities of water that might be removed from streams for municipal uses or irrigation. These provincial standards were developed initially in a partnership with the municipalities, especially the city of Calgary, but over time provincial-municipal relations took on a more adversarial, policing tone.[54] Together, these water quality standards, spelled out in various iterations of the Alberta Surface Water Quality Guidelines and other regulatory documents, became the performance standards by which municipal waterworks and sewage-treatment plants were held to account.[55] By the 1990s strict provincial regulations had placed firm caps on the amount of water that might be taken

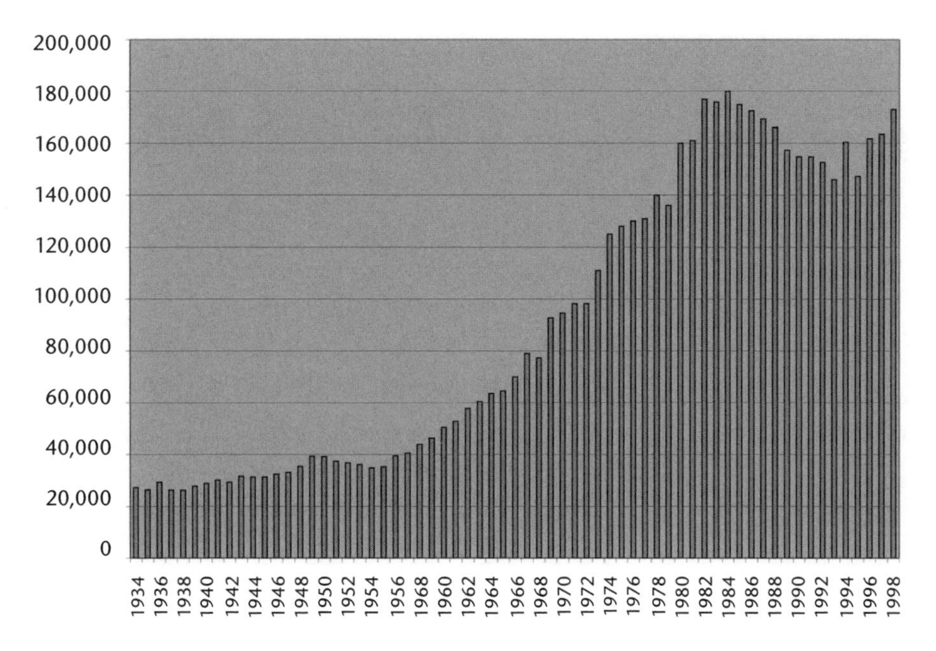

Figure 7.1 Calgary total water production in millilitres.
SOURCE: CCA, Engineering and Environmental Services, Annual Reports.

from the Bow River as well as the amount of organic material, bacteria, heavy metals, phosphorus, and phenols that might be returned to the river in municipal effluent.

Strict provincial regulation of water quality and in-stream flow, beginning in the 1970s but becoming much more effective in the 1990s, fundamentally changed the relationship between the city of Calgary and the Bow River. But it must be stressed that the city itself and its engineers, bureaucrats, politicians, and ratepayers developed an environmental consciousness of their own which also influenced municipal behaviour. For most of the twentieth century, as the city needed more water, it took more. Similarly, it poured its rising quantities of effluent back into the river with a minimum of treatment. Regulation restricted how much water might be taken and specified precise levels of contamination permitted in the return flows. From the 1980s onward Calgary and other cities lived in a world of caps. The linkage between population growth and water consumption would have to be broken if the city was to grow. Substantial civic pride, spurred on by provincial regulation, launched a major effort to come to terms with the pollution of the river that went well beyond minimum requirements. A civic environmental politics made the city aim

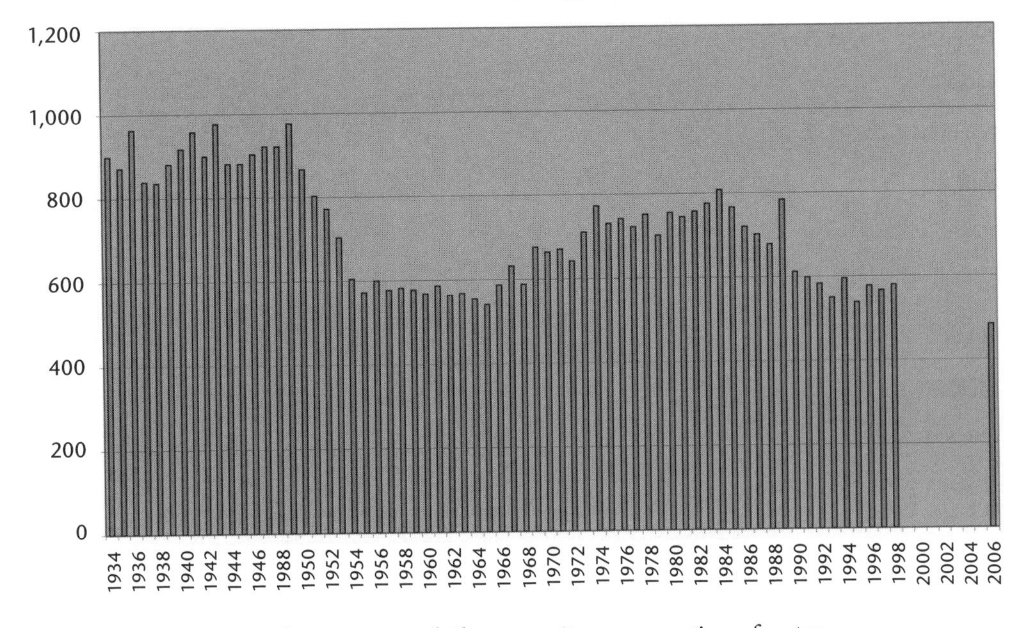

Figure 7.2 Calgary average daily per capita consumption of water.

SOURCE: CCA, Engineering and Environmental Services Annual Reports.

NOTE: This figure should be read for trends rather than precise readings per year. City populations have been estimated between census periods, and daily consumption computed from annual data in some cases, both of which introduce some degree of error.

to live within its water-taking limits and also to operate its sewage treatment at levels of effectiveness that not only met but significantly exceeded provincial goals.[56]

Cities require water, but how much water they need is a social and ultimately a political choice. As Calgary grew, it took more and more water from the river as figure 7.1 shows. While population was the primary driver of total consumption, per capita consumption rose and fell depending upon social practices, leaks in the mains, and the weather. As water was unmetered and relatively cheap for most of this period, price had little impact upon consumption. Throughout the 1930s and 1940s per capita consumption averaged about 200 gallons (over 800 litres) per day, a relatively elevated level in historical and comparative terms, probably on account of the local practice of leaving taps running in winter to prevent freezing. Figure 7.2 indicates that per capita consumption levels (both maximum and average daily) declined during the 1950s but then began to climb in the 1960s.

A population with increasing wealth demanded more water. Within the household taps, toilets and baths proliferated, especially after World War II.

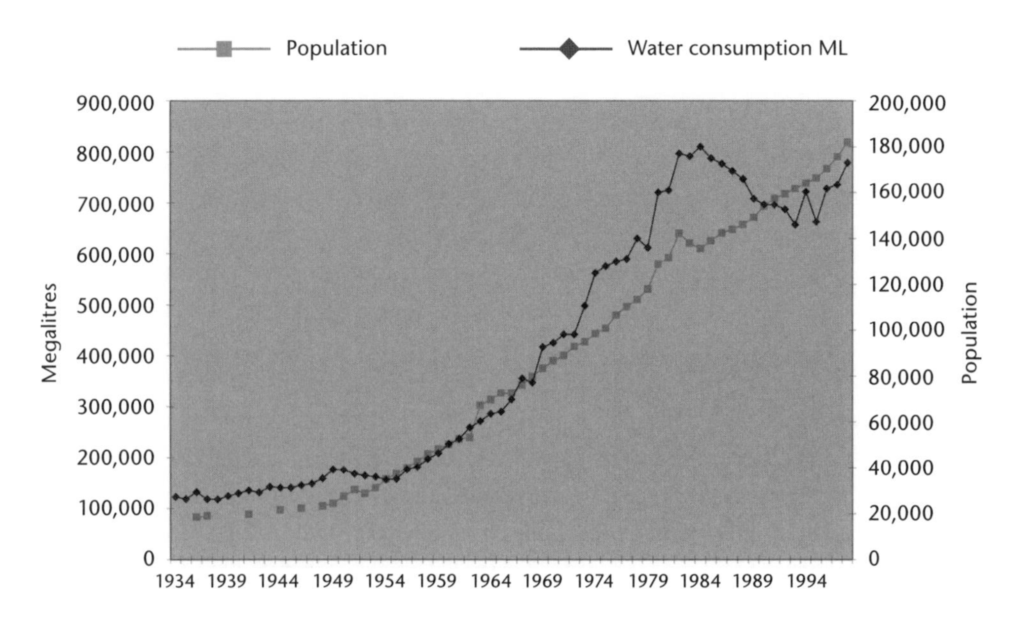

Figure 7.3 Population and water consumption.
SOURCE: CCA, Engineering and Environmental Services, Annual Reports.

Domestic uses expanded from kitchens, laundries, and bathrooms to watering lawns and gardens and washing vehicles. For most of the twentieth century the primary force driving water production was rising population, as figure 7.3 indicates. In the 1960s automatic washing machines, showers, dishwashers, swimming pools, central air conditioning in office towers, and larger suburban gardens pushed up per capita water consumption and concomitant wastewater production. Water consumption tended to peak in the summer months, when baths, showers, extra laundry, and outdoor use intensified. By the 1970s winter consumption stood at about 50–60 per cent of summer levels. The peak moved in accordance with shifting temperature spikes and dry spells. Maximum daily consumption, highly correlated with hot spells and heavy lawn watering, which from the 1930s to the end of the 1950s had averaged only about 50 per cent above daily averages, regularly amounted to double the daily average during the 1960s and 1970s, and water systems, like other utilities, had to be built to accommodate peak demand. Each year, it seemed, total water consumption increased with the population, augmented by the rising trends in average daily per capita consumption. This phenomenon could not go on forever. The province imposed an upper limit on Calgary's water-taking licences based upon a growing awareness of the in-stream needs of the river.[57]

With limited water sources, Calgary, in its relative affluence, developed a water-intensive culture. Even in the 1960s city engineers realized the impossibility of indefinitely increasing water production capacity to serve a rapidly growing population and turned their minds instead to devising ways of cutting back on consumption. Water metering offered the most obvious solution to this problem, but it was politically unacceptable to a population and their politicians inured to a flat rate. At length, however, a politically acceptable compromise was arrived at, one which in a rapidly growing city also served to temper consumption levels. Existing domestic users were grandfathered, but new construction would be metered, and existing homes would be transferred to metered supply when sold. A new regime of higher water rates in the 1980s further curbed consumption. By the year 2000 about 75 per cent of homes were metered; after 2002 meters became mandatory for all domestic users. An aggressive campaign of public education, combined with metered use and increased water rates, steadily reduced per capita daily consumption from about 750 litres per capita per day in 1986 to 451 in 2006.[58] Future growth of the city depended upon learning to use less water. Currently, the city aims to deliver the same amount of water produced to serve its slightly more than 1 million inhabitants to a projected population of 1.5 million in 2030. Limited by the province and pressured by its own conscience, Calgary at the end of the twentieth century had begun to adjust to a world of limited water.

Plumbing led most of this water back to the river. Not all of it, of course. Some water was consumed, transformed into steam, evaporated under the boiling sun, or lost through broken mains into the soil. On the other hand, snowmelt, runoffs from rainstorms, and infiltration of groundwater through porous mains added new water to the return flow. Variation in the monthly flow through the city sewers in general followed the pattern of water consumption. The rise in the summer was, however, more gradual on average because lawn water rarely reached the sewers. But sudden upward spikes in volume, occasioned by rainstorms, could dramatically load the system, momentarily pushing the city above the provincial pollution limits. A heavy rainstorm flooding the combined sewers in the older districts and pouring through stormwater outfalls into the river could lead to loadings briefly above the prescribed maximum. Both water consumption and waste-water production were, to a certain extent, sensitive to the economic environment. The recession of the early 1980s associated with the National Energy Program, for example, can be seen in these charts.

Calgary invested heavily in sewer construction and repair and in improved sewage treatment during the last quarter of the twentieth century. The renovations to the collection system cut down on the volumes of groundwater infiltrating the sewers through cracks and breaks, thus reducing overall volumes to be treated. Investments in more effective treatment equipment dramatically

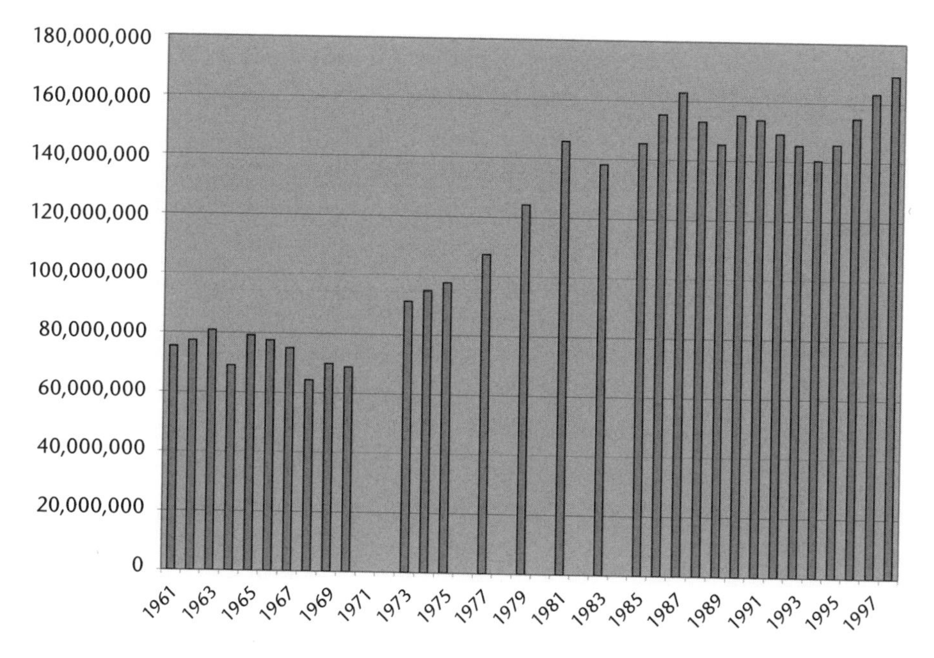

Figure 7.4 Calgary sewage treatment, all plants.

SOURCE: CCA, Engineering and Environmental Services, Annual Reports.

reduced river loadings. Throughout the 1990s the expanded Bonnybrook and Fish Creek plants removed 97 per cent of the biochemical oxygen demand, 92 per cent of the suspended solids through primary and secondary treatment and innovative tertiary treatment, and 88 per cent of the phosphorus from Calgary's waste water. Moreover, the daily load of a host of other nutrient and toxic chemicals had been severely cut back.[59] Total coliform counts and fecal coliform counts, which had dropped from the high levels recorded in the 1950s and 1960s, began to creep upward with greater volumes of effluent flow in the 1990s. Then advanced ultraviolet disinfection processes began to drive down coliform trends.[60] In the late nineties, as more efficient sewage treatment took effect, both the province and the city turned to the management of largely untreated stormwater flows into the river which, by some measures, accounted for up to 50 per cent of the total organic and inorganic loading of the river as it passed through the city.[61]

On every measure the city's sewage works performed much more effectively during the 1990s than required by provincial regulations. Moreover, the effectiveness of sewage treatment became a source of civic pride. The city operated

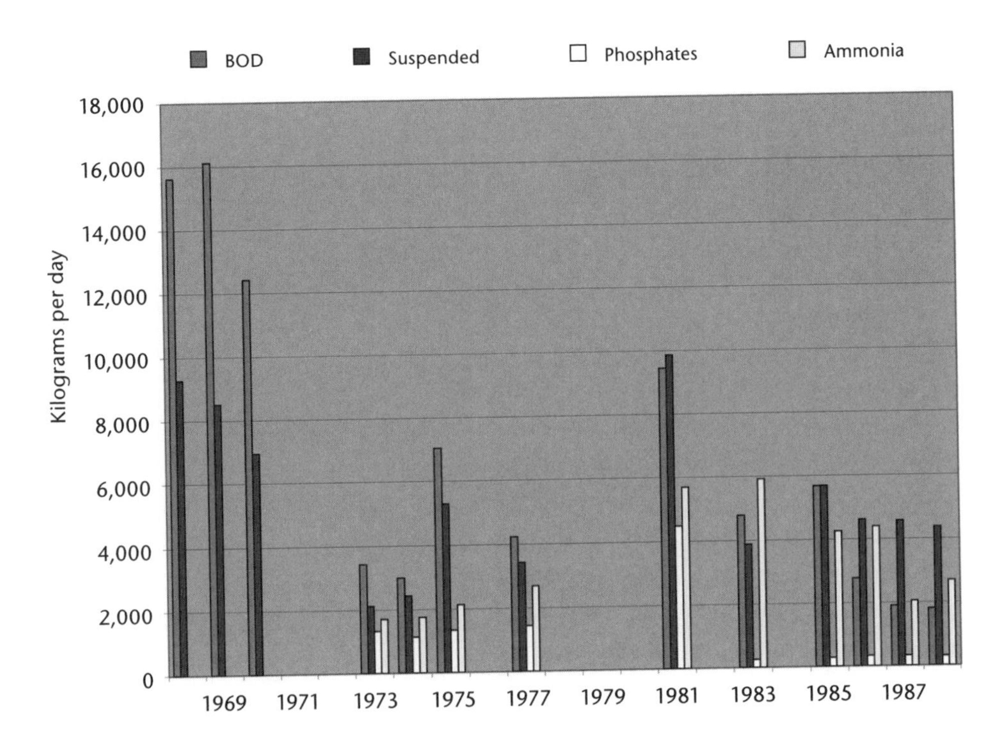

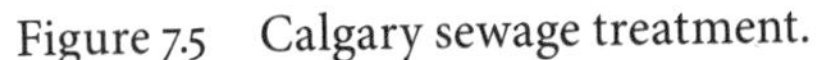
Figure 7.5 Calgary sewage treatment.

SOURCE: CCA, Engineering and Environmental Services, Annual Reports.

NOTE: This figure shows the steadily declining river loading from BOD (biochemical oxygen demand), suspended solids, phosphorus, and ammonia/nitrogen.

well under provincial guidelines in almost every category of measurement. For example, loadings of the Bow River in the nineties from Calgary sewage effluent averaged 15 per cent of provincially allowable BOD limits for Bonnybrook and 22 per cent for Fish Creek, and 39 and 49 per cent respectively for total suspended solids. If it chose to, the city could have poured many times more contaminants into the river, but instead it operated on a leadership policy well beyond mere compliance. In 1993 Calgary's sewage treatment efforts received a grade of A- in the Sierra Legal Defence Fund's National Sewage Report Card. Fishermen began to complain mildly that the fish in the Bow, less numerous and often smaller, were becoming harder to catch.[62] Nonetheless, waste-water return flows from Calgary as well as its untreated stormwater runoff still seriously compromised water quality in the river reaches below the city.[63]

The perverse logic of the sanitary imperative – that the city should be cleansed at the expense of the river – was largely broken by the end of the twentieth century. In the name of environmental protection and public health, the province decreed precise limits and exactly how much the city might be allowed to contaminate the river at the end of the pipe, a general policy largely supported by a more environmentally conscious urban community and government. Similarly, the expansionist policy of taking more water to serve the demand of a rapidly growing population also came to an end, to be replaced by a regulatory regime of finite limits. The intake could not be expanded indefinitely. A history of ever-growing volumes at the intake and the outlet was replaced by a future of fixed water withdrawals from the river and, by extension, waste water returns. Moreover, strict regulations governing effluent discharges limited the biochemical oxygen demand the city could impose upon the river, along with caps on suspended solids and a long list of bacteria, chemicals, and heavy metals. Growth could be accommodated only by using less water per capita and, accordingly, producing less waste water treated to higher standards. However, these requirements, imposed by provincial regulation but strongly supported by a growing civic environmental awareness, had yet to be tested. They were, after all, self-imposed limitations. What would happen when the city ran up against the maximum volumes permitted by its water-taking licence or its effluent reached or exceeded provincial guidelines were hard questions left for the future.

8 · The Fishing River

Take the Trans-Canada Highway southeast from Calgary any day in the summer, exit at one of several turnoffs, and find the river's edge. Push through the cottonwoods, and you will see men and women in wading pants casting lines across the water. Perhaps you will glimpse a large rainbow trout being pulled from the river writhing and splashing.

Whatever else it once was, the river below Calgary, as far east as the Carseland weir, has become a renowned recreational fishing area, heavily used by people in search of sport and play. This forty-six-kilometre section of the river is not the only part used by sport fishers, but it is the most popular site today, promoted in international sport magazines, trumpeted in books and guides about the fishery, and known in Calgary as *the place* to catch fish.

The fishing river exists, not in spite of development and change along the Bow, but at least in part because of it. While so many other North American rivers experienced precipitous declines of fish populations as settlement, damming, and pollution made their marks, on the Bow some of the effects of human use have produced mixed outcomes. The dams on the Bow's mainstem and tributaries have regulated water flows in ways incidentally favourable to some fish species. The sewage deposited in the river at Calgary provides a massive nutrient boost to the river's ecosystem, sustaining aquatic plant growth and insects important to fish. Over time, people have added species to the river, such as the rainbow trout, which have provided new opportunities for sport fishers. Humans have helped to make this fishery, usually without even trying or thinking through the implications of their actions. One fishing writer has described this outcome as a "magnificent accident."[1]

The fishing river extends beyond this prime section, but it falters in places and has ceased to exist in others. As on so many other North American rivers, formerly productive lakes and tributaries have been turned into the slack water of reservoirs. Unlike the remarkable section of river below Calgary, where a more productive fishery exists than ever before, in the upper and lower basin conditions are less robust. Dams have eliminated habitat, fish stocking has caused the redistribution and reordering of fish species, and changed flow conditions have produced harmful consequences. This is a story less often told. It is not a magnificent accident but a tale in which development trumped the protection of fish and fish habitat at every turn. Writing in 1956, Richard Miller, a prominent fisheries biologist in Alberta, described these changes using words such as "murder" and "destruction."[2]

The fishing river is thus a river of ironies and tragedies. Over time, diverse human actions have caused the spatial redistribution of fish habitat and species, the rise and fall of fish populations, and the reimagination of the river as a place for sport and play.

Beginnings

Fish have been living and changing in the Bow River for millennia. After the last ice age, fish reinvaded the Bow and broader South Saskatchewan system from the Columbia and Missouri-Mississippi systems. These fish, called glacial refugia, or species that survived glacial conditions in warmer climatic conditions, used various points of entry, sometimes through rivers that no longer exist. Mountain whitefish and bull trout possibly entered from the Columbia basin via a direct waterway between the Kootenay and Bow Rivers or via the Oldman River, which may have been connected to Columbia tributaries via the Crowsnest River.[3] The exact routes and periods of invasion are difficult to pinpoint. The fish communities that resulted interacted with a dynamic river system for thousands of years before humans actively sought to shape them. Before the Bow was a fishing river, it was a river of fish.

Different reaches of the river proved more favourable to some species than others. Although the causes of fish distribution are many, factors such as water temperature, stream-flow characteristics, bed materials, and water depths are important and differ broadly between the upper and lower basin. In the upper basin zone, relatively cold flows favoured mountain whitefish and bull and cutthroat trout. In the lower basin zone, where waters were much warmer and flows slower, northern pike, goldeye, mooneye, sauger, and walleye could be found. In between these two zones, a transition in water temperatures and flow types provided habitat for a mixture of species found in the upper and lower basins. Of course, fish move, and many species used different sections of the river at different points of their life cycle. For example, various trout species found primarily in the river's mainstem as adults might well have originated in tributaries characterized by cold and constant water flows and suit-

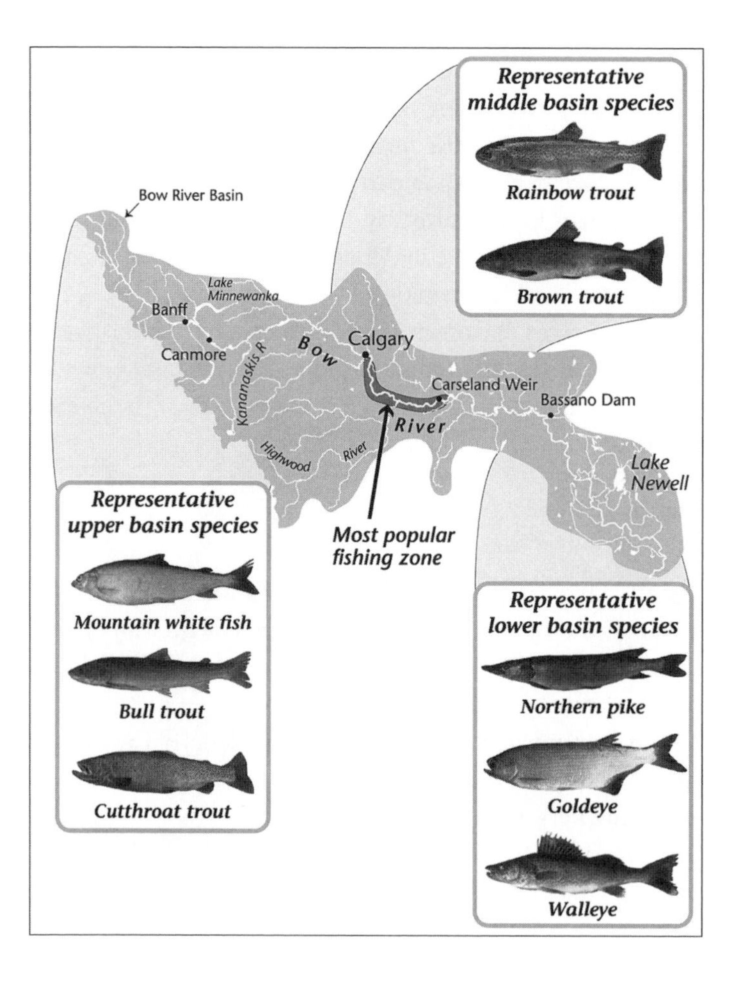

able gravels for spawning. Although fish distribution followed a few general patterns, environmental changes in the river or life history changes in fish could make new patterns. The river was a flexible habitat (see map 8.1).

Before European contact, indigenous peoples fished in the Bow River, but the extent of that fishery remains unknown. In the first chapter we noted the importance of the fishery to the Nakoda in the period of encounter and the absence of much indigenous fishing in the middle and lower basin. The ethnographic literature on the northern plains has traditionally emphasized the dependence of plains groups on the bison resource. Because fish provided limited energy values, while presenting considerable capture and preservation difficulties, they were rarely exploited. Nevertheless, anthropologist Brian Smith argues that fish may well have been an important food source for some northern plains groups in the distant and more recent past. Based on archaeological evidence from the Qu'Appelle valley and ethnographic evi-

Map 8.1 (*facing page*) Representative fish species in the upper, middle, and lower Bow basin. (Fish images are reproduced from Joseph Nelson and Martin Paetz, *The Fishes of Alberta* [Calgary and Edmonton: University of Calgary and the University of Alberta Press, 1992])

Illus. 8.1 (*right*) A good catch, 1912. (Glenbow Archives, NB 41-52)

dence from early contact documents, he suggests that some groups may have exploited fishing sites with weirs and traps within the seasonal round in order to provide variety in diet and to bridge lean times in the bison hunt.[4] One of the smaller bands of the Kainai bore the name "Fish Eaters," suggesting perhaps that they might have been unusual in that respect.[5] It is also possible that during the early nineteenth century indigenous fishing may have increased to supply fur trade posts, though this likely occurred further north, closer to points of consumption. Because the Bow was a marginal fur trade river, it is reasonable to assume that its fish were a peripheral element of fur trade diets. In the settlement era, when indigenous peoples were placed on reserves, some attempts were made to set aside particular tributaries or lakes for fishing purposes.[6] However, in the absence of more evidence, the local emphasis and particular pressure of indigenous fisheries before contact must remain vague. We should not, though, assume that fish were never important to bison hunt-

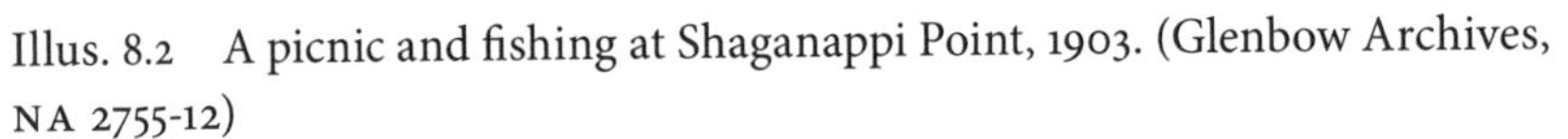

Illus. 8.2 A picnic and fishing at Shaganappi Point, 1903. (Glenbow Archives, NA 2755-12)

ers of the plains or that Euro-Canadians encountered an untouched fishery on the Bow.

In the late nineteenth century, Euro-Canadian settlers began to exploit the river for sport and sustenance. At the edges of newly established ranches, settlements, and work camps, men and women approached the river with makeshift poles, bringing assumptions about fish and sport from elsewhere. In the photographic archive of the Glenbow Museum, one finds evidence of this unorganized, occasional fishery. Men and women pose by streams, casting or kneeling with fish caught that day strung on a line. The accompanying photographs emphasize bounty, abundance, and pleasure (see illus. 8.1 and 8.2). The diary account of a transplanted New Brunswick farmer, Colin Crawford, gives a sense of an early fishing experience. In the course of a cattle drive up the Bow in 1883, shortly before the completion of the CPR's transcontinental line, he and his fellows camped near Calgary and spent an evening fishing. Crawford filled his diary with descriptions of the tugs on the line and "the pleasure of seeing a fine trout of some four pound weight push from the water glittering with the lines of gold and silver as he smartly waved through the air and landed safely at my feet."[7] He experienced little difficulty putting in a strong catch in one evening, and so did his comrades. The next day they feasted on a breakfast of trout and enjoyed the variation in their diet.

With that, they moved on. This sort of occasional use of the river's fish must have happened often in the early settlement period. Groups of people passing through caught fish, ate them, and left. Others fished streams in their locality as a leisure activity.

After the completion of the railway, fish in the Bow River came under the scrutiny and pressure of a new international tourist fishery. Almost all this activity focused on Banff and its surrounding area. A CPR promotional booklet published in 1893 advised tourists of choice fishing and shooting opportunities across the country. Its authors gestured vaguely towards fishing opportunities in some prairie lakes before highlighting Banff's "fine mountain trout fishing on the Bow and Cascade Rivers … [and] deep trolling for lake trout on Devil's Lake." The prairies were fine for hunting, the pamphlet suggested, but had little to offer anglers.[8] But within the Rocky Mountains National Park, an early tourist fishery emerged, complete with paid guides, local fishing lore, and an international clientele. On lakes and streams in the summer months, tourists could be found angling for trout. In the registers kept at the Lake Minnewanka Hotel and the Beach House Hotel, guests occasionally recorded comments about the quality of fishing. Some noted the catch: "July 9, 1889, Mrs. Gwyther Parker, Lincolnshire, England – 3 days fishing 22 fish 90 lb." Others, such as "George H.F. Nuttall, MD, PhD, etc, Johns Hopkins University, Baltimore, Md., USA," indulged in detailed fishing tales: "June 8, '91. Caught a 25 lb 39 inch trout off 'Gibraltar.' H.P. Panter of Beach House rowing. Had the latter not caught the fish under the gills with his hand we would have lost him as the gaff wouldn't hold." Still others made excuses. George W. Cassell of South Africa simply wrote, "No gottee whiskey no can catchee fish." Although visitors chose to represent their experiences differently in these registers, the catch was impressive. In 1892 some recorded well over fifty pounds of fish caught in three hours. With a man from the hotel to row and guide and ample opportunity to reward minimal effort or skill, the tourist fishery gained a considerable reputation and probably removed thousands of pounds of fish from the upper Bow basin each season.[9]

In addition to the tourist fishery, a food fishery to support workers from local mines developed in the upper basin in the late nineteenth and early twentieth century. During slack times and particularly during strikes, working men from mines in Canmore, Bankhead, and Anthracite would fish lakes and streams in and around Banff, including the Spray Lakes and Devil's Lake. They were after protein rather than play and used various methods to capture fish efficiently, including nets and dynamiting lakes. When large catches were made, workers preserved fish by salting them in barrels. All of this activity caused predictable outcries from fisheries officials and anglers, who believed that fish properly belonged to sport enthusiasts and should not be captured in

Illus. 8.3 A woman fishing near Banff in the 1890s. (Whyte Museum of the Canadian Rockies, NA 66-1359)

such quantities or by such methods, or perhaps even by such people. To preserve the sport fishery, anglers pressured the federal Department of Marine and Fisheries to regulate the fishery more conscientiously and to crack down on "foreign" workers who spoiled conditions for sport and tourism. The criticism of "unsporting" behaviour and the "foreign" nature of the workers tended to go together. Before long, signs stating legal catch levels in Banff and area carried their message not only in English but also in Polish, Danish, and Italian.[10]

In the middle basin, the fishery had a different focus. Locally based sport fishing was the primary use. Fishing focused near Calgary and extended south and east into ranching country on the Highwood River, where, as one angler put it, the trout were "as fat as butter."[11] As in the upper basin, this sport fishery operated without much regulatory oversight, and as early as 1907 anglers formed local groups such as the Southern Alberta Fish Protective Association to lobby government for stronger enforcement and stricter controls. When, in 1910, a fisheries commission toured southern Alberta in the course of a broader set of investigations in the Prairie provinces, witnesses in High River and Calgary complained about the unsporting ways of some fishers.[12] In the 1920s, following a surge in settlement after World War I, a raft of fish protective associations sprang up in towns and small communities to represent the concerns of older settlers against the incursions of supposedly undisciplined newcomers.[13] With the improvement in roads and access, some local groups along the Highwood River began to complain about day-tripping Calgary anglers.[14] The formation of associations signalled a greater organization of angling and the development of local claims of entitlement, but much fishing still fell outside the bounds of associations or government regulation. W.H. Bell, a boy in the 1930s, remembers fishing for suckers with bread off a crowded bridge over the Elbow River and chasing trout in shallow pools in the spring on the Bow.[15] Boys and girls climbing trees, rafting, and plopping fish into Mason jars also had their day on the river.

East of the Highwood River, the water temperatures on the Bow rise in the summer months as the river winds its way into some of the driest and warmest territory on the Canadian prairies. Such conditions did not support large trout populations or give rise to a strong sport fishery. When the 1910 fisheries commission held hearings at Medicine Hat, witnesses mentioned only a few sites in the lower Bow as worthy fishing holes: a reservoir behind the CPR dam at Bassano and the CPR irrigation project's main reservoir, Lake Newell. No mention was made of the Bow's mainstem. Fishing in the lower river focused on pike, pickerel, and goldeye, not the trout species favoured by anglers upstream. The lower river was more sparsely populated, was poorly integrated into existing transportation connections, and held few of the fish anglers favoured. It is notable that the only fishing sites mentioned had come into existence in the previous decade in the course of the construction of the eastern section of the CPR irrigation development. The lower river held a relatively new and marginal fishery built within the networks of irrigation infrastructure. It is revealing that one of the few articles in the local press treating the topic of fishing in the early twentieth century sought to boost the region's reputation for anglers by reporting the excitement of three Calgary fishers who were said to have caught a trout in the vicinity of Bassano.[16] Catching a trout on the lower Bow was news indeed.

"Mother Nature made no provision for the angler in her scheme"

Although advocates of fisheries protection often targeted other social groups, other fishing practices, and the lack of regulation as primary causes of fisheries decline, they also proposed a series of changes to the fishery that went beyond new forms of regulation and enforcement. Fundamentally, anglers wanted to make better fish. They wanted to introduce desirable game fish to the Bow from without and to expand the populations of those already in existence through artificial means. They sought, in short, introductions and improvements.[17]

The improvement agenda had roots extending beyond Alberta and Canada. In the late nineteenth century, across the Western world, fish had been introduced in a host of rivers and lakes as one facet of a broader acclimatization movement that sought to bring favoured species to places where they did not exist. In New Zealand, acclimatization societies had introduced trout and salmon from North America, as well as numerous songbirds and deer. Much of this activity was born of sport enthusiasts' desire to hunt and fish highly prized game, but it also signalled a longing for familiar landscapes and animals from homelands, particularly European and, in British settlement colonies, English and pastoral.[18] In Alberta, fish introductions would run parallel with these international developments. Sport enthusiasts would introduce birds such as partridges and pheasants as well as favoured trout in the interests of turning Alberta into a sporting paradise.[19]

Another aspect of the improvement agenda drew on international attempts to make fish – to reproduce them in hatcheries and stock them in lakes and streams. In the United States, fish hatcheries had become a fundamental plank of fish conservation policies.[20] On the West Coast, fisheries experts had established hatcheries to increase the productivity of native fish such as salmon or had stocked introduced species to replace depleted native ones. In Canada hatcheries had been established in British Columbia and the east to augment natural reproduction in heavily fished rivers and lakes.[21] Although most fishing interests and regulators acknowledged that activity had to be controlled in order to avert depletions, the possibility that fish culture techniques could increase supplies and relieve the stress on fisheries proved highly attractive. Given the growing pressures on fishing rivers in Alberta, concerned interests began to extol the possibilities of introducing the latest hatchery techniques to the province.

Fish introductions in Alberta preceded the organization of hatcheries.[22] Few records survive that explain how or when such introductions happened. Fishing enthusiasts moved fish from stream to stream, imported fish, and planted them near local fishing sites. Rainbow trout, one of the most successful introduced species, had probably been planted in several places by 1900 and would continue to be planted by different groups over the years. In the early 1920s a shipment of brown trout destined for another part of Alberta was dumped into a creek near Canmore when the vehicle transporting it broke down. Better, thought the driver, to save the fish by turning them into the river. Brown trout have since spread throughout the Bow basin. In the lower Bow, farmers and fishing enthusiasts stocked reservoirs with whitefish in the mid-1920s.[23] Other introduced species failed. Eastern black bass, planted in Banff National Park, did not take. Nor did Nipigon trout. Possibly, many other unrecorded introductions also failed. Some rare fish succeeded but occupied highly specific niches. In the Cave and Basin in Banff National Park, mosquito fish were introduced into the hot sulphurous waters in the late teens or early twenties to help control the local mosquito problem.[24] The fish may have done little to control mosquitoes, but they did apparently cannibalize and annihilate pre-existing species.[25] More recently, fisheries biologists have discovered other tropical fish in the hot waters, survivors of more than one domestic fish bowl dumped into the Cave and Basin pools at various times since the late 1960s.[26]

The pre-eminent institution for fish improvement in the Bow basin was the hatchery. As early as 1910, various groups in Alberta, including the Calgary Board of Trade, had campaigned for a hatchery on the Bow.[27] Before the fisheries commission launched its survey of fisheries in Alberta that year, the Alberta Fish and Game Protective Organization had sent a memorial to the federal Department of Marine and Fisheries stating that it would "be advisable from the stand point both of sport and business to stock the streams with suitable fish."[28] Local witnesses subsequently convinced the fisheries commissioners to include plans for a fish hatchery among its recommendations. Not only would a hatchery help to reverse noticeable declines in fish populations in the upper basin, the commissioners argued, but it would also grant Alberta a facility like many of the other provinces in Confederation. Fish and federalism could work together. In response to the report, the federal department sent officials to scout sites and establish operations.

In 1913 a hatchery was established in the Banff townsite. Part fish factory and part tourism spectacle, the hatchery drew water from the town main and displayed the intricate production process of fish from egg to fingerling before tens of thousands of grinning tourists each summer.[29] In its early years the hatchery struggled to procure eggs and milt and to produce enough fish for stocking. In the late teens, the hatchery received annual shipments of eggs and milt from Ontario, while in the twenties, supplies were imported from the United States.[30] In spite of these production difficulties, the hatchery became a regional fish production centre, which would gather available spawn from surrounding areas or import them and then raise and distribute fish to various areas within and without the park and the Bow basin. The production schedule and product line changed over time (more cutthroat trout this year, more rainbow the next), and so too did the sites of egg collection and distribution. In the process, fish stocks from different locations in the upper Bow system, and sometimes those imported from elsewhere, were thoroughly mixed up. Hundreds of small lakes in the Rocky Mountains, formerly devoid of trout populations, were stocked.[31]

Another, privately sponsored hatchery was created in Calgary in 1936 to supplement trout populations of the middle basin. Built and sponsored by J.B. Cross of the Calgary Brewing and Malting Company, the hatchery sought to marry the aims of publicity and production, not unlike the Banff hatchery. Apparently, the marketing connections between beer and small fry were stronger than one might have assumed. For several years, the hatchery operated as a curiosity and did not produce many fish for distribution. By the early 1940s, however, the provincial government had begun to operate the facility jointly with the brewing company. It expanded to carry about half a million eggs at a time.[32] In following decades the province assumed full responsibility for the operation of the hatchery, while the brewing company donated the use of the building and site. The facility continues today as the renamed Samuel Livingstone hatchery.[33]

By mid-century, hatcheries lay at the centre of a complex system of egg production and distribution for millions of fish in the Bow basin. Eggs were collected locally, traded with other hatcheries, and purchased from abroad. Once transported to site by tanker trucks, the young fish were reared in tanks with running water and fed on a diet of liver, bran formula, and vitamins. After the nursery stage, hatchery staff redistributed fingerlings and yearlings to various locations. The hatcheries had become a major force shaping the composition and existence of fish communities on the Bow.

Stocking rivers with fish had effects. Pre-existing fish communities could be reordered. In extreme cases, such as the Cave and Basin, native fish populations were annihilated by introduced species. In other instances, native and newcomer hybridized, as did the rainbow trout with cutthroat trout populations in various sections of the river. In some upper basin locations, bull trout faced competition for habitat from introduced fish. Across the basin, new species appeared, others hybridized, and still others disappeared from one location or became much reduced in numbers. Stocking the Bow to make it more of a fishing river had intended effects in many places: more game fish were available for more fishers at designated recreational sites. But stocking also produced new conditions that were neither anticipated nor desired by those who carried the banners of introduction and improvement. Before stocking, the upper Bow had provided habitat for eleven species of fish; today it contains twenty.[34]

In recent years, the logic of stocking has come full circle. Within Banff National Park attempts have been made to *reintroduce* formerly native species and to remove introduced fish.[35] An anti-stocking perspective has driven a new kind of stocking agenda. Lower in the basin on the Kananaskis system, attempts have been made to reintroduce and protect native bull trout on stretches of the river where they have been displaced. This new kind of stocking does not aim primarily or even secondarily at anglers' pleasure, as former stocking programs did. Now the goals are more abstract. Contested concepts such as restoration and ecological integrity provide loose goals for management policies that seek to atone for past environmental sins.

Dams

If stocking and hatcheries reordered fish communities in the Bow basin, then dams transformed the contexts in which those communities existed. Large dams on the Bow built between 1930 and 1955 interrupted the river's course and introduced a new set of ecological conditions. Lakes in the upper basin were impounded and flow conditions changed. On the mainstem, the seasonal patterns of flooding were altered to facilitate reservoir storage for power development and to eliminate flood threats in spring and winter. These new conditions precipitated changes in the composition of bottom fauna in lakes, deposited litter on lake bottoms, and introduced a new cycle of water levels according to power development schedules, not natural runoff.

The effects of dams on fish provoked interest and concern well before the 1930s. Since the turn of the century, dams of various sizes and types had been constructed on the Bow and its tributaries. These interrupted flow patterns, conditioned the movement of local fish populations, and produced reservoirs. Fishers and their organizations looked askance at them because they appeared to block fish movement. Fisheries officials judged that fish could sometimes pass dams and that it did not much matter if dams blocked the river. Both views, steeped in local experience, were based on rather vague notions of fish biology. In the early 1920s, for example, the case of the Eau Claire dam at Calgary attracted attention amongst anglers as an obstruction to fish migration that ought to be reconstructed with a fish pass. Various fishers and groups approached federal authorities to take action under the federal Fisheries Act, which empowered the Department of Fisheries to remove obstructions to fish in streams. As a result of the complaints, federal fisheries officials investigated the site, took photographs of the dam, and interviewed anglers.[36] Although the dam seemed to block fish, during high flows the river breached the dam, allowing fish to pass.[37] D.A. Richardson, the local fisheries overseer, was little concerned. He reasoned that below the dam, fish were mainly of the so-called coarse varieties not favoured by anglers. If the dam was removed or a fishway installed, the coarse varieties would invade the upper basin.[38]

Similar investigations took place in response to other complaints in the interwar period, without effect.[39] After anglers requested a fishway on the Bassano irrigation diversion dam, for example, federal officials visited the dam, interviewed people, and then decided that because the dam was so high, a fishway would be impossible. In any event, they claimed that the irrigation project had produced new opportunities for fishing in reservoirs. Lake Newell was "simply alive with fish, all pike, and affords sport for a large prairie district surrounding."[40] Dams did not equal damage in the minds of fisheries officials; as often as not, they were viewed as valuable additions to fish habitat.

The changes that had been observed by early anglers and fisheries officials paled before what was to come. In a period of twenty-five years, from about 1930 to 1955, dams were constructed across the Bow basin – for power, flood control, water supply, and irrigation purposes. In the headwaters, lakes were impounded to turn them into power reservoirs. On the mainstem, a new dam near the confluence of the Bow and Ghost Rivers turned a long stretch of the middle river into an extended and slack reservoir. The Elbow River near Calgary was dammed to supply the city with water. On the lower Bow new and reconstructed weirs diverted water for irrigation. At the end of the period, a dam was built at Bearspaw, above Calgary, to control flows through the city. All these dams had primary purposes. Some were regulated in order to coordinate with other dams in an integrated system of power production, flood control, and irrigation water supply. But following past practices, dams were not designed with fish in mind, and precious little was said or done to change or halt dams until after they had been constructed.

The new dams constructed on the Bow after 1930 did not face any greater regulatory scrutiny from fisheries officials. What did change, however, was the kind of knowledge produced about dams and their effects on fish. A new generation of university scientists interested in measuring the ecological effects of river development on fish were engaged by government departments to make sense of the transformed river. Most came from regional institutions – the University of Alberta and the University of Saskatchewan – and collaborated with government officials and staff. Studies of dams thus became more systematic in design, employed standard methodologies and instrumental recordings of river flows and fish counts, and engaged literature in ecology, limnology, and relevant disciplines. Starting in the late 1930s, investigators across the Bow basin sought to understand what had happened and what might happen to fish and their environments if large dams were placed in rivers, if fish were stocked in places where they had not previously lived, and if fish species and new dam-induced environments interacted.

Investigations of dams and their effects on fish first occurred as a by-product of the stocking agenda. In the late 1930s, in search of sites to rear fish from the Banff hatchery, provincial officials settled upon the Kananaskis Lakes as possibilities. They were isolated, not too far from Banff, and seemingly well suited to the purpose. The upper lake had also been dammed in the recent past, and the lower lake was slated for development. To inquire into the wisdom of the site as a rearing area, they engaged University of Saskatchewan fisheries biologist D.S. Rawson to conduct studies over two field seasons in 1936 and 1937.[41] Rawson was a specialist in the ecology of lakes and approached the matter not only with a view to the fish culture question but also with a wide array of queries about the biological productivity of the lake, its chemical composition, and the range of its water levels. He also fished the lake and examined the bodies and stomach contents of those specimens he caught. In general, he sought to understand the dynamics of lake ecology, how it bore on fish life in particular, and whether fish culture could work at the site.

Rawson concluded his investigations in 1937 and reported to government officials that the lakes could not support fish-rearing activities. The reason was not the location of the lakes, their size, or their chemical composition. It was that the lakes had been dammed, or soon would be dammed, and that the effects of damming changed the lake as habitat. In the upper lake Rawson found that forest debris now virtually covered the lake bottom. This material had been left when the dam was constructed. As the lake level rose and encompassed a wider area than before, it submerged riparian vegetation and forest. As a result, he concluded, aquatic plants could not root properly and shore fauna were scarce. A second problem with the lakes arose from their role as reservoirs, which could rise and fall in conjunction with power production needs. In the upper lake, water levels rose and fell each year within a thirty-foot range; proposed changes to the dam would increase that range to seventy feet. Fluctuating water levels, Rawson argued, lowered biological productivity in general and also disturbed fish-spawning patterns. After dam construction on the lower lake, he presumed, similar problems would arise. He therefore advised against fish culture in the lakes. "Power development, in this region at least, is largely incompatible with fish culture or as one might say – when Kananaskis lake was dammed for power development it was also damned for fish cultures."[42]

Rawson's study of the Kananaskis Lakes provided a kind of baseline. Although it began as a prospective assessment of the lakes as fish-rearing habitat, it introduced a number of other studies of the lakes that sought to measure changes to their environments under the impact of dams. In the next two decades, first Rawson and then other scientists continued biological studies of fish communities in the lakes. They aimed primarily to determine short- and long-term changes to environmental conditions caused by water drawdowns and the dynamics between fish species. Despite Rawson's early recommendations, lake stocking did occur under the aegis of government, but fish were probably also independently introduced by anglers. Changes to the lakes' ecology and the composition of fish communities therefore invited attention. After Rawson completed a final study of the lakes in the late 1940s, at least five other fisheries scientists analyzed the lakes at different times. The upper Kananaskis Lakes became, in short, a closely monitored case in which long-term patterns of change could be investigated.

In the mid-1960s Joseph S. Nelson, a zoologist at the University of Alberta, attempted to synthesize and summarize much of this work, completed over thirty-odd years.[43] Nelson described and analyzed a fish community that had undergone remarkable shifts in composition over a relatively short period of time. He began with an accounting of the fish community. Some species had virtually disappeared (cutthroat trout), some had decreased substantially (bull trout, brook trout), and other populations had remained relatively steady (mountain whitefish), while still others had expanded and succeeded in the changed circumstances (brown trout and longnose suckers). The reasons for these shifts, Nelson believed, had more to do with competition between species than with changes in the lakes' ecology, though the two were connected. Brown trout, for example, which may have been introduced into Carrot Creek before entering the Kananaskis Lakes in the early 1940s, expanded their range within the lakes and out-competed with other species for space and food. Brown trout were well adapted to fluctuating rivers. They succeeded in the dammed lakes environment at the expense of other species. By way of contrast, cutthroat trout had virtually disappeared over the thirty-year period. Nelson believed that they had probably hybridized with rainbow trout, though the incursions of brown trout probably also had adverse effects. Nelson described new natural spaces, carved from old, which operated according to a new set of patterns and processes. Some fish did well in these environments; others did not.

In different lakes in the upper basin and in sections of the mainstem, parallel processes occurred. In Lake Minnewanka, for example, following the construction of a large dam in 1941, water levels varied annually over forty-five feet. These changes affected the composition of bottom fauna, the accessibility of spawning gravels, and the interactions within fish communities. In 1954, after two decades of pre- and post-development assessments, zoologist

Richard Miller and fisheries biologist Martin Paetz summarized what had changed:

> The major sport fish was the lake trout, *Salvelinus nammaycush*, which fed on introduced cisco and on the native Rocky Mountain whitefish. Since the dam was built this fishery has degenerated very seriously. The trout do not reach sizes beyond about one pound; at this weight there is normally a change from plankton and bottom feeding to fish feeding. This change is not now made, and growth is arrested. The fluctuations in water level apparently result in an ecological separation of the trout and the forage fish on which they feed. An attempt to remedy this situation has been made by introducing the lake whitefish, *Coregonus clupeaformis*.[44]

Miller and Paetz and others who had studied Lake Minnewanka pointed to the same difficulties that Rawson had earlier highlighted: lakes controlled for power could not produce fish as they once had. Previous biophysical processes were upset. Only complex management strategies and stocking could maintain a recreational fishery. "Power projects on high altitude lakes and rivers," they concluded, "have proven to be, in Alberta, uniformly deleterious to the sport fishery."[45] Conditions were different but no better behind dams on the river's mainstem, Miller and Paetz claimed. Dam development on the Bow had changed the fishery and rarely for the better. This conclusion held for all of the dams on the Bow, except perhaps for the Bearspaw dam, built near Calgary in the mid-1950s.

Unlike the upper basin dams, which had transformed lake environments, the Bearspaw dam was built on the mainstem of the Bow at a point in the river not well known for its fisheries. Its primary function was to control flooding before the river passed Calgary. Most river regulation had to take effect during the spring, when flows were high, and during the winter, when floods triggered by ice dams and river surges posed a threat. This regulation schedule also accidentally produced favourable conditions for trout spawning downstream.[46] Before the dam was built, flood conditions had disturbed spawning grounds in the mainstem. A rapid torrent could churn up gravels and sediment on the riverbed and destroy redd sites. Furthermore, the seasonal fluctuations in flow meant that in winter the marginal flow of the river reduced the area covered by water, thus reducing habitat generally. Trout reproduced with greater consistency and productivity in middle basin tributaries, most notably in the Highwood River. But with the regulation of flood peaks, the dangers of surges and the disturbances they caused decreased. The river passing through

Calgary in the winter was also moderated in such a way that water cover was fairly consistent over the riverbed. This condition reduced risks of exposure. Within a fairly short time, the mainstem acquired the characteristics of a prime spawning habitat: cool, consistent, and moderated flows. To make matters even better for trout production, this stretch of the river was now one of the most biologically productive in the basin. Part of the reason was the contribution made by the City of Calgary's sewage system.

Sewage

All things are connected. When in the mid-1930s a toilet flushed in Calgary, its contents entered sanitary sewers and passed through the Bonnybrook treatment plant, where large grates collected the largest floating material and some suspended materials settled, before the liquid remains were piped into the Bow along with all of the other wastes that the sewerage had collected. Follow that imaginary toilet flush further, and its contents would mix and roll into the Bow's flow, be broken by the current, be changed chemically through the oxygenation process, and appear before long as an innocent input of nutrients and water. That toilet flush and thousands of others like it contributed to a massive boost in nutrients to the Bow River. Thousands of years ago no toilets flushed into the Bow; by the early twentieth century a growing urban settlement was contributing thousands, then hundreds of thousands, and now millions of flushes of material into the river each year. Fish do not eat sewage, but they eat insects that help to break down sewage, and they hide and swim amongst aquatic plants that have grown more plentiful because of the nutrients that sewage supplies. With each toilet flush, the environment that fish belonged to grew more biologically productive.

Nutrient inputs can smell, of course. Early in the twentieth century, sewage and other waste, from rotting carcasses to industrial wastes, entered the river untreated. The fisheries commission of 1910 decried the use of the river as a sewer and called on government to improve sewage treatment and on individuals to treat the river with more respect. "A good many people, living at or near streams, seem to think the main purpose for which they may be used is for the carrying off of manure, rubbish and other refuse, such as tin cans, old clothes, dead animals, etc."[47] In the following years, of course, sewage treatment was not measurably improved, while sewage and other waste inputs to the river increased. Government received complaints about sawdust being dumped in the river in large quantities.[48] Downstream communities charged that the Bow was contaminated for drinking water. After an oil refinery was established in Calgary in 1923, anglers believed that oil wastes were entering the river and spoiling fish. These complaints produced inquiries by governments but no resolutions to the real or perceived problems.[49]

Before the Bonnybrook treatment plant opened in Calgary in the early 1930s, millions of gallons of sewage entered the river daily. Below Calgary the visual and olfactory senses of citizens and anglers pointed to problems. In the early 1920s, fisheries officials interviewed several farmers who lived below the Calgary sewer. They offered the view that fish in the area were "not fit for food." "After the Calgary sewer was opened," said a Mr R. MacRae, "the fish caught in the river for miles below the mouth of the sewer had a bad taste something like gasoline." Mr Clifford, a rancher in the area, said that he warned "anglers not to fish here."[50] In 1926 D.A. Richardson, the fishery overseer in the Calgary area, reported to his superiors that Calgary's sewage did indeed contaminate fish and give them a bad, oily taste. The refinery was not the problem, in his view. Imperial Oil had added concrete holding tanks, charcoal beds, and waste sloughs to clean its waste water. The sewage, on the other hand, entered the river with little treatment. "The question naturally arises, that with hundreds of miles of pure waters to fish in, that anyone should choose this locality."[51]

Anglers were not alone in their concerns about pollution. As early as 1916, downstream water users in Bassano had tried to pressure Calgary with legal action to improve sewage treatment. But by the 1940s Calgary had still not improved treatment beyond the primary stage, despite several public promises to do so. Politicians in Bassano kept up the pressure on public officials in Calgary and the provincial government. Anglers proved ready to lend support. As W. Boote, proprietor of the Ogden garage in Calgary, put it to the mayor of Bassano in 1940, "I am interested in a Bow River fit for fishing, you are interested, I would think, in a Bow River fit for drinking."[52] At different times, anglers wrote letters to Bassano politicians explaining the failings in the Calgary sewage system and the wilful dumping of untreated sewage and manure. One letter writer, a city of Calgary employee and an avid angler, had reportedly advised superiors of problems in the sewerage and been told to "shut up and mind my own business."[53] Supplying Bassano with rhetorical ammunition proved helpful but did not force any action.

In the early 1950s an inquiry into the Bow River and its tributaries outlined the bizarre mixed effects of the sanitary river: the Bow below Calgary was dirty, smelly, and coated in places with a dark grey slime; on the other hand, the fish seemed to be enjoying this state of affairs. The stretch of the river between Calgary and Carseland had become remarkably productive. Word was out that this stretch of the river supplied large trout in high numbers. The inquiry offered the opinion that more trout were to be found in the stretch of river from Calgary to Carseland than in any comparable trout stream in Canada. In spite of the downside of this mixed outcome, the report's authors believed that the river could absorb yet more sewage from Calgary without inflicting problems on the environment or downstream users.[54]

Calgary only increased its sewage inputs in the coming decades. With a soaring population driven by the province's booming oil economy, new pressures were placed on the city's sewerage. In addition, changing patterns of domestic cleaning and gardening added to the load. As we have already seen in chapter 7, systematic study of the Bow in the 1960s by the Provincial Board of Health revealed a badly polluted river in which fish faced risks from high phosphate inputs and low dissolved oxygen levels. In addition, the role of phenols in the river in affecting the taste of fish became better understood. The resulting improvements to sewage treatment introduced by the city of Calgary at the behest of the Provincial Board of Health helped to turn around the situation. Calgary's sewage continued to provide the river with a massive nutrient boost, but only after some of its more harmful properties had been reduced through treatment.

Irrigation

Since early in the century, fish had expanded their range beyond the primary habitats of the Bow River and its principal tributaries. Following flows and channels, they migrated east, west, and south through weirs and diversion channels into complex webs of irrigation infrastructure. The extent of this colonization of new aquatic habitat was noted in 1910 when the fisheries commission pointed to the pike fishery in Lake Newell, the reservoir of the CPR's eastern irrigation section. As other areas of the semi-arid prairies were incorporated into irrigation systems in the early and mid-twentieth century, the networks of canals and reservoirs built to supply them also created additions to the potential range of fish. Lake McGregor, for example, constructed in the first decade of century, became, like Lake Newell, a catchment for fish that wandered into the Bow River Irrigation District's channels through its weir at Carseland. As water fanned out across the prairie to supply new irrigated farms, fish exploited new pathways into dry country. As noted in chapter 6, irrigation canals were poor habitats for fish, where they could become trapped and die.

Canal fish kills were feared by fisheries advocates from the earliest days of irrigation development on the Bow. The 1910 fisheries commission heard from witnesses in Calgary and High River who complained that irrigation canals had been improperly screened, leaving fish to enter them and die. However, as with the case of dams, irrigation interests were not forced to alter their diversion methods in the interest of fish and fishers. In 1894 the always blunt William Pearce wrote, "As to the matter of Fish versus Irrigation. If one or the other must suffer, it certainly would have to be Fish."[55] Although some weirs contained screens, they were rarely fine enough to stop fish entrants. If they had been, they also would have probably blocked debris and become inoperable. As a result, the problem of fish in canals became a recurring annual event in the lower Bow. Since the late 1990s the matter has become an important cause amongst regional environmental groups and fisheries advocates. Rather than endure more dead fish in empty canals, groups such as Trout Unlimited have organized weekend fish rescue operations involving over one hundred volunteers to scoop fish from canals and return them to the mainstem or tributaries.[56]

Irrigation systems posed risks for fish that entered them, but they also created some productive habitats. Those fish that did swim into major reservoirs survived in good numbers. From early in the century, self-sustaining populations of pike were found in Lake Newell. By the mid-1930s introduced whitefish were flourishing. One *Calgary Herald* article in 1938 suggested that forty-three fishers supported themselves by fishing whitefish commercially.[57] A year later the *Brooks Bulletin* reported that over the fishing season, commercial operators had hauled nearly twenty-five tons of whitefish from the lake. Packed in one-hundred-pound ice boxes, they were shipped to Edmonton before reshipment to US markets.[58] Similar small commercial fisheries operated later in the century on other irrigation reservoirs, such as Lake McGregor.[59] However, commercial fisheries such as these ran into conflict with farmers seeking recreational fishing. In the summer months, boating and fishing became popular recreational pastimes in the irrigation districts. During the coldest months of the winter, local farmers would gather around ice holes on Lake Newell and try to add a fish or two to dreary winter diets. As commercial takes increased in the 1930s, farmers' groups called for the fishery to be returned to farmers. The Cassis United Farmers of Alberta local, for example, passed a resolution in 1939 criticizing the scale of commercial fishing and insisting that farmers should have priority: "We feel we own these fish and should be allowed to use them for the people of the district."[60] In a slightly different vein, this was another conflict over the river's fish between advocates of recreation and those promoting use.

Irrigation projects affected fish without as well as within. On the river's mainstem, major diversion structures, such as the one near Bassano, introduced obstacles to fish movement and affected the composition of fish communities over time. A flowing river was replaced at particular points by slow-flowing lakes that favoured some species and not others. Fish sampling conducted in the reservoir behind the Bassano dam in 1947 suggested that the river at this point contained mainly common suckers and only a few pike. Fishing in the reservoir was very limited. In the tailwaters, however, more species were present (goldeyes, walleyes, whitefish, and suckers), and so too were anglers.[61] Just like power and flood control dams upstream, irrigation diversion structures and dams altered the environmental conditions of local fisheries. A more general change to fish habitats that is harder to measure occurred during the irrigation season when huge volumes of water were diverted out of the mainstem to feed the various irrigation districts. The lower stretches of the Bow between Bassano and the confluence with the Oldman River, which had never been a productive environment for fish because of high summer temperatures, could scarcely produce any fish in the twentieth century. Whereas upper basin dams had created steady water flows and winter cover as the river passed Calgary, below Carseland diversions produced an ephemeral summer stream not well suited to fish.

Gone Fishing

By the second half of the twentieth century, the river had been changed, and so had the fish. These shifts in turn led to different experiences fishing on the Bow. More people, from near and far, fished the river. They focused on different reaches from they had formerly fished. Anglers had to take out licences or be fined, comply with various rules, and observe catch limits. These were no longer the days of Colin Crawford pulling trout out of the river in whatever numbers as merrily as he pleased. In some sections of the river, catch-and-release rules applied. Fishers' actions were carefully monitored and regulated in a way that they had not been earlier in the century. Increased fishing pressure and changing concerns about environmental regulation and tourism prompted provincial authorities to manage and study the fishery on a continuing basis. State surveillance aimed to maintain fish populations and increase them, but they also generated statistics about the fishers as well as the fishery. On the basis of these regulatory statistics, it is possible to observe not only a remarkable shift in scale of recreational fishing pressure (which increased by leaps and bounds after 1970) but also clues about the social background of fishers and their preferences.

The first reasonably reliable censuses of the fishery began to appear in the mid-1970s. They suggested a local and international fishery on the Bow, emphasizing trout and fishing reaches near Calgary. A survey of anglers conducted in 1975 found that over 98,000 had taken out permits in the central administra-

Illus. 8.4 Ralph Klein fishing in Calgary in the fall of 1989, when he served as Environment minister. (Glenbow Archives, PA 1599-354c-83)

tive region of the province, including the Bow River and its tributaries. Most of the fishers taking out licences in the province were from Alberta, but fishers from most provinces were also active, as were thousands of Americans and hundreds more from elsewhere. Of these fishers, most favoured catching trout species to any others. The size and scale of the fishery had lately increased. In 1967–68, over 136,000 angling licences had been sold in the province. By 1975–76 the number had jumped to over 218,000. The reasons for this increase could be many, including better enforcement and regulation, but it also seems plausible that fishing had become more popular and that a rising proportion of the population was reaching for rod and reel. Men were overrepresented, but women made up a substantial presence on the river, purchasing one-fourth of the licences in the province. Fishing also attracted certain age groups more than others. The vast majority were between the ages of twenty and forty-four, though tens of thousands of men and women in their late forties and fifties were fishing as well.[62] The spike in fishing activity points to the marked shift in patterns of leisure, consumption, and lifestyle that had begun to affect so many aspects of river use, perception, and evaluation.

On the river the increased popularity of the sport and the diversity of the fishers became evident. Although many of the stretches along the Bow attracted attention and interest from anglers – from small lakes in the headwaters to the cherished Highwood River to pike fishing holes in irrigation districts – the river below Calgary had become the centre of gravity for anglers, pulling them in from near and far. They had been fed boosting articles in the US fishing press, and the place was well integrated into the highway system and close to an international airport. The river could be easily navigated by canoe or raft or fished from shore in waders. Small businesses sprang up which rented canoes and shuttled cars downstream, in addition to the outfitting and gear shops in Calgary. In the late 1970s, provincial officials estimated fishing pressure on this section of the river as "heavy" and judged that hundreds of fish were harvested annually per kilometre. They claimed that fishing pressure would likely only increase.[63] Jim McLennan, who began guiding on the Bow in the 1970s, witnessed remarkable growth in the fishery. "On a float trip in 1979 I mightn't have seen any other fly fishers and, if I had, I would have probably recognized them, for there weren't many of us out there … It's different today." Twenty-five years later the river was dotted with fly fishers he had never seen and would probably never see again.[64] A cherished place had become a fly fishers' thoroughfare. One newspaper photograph of Ralph Klein fishing in Calgary, taken in 1989, when he was provincial minister of the environment, shows clearly that he was not alone (see illus. 8.4).

Mixed Blessings

The conditions that provided for this popular and growing fishery were both natural and modified. The boulders and gravels that lined the riverbed had been dropped by glaciers and ground by flow and friction. They supplied superb spawning grounds for trout. The plants that fish swam amongst to rest and hide from predators were indigenous. The insects that fish ate were also. The Bow's flows had changed in recent years, but the water temperatures were moderated by climate, and the temperatures were cold enough for favoured trout varieties and warm enough for high productivity. Although many of the fish that reproduced in this reach were introduced species, they were naturally reproducing. Three important changes had been made to the river, however. First, the Bearspaw dam had been built above Calgary in the mid-1950s to protect against floods. It moderated flows advantageously for fish and particularly trout. Peaks and valleys in flow trends were reduced, and this mitigation helped to protect spawning beds from scouring during high flows and from exposure during low flows. Second, although sewage had provided nutrient inputs to the river for many years, enhanced water treatment installed by the city of Calgary in the early 1970s reduced the threat of fish kills in the lower river and helped to improve the perceived quality of the fish. Formerly, phenols in the water had given fish bodies an oily taste. But after enhanced treatment, this problem was reduced. Fish had all the benefits of increased nutrients and expanded biological productivity but no longer bore the taste traces of the sewage system. Of course, this improvement left them open to more human predators than ever before. Third, while fish had increased in some sections of the river, in other places they had practically ceased to exist. Some native species were crowded out, fished out, or found their favoured habitats destroyed or massively transformed. Stocking programs, dams, and irrigation all had diverse and often detrimental effects on fish populations, particularly in the upper and lower basins. As the settlement system along the Bow becomes more integrated and dense, problems of this kind will only increase. Bridges, roads, culverts, and straightened shorelines along residential properties can produce ill effects on fish habitat.

The fate of the river and its fish has become a matter of recurring public debate. In the 1980s a rising chorus of environmentalists in Alberta wanted to focus attention on the factors that still might alter or end the fishery in the future. Recent improvements were viewed as positive but hardly final. In the early 1980s the Alberta Wilderness Association published a glossy pamphlet that claimed that the river was "on borrowed time" (these words were emphasized in the publication). Pollution levels remained unacceptably high and threatened fish kills. Dams and diversions had to be controlled. Jet boats on the river posed an obnoxious threat to enjoyment. The association proposed that the river "should be established as a *recreational river* for the future enjoyment of all Albertans" (original emphasis).[65] The forward-looking and democratic quality of this language elided the fact that focusing regulation solely on the questions of the recreational river challenged and potentially threatened other river users: First Nations, cities and their sewage systems, power companies upstream, and irrigators downstream. Nor did complaints about the quality of fish and the spectre of oily tastes disappear. In the late 1980s, scientists investigated the effects of sewage on the river after complaints by anglers and downstream users. They found sewage to be properly treated and judged the complaints to be largely unfounded.[66] The recreational river remains a contested concept and a contested space.

9 · Overflow

In the middle of a bitterly cold night, 20 December 1938, the Bow River did something quite unusual, or so it seemed at the time. It flooded. Residents of the low-lying working-class neighbourhood of Lowery Gardens, three miles west of the city, were awakened by the sound of water rushing through their homes. Just after 4:00 AM in sub-zero temperatures firemen evacuated the neighbourhood. The waters retreated as quickly as they had come. The next day plates of ice hanging on tree trunks and cars locked in huge blocks of ice offered novel testimony of the extent and force of the flooding. This was the first time anyone could remember ice jams on the river sending frigid torrents spilling into a residential neighbourhood in winter. It would not be the last. Something strange seemed to be happening on the river.

Winter floods were new in Calgary. Long-time residents had grown accustomed to the occasional and sometimes spectacular inundation during the annual early summer runoff. But these had grown less frequent in recent years. Beginning in the late 1930s, winter floods began to replace summer floods as the main danger from the river. Again in 1939 and then twice in November and December 1940, floods drove the residents of Lowery Gardens from their homes.[1] For three winters running, the Bow had flooded. On 29 December 1941 Sunnyside was briefly inundated; the next year Lowery Gardens again found itself under water in mid-winter. In 1943 and again in 1944 winter flood waters reached Bowness. The winter of 1944–45 saw a serious flood affecting several blocks in the Hillhurst district. Virtually every winter some part of Calgary was underwater, usually upstream to the west but occasionally below the confluence of the Elbow. A summer flood eventually retreated, leaving a muddy reminder of its visit. Winter floods brought the mess but also drove shivering people out into bitterly cold temperatures and left behind huge panes of ice that remained until the next thaw, sometimes for weeks.

Engineers from the city, the federal government, and the power company first began tracking the winter flooding phenomenon in 1941. The Hillhurst flood in 1944–45 instigated a local ratepayers' insurrection, which led to the formation of a panel of experts to report on causes and remedies. Meanwhile, as ratepayers fumed, politicians created committees, engineers gathered data, and angry accusations flew back and forth over who was responsible, the river continued to overflow its banks in November, December, and January with frustrating regularity. Almost every year flooding and seepage, usually arriving swiftly in the dead of night, caused localized evacuations somewhere along the course of the river as it passed through the city and its outskirts. The rising winter waters, which regularly invaded working-class districts on the river flats, heightened social tensions in the community. This mysterious winter flooding also put professional management and scientific control on the defensive in the face of popular suspicions and vernacular knowledge. Under pressure to do something, the province of Alberta established a royal commission in 1952 to get to the bottom of the mystery and recommend measures to put an end to the nuisance.

If public pressure demanded action on winter flooding, public complacency stood in the way of managing the summer flood threat as it receded into distant memory. When the authorities began to understand the phenomenon of flooding better and took steps to protect the public, residents rose up in anger against what was perceived to be public vandalism against "their river." The apparent decline of summer flooding opened space for a complex process of political bargaining about risk tolerance, landscape design, and the limits of public liability. Floods raised tempers as well as waters.

Swept Away

Rivers flood. That is how they work. Irregular large-flow events (floods) and more routine full-flow events without inundations (bank full stages) are the two principal fluvial processes determining the shape of the river as it courses across its landscape. From month to month and year to year the flow of a typical river varies according to the volume and concentration of precipitation, the type of landforms in the watershed, and occasional obstructions. No one should expect a river to remain peacefully within its banks. Nor do rivers design themselves to contain their relatively rare maximum flows. That would be terribly inefficient.

Bank full stages work away incrementally on the cutbanks, gathering sediment to be deposited downstream in riffles and bends where the current slows down. This process gradually changes the shape of the river, adding to or reducing its sinuosity and depth. On the other hand, the occasional major flows of different magnitudes make quantum changes to the river by scouring the bed more deeply, carving out entirely new channels, cutting off old oxbows, and distributing gravel and soil widely across the flood plain.[2]

It should come as no surprise, then, that the Bow River would occasionally rise up out of its bed and spread onto the flood plain. But the bottom lands were appealing to both rural and urban communities: rich, flat, well-watered terrain. The site chosen for Calgary, a level, gravelly expanse by the river, ideal for railroading and real estate speculation, was in fact a flood plain. Long intervals separated major floods. So consciousness of the threat faded in memory, or it could be conveniently overlooked in comparison with the advantages of a riparian location. Just as humans were drawn to rivers by the amenities afforded by them, so too they tended to discount the flood peril. And as we shall see, there were ways of coping.

Historically, major floods on the Bow had usually occurred in early summer – June and July, more rarely May and August. The greatest flood probably occurred in the summer of 1879, according to the recollections of old-timers and the expanse of land covered in water, but there were no records of river flow kept at that time. On 18 June 1897, however, the town watched in horror as the Bow raged up over its banks, tearing out bridges, uprooting trees, dislodging buildings, and turning large parts of northwest Calgary into a swirling, silt-laden sea.[3] Three inches of rain had fallen in the previous day and a half. The swollen river drove more than sixty families from their homes on the flats southwest of the town. Floods weakened the foundations of the Langevin and the Calgary and Edmonton railway bridges and washed out portions of the CPR main line. Colonel Walker's house fell into the swirling waters and was carried away.[4] The camera faithfully recorded images of a vast turbulent lake, and the newspapers reported breathlessly on the ironies of the earth turned liquid. Judging by the high-water mark on a building close to where a recording gauge was eventually located, the water reached a maximum height of 14.9 feet. That meant that the peak flow probably exceeded 80,000 cubic feet per second (CFS). No subsequent flood in the twentieth century would come even close to this level and volume.

On the 4th of July 1902 the newly installed water level gauge on the Langevin Bridge recorded a height of 12.5 feet at the crest, which translated into a flow of 55,000 CFS. More importantly, it meant another major flood for the city centre. A wet June and a downpour early in July were the prelude to another destructive inundation. The rapid rise of the waters necessitated hurried evacuations and daring rescues from rooftops. This time Colonel Walker lost his barn. Washouts interrupted service on the CPR, and the approaches to the Calgary and Edmonton Railway Bridge were swept away. Once again

Illus. 9.1 Flooding in Calgary, 1902. (Glenbow Archives, NA 1086-1)

streets turned into torrents, basements filled up and had to be pumped out, and mucky debris left behind a huge cleanup job, along with the tiresome business of rebuilding damaged and displaced bridges. By then bridges at Centre Street had been demolished and rebuilt several times, as had the ramshackle Bow Marsh Bridge further upstream. As early as 1912, engineers had begun to get a more accurate notion of the Bow River's behaviour. "History goes to show that the Bow river is subject to very big floods," the chief hydrographer concluded. Nor could he be certain his measurements captured the potential maximum flows. He cautioned his colleagues that "in designing works such as dams and bridges a small amount at least should be added to the greatest known discharge."[5]

During the next major flood, on 26 June 1915, as the crest approached, the city engineer, George W. Craig, and the city commissioner, Garden, ventured out onto the temporary bridge built alongside the new Centre Street Bridge, then under construction. They were apprehensive about damage to their new structure. Suddenly the temporary bridge gave way beneath them, plunging them into the roiling current. Craig managed to cling to one of the fractured bridge timbers until he could be rescued. Garden, however, was swept away. Fortunately, two alert and courageous workmen with a rowboat set off in furious pursuit, snatched the commissioner from the water, and hauled him

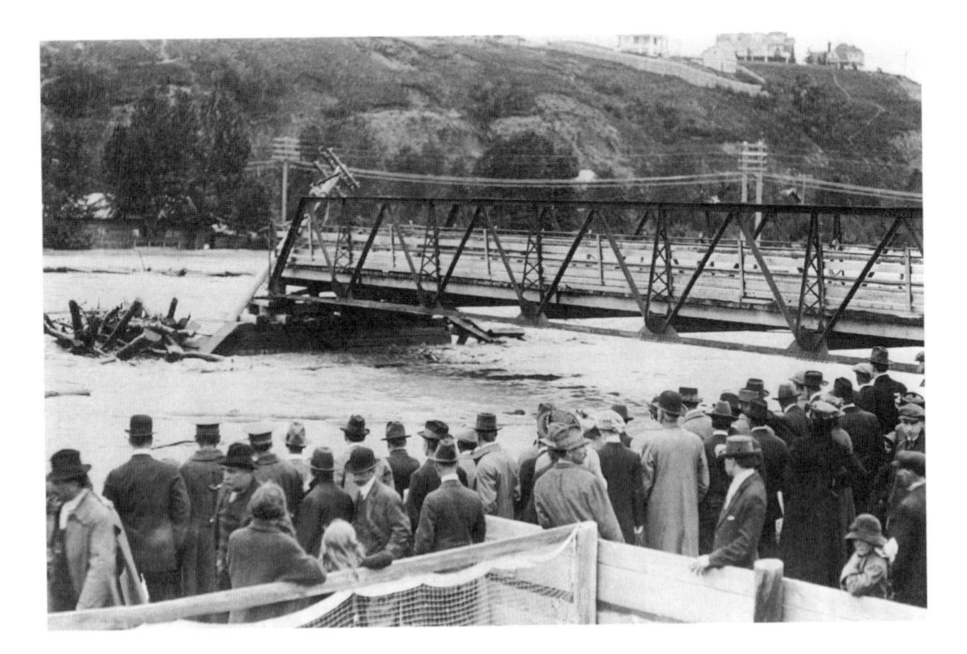

Illus. 9.2 The old Centre Street Bridge partly washed out by the flood of 1915. (Glenbow Archives, NA 671-8)

up onto St. George's Island, about a mile downstream. Despite this mishap, the only partially completed Centre Street Bridge had survived its first big test.[6] Elsewhere three lives were lost in this flood. Extensive property damage occurred in Calgary, and the irrigation weir at Carseland was destroyed.[7]

Close recording of river flow at Calgary began in 1911. From that date it is possible to measure with some degree of accuracy the levels reached by high water and the peak discharge of the river. Figure 9.1 and table 9.1 record the date, height, and flow of the major floods of the twentieth century up to the 1960s. What is clear from this material is that the biggest floods on the Bow River occurred early in the century: in 1902, 1915, 1929, and 1932. There were great differences in the height and flow of these major floods. The 1915 flood, for example, was only half the size of its 1897 counterpart, as measured by estimated peak discharge. In the next group of five floods – 1916, 1923, 1933, 1948, and 1953 – the peak diminished to only one-half the 1915 flow. There is a fairly strong tendency for flood peaks to diminish over time, though discontinuously and allowing for certain exceptions. The floods of 1948, 1953, and 1950 ranked ninth, tenth, and sixteenth in size respectively. By the same token, the floods of 1916 and 1937 were among the smallest recorded.[8]

Even these smaller volumes of water, however, could deliver devastating physical blows as buildings encroached on the flood plain and structures –

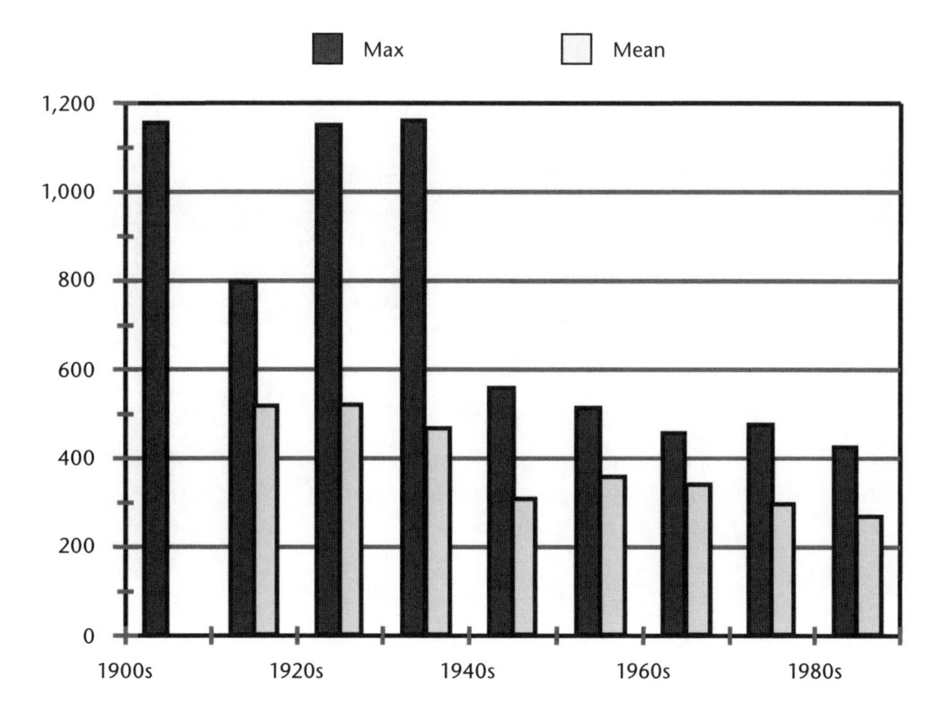

Figure 9.1 Maximum and mean flood volumes at Calgary in cubic metres per second (M3/s) by decade.

SOURCE: Data derived from the CD Rom South Saskatchewan River Basin Historical Natural Flows, 1912–1995, Version 2.02, Water Management Division, Alberta Environment.

dams, weirs, and bridges – were pushed across the river. The rambunctious nature of the river in part determined land values and uses on its banks. Storage yards, often verging on junkyards, warehouses, and light industrial uses predominated. Much of the area along the south bank was occupied by industrial sites, lumberyards, garages, and storage facilities, but on the north bank and in the western suburbs housing had encroached upon the riverbank. The Bow was not the only river running through Calgary. The Elbow River joined the Bow in the centre of the city and crossed a large residential area in its course. It was a much smaller stream, but it shared sources and characteristics with the Bow, flooding regularly in the early summer as well. And as the city grew southward into the Elbow's flood plain, the city was regularly visited by floods from it as well. The years 1923 and 1928 in particular witnessed extensive encroachments into newly built residential districts.

For the families living on the ranches and homesteads along the Bow, floods posed a frightening menace, especially the flash floods that occurred often during the spring breakup when water dammed up behind ice jams. High

Table 9.1 Major twentieth-century floods at Calgary

Year	Date	Gauge height	Peak flow in CFS
1897	18 June	14.9	80,000e
1902	4 July	12.5	55,000e
1932	3 June	12.5	53,600
1929	3 June	11.85	46,700
1915	26 June	11.2	39,800
1923	3 July	10.4	29,700
1916	21 June	10.7	28,600
1933	18 June		23,000e
1948	24 May	9.07	21,000
1953	13 June	8.20	20,400
1928	1 July	8.52	20,300
1927	11 June	8.28	19,900
1918	16 June		19,100e
1920	13 July	8.40	18,200
1934	1 June	7.96	18,100
1950	23 June		17,300
1917	3 June		17,200e
1952	23 June	7.92	17,200
1911	25 June		16,900e
1965	19 June	7.54	16,700
1938	23 June	7.94	16,200
1919	5 August	7.64	16,200
1912	10 July		16,100e
1963	30 June	7.30	15,900
1951	7 July		15,500e
1930	9 June	7.68	15,100
1936	3 June	7.57	15,100
1954	6 July	7.44	15,100
1913	12 June		15,000e
1947	12 June	7.58	15,000

SOURCE: CCA, Engineering and Environmental Services Department Records, series III, Unprocessed, 1970–77, Acc. CR 90-013, Montreal Engineering, *Flood Plain of Bow River in the City of Calgary*, 1: 12.

water flooded William Pearce's estate below Calgary several times. The McKinnons were flooded out in this way in the great flood of June 1897. The young rancher Lachlin McKinnon (J. Angus's father) and his family sought shelter on higher ground with their neighbour Bob Newbolt. They returned, cleaned up, and carried on, only to have history repeat itself the next year. McKinnon noted ruefully in his memoir: "The spring of 1898 arrived and with it another

Illus. 9.3 Flooding on the north bank of the Bow in Calgary, June 1932. (Glenbow Archives, NA 1044-3)

stab in the back by the Bow River.”[9] The sound of flowing water awakened the McKinnon family in the middle of an April night. Ice had jammed in the river, and the current now flowed around the house and outbuildings. Water eventually rose inside the house to above the cookstove. McKinnon drove his cattle up the hill in the dark, but two cords of wood floated away, and he lost two pigs to the river.

This second inundation almost broke his spirit, but he recovered, bought land a little farther downriver, and built a new place on a slightly elevated bench. The river threatened even this location. In 1902, after an unusually long period of rain, the river rose menacingly toward his property. McKinnon kept a team hitched to a wagon to take his family up the hill if need be. But the flood receded before he had to evacuate yet again. On this occasion he lost some fencing but suffered no serious damage. Life on the flats had its advantages for small ranchers, but the benefits came with some considerable liabilities. When the dominant form of agriculture shifted from ranching to wheat farming and it became more convenient to locate up on the prairie closer to the operations and the railway, the McKinnons and their neighbours left their old homes in the valley without much apparent regret.

At Banff, where the river became pinched between steep rock banks, late spring floods occasionally backed up into the parkland in the lower-lying sections of the town and washed out the bridge. But the hotels, business district, and most of the residential sections of the town were high enough to escape danger. In bad flood years the CPR lost some bridges, usually the smaller ones over mountain streams running into the Bow. Sometimes, though, flood waters weakened the bridges across the mainstem and washed out the main line.

The Bow experienced elevated flows up and down most of its length during these weeks in June and July. But it flooded mainly in Calgary. Certainly, the river frequently overflowed its banks, spreading itself over the low-lying lands along its meandering course and inundating forests, fields, and pastureland along its length. But to a certain extent floods are social constructs. It might be said with pardonable exaggeration that floods are caused by houses. They are highly correlated with built structures in towns and cities. In the countryside waters rise, run unimpeded in new courses, even rearrange stream beds in some stretches, but subside with little damage to property. This is not the case in the city. Ranchers, farmers, and indigenous people learned to build their houses and outbuildings on higher ground, frequently overlooking the Bow with a view of the valley but rarely close to the river. Horses, cattle, and wildlife could simply retreat uphill from the bottom lands for the time being. Crossing the river presented difficulties for both humans and animals, but not as much as the thin ice of early winter and spring breakup which made river crossing virtually impossible. Out on the prairie, when the oxbow turned into a lake for a few days, no one took much notice of it. In the foothills, when the raging current collapsed a mudbank and moved gravel and debris to new places, that did not inconvenience anyone very much. But in Calgary, when the river sought out a new course down city streets, floated away lumberyards, filled basements with mud, and swept away bridges, however thrilling the entertainment it might provide to awestruck spectators watching from a safe distance or boating in unfamiliar places, lives and property were in peril. That danger could not be tolerated, especially when it struck residential areas. High water in an urban setting sought a political outlet. Someone or some government must be held responsible.

By mid-century, summer floods had become a thing of the past. Each year the river would rise menacingly, but it rarely overflowed its banks. More often city works crews would be called to pump out basements infiltrated by water seeping through the gravel substrata upon which the city was built. Figure 9.1 charts the decline in both the recorded maximum flows during the decade from earlier in the century and the steady drop-off in mean flood flows per decade. After 1932, June and July brought simply higher levels and greater volumes in the river but no floods. This change inspired a popular belief that in this respect the river had been tamed. Yet the more engineers learned about the hydrology of the river system, the more they believed that such public confidence was misplaced.

Residents did not think of floods in terms of statistics. They had their own ways of recording flood events: memories of buildings being swept away and boats in the streets, lost bridges, drownings, heroic rescues, high-water marks on buildings, land gained and land lost. Thus the precision with which the height of water and volume of stream flow were measured requires an explanation. Who needed such information and who collected it?

Throughout the Western world in the twentieth century the science of hydrology, largely developed in the nineteenth century, was deployed by private companies and public organizations in the interests of navigation, flood control, hydroelectric development, and irrigation.[10] Rivers could only be tamed or used through knowledge of their fundamental behaviour. The basic science had been established by the mid-nineteenth century and standardized instruments developed to provide precise measurements of quantity, velocity, and level. Once the characteristics of rivers were known and their mechanics understood, remedial measures or economic uses could be designed. Typically, the state or agencies of the state underwrote this kind of inventory science in the interests of economic development and public safety.

Systematic hydrometric analysis of the Bow River began at the end of the nineteenth century with the investigations of irrigation possibilities inspired by William Pearce. Department of the Interior engineers considering irrigation works needed to know how much water would be routinely available at various points along the river. Rudimentary surveying by an agency called the Canadian Irrigation Surveys began in 1894. Stream gauges began to be installed to measure height and flow after 1902. Subsequent hydrometric surveys were conducted by the Forestry and Irrigation Branch after 1909, the Irrigation Branch of the Department of the Interior after 1911, and the Dominion Water Power Branch after 1920. Even after the Natural Resources Transfer Act in 1930, the federal government retained responsibility for measuring stream flow on the Bow and other rivers under a co-operative agreement with the province of Alberta. Though the province eventually assumed responsibility for administering its water resources in the 1940s, the federal government's water measurement role continued under various federal-provincial programs, and it does so to this day. Measurement of water by the federal government was thus one of the fundamental underpinnings of irrigation in the early twentieth century. Publication of the information generated by the surveys commenced in 1908 and continued into the 1990s under various titles such as *Surface Water Data* and latterly *Historical Streamflow Summary*.[11]

In addition to irrigation districts, power companies also needed precise measurements of stream flow to locate and design hydroelectric installations. As we have seen, power development on the Bow actually preceded systematic measurement. Had the precise nature of the Bow River been known to the developers, it is quite likely that hydroelectric development of the river would have taken a much different course. As it was, the power company put expensive dams and power plants on a river with a widely varying seasonal flow ill-suited to hydroelectric development. The result was a highly inefficient system. Thus began a decades-long struggle by a company with equipment in place to alter the river through upstream storage to make better use of its capital.

Belatedly, Calgary Power could make use of the data produced by the Department of the Interior hydrometric surveys. Federal government engineers conducted a special investigation of the Bow River in the years 1911–14 to locate additional potential water-power sites and investigate upstream storage possibilities.[12] As part of a special effort to publicize the hydroelectric potential of Canada, the federal government, under the auspices of the Commission of Conservation, surveyed the rivers of Canada, particularly those in the west.[13] From the 1920s onward, routinely generated data from automatic recording machines led to annual publications providing a lengthening picture of the hydrometric characteristics of the Bow River. These data gathered by government helped Calgary Power adjust its facilities, make strategic decisions regarding capital investment, and build the case for enhanced storage in the national park. Data were collected at public expense with intended uses in mind, irrigation and hydroelectric development primarily. The numbers were not neutral.

Beyond the needs of irrigation and power companies, federal officials also pursued a range of assessment activities in the hope of forecasting flood threats. The flood of 1915 served as the major catalyst. Under the leadership of the Meteorological Service, Water Power Branch engineers planted gauges above Calgary and determined levels that would indicate an impending flood downstream. On the basis of this monitoring, provincial and city officials would gain a jump on looming floods and prepare citizens and businesses for the worst. Although rapid runoff in the foothills was seen as the primary risk for summer floods, the Meteorological Service also spearheaded a new snow survey in the vicinity of Lake Louise to provide further predictive capacity. In time, it was assumed, a serial record of flood levels and snow conditions would develop to provide a stronger predictive capacity. By the late 1930s, four separate snow courses had been identified in the upper valley with ten sampling points located within each. Thus alongside the hydrometric surveys spurred by irrigation and power demands emerged an increasingly complex calculation of the river as a dynamic and flood-prone stream that needed to be measured to be understood and anticipated.[14]

What had caused the great floods of earlier in the century? Local residents held their own views, based upon personal experience and casual observation. They associated floods with heavy rains. It was widely believed that the storage dams constructed in the mountains to spread out the flow of the river in order to make the power stations more efficient over the entire year had an incidental beneficial effect of controlling floods. Engineers working for the power company, the irrigation districts, and the federal government came to a different conclusion.[15] Gradually, over the first half of the twentieth century, they began to piece together a picture of the dynamic hydrology of the Bow River watershed epitomized by the mid-1960s report on floods by the Montreal Engineering Company for the city of Calgary. They concluded that most of the water regularly flowing in the river came from melting snow high up in the mountains. But floods, it turned out upon close examination, were not caused by rapid melts; they are extremely rare in the mountain sections of the river. Meltwater certainly increased the flow in the river, but it did not cause the floods. As engineers gathered and examined stream-flow data, it became clear to them that flood events in Calgary did not coincide with peak flows moving down from streams above Canmore. Historically, stream flows in the mountain sections, draining some 1,500 square miles of the basin, were only beginning to rise when the flood crest had already passed through Calgary. Engineers determined that downstream floods were more closely associated with atmospheric events and hydrological conditions in the foothills. Rain rather than snow was the culprit. On those rare occasions when warm, moisture-laden air masses were pushed up from the Caribbean and met colder airstreams flowing east off the Rockies, heavy rains resulted on the foothills. High monthly rainfall was not in itself sufficient to cause a flood. What seemed to matter most was the degree of saturation of the ground by previous storms and the concentration of subsequent rainfall in a short period. Monthly rains totalling more than 9 inches in 1899, 1927, and 1954, for example, did not result in floods. The years of notable floods – 1915, 1929, and 1932 – recorded only 4 inches of rain or less in a month. It was the distribution of that rain that made all the difference.

During the flood of 1932 a complete system of stream-flow measurement was in place on the river. As a result, engineers were able to identify precisely the source of the flood waters. Late May had been quite wet. Heavy rains on 1 and 2 June dropped 2.6 inches on Banff, 5.3 inches on the foothills around Pekisko, and 3.3 inches on Calgary itself. The key measurement is the runoff in cubic feet per second per square mile of drainage. Some terrain held its rainfall much better than others. At some altitudes, for example, the rain became snow. Certain geological formations absorbed rainfall more readily. Other formations – rock, most obviously, but saturated porous soils as well – shed water as rapidly as it fell. In 1932, for example, water streamed off the foothills in the Ghost, Elbow, and Jumpingpound basins, but in the geographically much larger mountainous drainage areas on the Bow system, runoff was four times less. Moreover, the peak flows in the Banff region occurred a day *after* the flood had already hit Calgary. On the Kananaskis River, draining the eastern face of the front range, a transitional zone between the mountains and the foothills, runoff rates were about halfway between those of the mountains and of the foothills.[16]

The comparative absence of these meteorological and hydrological circumstances in the foothills regions is the primary explanation for the lack of a major flood in Calgary after 1932. Why this should be so has yet to be satisfactorily explained. But these data do show that human intervention had attenuated flooding only to a very small extent. The development of extensive upstream hydroelectric storage works did not play a significant role in flood control. The storage reservoirs held water from a huge area, but it was not an area from which high rates of runoff in CFS per square mile were usually recorded. The reservoirs were all fairly high up in the mountains. The source of major flooding occurred below these storage buffers. On the other hand, the placement of the Ghost dam, farther down the river and with reasonably large ponding capacity, meant that under certain circumstances that dam could absorb a good portion of the flood shock. In the 1920s the president of Calgary Power had often justified construction of this dam as a flood control measure. He recommended a similar dam on the Elbow for the same reasons. Note that in the 1932 flood, 44,000 CFS swept into the Ghost reservoir, but the engineers in control were able to hold back some of this volume, releasing only 38,000 CFS at the point of peak flow. The Ghost dam thus shaved a sizable proportion (14 per cent approximately) off the flood crest as it swept down the river. During the early 1930s the city of Calgary had begun construction of a dam on the Elbow to create a new water supply for the city. This dam was nearing completion in the spring of 1932; its reservoir had not yet begun to fill. When the Elbow went into flood, city waterworks personnel closed the gates, holding back more than half the stream flow. Whereas more than 25,000 CFS flooded into the reservoir, only 11,300 CFS spilled into the lower reaches of the river winding through Calgary to its confluence with the Bow. No flooding occurred in the residential districts. The Glenmore passed its first test as a flood control device. With a fuller reservoir, of course, the dam would be less effective.

Dams just upstream from Calgary – the Ghost, the Glenmore, and in the 1950s the Bearspaw – acted as shock absorbers when heavy runoff occurred in the foothills. The dams farther upstream – the storage reservoirs, in particular – were less of a factor. But the effectiveness of the dams depended in large measure on the extent to which they were empty or lower at the start of the flood cycle. After the early 1930s the Ghost dam on the mainstem of the Bow and the Glenmore dam on the Elbow could hold back some of the summer peak flow under certain conditions, but not indefinitely. The water finally had to be released cautiously without aggravating the situation. In the 1950s the Bearspaw dam just upstream from Calgary added a second buffer. But in the popular mind the upstream dams and storage reservoirs needed to make the river suitable for year round hydroelectric production had also brought summer floods under control. That was an interference with nature people could accept.

By the 1950s, then, summer floods were but unpleasant memories for residents of Calgary; perhaps they were gone forever. Yet the engineers worried about this complacency. The main cause of the decline, they concluded, was the relative absence of the conditions that created floods: concentrated periods of rainfall on previously saturated foothills watersheds, not the dams. These conditions could return, though no one could predict when or with what frequency in the future. A flood of the magnitude of 1879 or 1897 might recur perhaps once a century, and for floods of this magnitude, the city was ill-prepared and the dams would provide little aid or comfort. That such extreme events could occur was dramatically illustrated by the major flood that struck Fish Creek in the summer of 2004. Nevertheless, the residents of Calgary remained largely unmoved by the spectre of a hundred-year summer flood that might someday happen. They were, rather, more agitated about the distressing recent phenomenon of repeated *winter* flooding.

The Challenge to Expert Authority

Almost everyone had a theory about the unwelcome appearance of winter floods. Twelve times between 1938 and 1952 ice jams in the Bow flooded some part of Calgary or its suburbs. What was most puzzling – these floods occurred when the flow in the river was at its lowest levels. Early in winter, after several days of sub-zero temperatures, slushy ice floes would begin to appear on the open river. At bends and narrows, gravel bars, bridge abutments, and especially at the old Eau Claire weir as long as it stayed standing, floating ice would congregate and then congeal, forming a mass that eventually covered the river. Hardening into great chunks in the cold air, this ice pack would then build upstream as more slushy ice descended with the current, adding to the jam. Eventually the ice pack would compress and telescope in upon itself, hurling great plates up the banks and also forcing ice downward into the flow, partially damming the river. At that point water would begin to back up behind this ice dam, sometimes overflowing the riverbanks locally. The flood usually did not last long because the increased velocity of water forcing its way under and around this ice would eventually cut a way through the obstruction, releasing the pent-up waters.

When winter flood waters inundated the working-class Hillhurst district on 15 November 1945, damaging 220 houses, collapsing cellar walls, and filling basements with muddy water and ice, the angry residents demanded an explanation, compensation, and a remedy. The city of Calgary, the level of government on the front line, came under tremendous pressure to do something for poor, hard-working, and completely innocent ratepayers. It pumped out basements, grudgingly doled out $5,000 to help with repairs, and pleaded with the provincial government for assistance,[17] but most substantially, it set up a commission of engineers from the city, the Calgary Power Company, and provincial and federal governments to find out what was going on and recommend measures to prevent it happening in the future.

The Hillhurst Flood Report, issued in May of 1946, provided the first public analysis of the mechanics of ice formation on the river. The committee recommended the building of dikes to protect low-lying areas from surface flooding, the removal of certain natural and artificial obstructions in the bed of the river, and the maintenance of an even stream flow in the Bow below the Ghost dam. Ice-jamming through the Calgary reach could be reliably prevented, the committee believed, with the construction of a dam just upstream of Calgary at Bearspaw. This structure would trap the descending slush ice and at the same time release relatively warm water, which would deter ice formation through Calgary, delaying until farther downstream the accumulation of slush, where the resulting ice jams would overflow harmlessly into range-bottom land.[18] This report had the virtue of proposing some constructive, if expensive, remedies. But there were no guarantees that even with all of these preventive measures, ice flooding might not recur under certain circumstances. Seepage from high water, even if flood did not occur, would continue to infiltrate basements situated in the gravelly bottom lands adjacent to the river. The Hillhurst Flood Report largely ducked the most controversial questions: responsibility for the floods and the influence of varying the stream flow on ice formation.

The victims in low-lying regions "knew" the cause of these ice jams with a bitter certainty that only icy water regularly flowing through your basement can bring. It was the power company's fault. As the flooding became a regular occurrence, popular suspicion hardened into conviction. Calgary Power had been fiddling with the flow of the river. Opinion divided as to whether the relatively higher winter flow resulting from increased upstream storage or whether something more sinister caused the ice that backed the water up: sudden changes in water levels as the company turned its hydroelectric stations on and off to meet its peak power requirements. Either way, according to conventional wisdom, the power company was to blame for this new plague of winter floods.[19]

Calgary Power denied the charges, but it lacked a convincing counter-argument. As a local enterprise, it would do what it could to alleviate the situation without acknowledging any liability. In 1948 it entered a cost-sharing agreement with the city to raise dikes protecting some of the most vulnerable reaches.[20] Uncharacteristically for a confident corporation that normally had all the answers, it had to admit that it did not know very much about ice formation or even if the flood pattern of the river had really changed. It had begun serious study of the matter only in 1941; it had called in the leading experts in the world, but it still could not satisfactorily explain the flooding. It was learning along with the public. The company "knew," however, with a determination akin to a survival instinct, that modification in the flow of the river for power purposes was not the cause of the flooding.

Federal officials tended to regard ice jams as a perfectly natural phenomenon that had occurred regularly in the past. P.M. Sauder, the director of Water Resources, advised his superior in response to public complaints: "Ice jams have occurred on the Bow River above Calgary for many years, but when the adjoining land was only used for grazing, nobody took much notice of the flooding." He blamed the victims: "With plenty of land available, there is no reason why people should build so close to the river. I fear, however, that these people did not check up sufficiently before building."[21] Federal officials, guided by power company engineers, denied that upstream storage under licences it had granted bore any responsibility for the flooding. Rather, the problem arose from people living in the path of naturally occurring floods.[22]

Officials from the provincial government publicly maintained a similar facade of tactical skepticism. Ben Russell, the civil servant in charge of the Water Resources Branch of the provincial government, explained to his minister that while most people were convinced the ice jams occurred because of the increased winter flow from storage reservoirs for hydroelectric production and the fluctuation in the flow below the Ghost dam, as a hydrologist, he was not so certain. It was "not known to what degree these influences work to produce ice jams."[23] He held out some hope for some ice traps his department had devised, piers of logs and stone built in the shallows where, it was hoped, ice would be artificially trapped before it got to Calgary, harmlessly flooding vacant bottomland upstream. Addressing the problem, and even helping the flood victims, was further complicated by the spectre of legal liability. The town of Bowness and the city of Calgary, out of pocket for emergency relief measures, demanded that the provincial government assume some of the burden. The power company was, after all, operating under provincial licences.[24] The provincial government, not wishing to open itself to unlimited liability, brushed off these pleas for help in much the same way as the Bow River in the winter of 1948 had tossed its flimsy ice catchers aside.[25] No one would take responsibility. Meanwhile, each winter somewhere along its urban corridor the river would once again rear up a wall of ice, back up, and wander around for a short time outside its banks.

No one seemed to have convincing answers or know what to do to solve the problem, except to build higher dikes. Suspicion, rumours, and folk wisdom rushed in to fill the momentary vacuum created by the reticence of engineers, the stubborn agnosticism of the company, and the immobility of governments spooked by liability. Autodidacts suddenly knew as much about ice as engineers and bureaucrats. City officials, who patrolled the river watching helplessly as ice locked itself into jamming position, were suddenly inundated with suggestions from long-time river watchers. Thomas Stanley, an auto mechanic, proposed a giant V-shaped wedge be installed upstream to push the ice into empty fields like a plough. Christian Teske, a farmer, was one of many who proposed an ice-catching contraption be anchored in the river, "something like a fishnet, only on a bigger scale."[26] But there were more fanciful ideas. Lorne Saunders devised a scheme of "electrically controlled bombs" that could be floated under the ice, blowing it up before it jammed. And if the engineers scoffed at that idea, Saunders proffered an even better solution: melt the ice by setting the river on fire with underwater gas jets! In the postwar era, helpful residents thought that the artillery should be brought in to fire at the ice. In frustration a city engineer once startled the local RCAF base commander with a request that he prepare his aircraft to bomb a dangerous ice formation near St George's Island in the centre of the city.[27]

Convinced, like many others, that the variability of the flow in the river caused the ice to jam up, Olive Clare proposed a much less drastic solution than a flaming river or an aerial attack – power rationing. "I would infinitely rather sit in the evening by candle light," she told the mayor, "than be forced to continue in this present anxiety, and – if a flood comes – to be put to the additional expense of going somewhere else to live until the danger is over." Would it not be better for the city to use less power, thus reducing the flow in the river? Rationing would prepare people for greater emergencies, such as war.

Illus. 9.4 Ice buildup on the Bow River below the Langevin Bridge in Calgary, 1956. (Glenbow Archives, NA 5600-6641a)

"People in Britain have their electricity cut two hours a day," she reminded the mayor. "Surely Calgary business folk and citizens would not object to a little inconvenience for the sake of those of us in the threatened districts, living in this constant anxiety."[28] Here we can begin to sense the subtle social cleavages aggravated by the floods. Working people in inexpensive housing on low-lying land saw themselves the victims of better-off commercial and professional classes whose lights blazed confidently on higher ground. All social classes should, it was argued, share the burden. Reducing electricity consumption and thereby reducing the flow in the river seemed more equitable than periodically flooding poor people. Some people thought that winter power ought to be produced from coal instead. Dikes were not an entirely satisfactory solution either. Quite apart from the continuing seepage and backed-up storm sewers, those people with lots fronting on the river objected to a construction of huge walls of earth that cut off their view and lowered their property values. Surely it was better to address the root cause of the flooding, the surging flow induced by hydroelectric generation, than to build ugly barriers.[29]

Though the professional engineers withheld judgment on the impact of upstream storage and variable flow, citizens of Calgary and especially people who had lived near the river all their lives remained convinced hydroelectric releases were the culprit. In January and February of 1948, December of 1948, and then again in November-December of 1950 the Bow flooded. The Hillhurst Flood Committee had now morphed into a Permanent Flood Committee, advising the city on the removal of gravel bars and obstructions on the river.[30] In 1951–52, with the promise of the Spray Lakes storage development coming on line to augment winter flow even further, ice-jamming on the Bow all through the Christmas and New Year season flooded Lowery Gardens, Bowness, Hillhurst, and Sunnyside yet again. On this occasion the city buzzed with a rumour that someone in the know at the Ghost dam had warned friends in the city to watch out because the company was going to "flush out" the reservoir.[31]

It was a dangerous time. With the city demanding that the province be responsible for both flood damage and flood protection and with suspicion mounting that the actions of Calgary Power were really behind all of this, the province finally moved. To pre-empt passage of a city resolution formally blaming the province, the provincial government hastily convened a royal commission to investigate the causes of the Bow River floods and to propose remedies.[32] It was a measure that gave the appearance of action and bought some time. At the same time a royal commission might possibly condition public opinion, apply pressure on all parties to be more co-operative, and in general create a more accommodating alignment of public and private forces to avoid the bugbear of legal responsibility.

From 22 April 22 to 6 June 1952 the three-member royal commission – W.J. Dick, a professional engineer in the chair, D.W. Hayes, a retired engineer, and Angus McKinnon, a rancher from Dalemead southeast of Calgary – presided over a formal examination of witnesses and gathering of evidence in the Calgary Courthouse. J.C. Mahaffy acted as counsel for the commission, city solicitor E.M. Bredin for Calgary, and H.G. Nolan for Calgary Power. In this solemn judicial setting, lawyers for the city of Calgary and for the municipality of Bowness attempted to prove to the commissioners that the phenomenon of winter flooding was something quite new, that it was related to higher volumes of winter flow and open water generated by upstream storage and particularly to the rapid rising of the river when the Ghost dam opened up to peak hydroelectric production. The municipalities thus gave full legal voice to the popular conviction that power production caused the flooding. Residents argued from personal experience that in the old days the river, with lower levels of flow, used to ice over with board or sheet ice quite early and stay that way for most of the winter. With the increased flows of recent years, the river rarely froze over but frequently flooded on account of floating slush.

Thus hours of testimony were taken up with old-timers recollecting how they had once been able to skate in places where the water now remained open all winter. The sheer volume of this evidence showed that people had been paying very close attention to changes in the river over the years, its icing patterns, its changing volume, and its modulation of flow.[33] J. Ivor Strong, Calgary's city engineer, who presented a modest municipal brief largely describing recent floods, suggested that flooding during a low-volume flow in winter seemed to be more frequent and that a catchment dam at the Bearspaw site looked like the most promising means of addressing the issue. The city plainly lacked the technical capability of mounting a rigorous argument based upon systematic analysis of data and scientific research.[34]

Calgary Power, meanwhile, had been doing its homework and certainly was not caught off guard by the suddenly convened royal commission. While it was under attack from several quarters during the hearings, the company actually regained a good deal of lost ground as a result of the thoroughness of its presentation and the disciplined testimony of its expert witnesses. Calgary Power presented a massive submission to the commission in which it described its system and operations, itemized an extensive history of winter flooding from 1892 to the present, explained what it had learned about the science of ice formation in mountainous rivers in a cold climate, and suggested some possible remedies. This extensively documented brief set out to show that whereas company regulation of the river had been responsible for reducing summer floods, ice-jamming in winter was, in contrast, an entirely natural phenomenon, made apparently more noticeable recently as a result of the encroachment of buildings into the flood plain. Winter floods were just as numerous and, if anything, slightly less violent than in the past. Why, then, the widespread belief that the behaviour of the river had changed? Here the brief tiptoed delicately: "the incidence of winter flooding depends on the presence or absence of dwellings in close proximity to the river." As a result of recent urban sprawl, "high backwaters at the head of the icepacks no longer pass unnoticed, and reference to winter flooding became more apparent." Responsibility for the floods thus rested with those who allowed houses to be built in the path of such floods. Nature was simply taking its course: "the production of ice and winter flooding is fundamentally the result of low temperatures, and is a phenomenon found on many rivers in northern latitudes regardless of whether they were subject to man made regulation or not."[35]

Gradually the company had begun to understand the science of ice. There were no mysteries about the formation of board ice on still bodies of water. However, the dynamic formation of ice in rapidly flowing bodies of water had not been much studied until relatively recently. Research from Russia, the

United States, the Montreal Harbour Commission, and the company's own engineering work on the Bow formed the basis of its technical understanding, which in turn was a key element of its legal defence. It turned out there was more than one kind of ice. In shallows or slow-moving portions of the river, a sheet of ice formed right across the surface in the traditional and well-understood fashion, effectively insulating the water below. This was the ice that could be skated on. However, in extremely cold weather, in fast running open water, another kind of ice formed on the bottom of the river, especially during the extra cold nights. This anchor ice, as it grew up from the bottom, raised surface water levels and sometimes pushed the river over its banks. Company data showed how from 6 PM at night until midday, water levels in the Bow River rose with the formation of this bottom-growing ice. In the daytime this anchor ice was released as the sun raised water temperatures. It then bobbed to the surface, often carrying gravel, boulders, and debris with it. On the surface it mingled with another kind of ice, variously called frazil or frazzle ice. This kind of ice, formed when super-cold air lowered the surface temperature of the water, created ice crystals that coalesced into slush.

This combination of frazil ice and loosened anchor ice was carried downstream until it caught on some obstruction. Sometimes it built outward from the sheets of board ice formed in the slower-moving water along the bank. Sometimes ice formed where the water slowed down in deeper pools or excavations. More often it gathered around obstacles in the flow – boulders, a bend in the river, a gravel bar, a narrows around an island, or man-made structures such as bridges and weirs. In these places this amalgam of different kinds of ice began to build up and bridged the entire channel, leading to the telescoping, jamming, and often flooding already described. This phenomenon happened almost every winter somewhere along the Bow, the Calgary Power brief demonstrated, but it was not noticed because it occurred in the empty flats and bottom lands of the foothills. When a river overflowed onto empty rural land, that was not considered a flood. The recent unchecked spread of housing into low-lying areas along the river was, the company implied, the "cause" of the flooding.

The company also produced evidence to show how dams influenced the formation of this kind of ice. A dam obviously collected upstream ice in its ponds or forebays preventing it from passing downstream. But that was only part of the story. Board ice covering this slack water and the depth of the reservoir preserved the heat of the water. The water released below a dam was thus relatively warm – up to 34 degrees Fahrenheit – depending upon the depth of the reservoir and the level at which water was released below the dam. Flowing rapidly below the dam, this water slowly surrendered its heat, during which time the flow remained more or less ice-free. In the case of the Ghost dam, the water reached 32 degrees Fahrenheit within a distance of two to three miles downstream. The ice-free distance below a dam was, then, a function of the initial water temperature, the slope of the riverbed, and the air temperature. All other things being equal, a dam could actually "clear" a reach of river for several miles. But this pattern was by no means certain. That is why some engineers thought the only way to prevent ice jams was to line the river with dams so that the reservoir of the downstream dam backed up right to the footings of the next one upstream. That way board ice would form quickly and insulated the water. This was not a practical or particularly helpful suggestion. Under extreme conditions, frazil and anchor ice could form more quickly and, despite the engineering works on ice jams, could recur without warning. In this respect the Calgary Power brief in 1952 corroborated the findings of the Hillhurst Flood Committee in 1946.

On the one hand, these observations supported the arguments of the city and other critics that a dam upstream in the Bearspaw vicinity would likely have a beneficial effect. It would trap the ice descending, but its shallow reservoir would not have as much influence over exit water temperature as the Ghost. The company considered a dam at Bearspaw at best a marginal hydroelectric proposition; it was thus predisposed to believe "that even though the slush is held back at Bearspaw there will still be an ice pack formed downstream and that it will cause backwaters almost as high as those presently experienced. Hence the Bearspaw project would not provide a complete solution to the problem of winter flooding." Calgary Power therefore preferred an expanded program of diking instead.[36] The provincial government, in its largely innocuous brief, supported this view. Ben Russell, ever cautious, would only go this far: "there are so many contributing factors to the present conditions along the Bow River through Calgary, some of which are not yet even known, that one cannot be too confident that even a dam or an ice trap at the Bearspaw Site will be a permanent solution."[37]

But what of the public concern about the variability of the flow as a cause of flooding? The company readily admitted that it had significantly altered the flow of the river, and that with the coming on line of the Spray system, there would be an even greater storage effect. That, the company argued, had proven beneficial in shaving the peaks off summer floods. And there was no evidence that these higher winter flows had any influence whatsoever over the incidence or frequency of ice jams. The company readily understood that variation in the flow could cause alarm and might be thought to have some impact of ice jams. When the Ghost turbines cut in at 7 AM and ran full bore to 9 AM, that sent a wave of slushy water downstream which reached Calgary about 5 PM. "It was an unusual sight and it was natural that public opinion should associate it with the ice jamming phenomenon." However, the previous winter the company had run an experiment maintaining a constant flow in the river which disproved any association between variability and ice formation. Indeed, company engineers and other observers concluded that only slightly more ice formed with constant flow than with variable flow. Moreover, the company argued, uniform flow did not exist in a state of nature, where river flows change constantly with chinooks, the formation of anchor ice, and natural obstructions and surges of flow.[38]

In cross-examining the company's expert witnesses, lawyers for the commission and the municipalities tried in vain to get the company engineers to admit that there was a relationship between ice jams and company management of the river's flow. To no avail. J.K. Sexton, Theodore Shulte, and Thomas Stanley would only admit that it appeared winter flooding was becoming more frequent; that, yes, the company had changed the flow of the river; but that there was no proven relationship between flow modification and flooding.[39] On this point they were joined by the other professional engineers who testified: Ben Russell for the province and Owen Hoover, district engineer for the Dominion Water Resources Branch.

The rumour about "flushing" the Ghost reservoir would not go away, however. So the commission tried to get to the bottom of it. The story, as it turned out, had a complex genealogy. It had started with a letter from the Bowness activist Leonard Cooper, who had informed the mayor. Cooper had heard the story from a Mr Nesbitt, who had got it from his wife, who, when visiting the hospital, had heard from a woman friend that her daughter's boyfriend had been told by a Calgary Power employee "that Bowness had better 'look out' as they were going to 'flush' the dam at the Ghost on the night of December 30th, 1951."[40] Was there any truth to this story, the commission wanted to know? The company denied it but in the last week of testimony admitted that stop logs had been removed from the Ghost dam on the 27th of December and the reservoir had been lowered to repair a "rock blowout." The company engineer would only stubbornly insist: "I would not say it caused the flood. We let the water down. It may or may not have caused the floods down there."[41] So the rumour had some substance.

The royal commissioners composed their report with commendable speed,[42] and one of them, the rancher Angus McKinnon, also delivered himself of a mock epic poem chronicling the proceedings, which, mercifully, has remained unpublished.[43] The commissioners accepted the necessity of adjusting the flow of the Bow to make it an effective hydroelectric river. Commenting on the storage system, they observed: "The whole is a well conceived plan and if the Bow River had been left to its natural unregulated flow the whole of the industrial and domestic power outlook of Alberta would have been changed." As flooding had occurred both before and after the era of regulation, "the real problem still remains to be solved, namely 'How can winter flooding be prevented or reduced to such an extent to eliminate danger to life and property?'" Relying on the evidence of the Permanent Flood Committee report of January 1952, the commission rejected the most frequently proffered panacea, uniform flow: "The backwaters resulting from the uniform flows of 1951–52 were as high as the backwater stages observed during the fluctuating flows of the previous years. It appears that regulation from the Ghost plant is not a solution to the winter flood problem on account of its distance from Calgary and the many points in between where incipient ice jams may occur."

In its main recommendation, the commission supported construction of the Bearspaw dam. It fit into the overall system development plan of Calgary Power. It would reduce the amount of open water within which ice might form above Calgary from the current forty miles to ten. It would thus serve as an effective ice trap, and under normal circumstances, it ought to keep the river free of ice for from five to eight miles. Still, it was not a perfect solution. Under extreme conditions, ice could build up in this ten-mile reach, producing floods in Calgary. For that reason, the commission recommended a continuation of the dike program and the expropriation of those houses in Lowery Gardens and other low-lying areas exposed to extreme hazard. Even with dikes and Bearspaw, seepage would in all likelihood continue in some parts of the city together with ice buildup even if flooding did not occur. The commission proposed that a permanent control board ought to be established to continue the work of the former Permanent Flood Committee, advising governments on river improvements, the removal of obstructions, and the best methods to deal with seepage.[44] It was, in all, a good few months' work. The air had been cleared of suspicion, an impartial judgment rendered, and a clear course of action proposed. And professional authority based upon scientific analysis had reasserted primacy over lay knowledge.

The provincial government now faced the urgent necessity of ensuring that the Bearspaw project was built. In this it showed an almost devious cleverness. During the frantic campaign in the late 1940s to get on with construction of the Spray Lakes hydroelectric storage reservoir to meet looming power shortages, the province had in the end thrown its weight behind the power company's plan. Rather than go ahead with a provincially owned project, it had decided that it would be preferable for the company to proceed under provincial regulation. Provincial influence had been essential in persuading the federal government to complete the undertaking to carve the Spray Lakes district out of

Banff National Park. There was, it turned out, a little noticed quid pro quo. The provincial negotiators, mindful of the growing body of evidence about flood conditions on the Bow, had insisted upon inserting a clause that tied the Spray Lakes development to a requirement to build a dam at Bearspaw for hydro-electric and ice-control purposes should conditions warrant. In effect, the government in 1949 had made all future upstream hydroelectric development conditional upon the future construction of the Bearspaw project. In May of 1951, well before the royal commission, the government of Alberta reminded the power company of this clause when it refused the company permission to proceed with improvements to its Kananaskis storage facilities.[45] In September of 1952 the minister of agriculture wrote to the chairman of the power company more or less ordering construction: "Upon reviewing the situation to date and considering all of the conditions and provisions of the agreement and having referred the matter to a Royal Commission for a report, I am now of the opinion that the company should immediately undertake the construction of the Bearspaw Development."[46] The company wriggled and squirmed, and a public announcement had to be made before the company fell wholly into line, but in the end the Calgary Power agreed to build the dam. Thus through this clever bit of foresight and canny use of its political power, the provincial government managed to have this important flood-control project built without the expenditure of public money and with remarkable speed soon after the royal commission report.

The province then put up $25,000 to help the city of Calgary raise the dikes to the recommended 12½ feet. It helped financially with the removal of the old Eau Claire weir, dredging, and the removal of boulders and gravel bars from the river. It also provided $85,000 to facilitate the expropriation of properties in Lowery Gardens.[47] As predicted by the engineers, seepage problems continued each winter in the Hillhurst district.[48] Engineers remained skeptical, too, of the value of the dredging, diking, and weir removal projects. These were not solutions to the problem; they simply moved the flood to another location farther downstream. That, of course, was the point. It was not an elegant solution, but at least it displaced the high water to places where there were fewer houses.

The combination of the Bearspaw dam and channel improvement through the Calgary reach seemed, on the whole, to address the issue of winter flooding satisfactorily. With the Spray Lakes storage system fully operational by 1953, and with the addition of a third generator to the Ghost power station, Calgary Power was able to further increase the wintertime flow of the Bow River. In the 1930s, mean December stream flow averaged 30 cubic metres per second at Calgary. During the 1940s Calgary Power was able, with expanded

Minnewanka storage, to increase that to 44 cubic metres per second. With the Spray Lakes project fully operational, December stream flow averaged 61 cubic metres per second.[49]

December stream flow had effectively doubled in twenty years. And the vexing phenomenon of ice-jamming in the Calgary reach of the river gradually diminished. As the 1950s proceeded, the Bow River ice flood files of the provincial government and the municipal government begin to get thinner each year; they disappear entirely in the 1960s. On the closely watched river, winter flooding became a fading memory. The Bearspaw dam also provided another shock absorber in the river above Calgary to attenuate summer floods. With these additional works, the city seemed to be able to enjoy greater wintertime hydroelectric production from the river while escaping the most extreme floods from ice jams in the very centre of the city. Whether this change could be credited to this engineering solution or, rather – as with summer flooding – was the result of the relative absence in subsequent years of the frigid atmospheric conditions that created the ice jams is by no means clear. The ice could still rear up in an extremely cold winter, and the hundred-year flood still threatened but only in the minds of engineers.

Popular Assessment of Risk

By the mid-1960s Calgary had radically rethought its riverfront after a traumatic civic convulsion over the proposed relocation of the CPR tracks and an expressway along the south riverbank. After that plan had been rejected, the city with public support converted the riverfront from a decaying industrial eyesore into a park, a matter dealt with in more detail in a subsequent chapter. Planners and elites could not plausibly have envisioned beautiful parks along the bank if the river did not also co-operate by assuming a compliant character. By the late sixties it had seemingly given up its unruly nature. Summer floods were a thing of distant memory thanks, it was thought, to the Calgary Power Company's extensive system of hydroelectric storage reservoirs. At the same time the dikes and the new Bearspaw dam just upstream from Calgary, built in the late fifties, seemed to have solved the problem of winter flooding as a result of ice-jamming. The river was no longer a destructive force in the community. Tamed, it flowed gently between its banks, which, in the absence of repeated flooding, had become heavily treed. One could go down to the river without fear or foreboding and invest in expensive landscaping on the assumption that it would not all be swept away.[50]

Ironically, just as the public had begun to take the river for granted and think of it as an adornment, the city administration had been forced to take the possibility of flood disaster seriously and take action to minimize its liability in that event. Public reaction to municipal efforts to cope with potential floods indicates the extent to which both the placid river and the river park ideas had penetrated public consciousness, even in the face of the opinion of technical experts.

In the late 1960s, to ensure public safety and minimize public liability in the event of foreseeable disasters, the city of Calgary commissioned the Montreal Engineering Company to study the possibility of future floods and make recommendations to deal with them. In an extraordinary analysis of the hydrology and flood history of the Bow River, the consultants concluded that the possibility of catastrophic floods still existed. Upstream storage had been only a minor factor in flood control. Rather, the weather conditions that produced major floods – extensive wet periods followed by torrential downpours on the foothills – had simply been absent. An old-fashioned flood similar to 1897 of about 100,000 CFS had a return period, the report estimated, of between 100 and 150 years. The river had not been tamed; the frightening implications of this conclusion could be seen in the detailed maps of the areas likely to be inundated by a flood of historic proportions. The city could mitigate the problem, and accordingly reduce its liability for property damage, by clearing a flood plain, dredging the riverbed to remove obstacles, raising dikes, and introducing zoning changes that removed vulnerable buildings from the path. In one sense the Montreal Engineering flood study provided yet another argument for a river park. Only municipal parkland with no substantial built structures on it would be at direct risk in the event of a catastrophic flood. In response to this report, the city accelerated its riverbank acquisition program, renewed its efforts to widen and deepen the channel, and stabilized the banks of the river through the city. It also continued its riverbank beautification project, knowing that it had potential flood mitigation possibilities as well.[51]

Municipal attempts to act responsibly on expert engineering advice in the face of the likely risk of flooding therefore worked to reinforce the river park project until in 1973, in an effort to clear the channel of obstruction and open up the flood plain, city works crews cut down more than a hundred trees encroaching on the bed of the river near St George's Island. Then all hell broke loose. The hue and cry from the neighbourhoods echoed throughout the city. On an otherwise bald prairie, the river valley was the only treed area in the city. In an era of participatory democracy, citizens' groups demanded an equal say along with technocrats and bureaucrats in what could be done to *their* river. When confronted with the technical data about flood threats, many ratepayers' groups became alarmed at the maps of huge swaths of urban real estate declared to be in the putative flood plain and began worrying, not about remote floods, but rather about sinking property values.

Reeling from the uproar, the city and the province jointly commissioned a Calgary Bow River Study Committee, amply stocked with citizens' representatives, to make recommendations about what ought to be done in response to the Montreal Engineering Company report. The committee quickly recommended that the aggressive flood plain management plan proposed by the consultants not be implemented. Public meetings unanimously condemned a program that involved indiscriminate clearing of the floodway and rezoning of threatened districts. Popular opinion trumped expert advice. "Speakers on behalf of communities were emphatic in stating that the residents were prepared to take the risk of any flood damage," the committee reported in January 1974. They rejected intrusive excavations of the river or wholesale rezoning of property for flood control protection in favour of better flood forecasting, upstream flood control projects, and a flood insurance program. There were strict limits to what might be done to the river, even in the name of public safety. Certain aspects of the river could be tinkered with, but the dikes could not be raised to the point of obscuring views, property values threatened by regulations, or trees removed.

A study prepared for the Bow River Study Committee revealed the extent to which the idea of the riverbank as open space park had been firmly fixed in the public mind. "An overriding policy is required that gives priority position to the recreational resources of the Bow River Valley," the Lombard North Group consultants concluded after extensive discussions with community groups and individuals. The city should dedicate undeveloped land to parkland and develop a comprehensive system of parks along the valley floor throughout the metropolitan region. For years, the Calgary Local Council of Women had been a voice in the wilderness on this subject, Frances Winspear pointed out in presenting its brief. Within a decade its view had become the conventional wisdom: "We emphasise and reiterate the statement in this impact study that all city and provincial land owned in the flood area should be retained and dedicated to open land space, with no new buildings on these areas, but kept as future natural and recreational land." The long-term threat of floods, the Local Council of Women argued, need not interfere with the program of beautification it had long advocated.[52] The consultants noted that active public interest in the river was largely confined to communities alongside the river. However, in those districts the inhabitants were willing to accept what appeared to them to be a low probability risk of flooding in order to retain "the natural beauty of the river." The Lombard study emphasized the importance of preserving and enhancing the "visual zones of the Bow River through Calgary" by maintaining vegetation and natural features and rehabilitating "unsightly areas and eyesores."[53]

A follow-up survey conducted a year later reinforced these conclusions. Public opinion strongly supported the idea that the riverbank should be retained for park purposes, but it also insisted that it be cleaned up, landscaped, and improved for recreational purposes. Walking and enjoying the scenery were the two most popular activities that the river park should be designed to enhance. "Regarding future river bank planning, the need exists for the development of a policing body which will ensure more stringent river bank development compatible with the natural river environment," the survey concluded.[54] Afterwards a Calgary River Management Committee, with representatives from the city, the province, and various interest groups, was established to keep an eye on the extensive parks, recreation, floodway, and environmental enhancement projects under way in the metropolitan region to ensure the engineers were held firmly in check.[55]

The Bow River valley in Calgary was to be remade in accordance with a "designer" view of nature that privileged aesthetics and certain kinds of pedal recreation. This reconfiguration of the river and its banks depended in part upon engineering that mitigated the possibility of flooding. The extensive diking to prevent winter floods incidentally cleared a swath of land that could be conceived of as a park. But at the same time the heightened aesthetic values inscribed on the river placed strict limits upon the power of experts over it. A vigilant and aroused citizenry in the riverside neighbourhoods kept watch; an extensive consultative process had established a table of commandments as to what might and might not be done to the river. The urban river had been redesigned both with and against the advice of technical experts, and Calgarians expressed enormous pride in their accomplishment. They had established a new cultural compact with the river that would last, in all likelihood, until the unavoidable hundred-year flood returned.

10 · Building Banff

It is no coincidence that Canada situated its iconic national park where the Bow River performed its most spectacular tricks. In a relatively short span the Bow did all of those things one could ask of a river. It meandered gracefully but swiftly out of its fir-clad valley to pool in the Vermilion Lakes, providing a mirrored surface inverting in precise reflection the surrounding jagged peaks. It then wound its way through grassy meadows and savannah through a picturesque townsite that would grow up to serve visitors. There, forcing its way between cliffs and a sheer rock face, the Bow plunged in a breathtaking cascade, a sublime and terrifying descent seen close up, a pastoral picture-perfect composition when shot from below framed by mountains in the background. It foamed, roared, leapt and – its passion spent – subsided in a languid pool. After this spectacular show, the Bow retreated majestically down its scalloped valley, winding its way eastward through a forest vista to a distant horizon. Rocky Mountains National Park, colloquially and later officially known as Banff, was more than a river, much more, but it could not have been Banff without the Bow. The river was not incidental to the conception of the park but, rather, central to it.

It is one thing to focus upon the spectacular or the sublime and dignify such spaces as parks. It is another matter entirely to reimagine ordinary, day-to-day things as parks. The Bow was central to that process as well. Most people conceived of the river as a utility: an asset to be used to meet the shifting need for power, for irrigation, for sanitation, for transportation. But this mode of thinking had to contend with another imaginative construct, the idea of the river as a park, a place for pleasure and recreation set apart from the industrial domain. Parks were not simply there to be found; they had to be created, often in competition with more immediately utilitarian functions.

In connection with park-making we will meet yet again William Pearce, a man closely identified with the promotion of the economic uses of the waters of the Bow River for purposes such as irrigation and stock-watering. Yet from a very early date Pearce was also a keen promoter of parks, not only national parks but also urban green spaces. Initially he was a lonely voice. Gradually, however, popular enthusiasm for park-making gained momentum, especially during the third quarter of the twentieth century in the face of powerful competition for river resources until, over time, the idea of a river as park replaced and in some cases displaced other functional conceptions of the river. Farms became parks; junkyards and lumberyards became parks; buildings were demolished to make parks; until conceptually, and in large measure actually, the river became one long continuous park. But what kind of park? A park was not one thing; it could be many things – upstream as a national park and downstream as provincial and municipal parks, as we will see in the next two chapters.

The Park Idea

In western Europe the first parks had been created in the fifteenth and sixteenth centuries by monarchs or aristocrats as walled hunting preserves reserved for their exclusive use. In the seventeenth and eighteenth centuries there appeared domesticated versions surrounding castles and country houses, created and shaped by celebrated landscape architects. As cities expanded, they enclosed some of these unbuilt-upon areas; a number survive today even in places such as London and Paris. Gradually some of these parks were opened to the public and in time were added to by governments, local or national, which created green spaces from former military lands and newly laid-out grand avenues or along riverbanks and in reclaimed wetlands. The park idea was truly democratized in the United States, where the mid-nineteenth century saw the construction of spaces such as Central Park on the northern fringes of New York City. Parks came to be seen as important antidotes to the overcrowding and pollution of modern city life.[1]

At the same time a different kind of park also developed where city dwellers could amuse themselves in the evenings and on weekends. Vauxhall Gardens in London and the Tivoli in Copenhagen came to feature not only promenades with fantastic architecture among restaurants and dance halls but also boating lagoons and carousels. Transplanted over the Atlantic, amusement parks such as Coney Island in New York catered to young men and shopgirls, who could meet in relatively unstructured social circumstances free from older relatives or chaperones. In many cities around 1900 the new electric street railways built such amusement parks near the end of their suburban lines in order to drum up traffic during off-peak hours.[2]

With more high-minded motives, some Americans also sought to promote aesthetic pleasures by setting aside areas of particular natural beauty and protecting endangered landscapes by creating national parks. Yosemite came first in 1864, but equally significant was Yellowstone in Montana, which Jay Cooke of the Northern Pacific Railroad persuaded Congress to set aside in 1872. After languishing in bankruptcy for more than a decade, the railway finally constructed its spur line to a point about seven miles to the north of the park, closer to its extraordinary geysers, hot springs, and waterfalls. If the United States lacked ancient cities and timeless architecture, visitors might be lured to stand in awe at the majesty of the mountain landscape and nature's wonders. The "nationalization" of parks ensured that under public stewardship such wonders would not become the exclusive property of the rich but open to the people.[3]

Bathing and taking the waters gave rise to another strand in the genealogy of parks. Therapeutic bathing and medicinal drinks had their origins in antiquity, but their appeal survived into the modern era. Spas and hot springs were often located in picturesque surroundings where the well-to-do repaired to take the "cure," giving these retreats a reputation as restoratives from the stresses and strains of modern urban life of the same sort as the awe-inspiring mountain scenery of national parks.[4]

The idea of the "park," then, was a multi-faceted one. Green open spaces came to be distinguished from surrounding fields and forests at first glance little different in appearance. In the United States national parks preserved and protected (yet made accessible) some of the most picturesque features of the landscape, which it was believed would inspire awe and wonder to revive and improve the human spirit. Urban parks were deemed to have particular value as a curative or preventive for the ills of city life, and with the addition of entertainments and amusements, they became major attractions for a wide span of the population, not excluding young people and ordinary working folk.

A park could mean several things, foremost among them a natural place set apart in the city or the country, a health retreat, and a place of amusement. These ideas – of a park as a public preserve, containing notable geography or landscaped recreational facilities, as a place for rejuvenation and the restoration of health, and as a popular pleasure ground – all found expression along the banks of the Bow River as different orders of government and social organizations acted on the perceived need to build parks.

The Canadian National Park

In 1883 the Canadian Pacific Railway vice-president, William Van Horne, returned from a trip as far west as Lake Louise so impressed with what he had seen that he pressed William Pearce, the federal land agent, to reserve that part of the valley from applications for ownership. Following two personal visits in 1884 and 1885, Pearce drew up an order-in-council, approved in Ottawa on 25

November 1885, that designated a number of sections of land covering about ten square miles which contained "several hot mineral springs which promise to be of great sanitary advantage to the public." In order to ensure "that proper control of the lands surrounding these springs may remain vested in the crown," the government decree reserved them "from sale or settlement or squatting."[5] Though initially skeptical about the medicinal properties of the waters and the possibilities for future development, Pearce soon became a believer. By February 1886 he was writing enthusiastically about the experience of floating like a cork in water that was a "delightful temperature for bathing in." He also waxed poetic about the surrounding scenery, the fishing, and the hunting, concluding with the observation "With a small outlay [it] can be one of the finest parks in the world."[6]

Having paid off a swarm of opportunistic squatters, Pearce set to work drafting legislation "to make a large reservation taking in the best scenery in the neighbourhood" for inclusion in a permanent national park. In the process, the Nakoda band would also be dispossessed of some of its seasonal hunting grounds.[7] He laid out a rectangle roughly 10 by 26 miles (674 square kilometres) to become the Rocky Mountains National Park by act of parliament in 1887; the regulations covering the use of the parklands were largely based, said Pearce, upon the rules laid down in the United States for the Arkansas Hot Springs. So Canada's first national park originated as a spa.[8]

The spa model dominated early thinking about the park. Much was heard about the curative value of the springs at Banff in the parliamentary debate on the 1887 legislation establishing the park. One member of the House of Commons claimed to have seen invalids carried into the waters who were later miraculously able to walk unaided. Prime Minister John A. Macdonald, who had visited the region when the CPR began service to the coast in 1886, promised that, "carefully managed," the park would sustain itself financially from land leases and water rentals. Visitors would be attracted from across the country, from the United States, even from Europe. Care would be taken to prevent squatting and speculation, curb excessive commercial exploitation, and exclude the "doubtful class of people" frequently associated with spa society. Macdonald outlined his view of the future when he told the House of Commons that within the enlarged park, "There is beautiful scenery, there are the curative properties of the water, there is a genial climate, there is prairie sport and there is mountain sport; and I have no doubt that that will become a great watering-place."[9]

Following the model of the Northern Pacific in Yellowstone National Park, the CPR quickly took action to make Banff a tourist resort and spa by commissioning a New York architect to design a luxury hotel.[10] Work began on the foundations of the grand new hotel at Banff during the winter of 1886–

Illus. 10.1 The first Banff Springs Hotel above Bow Falls circa 1890. (Whyte Museum of the Canadian Rockies, NA 66-1351)

87. Van Horne personally redesigned and reoriented the plans so that guests could enjoy the spectacular Bow River scenery when it opened in 1888. The railway's doctor, Robert Brett, established a Hospital and Mineral Bath House at his Sanitarium Hotel, and numerous smaller hotels sprang up advertising the curative powers of hydrotherapy.

But relatively quickly the idea of the national park as a spa and health resort began to be subordinated to notions of more active recreation amidst spectacular mountain scenery. This shift was reflected in the early 1890s by the CPR's decision to open a mountain chalet above its station at Laggan, with views from the foot of Lake Louise towards the Victoria Glacier. Along with the Banff Springs Hotel, this destination became so popular that both buildings were frequently added to in the years prior to World War I. As Macdonald had intimated, the mountains, the scenery, the river and waterfall, and the wild animals and fish became the main attractions, and the park was ideologically and administratively reconfigured accordingly.

In the 1890s experienced American alpinists appeared in the Bow valley, eager to tackle the numerous unclimbed peaks in the Canadian Rockies. To cater to this new group of tourists and after several fatal accidents, the CPR hired two Swiss guides in 1899, making these adventures much safer for old hands and novices alike. In 1906 the railway gave its support to the formation of the Alpine Club of Canada (ACC), whose purposes included aesthetic as well as developmental goals: "The education of Canadians to an appreciation of their mountain scenery" and "the opening of new regions as a national playground."[11] Restyled by the CPR and mountaineers as "The Switzerland of Canada," Banff thus became the site of strenuous outdoor recreation – mountain climbing, trail riding, and hiking – as well as more sedate mountain pleasures such as tallyho rides, boating, and the novelty of bathing in the hot springs with breathtaking views.[12]

As the fame of Banff spread, assiduously cultivated by the CPR's aggressive advertising campaign, the number of visitors steadily increased each year. The Banff area underwent a burst of hotel building after 1900 as a modern town appeared complete with restaurants, souvenir shops, arc lights, a movie house, and a museum.[13] To encourage a stable, high-class clientele, as outlined by Sir John A. Macdonald, the park authorities set aside 180 substantial building lots on virtually perpetual leases with strict limitations on the type of dwelling to be constructed. Naturally, the most publicized project was the extensive additions to the Banff Springs Hotel itself. The original building became so overcrowded, with 3,890 guests registered in 1902, that the CPR built a new 200-room wing which basically duplicated the existing structure opened in 1888. As a result, 5,303 guests were accommodated in 1903 and 9,684 in 1904. Yet the park superintendent recorded that in 1903 at least 5,000 people were turned away, even though the hotel stayed open for an extra month that summer. Six-storey towers were added to both wings during the 1904–05 season, and new service quarters built over the next couple of years. Nevertheless, the hotel immediately filled to its capacity of 450 guests when it opened in the spring of 1907, and by 1910, 400 new arrivals had to be shipped back to the railway station to be accommodated in sleeping cars. With 22,000 registrations during the four-month season of 1911, the CPR began work on an eleven-storey central tower, which finally opened in 1914.[14]

Illus. 10.2 Early twentieth-century postcards from Banff. (Authors' collection)

Following its listing in the *Baedecker Guide to Canada* (1894) as one of the country's top five hotels, noted for its "good cuisine and attendence," Banff Springs attracted a blue-ribbon list of guests. (Lady Macdonald's enthusiasm for the area was such that she built a private cottage on an adjoining lot, which she used in the summers of 1887 to 1889.) When guests descended from the train at Banff, a dozen people were ushered to sit three or four abreast aboard the high-bodied tallyhos owned by the railway company (a service later franchised to Brewster Brothers). Behind a well-turned-out four-horse team, the journey from the station along Banff Avenue through the burgeoning townsite, across the bridge over the Bow, past the Sanitarium Hotel, and up the hill on Spray Avenue to the Banff Springs took about twenty minutes. Guests entered the hotel from a forecourt to the great octagonal lobby, four storeys high with its balconies above. Judging by the postcard images that have survived, the hotel itself, with its imposing central tower clad in limestone taken from a quarry on nearby Mount Rundle, was considered the most important attraction, scenic or otherwise, in the vicinity.[15] British visitor Douglas Sladen noted in 1895, "We tried both principal hotels, the Canadian Pacific and the

Sanitarium; the former cost nearly double as much as the latter, but then it is a palace hotel." Just after the turn of the century Bernard McEvoy grumpily reported finding the Banff Springs aswarm with American millionaires and their wives. "Confessing that they don't know what to do with their money now they've got it, they are going to every place they ever heard of, irrespective of expense or distance, and not much enjoying it either."[16]

From the windows of their hotels, "health and beauty pilgrims" could take in the spectacle of "roaring fall or piercing peak, glittering glacier or forest primeval," the 1904 pamphlet *Canadian National Park (Rocky Mountains), Banff, Alberta* advertised, all the while enjoying "almost every convenience of modern life." But the purpose of a national park went well beyond the superficial stimulation of the senses. Following a description of the wonders of "the Canadian National Park," the pamphlet affirmed: "It is one of the many beauty-spots on the American continent to which tired, over-worked people can come, with the reassurance that they may soon return invigorated both in mind and body, and the remembrance of this visit will never fade, but will deepen and quicken into a stronger feeling, not alone the sense of beauty, but also the sense of patriotism, for what son of the Dominion can fail to realize, after visiting such scenes, that a country of such beauty is a good country to live in, and to live for?"[17]

The focus was always on the well-to-do tourist. First on anybody's list should be a visit to the restorative wonders of the minerals springs. The CPR hotel always had a pool of water piped in from the Upper Hot Spring, and once its new central tower was completed in 1914, there were indoor and outdoor sections with a spectacular view eastward down the Bow valley. Then there would be time to appreciate the setting; as Bernard McEvoy put it, anybody would, "of course, want to visit the various springs, to cross the Spray and to walk down the Bow until he can look backward between Tunnel and Rundle Mountains; he will want to study the [Bow] falls from every accessible point of view, and taking the little steamer or a canoe, ascend the Bow, push into the Vermillion [*sic*] Lakes, and construct for himself varied pictures of the mountain."[18]

To add to the appeal for visitors, entertainment and pageantry based upon the traditions of the ranching and pioneer past of the "Old West" were grafted onto the call of nature. Gradually the conception of a park as a place of amusement and entertainment crept in, supported at first by park officials eager to attract customers in accordance with "the doctrine of usefulness" to justify their expenditures and labours. Local businessmen began to sponsor the week-long "Banff Indian Days" in 1907, a theme later more fully developed by the CPR. Like Calgary with its Stampede, these boosters recognized the appeal

Illus. 10.3 Early twentieth-century postcards from Banff. (Authors' collection)

of frontier days. Ironically, the indigenous people, who had been systematically displaced from the park as hunters, were invited back in (temporarily) as performers. The Nakoda band moved from its reserve near Morley to set up a teepee camp in town. Each day there was a mounted parade to the Banff Springs Hotel, followed by races and other competitions.[19]

Of course, the river itself always figured prominently amongst the natural wonders of the area. Artists and tourists with their box Brownies sought out the reflective surface of the Vermilion Lakes. Rowboats and canoes paddled languidly along river reaches, and steam launches puffed around the shores of the expanded Lake Minnewanka. Thrill seekers shuddered at the spectacle of the thundering waterfall. For hotel guests on the terrace and in the dining room, the CPR strategically appropriated the panorama of the Bow as it meandered eastward, a ribbon of blue in the deep green and rock grey mountain valley. The lake and river waters attracted anglers and afforded smooth surfaces for boating and canoeing.[20] More prosaically, the Bow system also provided water to the community, a convenient swift-flowing receptacle for its sewage, and eventually even hydroelectricity.

The provision of water in the town of Banff closely followed the Calgary experience with one important exception: the system was financed, built, operated, and maintained by the federal government. Creeks, springs, and wells had at first satisfied the needs of the hotels and the townsfolk. But by the turn of the century the groundwater and many wells had become contaminated. Concern for the health of the summer tourists forced action. To serve this summer peak demand, Banff required a system suitable for a town many times its size. In the fall of 1904 the superintendent of the park advised head office that "the time had arrived when it was absolutely necessary from a health point of view to provide Banff with a water works system and a proper sewage system." In the spring of 1905 a consulting engineer designed a system drawing its water from the upper reaches of Forty Mile Creek, which tumbled down from the mountains north of the town. From this height abundant clean water could be delivered to the town mains by gravity with sufficient pressure to meet the fire protection needs of the town and the hotels ringing the base of Sulphur Mountain. With commendable speed, the appropriations were arranged and construction begun. The system commenced operation in 1908, serving the CPR hotel and station, the other resorts and sanatoriums, and the village.[21]

Operated as a division of the national park, the water system served the community and its seasonal visitors more or less invisibly. Occasionally, however, it popped into view. Spring freshets in Forty Mile Creek would from time to time wash out the dam, silt up the intake, or undermine and burst the pipeline to town, which had been laid along parts of the lower stream bed. These momentary interruptions were rapidly repaired. From the beginning, local residents complained about the cost of service, comparing their water rates with much lower prices prevailing in other Alberta communities. The Parks Branch countered that rate comparisons had to take into account water rates and taxes; residents of Banff paid no municipal taxes. When local taxes were factored in, Banff water was a bargain. Permanent residents only paid half the true cost of their water.[22] Large water users, such as the hotels, initially reduced the burden on ratepayers. In time, however, as the hotel contracts remained fixed and their water demands rose, the cross-subsidy shifted in the opposite direction.[23] It took until the 1950s to renegotiate the original 1907 contract.[24]

Most of the time residents were extremely proud of their cold, hard, sparkling mountain water. Annually, however, they had to grin and bear it. With the spring breakup, grit and gravel carried down Forty Mile Creek silted up the intake dam. For two or three weeks in June the taps and toilets of Banff poured out a gritty, murky substance. There was so much silt in the water that the meters originally installed in 1908 all clogged up and had to be removed from the system. Townspeople, for the most part, considered this annual discolouration a mere inconvenience, just another aspect of life on the frontier. Hotel guests, however, objected to sand in their drinks and muddy bath water. Eventually the Banff Springs Hotel had to install a holding tank, to let the silt settle, before passing water on to the kitchens, laundries, swimming pools, and guest rooms. The rest of the town just put up with the problem.[25]

At the other end of the pipe Banff provides the best example of faith in and the relative effectiveness of the dilution principle. At the same time as the federal government built the Banff waterworks in 1906, it installed sanitary sewers. The superintendent of the park reminded his head office in 1904 of the urgency and especially the need for waterworks and sewerage to be provided at Banff to protect a certain image: "It is absolutely necessary if Banff is to keep up its reputation as a health resort that this matter receives attention. There are now three hotels on the South side of the River and over fifty dwellings and there is even no surface drainage for any of the sewerage. The hotels are using cesspools and when the soil gets impregnated with this it will be sure to breed fevers and other diseases."[26] Within two years a grid of sewer mains drained the village. The sewers, cleverly, were ventilated through hollow metal electric light standards. The combined cost of the complementary water and sewer systems reached $150,000 when the bills were added up. Construction had proven unexpectedly difficult on account of the high water table, quicksand, and extensive rock excavation. But by 1909 the two systems were in operation.[27]

Over the years, as the town grew, more outfalls poured sewage into the Bow River at Banff. The Banff Springs Hotel, for example, never connected with the town system. It built its own sewer, which conveyed hotel waste water to an outfall in the middle of the river just below the confluence with the Spray River. Eventually five sewers, draining different portions of the town, emptied into the Bow at Banff as the system expanded in 1916, 1937, and the 1950s. Storm sewers added their burden of waste water in the late 1940s. At some point in these modifications the main sewage outfall was relocated just upstream of the falls. This placement led to surely one of the greatest ironies in Canadian parks history. The most photographed waterfall in Canada, Bow Falls, had in effect been turned into an sewage aerator.

At first, of course, when Banff was small, the system added only a tincture of sewage to the river. As it turned out, in the winter, when stream flow was low, much of what came out of the sewers was heavily diluted with water. In an inversion of the normal state of affairs, water use in Banff was heaviest in the wintertime. To prevent freezing pipes, residents left their water running. Winter water usage in Banff sometimes reached 5 million gallons per day; in the summer, when the population rose, water use fell to between 3 and 4 million gallons per day.[28] In either circumstance sewage would have constituted only a minuscule proportion of the flow of the Bow River. Short of testing the water, no one could tell that Bow Falls had been turned into a fountain for sewage.

When Calgary Power came knocking in 1911 with a scheme to turn Lake Minnewanka into a hydroelectric storage reservoir, park authorities saw an opportunity rather than a threat. The dam would raise the water about 12 feet, eliminating some unsightly shallows, improving the lake for boating, and extending it deeper into the mountain canyon. The proposal had the additional virtue of creating the possibility for the park to generate its own power. Up to that point the townsite and the hotels had depended upon the power station at the Bankhead coal mine for electricity. The Lake Minnewanka storage reservoir project went ahead with the blessing of the park authorities on the condition that a "thimble" be built into the dam to allow a penstock to be installed at a later date to feed a power station farther down the Devil's Creek valley. Eventually, during the 1920s, the park built its own hydroelectric station, liberating the town from Bankhead thermal power. Banff National Park thus became the home of not only a hydroelectric storage reservoir but also a generating station. That power station would supply Banff until in 1942, under the War Measures Act, Calgary Power received permission to build a much larger storage hydroelectric project and a generating station prominently situated right beside the road and railway.

The river's edge also offered a site for a new recreational activity: golfing.[29] The game was seen as an important draw for the "carriage trade," which both the park officials and the hotel were keen to attract.[30] Golf, as it turned out, was important to the river too. As we have seen in previous chapters, golf and the CPR were the only things standing between Calgary Power and the virtual disappearance of the Spray River. But beyond that, a golf course, in the modern aesthetic, existed in harmony with the natural landscape – indeed, represented in some ways an idealized improvement upon it. The golf course was the most visible aspect of constructed nature at Banff. It involved blasting rock, carting in soil, planting trees in strategic places, trimming the forest, and shaping the land to please the eye. But those obvious efforts on the golf course could just as easily be taken to symbolize other less intrusive park-making policies that produced the manufactured environment experienced by millions of tourists.[31] The CPR gave the course special attention in its brochure entitled *The Challenge of the Mountains: The Canadian Rockies, the Playground of America*, noting that at 5,000 feet above sea level it was the highest in North America, combining scenic beauty with the "exhilaration of the mountain climate."

The town, the hotels, and the golf course were obviously constructed. But so to a very large extent were many aspects of "nature" that tourists came to admire. Wildlife, fire, even insects were vigorously managed to produce a desired outcome, a park designed by policy to meet the expectations of tourists. Viewing wild animals in their native habitat could be at one and the same time an aesthetic experience and a popular entertainment. But the fauna in the wildlife preserve in which a tourist town was situated often failed to play the Disneyesque roles scripted for them. The history of wildlife management in the park is a subject on its own beyond the scope of this book, but the experience of one Bow River denizen, the beaver, deserves brief comment to demonstrate the larger point that wildlife in the park became dependent upon human regulation. Bears, cougars, elk, and buffalo could just as readily serve as examples. Beaver were extremely popular with tourists.

Beaver were not initially indigenous to the upper valley. Legend had it that the beaver population grew from a pair that escaped from a Banff petting zoo. The truth was more prosaic. A fire in 1904 changed the forest ecosystem by destroying the cover of spruce and fir and allowing large stands of aspen poplar to become established along the river. Beaver appear to have migrated upstream from the foothills to this new food source. Thereafter beaver ponds multiplied, populations rising and falling in particular locations with the supply of fodder. Going about their work at dawn and dusk, the beaver incidentally entertained the tourists. But they also posed a health hazard.

Banff's water was not only unfiltered; for many years it was also untreated as well. The water came right down from snow-capped peaks through alpine meadows and a little visited valley. It was as pure as nature intended. Still, the medical officer of health became alarmed when beaver recolonized the region in 1938, building several lodges upstream from the water intake. Fearing an outbreak of disease from the mammalian invasion, he ordered the removal of the beaver. A struggle ensued over the forced evacuation, won initially by the beaver and the park naturalists. The beaver stayed, but the waterworks had to install a crude chlorination system in the late 1940s, which townsfolk deplored, preferring the crisp taste of their mountain stream, beaver notwithstanding.[32]

Beaver scarcity was just as much of a problem. In 1947 Special Warden Hubert U. Green reported a general decline in the wild beaver population in the Bow valley; he had trapped enough animals elsewhere to furnish a site at the Vermilion Lakes just west of Banff, where they could be viewed every evening to the delight of large crowds. By this time the biologists in the Wildlife Protection Branch of the Department of Mines and Resources had begun to have their reservations about such interference with ecosystems, but they

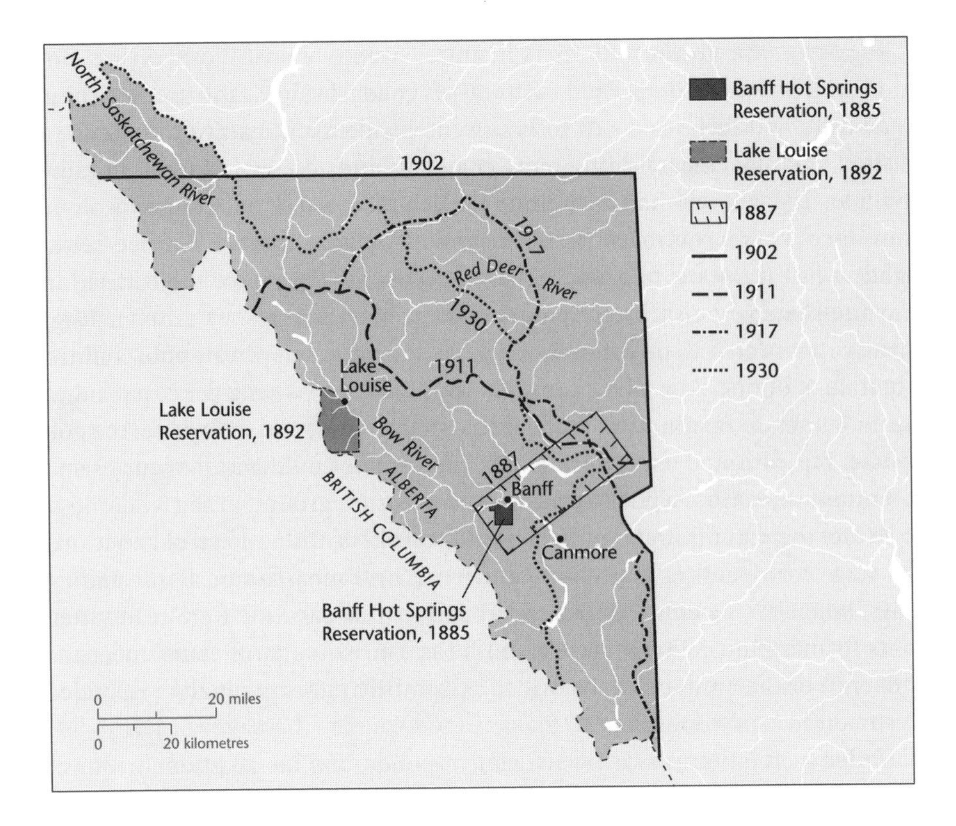

Map 10.1 The changing boundaries of Rocky Mountains National Park, 1885–1930. (Adapted from Geoffrey Wall, "Recreational Lands," in *The Historical Atlas of Canada*, vol. 3, ed. Donald Kerr and Deryck W. Holdsworth, plate 36)

bowed to the fact that visitor expectations deserved to be met: "It is not generally desirable to maintain wild animals by artificial feeding in a National Park, but an exception may well be made in this case, because of the publicity value and tourist influence attendant upon the existence of free-living beaver at a place near Banff townsite." When an American visitor complained in 1951 that the beaver pond at the Vermilion Lakes had been abandoned because of lack of poplar trees, Controller Smart asked the park superintendent whether there was some other site to "be developed where the beaver could be induced to perform for the public as in the old pond." As a result, poplar was cut and hauled to an area just west of the previous site, in hopes that the beaver would stay around in 1952, as any other occupied beaver ponds were too distant from the roads for motorists to bother venturing out to see them.[33] The beaver around Banff were effectively put on welfare to meet tourist expectations.

As a pleasure ground, the park required improvement. Scenery was not enough. Besides the hotels and bathhouses, roads, bridges, and scenic drives, a townsite had to be laid out to house the residents and serve the needs of visitors. To make the wildlife a more enjoyable and accessible spectacle, game (buffalo, elk, beaver) had to be imported, predators (lynx, cougar, wolves, "nuisance" bears) routinely destroyed, and as we have observed earlier, better fighting fish (rainbow trout) added to the lakes and river. For the edification and amusement of visitors, the government even built a geological and natural history museum.[34] To provide more diversion, public and private efforts added a buffalo paddock, zoo, picnic grounds, livery stables, playing fields, pavilions, steam launches, boathouses, bandstands for concerts, and, of course, the golf course. Businessmen opened museums and galleries selling Indian curios and mountain souvenirs. Under the rubric of a pleasure ground in the wilderness, to attract the maximum number of pleasure seekers, the national park became the stage for operas, a Highland Gathering, and most famously the Indian Days, when for a period the Nakoda band in full costume were readmitted to its former hunting territory to amuse the tourists.[35] Banff came under the influence of the amusement park idea as administrators gradually responded to perceived tourist interests.

Under such latitudinarian management policy, and the additional requirements that the park should be useful and largely self-supporting, many economic activities were deemed to be compatible within the boundaries of the park. Nor was there anything sacrosanct about the boundaries. The original 10-square-mile area, which had grown to 260 square miles in 1887, was again increased in 1892 with the addition of a satellite park around Lake Louise. In 1902 it was expanded to a huge park and forest reserve covering a vast triangular area between the foothills, the BC border, and an arbitrary line well north of the Red Deer River. Of course, this space contained many private properties and resource developments. It also brought into the park the upper reaches of the Bow, including its source, Bow Lake and Bow Glacier. In 1911 that area was substantially reduced to ensure unqualified access to potential mineral deposits and forest resources by confining the more regulated park space to the core Bow Valley region. Additions were made in 1917 in the Lake Louise sector, and reductions occurred in 1930, reductions that established essentially the present-day limits of the park.[36]

Over time, as it developed, the national park accumulated a heritage of ad hoc decisions that collectively added up to incoherence. The policy permitted multiple uses, especially if they attracted visitors and revenue to support the activities. Even within the core area of the park, coal mines functioned at Bankhead and Anthracite. Loggers operated on limits within the park into the 1930s. Lake Minnewanka had been converted into a hydroelectric storage reservoir, and a publicly owned power station installed at its outlet.[37]

In so many ways Banff was built. Moreover, it was built around the river, as an attraction and a utility. Banff was maintained as a wilderness, a sanctuary of extraordinary natural beauty, through the application of strict human control. Nature was improved upon with introduced species and wildlife management, diversions, and amenities to attract sufficient visitors to justify the upkeep, occupy them while they visited, and provide creature comforts to take the edge off roughing it in the bush.

National Auto Park

Until 1911 almost everyone arrived at Banff by rail. That, of course, was what the CPR intended. But the railway necessarily restricted visits as well. The number of trains limited passenger volume, and the high cost of train fare and hotels, not to mention the necessary leisure to travel, restricted access to the relatively well off. Again, that is what government policy initially intended. The automobile would bring mass tourism to Banff, and auto tourism would redirect park-making in the process.

The story of the relationship of the automobile to Banff National Park follows a familiar narrative arc: from rejection to acceptance to an enthusiastic embrace. The park then had to be redesigned to accommodate the influx of automobiles and tourists. This stampede of humanity and machines in turn engendered belated resistance and more stringent regulation. Of course, the quantum leap in visitors and pass-through transients had tremendous implications for the relationship of the park to its river.

Automotive tourism required increasingly intrusive road construction to handle large volumes of traffic. As with the railway, the broad gentle slope of the Bow River valley provided an ideal location for roadway construction and agreeable river views for the drivers as well. The alluvial gravels and nearby concrete plants even provided convenient road-building materials. In the park, cars and people required additional accommodation, parking, and service facilities. The mass-produced automobile effectively changed it from a national and international destination to a holiday retreat for British Columbians and southern Albertans. All-season access opened opportunities for new recreational activities, particularly winter sports. Banff had to be remade in the image of the automobile, a radical transformation that in time nurtured a counter-revolution.

But cars and roads not only increased the accessibility of Banff; in time they transformed the park from a destination into an extended – though dramatic – river parkway. Banff became an incidental stopover in a mass migra-

Illus. 10.4 Motoring along the banks of the Bow River circa 1911. (Glenbow Archives, NA 2414-4)

tion of autos, buses, and trucks east or west across the Trans-Canada Highway. The highway and its through traffic turned Banff into a major transportation corridor, even as it taxed its facilities. This vast increase in the number of mobile visitors also brought the river more directly into view.

At first Banff resisted the automobile. An application as early as 1903 to ship in an automobile by rail, to be used for tours along the trails that then linked up such sites as Lake Minnewanka and the Loop Drive, not only aroused strong opposition from the CPR and other livery operators but also drew the ire of the influential William Pearce. Accepting claims that cars were likely to frighten the horses, endangering the lives of tourists, the Department of the Interior banned them from the park the following year. When the province of Alberta was created in 1905, a "coaching road" was promised from Calgary to the eastern boundary of the Rocky Mountains National Park. This route ran along the north shore of the Bow, despite a steep and dangerous hill at Cochrane; the federal government agreed to improve an existing trail from the boundary to Banff, but it was assumed that it would be used only by horse-drawn vehicles.

Still, motorists continued to lobby, and in 1908, with road construction proceeding, hopes were expressed by some Parks Branch officials that it might be possible to drive from Calgary to Banff in another couple of years. In the summer of 1909, however, one intrepid Calgarian negotiated the unfinished road over eight and a half hours, only to be charged with violating the ban on cars in the park. Nevertheless, others soon followed, and by 1911 more lobbying had convinced officials to allow people to bring their cars to Banff, provided they were then parked at the Royal North West Mounted Police barracks. The Calgary Automobile Club, which included a number of keen golfers, chose the formal opening of the Banff Springs golf course on 15 July 1911 to make its first group outing to Banff. Not only did the club show up in force, but in a mild act of civil disobedience, some people drove across the bridge over the Spray River and along the Loop Drive to attend the opening of the golf course.

Previously, tourists had disembarked at the railway station in Banff and journeyed around the national park either on or behind one of the horses owned by the Brewster Brothers or another of the livery operators, before reboarding the train and heading out of the Bow valley to the east or west. Now motorists might come and go as they pleased. One hundred and fifty cars were reported to have reached Banff in 1911 alone, though a lobby by the local Board of Trade to allow driving around inside the park was unsuccessful. In 1913 car owners, including Senator James Lougheed,[38] a summer resident, did secure permission to drive on the town streets and up to the Banff Springs Hotel during daylight.

This accommodation with the automobile undoubtedly reflected the views of the new commissioner of dominion parks. J.B. Harkin shared none of the hostility of his predecessors to cars operating within the national parks, and in his first annual report in 1913 he waxed enthusiastic: "[W]hat motorist will be able to resist the call of the Canadian Rockies when it is known he can go through them on first-class motor roads? And what a revenue this country will obtain when thousands of automobiles are traversing the parks."[39] At the same time William Pearce also changed his mind about automobiles in parks, setting his hand to mapping out scenic routes throughout the mountain west. In 1914 he started badgering the government in Ottawa to promote such road-building, drawing upon American evidence and statistics about automobile tourism probably supplied by Harkin. Pearce even invoked John Muir of sainted memory in defending automobiles as democratizers of parks: "If recreational facilities are important for a nation, it is obvious such facilities should be made available for all the people."[40]

By 1915 the auto had triumphed. The authorities lifted most restrictions on cars on the streets of the Banff townsite, though horse-drawn vehicles and riders still retained the right-of-way. Enemy internees working on the Banff golf course were redeployed a few miles west to a camp at Castle Mountain to build the Banff–Lake Louise road, though this would not be completed until 1920.

Illus. 10.5 Calgary tourists visiting Banff, August 1913. (Glenbow Archives, NA 2788-84)

Early on, parks authorities had envisioned Banff as linked to a web of roads drawing motor tourists from the western half of the continent. The Banff-Windermere highway, conceived in 1913, would run south from Castle Mountain over the Vermilion Pass into the valley of the Columbia River, permitting a loop eastward over the Crowsnest Pass to Lethbridge and Calgary or a journey southward into the United States. Wartime budgetary restrictions stalled progress. Only in 1919 did the federal government promise British Columbia funds to construct the Windermere road over the next four years. A map prepared by the Calgary Good Roads Association in 1922 laid out a "Blue Trail," including both a "California to Banff Bee-Line Highway" and a "Banff to Grand Canyon Road" which would eventually permit motorists to make "a great circle tour connecting up all the great scenic attractions of western America." At the formal opening of the Banff-Windermere Highway on 30 June 1923 an invited audience (including William Pearce) heard J.B. Harkin celebrate the achievement with a comparison to the driving of the last spike on the CPR in 1885. More than 4,500 cars travelled the new road in 1924. Two years later the road from Lake Louise up to the Continental Divide and on to Field, BC, in Yoho National Park was completed, partly using the railway right-of-way over the Kicking Horse Pass abandoned when the Spiral Tunnel was completed, and the next year this "Kicking Horse Trail" was extended to the western edge of Yoho, where it linked up with a provincial highway at Golden, BC.[41] Roads not only led to Banff; they also permitted travel through it.[42]

The arrival of substantial numbers of motorists in the national park had an immediate impact. By 1915 park officials had created an auto camp on the Loop Drive; 73 permits issued in 1917 had increased to 2,388 seven years later, and sightseers on the drive were causing traffic jams. The CPR began building "bungalow camps" for motorists, and by the end of the 1920s the company could publicize the fact, "The comprehensive programme of road construction carried on by the National Parks Department of the Canadian government during the past few years has rendered easily accessible some of the most magnificent scenery in the Canadian Rockies." Rocky Mountains National Park was renamed Banff National Park in 1930, and the following year work began on the extension of the road to Lake Louise northwest towards the Columbia Icefield as a scheme to provide for the unemployed. In an effort to attract more tourists during the Depression, auto camps were established at both Banff and Lake Louise.

The opening of the Banff-Jasper Highway in the summer of 1940 was expected to create more traffic, as was the Big Bend Highway in British Columbia arcing along the Columbia valley from Golden to Revelstoke and opening a (roundabout) route west to Vancouver. Though the fall of France led to the cancellation of the formal ceremonies opening the two new highways, bus passengers from the Pacific coast were soon disembarking in Banff. Only the imposition of gasoline rationing for private cars in 1941 (and the banning of sightseeing buses the following year to conserve fuel and tires) cut sharply into the flow of tourists. The number of automobiles entering the national park fell from 69,798 in 1941–42 to only 36,332 in 1942–43.[43]

With the return of peace, visitor numbers quickly recovered beyond prewar levels. As a result of shortages of accommodation, three new camps were leased right after the war, and eventually a number of motels were constructed at the north end of Banff Avenue and elsewhere during the 1950s. The number of arrivals by railway entered a steady and irreversible decline, as the national park became a prime destination for auto tourists. That influx in turn produced growing pressure for more road improvements.[44]

Up to the end of World War II automobiles and their attendant roads acted as incremental change agents, the full force of their influence moderated by wars, depression, an evolving technology, and gradual market penetration. The postwar explosion of automobile ownership, leisure travel, and especially the routing of Canada's major transcontinental highway through Banff led to vast changes within the park. Parks officials eagerly encouraged these developments. In 1947 the chief engineer of the Parks Bureau, W.S. Mills, emphasized,

"Our National Parks constitute a major attraction for tourists," and he urged the federal government to get going at once on a broad highway improvement program so that visitors would not have to travel on lower-grade roads inside the parks. From the Banff Information Bureau, H.C. Wilson reported two years later, "Personal contact with thousands of tourists during the past two summer seasons has convinced me that there is a real necessity for a first-class Trans-Canada Highway. Americans have said that they would prefer to travel a Canadian route to see Canada, if only there was a good highway. We are also losing valuable dollars because many Canadians travel via the United States instead of travelling a Canadian route."[45]

As part of a postwar reconstruction program, the federal government negotiated the construction of a transcontinental highway with the provinces at a dominion-provincial conference in 1948, which agreed that such a road could be built in eight years at a cost of about $266,000,000.[46] The effect of a new highway upon the Bow valley depended, of course, upon the route chosen. The federal government left the final choices to the provinces, provided that they adopted the "shortest possible East-West route" consistent with both national and provincial needs, though Ottawa would bear the entire cost of the project inside the boundaries of any national park. Some Albertans preferred a northerly route through the Yellowhead Pass at Jasper, as did the military for security reasons. Unfortunately for these lobbyists, Alberta premier Ernest Manning had already opted for the southerly route through Medicine Hat to Calgary and Banff at the secret dominion-provincial conference at the end of 1948. To deflect criticism, Manning later argued that once Saskatchewan and British Columbia had made their route selections, he was left with little choice in the matter.[47]

It is not possible here to go into details here, but the alignment and construction of this highway would have a major impact upon the park and the river, as would the vastly increased traffic. A new road would be built from the east gate to Banff; 3,600 feet of the CPR's main line would be shifted at the expense of the government.[48] Beyond Banff the highway followed the south bank of the Bow River, a controversial decision as it effectively cut off the businesses and attractions established along the old road to the north.[49] Seeking a line of least resistance with good visibility, easy grades, and sweeping curves, the highway route closely followed the course of the Bow River. A dispute with the Nakoda over compensation for reserve lands taken for the road alignment delayed the project. Nonetheless, the route over Rogers Pass was formally opened with great ceremony in September 1962, unleashing a flood of travellers, truckers, and tourists upon the Bow valley from east as well as west.

Easier all-season access to the national park via the Trans-Canada Highway helped fuel schemes for further development. Alpine skiing in the park had begun in a small way in the 1920s at Mount Assiniboine and at Skoki near Lake Louise. When local Parks Branch officials consulted Ottawa in 1930 about permission to construct a ski lodge as "a further attraction to the Banff National Park as a winter playground for the tourist," J.B. Harkin displayed little of the enthusiasm he showed for road construction. Eventually, however, his resistance was overcome, and Cliff White of Banff was granted a licence for his Ski Club of the Canadian Rockies. This operation soon fell upon hard times, but it was taken over and extended by a wealthy Englishman, Sir Norman Watson, in 1935. Afterwards the Brewster family developed the Sunshine area, and runs were cut on Mount Norquay, within sight of Banff, where the national ski championships were held in 1937.

During World War II many service personnel proved to be keen skiers who entered competitions at Mount Norquay and helped to keep the lodges at Sunshine and Skoki in business. In the postwar period new chairlifts, improved facilities, and longer runs attracted thousands more skiers.[50] The number of winter visitors to the national park increased at three times the summer rate in the 1950s.[51] Skier-days reached 160,000 in 1965–66 and rose to almost 580,000 a decade later. New hotels were built near each of the resorts, and in 1969 the CPR decided to keep Banff Springs open all winter, followed by Chateau Lake Louise in 1973.

Winter and summer the Bow valley remained a highly popular destination for automobile and bus tourists, mainly drawn from inside Canada. Statistics taken at the east gate of Banff for the years 1960–65 show that the annual number of passengers in Canadian vehicles rose from about 800,000 in 1960 to almost a million and a half in 1965, compared with an increase from 60,000 to just over 100,000 in foreign (US) vehicles.[52] By early 1964 the director of the Parks Bureau was pressing his deputy minister to approve four lanes of the highway at least over the fourteen miles from the gate to Banff, as the number of vehicles per day on that part of the road during the summer had increased by one-third over four years. A similar increase was expected by 1967 to 10,000, double the rated capacity of the two-lane highway, so that average speeds were already declining.[53] The steady increase in the number of visitors to the area, along with the rising importance of transcontinental truck traffic, ensured that widening the Trans-Canada Highway was only a matter of time.

Policy Shift

During the 1920s, as professional managers took hold of park development and influential conservation interest groups formed,[54] the mission of the national parks had begun to shift once again. In part because of the remarkable success at constructing a popular, accessible pleasure ground, this approach to park-making met critical resistance in the form of the notion of a park as a sanctuary. Inspired largely by the preservationist philosophy of the naturalist John Muir, especially his notion that in wilderness lay the salvation of humanity, parks were deemed to serve both a practical and a moral purpose. In most parts of the country, economic motives and utilitarian values maximized wealth and well-being and transformed the earth. Amidst this roaring tumult, parks were to be untouched sanctuaries, archives of the primitive, unique protected reservoirs of the wild. In parks, scenes of "virginal beauty" were to serve as a reminders of what had once characterized the entire country. This was a duty not only owed to the present but also to be passed on to future generations. "To preserve out of its vast area, for the generations to come, a certain share of primitive nature is the meaning of the national parks movement," J.B. Harkin, director of the Parks Branch of the Department of the Interior, argued in 1924. Capitalism might be unbridled elsewhere, he affirmed, "but in the national parks at least primitive beauty may remain untouched and unscarred by the hand of man." Parks were places to find refuge from the press of civilization, to be revived through encounters with nature, and in the cathedral of nature to renew the spirit.[55] At least that is what Harkin, in his most lyrical moments, and the preservationist lobby hoped.

This line of thinking influenced the revision of the National Parks Act in 1930. In the preamble parks were justified in these terms: "The parks are hereby dedicated to the people of Canada for their benefit, education, and enjoyment ..., and such Parks shall be maintained and made use of so as to leave them unimpaired for the enjoyment of future generations."[56] But note, the revision did not make the preservation of nature the prime objective; rather, it required that use and enjoyment be regulated by the demands of unimpaired transmission. The act mandated a balancing of use and enjoyment against preservation and stewardship. The new order would find certain uses to be inappropriate – logging, mining, hydroelectric generation. But that change was a double-edged sword. Where those uses were deemed to be paramount, such places could not be parks, and park boundaries were necessarily adjusted, as was the case when a small section was carved out of the park in 1949 to accommodate the Spray Lakes storage reservoirs. Nor could a preservation policy be made absolute. Some uses were necessarily grandfathered (Lake Minnewanka storage); some were governed by long-term leases (the building lots); and some theoretically non-conforming activities, such as the location of a transcontinental railway in the park, predated the formation of the park. In Banff the archive of past policies could be read on the landscape.

The dramatic changes wrought by new highways, the rapid increase in summer visitors, and winter recreation possibilities began to expose contradictions in this park policy during the 1960s. To some close observers and to the park officials themselves, use seemed to be overwhelming the stewardship mission. Some uses could be mitigated, but others had to be resisted. Water provides a good example of conscious remediation.

The additional hotel and motel rooms and mass tourism overwhelmed the rudimentary waste-water disposal system. By the 1950s, with the growing all-season popularity of Banff, the function of Bow Falls as the park's sewage treatment plant could at certain times be readily detected by casual observers. Engineering consultants retained by park officials in the mid-fifties to evaluate the waterworks and sewage systems tiptoed around what was obviously a delicate issue. "Since sewage is discharged untreated to the Bow River, the River's ability to carry away sewage solids without nuisance and its capacity for dilution and self-purification are of great importance. Fortunately the spring freshet coincides with the start of the tourist season and high river flows are maintained while the population of Banff (and hence the discharge of the outfalls) is increased." Nevertheless, the engineers discovered "stranded solids" along the river in June. None could be detected in July, however. Moreover, as a natural sewage-treatment plant, the river worked extremely effectively. Near the outlet the river was badly contaminated, but tests showed that pollution diminished rapidly downstream. This pattern demonstrated the remarkable self-purification powers of the cold, turbulent river as it flowed briskly away from the town.

Even though the falls and the river continued to digest these much larger volumes of sewage, the practice touched an ideological nerve. The consultants reassuringly pronounced the sanitary sewers "adequate for many years to come." The outfalls might be pushed further out into the middle of the flow, and perhaps the sewage could be ground up before being flushed into the river, but essentially the system functioned effectively. Nevertheless, the consultants elliptically questioned the policy of pouring untreated sewage into one of Canada's most scenic rivers in the heart of a famous national park, however efficiently the river performed its new function: "there is undoubtedly some risk to public health and amenity, particularly since visitors have unrestricted access to the left bank of the River from Tunnel Mountain downstream. Furthermore there is a nation-wide movement to abate the pollution of rivers by sewage, and it appears most desirable that the Government should, in so famous a beauty spot as Banff, set an example for other authorities."[57] Even if the tourists were oblivious and the residents were not complaining, the consultants thought that the Bow River at Banff should not be turned into an open sewer. It should be said, too, that at Banff the National Parks Service took a new view of its responsibilities in the 1960s. The practice of dumping untreated sewage into a river in Canada's most famous national park was no longer publicly defensible. In the mid-sixties, at federal government expense, the town acquired a modern sewage-treatment facility tucked discreetly away beyond the nether reaches of the golf course. In Banff sewage treatment was driven not by health concerns or a public scandal but, rather, by the influence of the new environmental conscience of the 1960s upon public policy.

The new stewardship sensibility animating parks policy could not stop the four-lane highway juggernaut, but it could and did halt plans for the Winter Olympics and attendant development. Talk of such a bid was also quickly followed by the birth of a National and Provincial Parks Association of Canada, concerned to preserve parklands from overdevelopment. In 1964 the new Liberal minister of Northern Affairs and National Resources, Arthur Laing, declared that the time had come to provide "safeguards against excessive or unsuitable types of development and use." He proposed the creation of long-range plans for all the national parks, which would divide them into zones ranging from wilderness areas to permanent townsites. Laing gave no great encouragement to Banff's Olympic bid, and the following spring he announced that, while he wished to see as much skiing at Banff as possible, he favoured the full utilization of existing resorts before any new areas were opened up. An Imperial Oil proposal in 1970 to build Village Lake Louise with rooms for 6,000 visitors, staff housing for 2,500, and a huge parking lot right at the bottom of the hill brought this issue to a head. Public pressure effectively forced the governments of both Canada and Alberta to withdraw their support for the project. If the Winter Olympics were to be staged in Alberta, it would be outside the boundaries of the national park.[58]

Organizations such as the National and Provincial Parks Association of Canada succeeded in arousing concerns about environmental impacts of highways in the park, and the public hearings on the highway in 1971 provided a forum for critics of the four-lane scheme. Opposition focused upon the destruction of the narrow "montane ecoregion" along the Bow River and the elimination of wildlife habitat. But the perceived transportation needs of the country prevailed. When federal officials released a report recommending the twinning of the first section west of the park gate in 1975, groups such as the Bow Valley Naturalists attacked Parks Canada for promoting the scheme instead of protecting the integrity of parklands. These objections delayed matters until 1978, when Public Works announced its intention to submit the proposal to a formal environmental assessment. Parks Canada now criticized certain aspects of the construction plans, particularly for the sensitive Vermilion Lake wetlands just west of Banff. Seeking to convince the government that widening the road was not really its responsibility, the Parks Branch rewrote history in 1968 with the claim that it had only reluctantly agreed to approve a "major intrusion" into the national parks in the broader national interest. But public hearings showed that powerful local interests such as the Banff–Lake Louise Chamber of Commerce supported the plan. In 1979 the federal environment minister approved the project, conditional upon the highways being fenced to keep out large animals and tunnels being provided under the roadway to facilitate migration.[59]

Banff would get a four-lane superhighway but not the Olympics. The prospect of much more extensive recreational and commercial development inspired a counter-insurgency drawing upon growing pools of environmental sensitivity, which will be discussed in subsequent chapters. But the automobile and mass tourism, like the proverbial camel, had occupied a good part of the tent and could not be dislodged.

On its upper reaches the Bow River contained the classic example of a certain park ideal, the wilderness-protecting national park. Yet for most of its history Banff National Park has been a bundle of diverse objectives and logical contradictions, a selection and modification of past practices only partially aligned with current policy. Its purpose and meaning shifted several times over the twentieth century and would continue to do so. Though the mountains supplied the dramatic backdrop, the Bow River served not only the scenic but also the sanitary, power, and transportation needs of the transient community.

Night and day, winter and summer, transport trucks, buses, and cars carrying travellers and tourists raced along the four-lane, controlled-access highway running right alongside the Bow River from Seebe through Canmore and Banff to Lake Louise and over the Continental Divide. Some people stopped en route, while others never even slowed down. Though the Canadian Pacific Railway had long since surrendered its place as the prime means of access to the national park, its main line still ran beside the Trans-Canada Highway, rumbling with heavy freight trains snaking their way through the mountains at all hours. With hundreds of hotels, motels, restaurants, campgrounds, stores, and gas stations to serve travellers visiting the golf course, the ski lifts, the pack-horse trails, and the boat ramps, John A. Macdonald's wish that Banff National Park would become "a place of great resort" was realized in every modern sense of that final word.

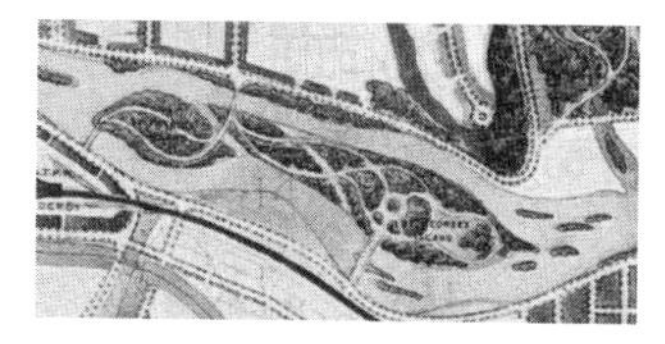

11 · Greening Alberta

The city of Calgary and the province of Alberta played a major role in the transformation of the Bow River into virtually a continuous park during the twentieth century. These two orders of government responded to different impulses, moved according to a different timetable, and constructed parks in their respective jurisdictions in an uncoordinated fashion, though it must be said that increased leisure time, a taste for outdoor recreation, and the democratization of mobility underlay all the park-making programs. While the presence under federal control of Canada's major national park in the region obviously influenced thinking and policy in some respects, the park-making impulse at the municipal and the provincial level served different needs and produced different kinds of parks.[1] Because the Bow River lay in a densely populated, much travelled portion of the province and because of its natural attributes, it became one of the major sites in what might be thought of as the "greening" of Alberta.

As Calgary grew during the first decade of the twentieth century, the city council had more pressing priorities than parks, such as laying streets, water pipes, and sewers while providing schools, police, and fire protection. Nevertheless, at the turn of the century, Calgary found itself manoeuvred into the park-making business by a now familiar figure, William Pearce. He had played a key role in the establishment of the Rocky Mountains National Park; he would also be the moving spirit behind Calgary's early urban park movement.

The province of Alberta entered the business of creating parks comparatively late compared to the federal government and the city. In the United States the process of national park development and the establishment of parks at the state level had gone on in parallel since New York State had created a reserve at Niagara Falls in 1878.[2] Similarly in Canada, Ontario led the provincial park movement with the creation of the Queen Victoria Niagara Falls Park in 1885 and Algonquin Park three years later. The former was driven by a desire to preserve an international icon from excessive commercial exploitation; the latter by a diverse conservation and recreation movement that joined in a desire to protect highland watersheds, conserve the forest, establish a wildlife sanctuary to replenish the province's depleting stocks of game, and reserve spaces for outdoor recreation.[3] Quebec, too, created large hinterland wilderness parks with Mont Tremblant (1894) and the Laurentides Park (1895).

When Alberta acquired control over its Crown lands from Ottawa with the Natural Resources Transfer Act of 1930, the provincial government had just begun to take the first steps towards creating provincial parks. Premier John Brownlee had returned from a trip to England in 1928 to recruit agricultural settlers profoundly impressed, like many North American tourists, with the park-like beauty of the English countryside. His nostalgic desire to create a green and pleasant land in his home province provided the initial impetus for the establishment of provincial parks in Alberta.

Inventing the City Park

William Pearce believed in the imperative of employing land in its highest possible use by following its possibilities and improving it wherever possible, hence his enthusiasm for irrigation. He also had a passion for trees. On the bald prairie, trees were the mark of civilization. Vertical against the dominant horizontal, they indicated culture working with nature. Trees, he believed, had the power to remake the climate. In urban environments he further believed in the importance of taking advantage of geography and natural possibilities to create a civic sense. Cities had to be more than a rude scramble for wealth. They needed to be softened by parks, monuments, treed avenues, and public squares. A great city, in Pearce's view, had to be more than the sum of its private interests. Beauty, grandeur, monumentalism in a cultivated natural setting – these things nurtured a sense of enlightened citizenship. Parks expressed that uplifting collective spirit. In parks his vision of the highest use of land and his passion for trees came together.[4]

As early as 1890 Pearce engineered the transfer of the islands at the confluence of the Bow and the Elbow, named after St Andrew, St Patrick, and St George, from the federal government to the fledgling city of Calgary for $1. In return the city promised to spend $100 per year improving the islands for park purposes. For Pearce, ever the micromanager, that meant planting native spruce trees to stabilize the soil and create a "serpentine" tree-lined drive affording views out over the river. These islands, he claimed, were the best places in the region to grow trees. He insisted upon choosing native spruce rather than deciduous trees because, based upon his experience with seedlings on his nearby farm, he knew that spruce trees would thrive in this climate and moist earth. But for many years the city reneged, claiming it did not want to waste money on land that was frequently flooded. Pearce retorted, somewhat unconvincingly, that in twenty-five years he had known the islands

Illus. 11.1 Entrance to the auto campground on St Patrick's Island in the Bow River about 1911. (Glenbow Archives, NA 2365-40)

to be under water for more than twenty-four hours just three times. Notwithstanding these insignificant inundations, a frustrated William Pearce regularly pestered the city council to keep up its side of the bargain to create a "pleasure resort" unparalleled on the prairies for a nominal sum.[5]

Only in 1908 did the city begin to act on Pearce's gift. Thousands of spruce trees were planted, drives laid out, and rustic bridges and shelters constructed. The now accessible and treed islands provided Calgary's citizens with fishing, boating, picnicking, and other recreational opportunities in the heart of the growing city. Automobile clubs used this parkland for camps and rallies. In the late twenties the islands, reduced from three to two by erosion and infilling, became the site for the municipal zoo, a purpose they continue to serve to this day.[6]

About the same time Pearce also managed to reserve for park purposes a 200-foot strip on the north bank of the Bow River through the city. Originally, the CPR had intended to use this land for its main line, but when the railway rerouted its tracks across the more spacious and superior real estate opportunities to the south, this land reverted to the Crown. Pearce saw in it possibilities for a shady avenue or treed parkway along the river and arranged a transfer of the property to the city for that purpose. Once again nothing happened, despite his repeated promptings. Just as the city began to take an interest, World War I intervened and plans had to be scaled back. After the war, civic-minded service clubs joined in, planting trees along what became

Memorial Drive. Thus in the 1920s a curving parkway along the north bank of the river, widened and enlarged in 1931, became an automotive arboretum to the memory of the fallen.[7]

Such grudging, piecemeal efforts rankled with Pearce. He wanted to make a much bolder statement than the city of Calgary seemed capable of imagining. In the glow of the real estate boom before the war, Pearce had seized an opportunity to have his vision given the stamp of approval by an internationally renowned expert. The British town planner Thomas Mawson, passing through on assignment from the federal government to draw up a town plan for Banff, accepted an invitation to draft a design for Calgary. As chair of the largely ceremonial city planning commission, Pearce guided Mawson around town, taking him to strategic points and overlooks while impressing his own views upon him.[8]

Mawson prepared a plan that applied the principles of the City Beautiful movement to Calgary. Town planners of this persuasion hoped to counterbalance the individualism, anomie, commercial clutter, and haphazard development of industrial societies with architectural grandeur designed to inspire civic engagement and maintain public order. Imposing public spaces, fountains, tree-lined avenues, parks, vistas leading the eye to statuary, monuments, and dignified public buildings were the instruments whereby a proud citizenry could live a better life and contemplate a higher destiny.[9]

Mawson deployed all the elements of the City Beautiful vocabulary in his proposed remake of Calgary into something resembling the capital of a theatrical middle European principality. He realigned the main east-west axis of the city along the river, drawing the civic core of the city northward into the curve of the Bow. At the intersection of the north-south and east-west axes he placed a huge roundabout – a Trafalgar Square, a Place de la Concorde – centred on a plinth. To the west he situated a domed city hall, facing east onto a long mall, liberally adorned with state buildings, that led to another large traffic circle at Centre Street, where a huge municipal auditorium echoed the government buildings at the opposite end. Government (city hall), history (museums), and the performing arts (a concert hall) anchored the three cardinal points of Mawson's triangular core; he connected these points with grand avenues, ceremonial spaces, and, on the hypotenuse, parks along the river.

Above all, as Pearce hoped, Mawson's plan turned the river into a major civic asset. Quays along the south bank lined by neoclassical building facades lent a suggestion of London or Paris to the Bow. An elegant bridge carried the main north-south artery across the river, between two grand museums on Prince's Island. As the road entered the city between towering gates, across massive squares, and under an Admiralty Arch, two scalloped lagoons carved out of the south channel served as elegant reflecting pools to the surrounding

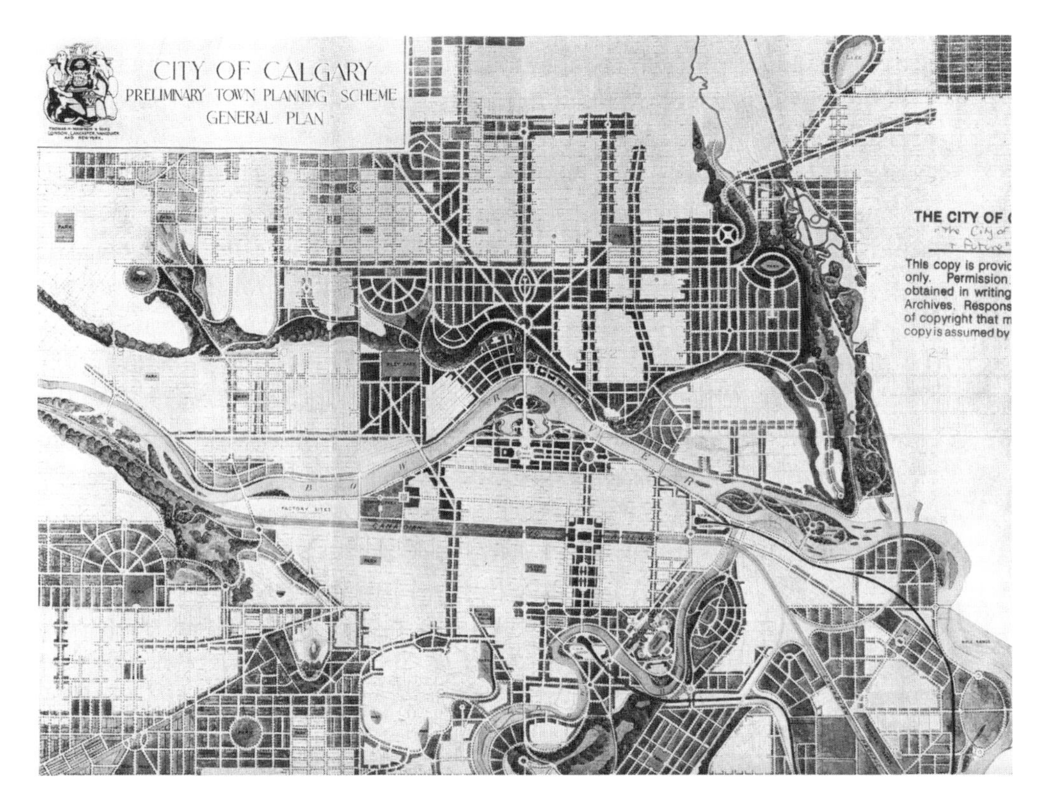

Illus. 11.2 Detail from Thomas Mawson's plan for the city of Calgary. (Calgary City Archives)

architecture and boating basins in the summer. Mawson lined the Bow with green space and elegant driveways on both banks.

In the eyes of history, Mawson's plan is little more than a designer's pipe dream.[10] The timing could not have been worse: the collapse of the real estate boom in 1913 put a damper on creative thinking about urban design; the war stamped it out entirely. After the war the imperial triumphalism of Mawson's design rang a little hollow. Beautiful and awe-inspiring his plan might be, but it seemed obviously unsuited to a city this size, out of step with public taste, the western ethos, the climate, and especially the turbulent and unpredictable behaviour of the river.[11] Major floods in 1915, 1923, and 1928 put a good deal of the lands for Mawson's plan under several feet of muddy water. But from Pearce's point of view, through small-mindedness and a lack of vision, the city had once again missed out on a magnificent opportunity.

While the city rejected the Haussmann-esque fantasies projected onto the river valley by Mawson and Pearce, it did respond to other park-promoting instincts. By the river on its western outskirts, the city presided over the development of a classic streetcar pleasure garden during the first third of the twen-

Illus. 11.3 Boating in Bowness Park, August 1930. (Glenbow Archives, NA 1604-66)

tieth century. It began in the traditional way, as a real estate play. John Hextall, a developer, donated a long, low-lying upstream island in the Bow River to the city for a park in return for the city agreeing to extend its streetcar line into his adjacent subdivision. The municipal streetcars began running to the park and the western suburb in 1912. After World War I Bowness Park's lagoons and winding waterways attracted boaters and swimmers in summer and ice skaters in the winter. The picnic grounds, sports fields, and occasional carnival amusements attracted flocks of young people, couples, and families. A streetcar ticket away, cheap, alternately a dreamy waterside rendezvous and the site of gaudy amusement, Bowness played the role of Calgary's Coney Island, making allowances, of course, for differences in scale. Like so many parks of its type, it flourished in the 1920s, languished in the Great Depression, and struggled during the post–War World II era with the disappearance of the trolley and the rise of mass automobile travel, which brought more distant and attractive recreational possibilities into reach.[12]

Pearce's vision of a serpentine of parks strung along the river through the centre of the city would not be realized in his lifetime. To his disgust, municipal authorities seemed more interested in amusement parks and flower beds. Despite disappointment, he did not lose faith. Indeed, he arranged to leave his east-end estate, where the Bow bent majestically southward, to the city. On this land he had experimented with ornamental tree plantings, scenic drives, and prototypes for irrigation projects. The woods, gardens, and shrubbery of the Pearce estate stood out in marked contrast to the bald prairie knolls on the other side of the river. This evidence of human improvement and the bounty of the land gave Pearce enormous satisfaction. Occasionally, however, flood waters still swept across this low-lying land, even threatening his house and leaving a thick deposit of silt in their wake. As a final public bequest, he willed the property to the city as a park.

After he died, the property descended into a sorry state. The city could not afford to keep it up, and it became neglected and somewhat derelict, a forgotten piece of land on which much later the city plopped a distinctly unlovely fish hatchery named after someone else. In a cruel ironic twist, the Pearce estate gradually reverted to wetlands, the kind of unimproved wasteland he would have deplored. Now mainly a marshland being restored for waterfowl, the Pearce estate offers no evidence of its former self.

The property of his friend and neighbour Colonel James Walker enjoyed a similar fate, though by intention. As a private philanthropy, Walker's son, the naturalist and park enthusiast Selby Walker, opened the property to the public as the Inglewood bird sanctuary and wildlife preserve during the 1930s and 1940s.[13] The city acquired this much reduced and overgrown property during the 1960s. In this way two of the most prized riverfront properties in the region passed into the hands of the city to be used for park purposes, in this case wildlife sanctuaries.

But as Pearce lamented, no coordinated vision of urban parks guided these acquisitions or their development. Up to World War II Calgary's parks were in a sense an afterthought, something a city knew it should have but with no plan linking them to other civic functions.

Parks for the Province

At the provincial level, thinking about parks began with Premier Brownlee's trip to England and the transfer of Crown lands to provincial control in 1930. Convinced that an effort to "beautify" the landscape of his own province would "make rural life more attractive," Brownlee introduced a Town Planning Act (1929) and a Provincial Parks and Protected Areas Act (1930) to promote that objective. Even before this legislation was in place, he appointed a special committee to advise on park development. The deputy minister of works canvassed his district engineers to identify likely bathing beaches, popular picnicking grounds, fishing holes, and "natural beauty spots adjacent to the provincial main highways" that might be suitable "auto parks" for tourists. Guided by the district engineers' responses, suggestions from local United Farmers of Alberta organizations, and the Edmonton Local Council of Women, the special committee recommended the creation of a long list of parks scattered about on the lakes and rivers of the settled parts of the province.[14]

During the depression of the 1930s, Premier Brownlee's hopes of greening Alberta evaporated. However, the more modest notion of creating recreational parks for auto tourists did take root. Even in depression-ravaged 1935 the secretary of the Parks Board, set up to oversee the policy, argued that social and economic change still justified an aggressive park-making program. In the old days pioneers had lacked the leisure and the habit of mind to enjoy the contemplation of nature; then they were too busy trying to wrest a living from the land, sometimes wantonly destroying its beauty and wildlife in the process. Settlement drained ponds, levelled forests, depleted game. Economic progress in the towns, cities, and farms created a population with the leisure and means of seeking outdoor recreation, but with limited and often unsatisfactory commercial recreational opportunities available. Provincial parks would serve needs between urban parks, on the one hand, and national parks, on the other.

Provincial parks, "while not attaining to the dignity and expansiveness of a Federal Park, will preserve for our citizens areas of outstanding beauty and create where necessary areas for outdoor recreation and enjoyment within easy access of every community." Besides, the federal mountain parks were too distant and expensive for most Albertans. A comprehensive park program like that of the state of Iowa would help reduce the "monotony of farm life," counteract "the pull of the cities," and by protecting natural areas, serve the utilitarian function of maintaining water tables. Parks would even be self-financing. Increased tourist traffic would boost gasoline tax revenues. Motorists and their families, enjoying weekend outings in the summer, needed nearby park destinations. Must a citizen of Alberta "be compelled to spend his short vacation travelling along the dusty highway, lounging in the crowded areas of urban parks," the secretary of the Parks Board asked rhetorically, "or will some of the more enticing areas of lake and woodland, with their healing and invigorating powers of beauty, fresh air and sunshine, be preserved for his use before it is too late?"[15]

The Parks Board, mandated to provide recreation facilities for the enjoyment of the inhabitants, to propagate and protect wildlife and vegetation, and to preserve objects of geological, ethnological, or historical interest, created nine public parks in 1932. Most were essentially public bathing beaches and campgrounds situated on lakes scattered about the province, such as Aspen Beach on Gull Lake near Red Deer. The biggest, at 534 acres, was situated on the Bow River, upstream from Calgary, on land donated by the Calgary Power Company. This park effectively opened access to the reservoir then forming behind the Ghost dam. The terms of the gift required that it would have its own dedicated management board within the parks administration. In this first instalment provincial park-making implied creating access to water for fishing, boating, swimming, picnicking, and camping. Wildlife preservation and spiritual uplift from nature took second place to aquatic recreation.[16]

Soon after these achievements the economic crisis stalled further action, and only two more parks were added during the 1930s. Death removed the program's two main champions, the deputy minister of public works and the director of town planning. The Depression devastated the provincial government's budget, and funds for luxury expenditures like parks dried up, but it also effectively killed tourism, reduced automotive travel, and wiped out vacations. This is a brutal reminder that whatever the emotional attachment to parks, the necessary conditions for their creation are leisure and money. Park-making languished for the duration, and World War II redirected resources to other priorities. The province of Alberta thus emerged into the postwar era with one of the smallest provincial park programs in the country. Quebec and British Columbia, with approximately 14,000 square miles each, accounted for about four-fifths of the 34,000 square miles of provincial parks in Canada in 1950. Alberta, by contrast, accounted for less than 11 square miles, an insignificant proportion. In the late 1940s the annual budget for park development amounted to only $3,500. The Parks Board did not even provide maps or brochures directing people to its parks.[17]

The return of prosperity, along with the discovery of oil, the proliferation of automobiles, and rapid urbanization and population growth of southern Alberta in the 1950s, restored the necessary conditions for a renewal of park-making and directed its focus to the Bow River valley. Writing in 1953, fishery officer R.A. Wileman pointed out that urban growth had overwhelmed many traditional recreational areas in the region and at the same time increased the demand for accessible outdoor activity space for the growing population. "The most urgent problem," he noted, "is that one regarding parking and picnic spots within a short drive of the city, to which the average citizen may repair for an afternoon or evening of fishing, swimming, picnicking, berry-picking, etc., without being put to much travelling expense."

Wileman lamented the survival of certain pioneer attitudes in rural areas: "looking at the broad picture, is it wise to press every available acre of soil into farm production? Should we not attach more importance to the intrinsic values of the land? Should the countryside surrounding our urban centres be allowed to become regarded as alien soil where no one but a handful of landholders may walk and enjoy the benefits[?]" Fences and gates were going up, even on public road allowances that led to watercourses, and "No Trespassing" signs were sprouting. For instance, at Twin Bridges on the Elbow River south of Calgary, long a highly popular place for fishing and picnicking, this had occurred even though the land "used here by the public has absolutely no agricultural value, being an area of silt and gravel thickly overgrown with willow and other shrubs. I cannot imagine a more suitable area for public recreation, where fires will not run, the water is clean, and there is nothing that could possibly be damaged." Wileman wanted the area expropriated and "declared common land for all time, as was done with many areas in Britain where the countryside is dotted with such open lands, where no one may fence and build ... Two or three of the landowners may be pretty jealous of this little bit of natural parkland at their back doors, and naturally opposed to public use, but we could rest assured that we would win whole-hearted support from a considerable portion of the populace by bequeathing such an area to the people."

The Lands and Water Resources Branches of the Department of Public Works should be prevailed upon to draw up new grazing and hay leases for Crown lands that would open access to watercourses, Wileman argued. Everywhere a bridge crossed a stream of any size, an acre or two could be reserved as an angling or picnicking ground. Nothing less than a full-scale inventory of the territory along the foothills ought to assess what was needed for parklands before any more Crown land was alienated. And should the landscape be deficient, "If the Parks Branch were ever to be casting about for ways to spend a large sum of money for a worthwhile recreational project, they might look into the matter of creating a lake in this conspicuously lakeless area, choosing a picturesque site."[18]

Officials such as Wileman were joined by the recreational fishing lobby in spearheading a movement to develop a comprehensive parks system. In the mid-1950s the government began to increase annual appropriations for parks from $30,000 to more than $200,000. Several new parks were opened on various lakes and irrigation reservoirs about the province.[19] The Calgary Fish and Game Association lobbied for more parks and compiled an inventory of suitable locations on the Bow. Government did begin acquiring and reserving "desirable spots along rivers and lakes," though its main priority for most of the 1950s was the development of campgrounds along the main highways to serve motorists.[20]

The need for more provincial parks along the Bow Valley corridor, with its rapidly growing urban population and heavy tourist traffic, was now seen as particularly acute. In 1957 the forest officer at Canmore pointed to the need for more extensive public facilities for tourists, weekenders, and school and church groups unable to find accommodation within Banff National Park. When the facilities in the national park filled up, disappointed visitors simply camped along the highways outside the park boundaries. Moreover, he pointed out, "the scenic attractions of the Bow corridor with access to recreational areas in the Spray and Kananaskis valleys rivals [*sic*] Banff Park, and even surpasses Banff in popularity in some respects."[21] As park visitors were now more likely to come from Alberta than the other provinces or the United States, locals, taxpayers, and voters were the ones being inconvenienced, rather than visitors from other jurisdictions. The language here reminds us that roads and traffic had turned the upper Bow valley into a "corridor" full of moving people with needs.

Fishing enthusiasts may have been the most articulate and organized proponents of parks, but the masses of newly mobile urban residents seeking camping holidays in serviced campgrounds provided the main driving force behind provincial parks policy. In response, the province consciously strove to create a new kind of park situated somewhere between the iconic, high-end, reverential national park, on the one hand, and the playgrounds and formal landscaped urban parks in cities, on the other. In between these archetypes the provincial mission was to build and maintain rustic, rural, recreational properties, but suburbanized and serviced with sanitary facilities and other amenities to accommodate large numbers of Albertans with leisure, mobility, and camping equipment.

Responding to the tremendous growth in interest in picnicking, camping, and hiking, the government began an intensive search for a major park location along the Bow River. Park planners poured over aerial photographs searching out suitable locations within a fifty-mile radius of Calgary. Fish Creek, just south of the city, was ruled out because the privately owned land seemed too expensive. The area around the Western Irrigation District weir was accessible but proved extremely treacherous on account of steep slopes and the drowning danger from the undertow. Locations along the eastern slopes of the Rockies could not be considered because of the potential fire threat posed to the forests there.[22]

Eventually, in 1959 the government announced plans to create a major park at the junction of the Kananaskis and the Bow Rivers. The 2,400-acre Bow Valley Provincial Park would relieve some of the pressure on the Canmore corridor by accommodating overnight campers, hikers, and picnickers. The rocky, spruce-covered site on the Bow surrounding Chilvers Lake was ideally located near the Trans-Canada Highway and readily adapted to motorized camping. The only disadvantage came from the nearby lime plant, whose faulty dust-control equipment regularly coated the park with a fine layer of white powder, a problem that persisted until the plant moved in 1973. Within a decade, the park was receiving almost 70,000 visitors a year. Bow Valley Provincial Park's fame was later enhanced by its use in 1983 as the campground for 20,000 Boy Scouts attending an international jamboree.[23]

Nature in the City

During the first half of the twentieth century, the city of Calgary did acquire several isolated properties along the Bow River but without any clear plan or purpose. These green spaces contained the germ of a system of river parks, but for the most part the banks of the river in both town and country remained in private hands, often somewhat neglected on account of frequent flooding of low-lying areas. The state of the south bank of the Bow River in the centre of Calgary in the 1950s provided dramatic evidence of the failure of Pearce's ambitions (and Mawson's plan) to be realized. Plainly put, this strip was an eyesore, a decaying and derelict commercial site. A greasy zone of light industry separated the downtown from the river. Warehouses, auto body shops, garages, parking lots, and the abandoned Eau Claire sawmill, the last a painful reminder of the transience of resource-based industrialization, blocked approaches to the river in the urban core. Weeds, shrubs, poplars, and cottonwood trees had haphazardly recolonized Prince's Island. Seen from across the river, the south bank presented a grim, vaguely fortified brow of hastily thrown up dikes, bulldozed from the scoured riverbed in the mid-fifties, casually surfaced with broken pavement from city streets, and punctured occasionally by storm sewer outfalls. Chain-link fences topped with barbed wire lined the dikes guarding littered storage yards and squat cinder block buildings. Literally hundreds of properties abutted the riverbank along the south shore, a miscellaneous collection of residential lots and commercial establishments, all resolutely backing rather than fronting on the river. Not exactly out of sight but certainly out of mind, the riverbank was Calgary's "out back," a place to pile rubble waiting for another day or toss aside broken pieces of equipment, such as the wrecked car captured in a 1955 photograph teetering on the dike as if about to tumble into the water.[24]

The strongest public pressure for riverfront redevelopment came from the Calgary Local Council of Women. As part of a larger campaign to improve social services and public amenities in the city, the women petitioned city council to create a ribbon of parks along the south bank of the Bow River from 14th Street in the west to the Cushing Bridge in the east.[25] Council

Illus. 11.4a Calgary seen from the north side of the Bow River during the 1930s. Note the city turned away from the river. (Glenbow Archives, NA 3182-14)

Illus. 11.4b South bank of the Bow seen from the Centre Street Bridge in 1955. Note the backyards along the river. (Glenbow Archives, NA 5093-205)

Illus. 11.4c Junk on the banks of the Bow in Calgary, 1955. (Glenbow Archives, NA 5093-206)

accepted in principle the idea of a "River Bank Development Scheme" in 1955 and even took some small steps in the direction of carrying out the plan. But the city lacked the will and accordingly the resources to tackle a larger river park project seriously. The City Planning Department and the Technical Planning Board estimated the total cost of even a modest riverbank development scheme at $852,486.

By 1958, in an effort to stem the flow of residential development toward the suburbs, council rezoned the area between the city core and the riverbank from light industrial to high-density residential. This zoning change, it was hoped, would revitalize a decaying area, replace declining industry and deteriorating single family houses with modern high-rise apartment buildings, complement the commercial core with a nearby residential area, and increase tax revenues to pay for improved services, including parks. At the same time the city began to engage the many interest groups, public agencies, and its rudimentary planning capabilities in a process of thinking about a more comprehensive development plan for the downtown that would cover both the central business district and the river.[26] Despite good intentions, progress on the river park stalled when it came to the cost.

Appropriately, in a petroleum city, the automobile drove the next phase of the process. The corollary of suburbanization was, of course, traffic congestion entering and leaving the city centre. There were effectively three barriers to smooth traffic flow in Calgary: the river, the railway tracks, and the narrow downtown streets. In the eyes of transportation planners, the city needed more bridges across the river to connect with the northern suburbs, more access points through the railway lands to the south, and a major east-west artery through the central business district to distribute and collect traffic. Of course, parking places also had to be created for the thousands of cars pouring into the city each day. Beyond this narrowly focused thinking lay a broader modernist concern: an old, worn-out, and largely wooden two-storey cattle town needed to be completely refurbished with roads, office buildings, and apartment towers to make its proper mark on the skyline of a booming oil economy.[27]

At about the same time the CPR began to consider ways of maximizing the real estate potential of its underutilized rail lands in many cities, including Calgary, where it occupied a prime 107-acre tract running right through the heart of the central business district. The company developed a plan to build a string of office buildings, a hotel, a convention centre, car parks, and a transit hub on the site of its former main line. From the city perspective, this proposal would convert tax-free rail land into taxable commercial real estate and open up more roads through the rail barrier. But the tracks would have to go somewhere. Here, too, the city's and the company's interests seemed to coincide. The city would acquire land running along the south bank of the Bow where the CPR main line could be routed. Parallel to the railway the city could build a four-lane crosstown expressway. This mammoth project would in one fell swoop create a modern city of shiny office towers and expressways, greatly increase the tax base, smooth the flow of traffic in and out of the core, reroute the railway, and locate the highway on largely derelict and underutilized lands by the river.

Thus the city and the CPR almost buried the idea of a river park along the Bow under ballast and asphalt in the 1960s. But after a classic urban resistance movement and political debacle, the park idea resurfaced, now stronger and more ambitious than ever. The details of the controversy over the CPR development and the rerouting of the rail line need not detain us here; they have been dealt with elsewhere.[28] Even though the predictable "progressive" forces (the Chamber of Commerce, business groups, organized labour, real estate promoters) lined up behind the joint modernization project of the city and the CPR, it nevertheless failed. Much of the credit for this outcome could be claimed by the vociferous voices raised in opposition – the women's move-

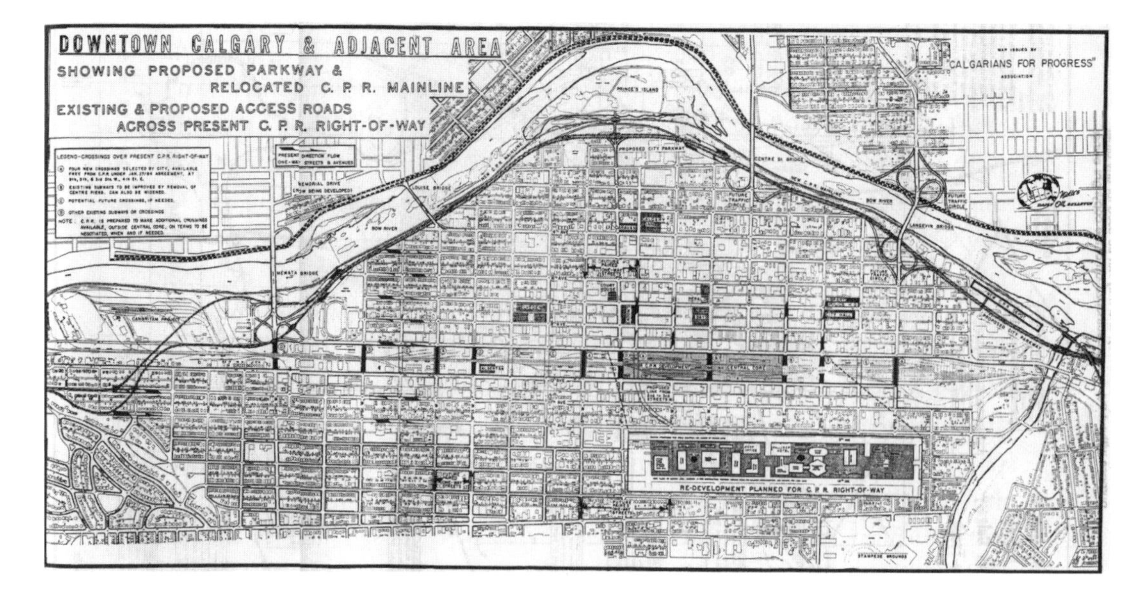

Illus. 11.5 Plan for the CPR lands development and track relocation. (Glenbow Archives, Barron Fonds)

ment, ratepayers in the way, disgruntled interest groups, academic critics, and congenital enemies of the railway. But the project also failed because an over-eager and inexperienced municipal bureaucracy negotiated a poor bargain with its CPR partner that exposed the city to all the potential risks and cost increases associated with land acquisition. When the incompetence of the city negotiators became known and the terms of the one-sided bargain became clearer, the whole scheme collapsed like a house of cards.

The fight against the rail corridor–expressway project in Calgary formed part of the remarkable urban political insurgency arising at the same time elsewhere in Canada and the United States in the face of major highway and redevelopment schemes. The intense political spasm caused by the scheme left the city exhausted and created enemies for life entrenched behind memories of what went right or wrong, what might have been or mercifully never was.

This intense controversy decided that the riverbank would not become a transportation corridor. It did not, however, determine that the riverbank would become a park. It would take a realignment of social, economic, and political forces to move from the negative to the positive position. Briefly, the separate forces converging after the CPR debacle could be enumerated as follows: the rise of professional planning, the creation of an effective park-promoting coalition, the apparent taming of the river, the development of a popular environmental consciousness, public affluence, and private philanthropy.

Planning received a big boost from the controversy, and planners and ratepayers placed more value upon parks than developers or politicians. The need for a more comprehensive concept of the urban core, rather than a piecemeal approach to individual developments, had been amply demonstrated. Planning had not yet acquired much authority or professional influence in Calgary. The city did not have an official plan at the beginning of the CPR negotiations.[29] Soon after the collapse of the project, however, it launched a major planning exercise under the direction of a Planning Advisory Committee. The committee's report, *The Future of Downtown Calgary*, published in 1966, featured a stinging critique of the "drab and depressing" downtown, the "ring of blight" surrounding it, the worn-out residential, commercial, and retail space, and "the lack of downtown parks and breathing space." Two of its recommendations had implications for the riverbank. In the first place the committee displaced the crosstown expressway, formerly designated for the south bank, farther south to 4th Avenue. At the same time it deplored the shortsightedness and lack of proper zoning regulation that had allowed the "numberless beauty spots" along the rivers to become blighted. Only "comprehensive redevelopment" would restore the bank to its rightful purpose and former beauty. The thirty-eight acres of Prince's Island presented an excellent opportunity to create an amenity as handsome and useful as Vancouver's Stanley Park in immediate proximity to the downtown. The committee plan envisioned riverbank redevelopment on either side of the main island park which would afford "leisurely, landscaped walkways as an integral part of the Downtown pedestrian system."[30]

As the new planning bureaucracy embraced the park idea, some veteran advocates of a riverfront green strip continued their campaign and gained powerful new allies. The Local Council of Women's campaign for riverfront beautification began to involve some influential urban groups, including such elite male domains as the Ranchman's Club and the local golf clubs. In 1969 Mr Justice Colin McLaurin's newly formed Bow River Beautification Committee joined forces with the Local Council of Women. McLaurin, a respected judge and chancellor of the university, having grown up in Calgary skating on the river, deplored the fact that the river had become the city's "back alley." After years of neglect, the moment had come to redeem the situation and adorn the city with the continuous river park it ought to have enjoyed all along. Following three years of arm-twisting, McLaurin formed a classic elite pressure group of forty-two men and women. These people knew how to work the political system. First, they put up $36,000 to fund a professional feasibility study proposing a design for the keystone of a river park system. A renewed Prince's Island would in turn form the backdrop for major urban renewal in the Eau Claire area.[31] They also knew that existing shared-cost programs reduced the city portion to only 20 per cent of the total cost. The federal government would put up 50 per cent and the province 30 per cent.

Thus in the late 1960s a powerful lobby group presented a convincing proposal to the city, consistent with the emerging urban plan, along with an attractive proposal to finance it. As it turned out, the city did not need much convincing. City officials, working in parallel with this private lobby, had meanwhile developed their own master plan for the island. In 1968 the city would launch a five-year, $1.5 million Prince's Island and Bow River Bank Beautification Program designed by landscape architects Man, Taylor, and Muret.[32] In just a few years a river park had become the conventional wisdom commanding social, bureaucratic, and political support. And all of this occurred without so much as a word being said about the possibility, raised only a few years earlier, of the riverbank becoming a transportation corridor.

When Calgary adopted a general city plan in 1973, the river park had become firmly embedded in municipal thinking. Maps accompanying this report showed continuous green space along the Bow and much of the Elbow. This report demonstrated that parks had moved from municipal luxuries to necessities in two fundamental ways. The planners pointed out that the growing leisure accompanying economic and social change created new responsibilities for municipal governments to provide more parks in order to incorporate recreational activity into daily life of the citizens. At the same time well-placed and properly designed parks enhanced the visual quality of city life. Parks thus filled two roles: as places for recreational activity and as aesthetic elements in the urban form. Accordingly, the 1973 official plan proposed that the river valleys be developed as open spaces which offered continuous public access to all the riverbanks at all times and connections to public footpaths throughout the city. The park areas should provide protection against air, water, noise, and visual pollution. They were thus a type of wearable art for cities, providing functional amenities and beautiful forms.[33] Meanwhile, oil-based affluence filled the municipal coffers. Just as important in rapidly growing budgets, these relatively small expenditures could be easily accommodated.

The river itself seemed to have settled down between its banks; it had shed its rogue image, appeared more benign and tranquil, and as such, it became something to be adorned and admired rather than feared and shunned for its floods and its burden of sewage. Planners and elites could not plausibly have envisioned beautiful parks along the bank if the river did not also co-operate by assuming a compliant character. By the late sixties it had seemingly given up its unruly nature, no longer lashing at its cutbanks with such fury, making islands and depositing debris in awkward places. Banks had been stabilized by vegetation. Summer floods were a thing of distant memory, thanks, it was thought, to the Calgary Power Company's extensive system of hydroelectric storage reservoirs. At the same time the dikes and the new Bearspaw dam just upstream from Calgary, built in the early 1950s, seemed to have solved the problem of winter flooding caused by ice jamming. The river was no longer a destructive force in the community. Apparently tamed, it flowed gently between its banks.[34] One could go down to the river without fear or foreboding and invest in expensive landscaping on the assumption that it would not all be swept away.

A broader cultural shift occurring across the continent heightened public awareness of the environment. In this new ethos the river and its valley represented nature in the city, something to be protected and enjoyed. This shift in popular consciousness can perhaps best be illustrated by the surprising furor that arose when, as we saw in an earlier chapter, the city proposed to cut down some trees around St George's Island to reduce the risk of flooding.

A certain kind of environmental sensitivity thus pervaded public consideration of the Bow River in the late 1960s and early 1970s. The river and its banks could be both objects of beauty and a welcome natural presence in the urban environment. But this sensitivity involved a highly selective view of nature. Not everything about the river was beautiful, and nature was valued most when it served human goals. In the consensus view nature should not be interfered with, but in other respects it needed to be improved upon. Minor adjustments to contours and banks could be permitted. Lagoons and reflecting ponds could be scooped out and the channel reamed, but dikes could not be raised too high or trees removed wholesale. Indeed, it was not sufficient just to leave the riverbanks as they were; certain kinds of improvements, some quite expensive, needed to be undertaken, such as planting, landscaping, accommodating picnickers, and building cycling and foot paths, for the river to achieve its full potential. Grass had to be planted and cut, flowerbeds and rockeries tended. On the treeless prairie the poplars and cottonwoods growing alongside the river in protected valleys were especially valued. Trees were to be encouraged, planted where necessary, and protected not only from human interference but also from the increasingly abundant beaver drawn to the city by these very trees.

Even within this popular view of nature, contradictory elements coexisted; it emphasized some things and ignored others. The Bow River valley in Calgary was to be remade in accordance with a "designer" view of nature that privileged aesthetics and certain kinds of pedestrian recreation. Nonetheless, in less than a decade the green strip, once the province of the Local Council of Women alone, became a highly popular policy, one closely watched by a vigilant and aroused citizenry in the riverside neighbourhoods. The city set in motion an extensive program of public works within the framework of an official plan and with strong backing from urban elites. This program retained broad popular support as long as certain rules were observed, a table of commandments spelled out in considerable detail by the Bow River Study Committee. Afterwards a Calgary River Management Committee, with representatives from the city, the province, and various interest groups, kept track of the extensive parks, recreation, floodway, and environmental enhancement projects under way in the metropolitan region.[35]

One more element was needed in order to bring a system of connected river parks into being, private philanthropy. Calgary's first family and largest fortune offered to match municipal expenditures to ensure public pedestrian access to a continuous trail along the riverbank throughout the city. In 1975 Donald Harvie's Devonian Foundation, which had generously endowed the city with the Glenbow Museum and made many other charitable donations, offered to put up $330,000 towards the construction of a fourteen-mile hiking and biking trail if the city would commit $115,000 to the project.[36] The city readily accepted. In subsequent years both the Devonian Foundation and the city increased their contributions. The partnership expanded to include the provincial government, which contributed both land and money. More than a million dollars eventually would be poured into this aspect of the river park alone. As a result, a park and trail system expanded many miles upriver to the west towards Bowness and downstream to the southeast towards the Pearce Estate and Walker's Inglewood Bird Sanctuary, the products of an earlier generation of philanthropy.[37]

The word "trail" had a special resonance in a Calgary setting. Aboriginal trails through the mountains and across the prairie and the trails over which cattle were driven were part of popular folklore. When the city built suburban expressways, it called them trails. Thus the Harvie gift and the government response also inscribed the river valley with another evocative meaning. The river became both a park and a trail. This string of public lands through which a four-foot-wide trail had been laid was thus crafted into a symbolic space, free of autos, commerce, and buildings, for people to recreate a uniquely western experience of moving across the land on foot (or perhaps bicycle) under a big sky. There was no historical justification for a trail parallel to the river; aboriginal travellers took more direct routes.[38] But this was a postmodern trail, not leading to or from anywhere but, rather, along and through something, a space to pass through without immediate purpose, simply for the joy of moving. A city being refashioned by oil money, which gloried in its anachronistic Stampede, added a river trail through parkland to its image.

Thus in the 1970s and 1980s, strongly supported by planners and public opinion and abetted by private philanthropy, the city of Calgary became a major force for park-making along the river. Indeed, within the city the river became, in conception at least, a continuous green-space trail dedicated to recreation rather than private appropriation or industrial uses. It became a civic space and a natural space, though natural in a carefully contrived way.

Park Province

During the 1960s, as the city of Calgary experienced a convulsion over riverside development ironically resulting in a program of extensive park building, the province of Alberta also went from being a laggard in the development of provincial parks to one of the leaders. By 1970 the Alberta provincial parks system covered almost 900 million acres. Government policy defined parks as tracts of land providing outdoor recreation or areas of natural or scenic beauty or historical significance. In the process of expanding park space, the stated goal of parks policy shifted from the original people-centred aim of providing "active and passive recreation in a setting of natural beauty for the population of the province and visitors" to the more environmentally sensitive objective of providing "recreational facilities in harmony with the preservation of specific geographical, geological, biological or historical features but always so as to minimize impairment."[39]

During the 1970s a "quiet revolution" occurred in Alberta, a sea change in provincial politics that deserves more extensive study.[40] Though it has been little noticed, parks policy formed a central element of the new province-led strategy of economic and social transformation. Following the election of Peter Lougheed's Conservative government in 1971 and a major revision of the Parks Act, parks policy became part of a conscious province-building strategy. A restructuring of government led to the integration of recreation resources and parks into a broader development plan for the province. In response to a growing concern over environmental issues, Lougheed created a new Department of the Environment to bring focus to the previously scattered government initiatives. Beyond that he created a high-profile Environment Conservation Authority to advise the government on land use, resource development, and planning, particularly on the eastern slopes of the Rockies. Even more broadly, the provincial government and its bureaucracy began to think in terms of a comprehensive strategy for the province in which economic, social, and cultural development planning would be coordinated with resource use, infrastructure creation, environmental protection, and wildlife management. According to this new way of thinking, water, parks, outdoor recreation, and environmental protection became as important as oil in a more nuanced and sophisticated provincial development strategy. This shift emerged as part of a changing conception of the process of policy formation, along with the professionalization of the bureaucracy in the provincial order of government.

With respect to parks and recreation, a powerful impetus was provided by the controversy in 1972 which led to the decision not to expand winter sports facilities in Banff National Park to accommodate a proposed bid for the Winter Olympics. Sporting activities on that scale and the infrastructure needed to serve them were deemed incompatible with the mandate of a national park. But a newly rich and ambitious Alberta wanted the Olympics, wanted more growth, and at the same time wanted to be a responsible steward of the environment. The imbroglio over the Winter Olympics drove the message home to the provincial government that it would have to reconcile these competing objectives according to its own plan for land, over which it possessed exclusive control.

Federal control of a key asset – outdoor mountain recreation in the Bow valley corridor – seemed to be standing in the province's way. Not only did Ottawa control the lands lying within Banff National Park, but by virtue of its involvement in the Eastern Rockies Forest Conservation Board, it also possessed great influence over development covering the entire eastern mountain corridor. As we have seen, the ERFCB had been created in 1947 by an agreement between the federal government and the province of Alberta to control fire and conserve forests in the headwaters of the prairie rivers with a view to maintaining downstream water levels. As time passed, however, the board's obvious bias against recreational uses of the eastern slopes in favour of fire control alienated Alberta, which gave precedence to recreation. Alberta withdrew its support from the federal-provincial agency in 1973 and replaced it the following year with a body under sole provincial control, the Environment Conservation Authority, which had a broader mandate, including recreation, tourism, economic development, and wildlife management.[41]

This new perspective permitted an imaginative reshaping of the river and its valley. As parks became integral elements of a provincial development plan, the Bow watershed became the focus of intense public scrutiny. For example, when the new Environmental Conservation Authority set about in a rational, bureaucratic way to develop a comprehensive policy framework for the entire eastern slope of the Rockies, it quickly directed its analytic energies to land use and resource development in the Bow River basin, which was under the greatest immediate pressure. In its report the authority noted that the Bow valley had experienced more "human intrusion" than any other area of the province, so that the increasing demands of recreational users, natural resource companies, agriculturalists, utilities, and transportation networks would have to be "accommodated to the traditional land use priority for the Eastern Slopes region, which is watershed preservation." In short, priority for any single use would have to give way.[42]

The Environment Conservation Authority did not so much produce a new policy as it prepared the ground for a policy shift. In its hearings and reports it gathered, aggregated, prioritized, and to that degree legitimized the views of various interest groups. Thus it prepared public opinion for changes even while it informed government of the needs and concerns discovered through polling, from written submissions, and at public hearings. For example, the 108 briefs received concerning the Bow identified recreation and tourism as the leading area of concern, followed closely by planning and further down by conservation. Since the mission of Banff had shifted to "landscape preservation," the unstated grounds for the rejection by the federal government of the Winter Olympic bid, the report concluded that the accommodation of outdoor recreation should be given high priority outside the boundaries of the national park on the provincial public lands within the Canmore corridor. A public opinion survey conducted as part of the broader eastern slopes study indicated that 65 per cent of respondents had visited that region within the past year, that Calgarians came to the region most often, and that the most popular activities were sightseeing (67 per cent), camping (38 per cent), hunting and fishing (20 per cent), hiking (18 per cent), and skiing (15 per cent). The topics of concern most frequently addressed in both the eastern slopes and the Bow basin hearings were "the protection of rivers and streams, the protection of wildlife, the development of recreational areas, the development of natural areas, the development of natural resources, and urban development." When participants were asked to rank these priorities, "most people felt that the protection of rivers was the most important concern, and that wildlife preservation was a close second." Accordingly, whereas both reports recommended a policy that "preserves and takes advantage of the natural endowments of the foothills and mountains, so as to retain the whole region as a renewable resource for the province," they quickly added a caveat that the highest priority be given to "the maintenance of the integrity of the watershed basins" to regulate seasonal flow and ensure water quality for industry, energy production, agriculture, food production, and urban populations.[43]

On the basis of these and other reports, the provincial government formed an Eastern Slopes Interdepartmental Planning Committee to prepare a zoning plan to implement these priorities and regulate conflicting land use needs within the Canmore corridor. Surprisingly, out of this bureaucratic swamp of quasi-independent agencies, stakeholder consultations, participatory democracy, task forces, and cabinet committees that had come into vogue in the 1970s, a policy for the Canmore corridor actually did emerge.[44] The most significant element of this plan with respect to new parks on the Bow River was the creation of a major public park covering most of the watershed of the tributary Kananaskis River. On the eastern face of the Rockies, within easy reach of Calgary, the government in 1977 set aside a large backcountry area in the upper zones of the river system for a provincial wilderness park and a large recreational area downstream for more intensive development of hotels, ski areas, and townsites. Opening up these areas previously dedicated to forestry, hydroelectric development, and watershed protection to various forms of recreation would take the pressure off Banff, while providing ample outdoor recreation opportunities of diverse kinds to the burgeoning population of Calgary and southern Alberta. Confining the two main recreational activities to dedicated zones – backcountry hiking and camping to the provincial park and hotel-based tourism and skiing to the recreation area – separated conflicting uses.[45]

Like Banff National Park, which it abutted, the Kananaskis Provincial Park would be far from "pristine." It had been burned over many times and logged since the late nineteenth century. The Spray River had been largely diverted to flow out of the reservoir in the Spray Lakes basin into the power canal leading to the hydroelectric plants above Canmore. The Kananaskis River itself had been rerouted through penstocks and turbines, and the exposed river canyon turned into hiking trails. But even as a used and redesigned wilderness it more than matched expectations as potential parkland with its mountain vistas, scenic lakes, and abundant wildlife. The recreational area opened up space for commercial recreation involving lodges, hotels, retail operations, and alpine skiing. By permitting multiple uses in different zones, the new provincial parks policy virtually restored to public control the area within the original 1911 boundaries of the old national park south of the Bow River. If greater ideological consistency about wilderness preservation was now being pursued inside the national park, a pragmatic zoning system was imposed on the provincial lands outside. Thus on one of the main tributaries of the Bow the province created both a showpiece "wilderness" mountain park and a five-star mountain resort fully capable of accommodating Olympic ambitions.

Through a parallel and virtually simultaneous process, the provincial government determined that Calgary should be endowed with a multi-purpose urban park. Curiously, this need seemed most keenly felt from the legislature in Edmonton, which overlooked the thousands of acres of recreational land covering the floor of the North Saskatchewan River valley in that city. Planners had begun thinking about Calgary's growing recreational needs as far back as the late 1950s, but several obstacles stood in the way of action. In the first instance, there was the cost of suitable land relative to the resources available. Secondly, the Parks Act made no provision for provincial parks within municipal boundaries. Until the requisite land, purchase money, and enabling legislation could be assembled, the occasional public campaigns in support of a major park in Calgary went nowhere.[46]

When the opportunity to acquire over two thousand acres on the southern limits of Calgary arose in the early 1970s, the Lougheed government sprang into action. It hastily allocated $15 million in the provincial estimates, made plans to amend the Parks Act to permit metropolitan parks, and quietly bought a strip of land about six miles long by a half-mile wide in the shallow grassland valley formed by Fish Creek where it joined the Bow River (land that had earlier been deemed too expensive to acquire). In February 1973 Premier Lougheed, with Mayor Rod Sykes at his side, announced creation of what they claimed was the first urban provincial park in the country.[47] In a further populist gesture the province appointed a Citizen's Advisory Committee of three men and three women with a mandate to consult the public to ensure that "one of the world's great parks" would be designed by the people themselves. Such a consultative process also served the delicate political purpose of brokering the highly polarized public opinion that had already formed on the subject of what kind of park that it should be.

The saga of the construction of Fish Creek Park is too complex and in some ways comic to be recounted here. No sooner did the province announce plans for the park but the city unveiled plans to drive a major expressway through it. The process of public consultation led predictably to something for everyone. The eventual defining theme of the park ultimately arose not from public consultation but, rather, from its location. Cattle still grazed the lush valley grassland during these deliberations. The proposed park occupied land that had once been part of the Bow Valley Ranch. This was not just any ranch but one that had connections with the some of the most memorable names in Alberta history. One of the earliest settlers, John Glenn, had located his cattle operation at the confluence of Fish Creek and the Bow River in the late nineteenth century to supply beef to the Mounted Police and provisions to Indian reserves. William Roper Hull expanded the operation after 1896 to include over 11,000 acres of deeded land in the Bow valley. After the turn of the century Hull hitched up his western Gothic city house to a team of sixteen horses and hauled it to a new location nestled in the lee of the bluff on the north side of Fish Creek. The industrialist Pat Burns later acquired the property and included it in the portfolio of ranches and farms connected to his integrated meat-packing business. At one time the property extended north to the Calgary Stampede grounds. Bit by bit the Burns family sold off parcels for suburban development. In the postwar era only the much reduced Fish Creek operation continued as a working ranch and farm finishing cattle on the southern outskirts of the city. The Bow Valley Ranch thus had associations with the very beginning of ranching in Alberta, with the epic period of ranching, and with one of the legendary Big Four ranchers who had launched the Calgary Stampede.[48]

The site in a sense recapitulated the history of ranching in southern Alberta. Fish Creek Park therefore became an ideal site close to the city for the province to celebrate its ranching heritage by preserving an iconic architectural landmark and pastoral setting. Thus the design committee recommended that, beside its natural and recreational aspects, the park should also emphasize these historic associations by restoring the buildings to their 1896–1912 appearance and mounting an interpretive program. The Burns Ranch buildings could be restored and the main house converted to a tea room and restaurant.[49] The management committee initially rejected these expensive proposals in 1978 but returned to them in the early 1980s.[50]

Thus with popular input and recognition of heritage activities, Fish Creek Park, declared to be the first urban provincial park in Canada, presented a somewhat hybrid design after it officially opened in 1975. Perhaps because of its thematic incoherence, the park proved immediately popular, especially with the burgeoning population of the southern Calgary suburbs. Off in one corner a group of ranch buildings mouldered and then underwent reconstruction. Expressways and sewers might be deemed antithetical to park uses, but that did not mean that open space, left in its "natural state," would suffice. Pathways were cut across the lands for biking and hiking. Picnic facilities, shelters, and other structures were built. Low-intensity recreation justified the excavation of a ten-acre artificial lake for swimming, an amenity that proved very popular. Fish Creek itself needed improvement to meet park expectations – it often went dry in the summer. So upstream storage reservoirs had to be built to ensure "that the creek would look like a creek."[51]

In the postwar period, then, as part of a broader economic and cultural program, the province of Alberta created parks along the tributaries and the mainstem of the Bow River, upcountry, and inside the Calgary metropolitan area. It did so to meet the perceived needs of an increasingly affluent, urban, motorized population seeking outdoor recreation. By providing access to mountains, lakes, hiking trails, ski hills, picnic tables, and campsites, Alberta was not only serving citizens but also branding a jurisdiction. The river and its valley became spaces upon which government could imprint its ambitions for the region. An increasingly urban, industrial society nestled comfortably

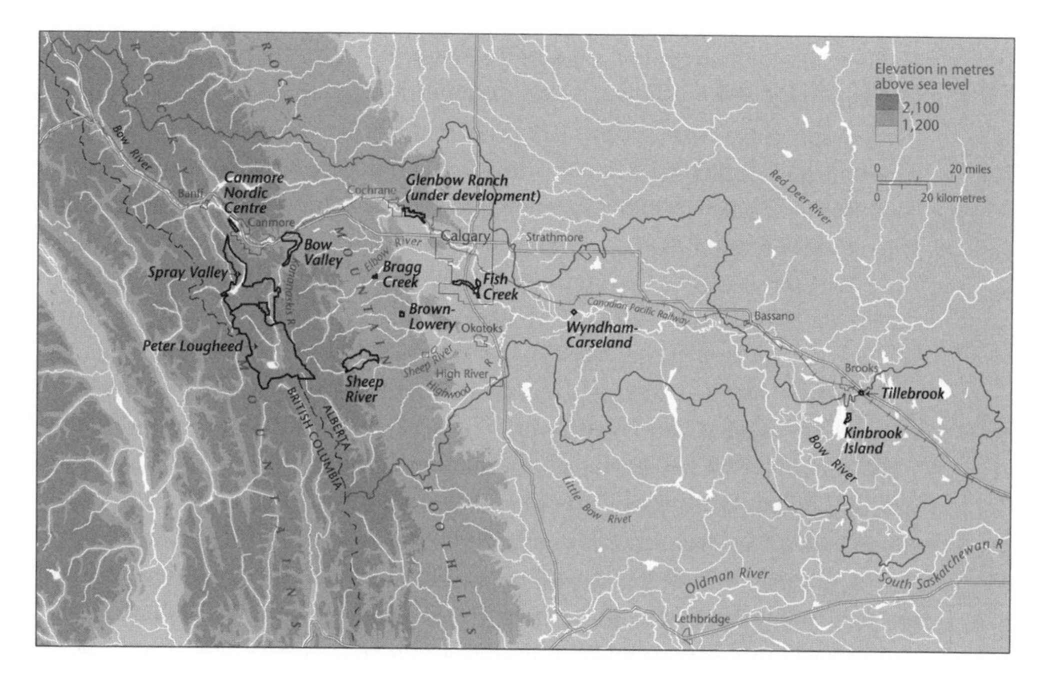

Map 11.1 The Bow River's primary provincial parks in 2008: Alberta parks and protected areas. (Adapted from Alberta, Parks and Protected Areas Maps: http://www.tprc.alberta.ca/parks/landreferencemanual/docs/pasites_pdfmap.pdf; http://tpr.alberta.ca/parks/landreferencemanual/docs/provpark/glenbow_oc.pdf; both accessed 10 November 2008)

and without contradiction upon the bosom of nature. Following a more latitudinarian parks doctrine, these provincial establishments situated themselves somewhere between urban playgrounds and federal sanctuaries. There were parks in which the Winter Olympics could and would be held. Here, too, in a society hurtling towards modernization, aspects of provincial history could be preserved in an explicit Heritage Park in the middle of Calgary, on the rolling bank of its water reservoir, and on Fish Creek on the old Burns property. Later the province would also acquire the abandoned McKinnon Ranch, where McKinnon Flats Provincial Park would become the main public access point to the downstream Bow River fishery.

In one sense Premier Brownlee's earlier vision of making Alberta a park within which its business could be conducted on a more agreeable and human basis was finally implemented by the Lougheed government after 1971. To give it an unfamiliar connotation, the oxymoronic phrase "industrial park" perhaps best captures the underlying provincial objective of embedding a highly sophisticated, modern, metropolitan society within a purposefully designed and zoned natural world that permitted multiple uses.

Three orders of government in Alberta followed different political imperatives in making parks along the Bow valley. The water and geography of the river were necessary central conditions to all three social movements. Thinking about the national park followed a trajectory roughly from spa, through mountain playground, to nature reserve and environmental sanctuary. By the late 1960s, thinking at the provincial level had undergone a profound change, a shift from seeing parks in isolation as service centres towards a comprehensive vision of southern Alberta with a variety of parks serving differing purposes, in which an event as large as the Winter Olympics could comfortably be staged. Only during that same decade did the city of Calgary move much beyond a cluttered vision of parks as playgrounds, gardens, and amusements. Eventually, conflicting schemes for future urban development led to the emergence of a strong public-private consensus formed around the notion of the river valley as a horizontal continuum of greenery winding through the gleaming towers of the city, connected by a sinuous pedestrian trail. Contradictions prevailed in all three evolving policies. Park-making did not consist of pure thought being realized at any level. Traditional uses, inconsistent goals, lack of resources, entrenched interests, and other awkward attributes of humanity often got in the way and compromised policy implementation.

Thus, even as the river was being used for economic purposes – forestry, hydroelectric generation, water and sewage, irrigation – it was at the same time being incorporated into culture for its aesthetic possibilities, which changed, in turn, with the changing eye of the beholder. Use as a park could mean many things, but it presupposed a certain kind of benign river, public management, and conscious design permitting and emphasizing some things and excluding or regulating others. Like other uses, parks made their own impacts upon the environment and ecosystem of the river. But these influences, seen through the lenses of the environmental consciousness of the late twentieth century, were deemed to be desirable.

12 · Water Powers

Nature united the Bow River. Humans put it asunder, drawing lines in the water, dividing it for purposes of control and exploitation. In law the bed was separated from the banks, the banks from the flow, the fish from the water. Different orders of government asserted jurisdiction over parts of the river; indigenous peoples, riparian owners, water-taking organizations, and corporations acquired rights to different aspects of it. The instruments of division were science and law, separate spheres in an imaginative conceptualization of nature. The river and its valley were visibly one thing; government imposed a multi-layered grid of invisible jurisdictions and rights over it.

Law attempted to part the waters. But the river, in its own relentless way, kept reasserting its essential unity, connecting those things that had been disconnected. Floods, droughts, deteriorating water quality, and environmental change forced an intellectual reintegration of this ownership patchwork. In time, humans would want to reunify the river for purposes of regulation, but what had been cut into so many pieces could not readily be put back together again. This chapter examines the evolving tension between the human desire to divide, appropriate, and control the river's attributes and, conversely, the wish to connect the separate pieces and reintegrate the river in order to permit collective action across multiple jurisdictions.

Division: Uncommon Law

Water is indispensable to humankind, but it is impractical for an individual to stake a claim to any particular volume of the liquid, which unused drains away or evaporates. As a result, water supply and use become a matter of the right of access to a given source. Yet such rights may conflict, and rules and regulations become necessary. The fluidity of the resource creates challenges for lawyers: try as they might, they have had many difficulties in translating such rules into an adjunct of real property law. In Canada, which drew upon several varying legal traditions (American and British, including Australian), cases and precedents in water law have not been numerous, nor are they always helpful in trying to comprehend the settlement of particular issues.[1] In fact, administrative regulation has been the principal means by which water rights have been divvied up, at least in the valley of the Bow.

The first step in the process of conceptual division involved the separation of the river from the land around and under it. This occurred as an incidental aspect of the incorporation of the North-Western Territory into the Canadian state and then as a purposeful act of public policy. As a part of Rupert's Land, the Bow River had been a possession of Great Britain since 1670, administered by the Hudson's Bay Company. But as we have seen, the region remained under the de facto control of indigenous peoples. No single cultural group or confederacy maintained dominance over the lands and waters of the entire Bow basin, as different indigenous groups prevailed in various places, and these groups changed over time. Their migrations, fortunes of war, and politics vis-à-vis newcomers determined who had access to the river and on what terms.

The transfer to the Dominion of Canada of the Hudson's Bay Company lands in 1870 extended to that area the prevailing Canadian (and English) common law of riparian rights. By this doctrine, rights in the water appertained to the land adjoining it. Owners of bank and bed acquired certain rights to water for domestic and agricultural purposes, to fish, and to use water in passage (the right of usufruct), as long as these activities did not impair the flow downstream or noticeably degrade water quality. Riparian rights holders owed certain obligations to downstream users and in turn were protected by common law against the adverse actions of landowners upstream. In the event of impairment of the flow by diversion or of the degradation of water quality, riparian owners had the right to sue and in turn were subject to claims of nuisance should they act unreasonably. Over the years the courts developed tests of reasonableness regarding flow or water quality impairment: essentially, if changes were imperceptible, no nuisance had occurred.[2] This regime of riparian rights, developed in regions of abundant water flow, would be put to the test in the semi-arid Canadian west.

The situation of de jure control and de facto possession remained in place after 1870 until the signing of Treaty No. 7 in 1877. At that point indigenous peoples surrendered their ownership claims to the land and its rivers based upon residency and control over access. In return they received reserve lands for their own exclusive use, per capita payments, ongoing annuities, and promises of educational, medical, and agricultural assistance. In addition, Treaty No. 7 granted indigenous peoples continuing rights to hunt (and presumably fish and trap) not only on their reserves but also on their "traditional lands," subject to regulations that the government might subsequently make.[3]

Thus the first lines across and through water defined the Indian reserves. The main Siksika reserve, as described in Treaty No. 7, consisted of two strips of land, one four miles deep running along the north shore of the Bow River and the South Saskatchewan River from a point west of Bow River Crossing to the junction of the Red Deer River and the other one mile deep on the south shore. The actual reserve as laid out in 1880s, however, was a roughly square block of land covering both sides of the river running much farther to the north and the south but not as far east. The Bow thus passed through the middle of this reserve.[4]

By the same treaty the Kainai and Piikani peoples received unspecified lands on the Oldman River, while the Nakoda were promised a reserve on the Bow in the vicinity of their main centre of activity around Morleyville. The Nakoda reserve as subsequently surveyed differed from these other allocations in that it consisted of three contiguous but separate parcels of land to accommodate the location of the distinct Chiniki, Bearspaw, and Wesley subgroups. These lands were legally described as running from the bank of the river to a line joining survey points in the uplands.[5] When the Tsuu T'ina also asked for a reserve of their own separate from the Siksika, they received a block of land south of the Bow River through which the Elbow River flowed. The federal government later insisted that even though the boundaries of the reserves were set forth as running along the two banks of the Bow, the reserves included ownership of the riverbed. Shortly after the Natural Resources Transfer Agreement of 1930 (handing over the unalienated Crown lands and resources to the province), Alberta claimed ownership of the riverbed, but the Justice Department insisted that the Treaty No. 7 reserves on the Bow included "the bed of the river at the sites in question." As trustee for the indigenous people on the reserves, the federal government retained ownership.[6]

The rights of riparian owners were also subject to certain other restrictions. The British North America Act (section 91 [10]) granted jurisdiction over "navigation" to the federal government. Thus on navigable rivers there existed an "easement" or "servitude" which permitted Ottawa to protect against any obstruction or interference and trumped other competing claims. In the late nineteenth century the government of Canada jealously guarded this jurisdiction over navigation. Federal powers expanded more broadly along this axis than under Ottawa's authority over trade and commerce, and navigation became a paramount consideration in the judicial interpretation of the law pertaining to waters in Canada. The public interest was deemed to be served by encouraging the flow of commerce. Therefore the courts strictly subordinated the rights of riparian owners who, in the pursuit of their private interest, might attempt to monopolize a river or erect structures that would hinder passage. Thus federal law and the courts regarded navigable rivers as shipping lanes that were common property, to be kept open for public use.[7]

Navigability in law was, and is, not necessarily identical with or determined by actual boating practice. Western rivers had historically been important transportation routes for the fur trade. In the late nineteenth century steamboats plied several western rivers, including the South Saskatchewan and parts of the Oldman; the shallower Bow, with its variable flow, resisted commercial steamboat development. Yet it was enough that the river *could* be used for communication by boat or vessel continuously between two points of public access for it to be deemed a navigable river. In the early years of settlement, transportation and communication in the region actually did rely to some extent upon the river. Rowboats carried goods and people; loggers drove their timber downstream to Calgary each spring. Eventually the railway and roads displaced the river as a practical transportation route, but in law it retained its theoretical navigability. As late as the 1940s the federal Department of Public Works still classified the Bow as navigable all the way downstream from the Louise Bridge in the west end of Calgary to its mouth.[8] Subsequent judicial interpretation has continued to stress the potentiality rather than the actuality of transportation use. As a government of Ontario legal opinion concerning a recent judgement (*Canoe Ontario v. Julian Reed*) concluded, "In essence, the test of navigability developed in Canada is one of public utility. If a waterway has real or potential practical value to the public as a means of travel or transport from one point of access to another point of access, the waterway is considered navigable."[9]

Control over navigation by the federal government could limit the rights of private riparian owners on navigable rivers as it could not on non-navigable rivers and streams. Ottawa had full power to regulate the erection of piers, weirs, bridges, dams, or obstructions of any kind that might hinder navigation. Once the Dominion Lands Act had been passed in 1884, ranchers acquiring property through leases or freehold also acquired rights of access to the shores of rivers, to drive cattle across them, and to use their waters in passage for domestic purposes. They could not, however, build unapproved structures in rivers or divert them and moderate their flow unreasonably. The beds of the river flowing through both the Nakoda and Siksika reserves remained Crown land. For further emphasis, Treaty No. 7 stated that control over navigation on the Saskatchewan and Bow Rivers as they passed through territory reserved for the Siksika remained in the hands of the federal government. In eastern Canada a raucous dispute arose between Ontario and the federal government as to who owned the bed of navigable waters, a conflict eventually resolved in favour of the province.[10] In the prairie west, however, Ottawa retained control of Crown lands and natural resources to promote settlement; so these resources remained a federal responsibility after the formation of a territorial government in 1894. Even after the creation of Alberta in 1905 the unalienated Crown lands and resources of Alberta stayed in federal hands until the Natural Resources Transfer Agreement of 1930.

Thus ownership of the bed of the Bow within the Siksika reserve continued under federal control for two reasons. In the first place, there was the deemed navigability of the river at that point. And second, along with the other Indian reserves of the Nakoda upriver at Morleyville and the Tsuu T'ina, whose reserve was situated astride the non-navigable Elbow River, the federal government continued to claim ownership of the riverbeds as trustee for the indigenous peoples. This complex legal situation may seem a small and perhaps pettifogging point to make, but it does underscore the much larger point that law and regulation divided the waters by granting different kinds of rights to various groups of people while reserving major powers to the Crown. The navigable waters doctrine, subsequently enshrined in the federal Navigable Waters Protection Act, stripped away significant riparian privileges from landowners along the Bow, including ranching leaseholders, homesteaders, and possessors of freehold title. Riparian landowners did, however, retain their usufructuary rights to take river water for domestic and agricultural purposes.

The application of riparian rights to water in a dry land necessarily "sterilized" much of the prairie in agricultural terms. Only those owners whose land abutted rivers or streams had access to water; a few strands of barbed wire could exclude others. Without water, ranching collapsed. Dry farming techniques left farmers dangerously exposed to the whims of the weather. However, if the water could be moved from the rivers to the inland properties, irrigated agriculture could bring the land into its highest productive uses. Thus a new pattern of water allocation emerged in the west, epitomized by the North-West Irrigation Act of 1894, which subordinated riparian rights to the public interest of watering a much broader region.

The North-West Irrigation Act, as discussed earlier, drew its principles from the water law of the American West and the Australian state of Victoria. From the earlier discussion it will be recalled that from Australia the Canadian legislation took the concept of Crown reservation and licensing of rights to water for irrigation. From the United States it borrowed the concept of prior appropriation to regulate the allocation of water in time of scarcity. These two doctrines – government ownership and licensing, on the one hand, and "first in time, first in right," on the other – became the cornerstones of subsequent water policy in western Canada.[11] This legislation and its subsequent amendments and regulations were administered by the federal Department of the Interior until 1930 and subsequently by the province of Alberta under its Water Resources Act.

With the North-West Irrigation Act and its subsequent amendments the federal government may have set out to accomplish the "total suppression of riparian rights" in order to create a much more efficient and equitable means of allocating water in an arid territory. But the legislation did not in fact achieve these purposes, leaving riparian owners with their common-law rights to take water for domestic purposes, which presumably included such things as stock-watering and gardening.[12] However, the legislation did clearly vest the right to regulate water use in the rivers and streams of the North-West Territories in the Crown, and by an amendment the following year it made the water itself federal property.[13] The Department of the Interior could now decide who had the right to use water, for what purposes, in what quantities, according to what priority, and for how long. A federal bureaucracy thus received and evaluated applications, drew up licences, established the priorities, and then regulated operations.[14] As the act's title would suggest, government had the promotion of irrigated agriculture squarely in view, but the legislation regulated all other uses as well. Sawmills needed federal licences for flumes, dams, and industrial uses.[15] The government licensed railways to withdraw water at various points for their locomotives.[16] Withdrawals for town water systems required federal permission. The act facilitated diversions for large-scale irrigation by the CPR and later organized irrigation districts, but it also regulated withdrawal and diversions on a very small scale for domestic and agricultural purposes.

By virtue of its comprehensive language, the North-West Irrigation Act also seemed to regulate hydroelectric undertakings. In 1903 the Department of the Interior received an application for a licence to the water power of Horseshoe Falls on the Bow. The negotiations for this licence brought into sharp focus some of the issues raised by the division of rights in the river and in turn showed that even the bureaucrats were confused by the multiple layers of jurisdiction.

At the outset, Edmonton lawyer Frank Oliver wrote to the minister of the Interior, Clifford Sifton, on behalf of some clients about securing power for a sawmill and flour mill at Horseshoe Falls on the Nakoda reserve.[17] The applicants would have to negotiate the purchase of reserve lands to reach the falls and construct their dam, power station, and transmission facilities along the river, as well as flooding land with their reservoir. Oliver was told that no water powers had yet been granted within Indian reserves in the North-West Territories, the implication being that a permit would not be forthcoming. Privately the departmental counsel agreed that a land sale might be arranged but doubted whether the Interior department had the authority to issue permits for a water power on what was deemed a navigable river. Ironically, as Interior minister himself after Sifton's resignation in 1905, Oliver would have to deal with the matter. He ruled that applications for diversions of water for power development could not be licensed under section 8 of the North-West Irrigation Act, which permitted grants for "domestic, irrigation or other purposes," on the grounds that water for power did not usually alter the quantity or quality of a river's flow, making it different from domestic or irrigation uses.[18] When the Nakoda people themselves were approached directly by the power developers, some younger band members approved selling land for water-power development, but Chief Chiniquay and his family disapproved of parting with such a large piece of the reserve.[19]

In April 1906 the CPR approached the Department of the Interior about acquiring rights to Horseshoe Falls and another water-power concession just upstream at Kananaskis Falls. In confusion, departmental bureaucrats turned down the railway's application on the grounds these water powers lay in what the department now considered "unnavigable waters within an Indian reserve." Responsibility for such matters was said to fall within the jurisdiction of the Department of Indian Affairs. In effect, the Lands Branch authorities appear to have been under the impression at this point that the Nakoda possessed riparian rights in the falls.[20] They would subsequently change their minds.

Indian Affairs began a delicate process of dealing with the Nakoda, using the Reverend John McDougall of Morleyville as an interlocutor, to find out what terms they would ask for should the CPR wish to press ahead. The Nakoda expressed some frustration that the company had previously conducted surveys without their permission and that its railway operations had often set fire to their pasture lands without compensation. McDougall was instructed, nonetheless, to continue the dialogue, making certain to separate the issue of land sales from water-power sales: "It is to be borne in mind in connection with the disposition of the water powers that there is serious doubt as to whether these are the property of the Indians, on account of the nature of the Bow River, and it is important that the question of the value of the land should be determined apart from the value of the water powers." Further discussion in the summer of 1906 indicated that the Nakoda might be willing to part with 3,000 acres for $10 an acre, a per capita payment of $20 ($13,000), and 300 head of cattle (worth $9,000). At this point the CPR dropped out of the negotiations.[21]

Soon afterwards W.M. Alexander and W.J. Budd applied to the Department of the Interior in December 1906 to obtain the water-power rights to Horseshoe Falls. On the strength of a previous legal opinion, the applicants were told by the department to negotiate a purchase agreement with the Nakoda for their works (a deal that would have to be approved by Indian Affairs). But the Lands Branch bureaucrats had now decided that though the river was navigable, the bed of the Bow belonged to the new province of Alberta; they also believed that a water-power use permit must be negotiated with their office, though that remained open for ministerial decision.[22] The land administrators decided that once the developers had acquired title to the necessary real estate, an interim licence to divert water could be issued, subject to conditions. Once development had been satisfactorily completed, the department would grant a permanent licence.[23]

Alexander and Budd therefore approached the Department of Indian Affairs to purchase the necessary land. When presented with this proposal, departmental officials explained to the applicant's engineers that they lacked the power to "grant land under water or permit diversions," but that the department could assist with the land transfer.[24] On 12 March 1907 Messrs Alexander and Budd signed an agreement to purchase the 1,000 acres of land from the Nakoda reserve. The band surrendered what it considered "gravelly land" unsuited to agriculture "upon such terms as the Government of the Dominion of Canada may deem most conducive to our welfare and that of our people." The power developers got the property and whatever rights that went with it for $10 an acre, a one-time payment of $3,350 (distributed as follows: $5 per capita, $15 for headmen, and $25 for chiefs), the purchase of 50 brood mares, and the promise to fence their property. Additionally, the businessmen agreed to pay $1,500 to the superintendent of Indian Affairs annually as a water rental. An interim agreement with Indian Affairs was subsequently drafted in the name of the Calgary Power and Transmission Company.[25]

J.B. Challies, the water-power officer in the Lands Branch of the Department of the Interior, strenuously objected to his bureau's exclusion from the preparation of this interim lease. He believed that neither the indigenous people nor Indian Affairs had the right to lease the water power under Treaty No. 7 or as riparian owners. They might sell the lands but not the right to the water power; so he argued the annual rental was being paid to the wrong party. His superiors agreed that the Department of the Interior alone (not Indian Affairs) had the right to issue a permit for the use of the water. Why they believed this is unclear from the documents. Section 92 (24) of the British North America Act granted the federal government jurisdiction over "Indians and lands reserved for Indians," exercised by the superintendent general of Indian Affairs.[26] Yet the boundaries of the Nakoda reserve lands were described as running along the banks of the Bow; so the administration of the riparian rights to the river-

Illus. 12.1 Penstocks of the Horseshoe Falls power station. (Authors' photo)

bed may have been thought to lie with the Department of the Interior rather than Indian Affairs. In any event, the water-power bureaucrats in the Lands Branch (who were in the process of seeking their own divisional designation as the Water Power Branch) insisted that the lease should stipulate the conditions under which the power might be used, the term, the obligations to continuous development owed by the lessee, the authority of government regulatory oversight of rates, and the conditions under which the licence would be suspended and the property recovered.[27]

Likely as a result of these negotiations and to ensure greater certainty and clarity, the federal government amended section 35 of the Dominion Lands Act in 1908 to assert its rights to control and license the use of water for power purposes, establish a regulatory framework, and create the Water Power Branch in the Department of the Interior to identify locations, conduct hydrographic surveys, issue licences, and regulate the construction and operation of hydroelectric works. The minister of the Interior was specifically granted authority to make regulations governing "the diversion, taking or use of water for power purposes" and "for fixing the fees … to be paid for the use of water for power purposes, and the rates to be charged for power or energy derived therefrom."[28] The Calgary Power Company, which took over the Calgary Power and Transmission Company licences, had to agree to the retroactive

Illus. 12.2 Kananaskis dam spilling water, 2005. (Authors' photo)

application of this legislation to their operation before receiving an interim permit in 1909. All the suggested provisions were subsequently included in the final lease with the Calgary Power Company.[29] This legal framework and regulatory regime of the amended Dominion Lands Act were later consolidated with the passage of the Dominion Water Power Act.

The subsequent application by Calgary Power to dam the Bow at Kananaskis just upstream from Horseshoe Falls raised many of the same jurisdictional questions since the proposed dam and power works lay inside the Nakoda reserve while the upstream reservoir was inside the Rocky Mountains National Park. Once again lands were purchased from the Nakoda along with an annual water-power rental fee. In the late 1920s the company sought to develop the Ghost-Radnor site, which would flood lands near the easterly edge of the Nakoda reserve. Each case revealed the complexity of the jurisdictional issues at play.[30]

Clearly, the Dominion Water Power Act, along with the Irrigation Act, placed virtually all water allocation decisions in the prairie provinces in the hands of the federal government. Private riparian rights, to the extent that they existed at all, remained restricted – confined to withdrawals or diversions for domestic use only. The creation of a national park in the headwaters of the Bow with steadily expanding boundaries to 1911 further extended federal authority over the river. And to conclude this litany of the plenitude of powers held by the federal government, its constitutional jurisdiction over fisheries, defended in the courts against provincial incursions and codified in various versions of the Fisheries Act, conveyed sweeping authority over the licensing of fishing in western rivers and lakes, interference with fish habitat, and ultimately control over water quality.

The federal government amassed almost total control over land and water during the process of colonization and settlement on the prairies. Alberta, by virtue of provincial jurisdiction in the field of health after 1905, did assume responsibility, through its boards of health and subsequently its Department of Health, for the licensing and regulation of municipal water systems, including those drawing from rivers. Under pressure from downstream users, the health authorities also extended provincial powers over waste-water deposits into rivers, although as we have seen, they were reluctant to force expensive sewage-treatment systems upon the municipalities until quite a recent date. The province also began to license fishing and hunting. Thus gradually during the first decades of the twentieth century a further division of the river into parts occurred as the provincial order of government started to exercise its legislative authority as well.

As the prairie provinces became more developed, they chafed at the increasingly anomalous situation of federal responsibility for areas that in eastern Canada lay within provincial jurisdiction, such as natural resources. Federal control over resources involved loss of provincial revenues from licensing extraction and development, although the provinces did receive a federal grant in lieu of lost revenues. Eventually a transfer of jurisdiction was negotiated which brought the prairies more fully into jurisdictional equivalence with their eastern provincial counterparts. In 1930, with the passage of the Natural Resources Transfer Act, the federal government conveyed responsibility for lands, forests, waters, and minerals to the prairie provinces.

In theory, the federal government surrendered its authority in 1930, and the province of Alberta took its place. Except, of course, the transfer of powers was far from complete. Indeed, some concern remained that water had not been explicitly conveyed in the transfer agreement, a matter rectified in 1938 by an amendment. Even though the provincial government assumed the responsibilities of the federal Department of the Interior for regulating water-taking, diversions, and hydroelectric development with the passage of its own legislation and the creation of its own bureaucracy, substantial areas of federal jurisdiction remained intact. Federal regulation of navigable rivers continued, though with the resources transfer the beds of navigable rivers in Alberta became the property of the provincial government, yet still subject to the control of navigation by Ottawa. Inland fisheries, too, remained a federal responsibility, which entailed some degree of authority over water quality. National parks and Indian reserves, which stretched over substantial reaches of the Bow River, persisted as areas of exclusive federal jurisdiction. And, indeed, during the dark years of the Great Depression immediately following the resources transfer, the federal government reinserted itself in areas of provincial jurisdiction, particularly water management, under its Prairie Farm Rehabilitation Administration, designed to restore drought lands and assist irrigation.

Thus even after 1930 the Bow River remained divided as before between federal and provincial authority, a complex shared jurisdiction in which two orders of government in different degrees controlled various aspects of the river. Moreover, from these two orders of government many individuals, businesses, municipal corporations, railways, electric utilities, irrigation districts, indigenous peoples, and other groups such as recreational users acquired sometimes overlapping and conflicting rights in the river and to its waters. The Bow River became a zone of federal-provincial jurisdictional entanglement. But together, governments and those acting on their authority largely stripped away control over the river waters from the property owners and residents on the banks. The river became a divisible bundle of utilities, access regulated by different orders of government and their agents.

Division of the river for a myriad of single purposes created conflicts. Upstream users diminished the flow at certain times and contaminated the water, rendering it unsuitable for use. Adjustments to the flow seemed to produce mysterious winter floods, even as variations apparently reduced summer inundations. The river connected, linking different single-purpose users through intended and unintended consequences. These differences were not matters to be settled by the courts in lawsuits but, rather, by negotiations between federal, provincial, municipal, and corporate authorities. Often the jurisdictional lines determining who was responsible for what remained unclear or disputed.

Interlude: Co-operative Federalism

The Great Depression and World War II left a powerful imprint upon the character of Canadian federalism. Prairie governments, which had notionally just graduated from the tutelage of the senior level of government, found themselves near bankruptcy with impoverished tax bases that left them unable to afford to act in their areas of jurisdiction. By comparison, the war endowed the federal government with greatly expanded taxing and spending power, a much more powerful bureaucracy with a reputation for managing economic and social forces for grand national purposes, and a mandate to use the state to develop the society and the economy. To achieve its goals in health, education, welfare, housing, transportation, and infrastructure – key areas of provincial jurisdiction – the federal government deployed its spending power to build a host of federal-provincial cost-sharing programs. Through the instrument of co-operative federalism, the federal government thus led the provinces, some more eagerly than others, into the age of the welfare state. Nor should it be thought that these projects were all driven from the top down. As we shall see, the provinces invested considerable energy (though at the cost of some self-esteem) in drawing a sometimes reluctant federal government into expensive programs within the provincial domain.

These joint federal-provincial programs had emerged in the 1920s and were then deployed with varying degrees of success during the 1930s. The most visible and effective program of this kind in western Canada, the Prairie Farm Rehabilitation Administration (PFRA), channelled federal funds to rebuild infrastructure, farms, businesses, and communities devastated by drought conditions of the 1930s. Alberta was a major target of this program, and irrigation infrastructure a prominent beneficiary.

This ethos of federal-provincial co-operation in areas of provincial jurisdiction would be expanded in the postwar era. More particularly for our purposes, federal-provincial efforts aimed at reconstruction and economic development profoundly affected control over the Bow River. Following the transfer of jurisdiction in 1930, the province of Alberta had slowly begun to build up its own statute law and regulatory bureaucracy in areas such as water resource management, irrigation, and wildlife control, in some cases simply by re-enacting federal legislation and hiring former federal government employees to run the provincial operations. But on the ground the transfer occurred much more slowly than expected. Fiscal problems during the Great Depression and then personnel shortages and conflicting priorities during World War II interfered with the province-building process.

During the era of postwar reconstruction the coordinated federal-provincial cost-sharing programs filled the gap. In the field of environmental management, two major intergovernmental agreements significantly modified what might have been a straightforward story of devolution of jurisdiction from the federal government to the province of Alberta. With the aim of conservation, orderly development, and the avoidance of interprovincial conflict, two supra-provincial agencies exercised authority over the Bow River and its waters, these bodies being the Eastern Rockies Forest Conservation Board (ERFCB) and the Prairie Provinces Water Board (PPWB). This form of multi-level governance may initially have seemed a satisfactory arrangement for a relatively poor, politically conservative jurisdiction facing a determined, well-financed central government intent upon social and economic planning for the nation, but it certainly did not fit the circumstances of what became an oil-rich, ambitious, rapidly growing province resentful of outside interference. As the structural transformation in the relationship of the two orders of government occurred between the 1940s and the late 1960s, it gradually undermined the foundations of this era of co-operative federalism on the river.

We have already dealt with the origins, operations, and ultimate demise of the Eastern Rockies Forest Conservation Board. It would be useful to keep in mind that episode, while here we shall concentrate on a parallel federal-provincial initiative affecting the Bow River, the Prairie Provinces Water Board. A joint venture of three western provinces (Alberta, Saskatchewan, and Manitoba) and the federal government, the PPWB was established after the war to ensure "integrated development of these interprovincial waters in the interests of all concerned."[31] The ERFCB and the PPWB, operating under concurrent federal and provincial statutes, represented an important delegation to an intergovernmental body of provincial responsibility over the Bow River and other resources.

The water board traced its origins to intergovernmental discussions during the Great Depression. Recognizing that there should continue to be a single registry of water-use plans, provincial officials had already agreed upon the establishment of a Western Water Board, which would centralize record-keeping. The federal cabinet, however, failed to approve this takeover of the functions previously exercised by the Department of the Interior. A prolonged drought in western Canada and associated rural distress led the federal government to establish the Prairie Farm Rehabilitation Administration in 1935, which encouraged farmers to undertake small irrigation projects. Yet Ottawa still declined to participate in the creation of a water-use agency with more wide-ranging powers for fear of being drawn into expensive, open-ended projects.

During World War II Ottawa's Advisory Committee on (Postwar) Reconstruction recommended "the setting up of a Western Provinces Water Board ... for the purpose of advising the respective governments on the most equitable and advantageous use of the limited water supplies."[32] After the war PFRA staff member W.M. Berry drew up a report entitled "Notes on Water Resources Development on the Canadian Prairies,"[33] pointing out the need for intergovernmental collaboration to avoid large development schemes being commenced without any prior technical analysis. In 1946 the three provincial governments took the initiative when their senior hydrological experts met to form the Prairie Provinces Advisory Water Board (PPAWB). Initially suspicious of provincial efforts to interfere in areas of federal jurisdiction, Ottawa eventually consented to send representatives from the PFRA and the Department of Mines and Resources to a second meeting of the PPAWB the following year. These bureaucrats convinced the federal government of the need for a co-operative regulatory agency covering the Saskatchewan-Nelson basin.

Within a few months a proposal had been drafted by the federal government for a new Prairie Provinces Water Board. Despite initial provincial concerns that the new body would become simply an arm of the federal PFRA and anxiety that its decisions might be binding on the provinces, the PPWB agreement was finally arrived at between Ottawa and the three provinces in 1948, creating a five-member body. Two members of the board would be appointed by Ottawa, one being the director of the Prairie Farm Rehabilitation Administration and the other a federal civil servant from the hydrology branch of the Department of Mines and Resources, while each of the provincial governments appointed one representative. Costs were borne one-half by the federal government and one-sixth each by the three provinces. The director of the PFRA became the first chair of the PPWB, which was also had its headquarters in Regina.

The mandate of the new body clearly reflected the concerns of prairie politicians. Alberta's agriculture minister, D.B. MacMillan, had warned against "clothing a body of civil servants with judicial power." Any one of the four governments could refer an issue to the PPWB, which was charged with the task of "recommending the best use to be made of interprovincial waters" and "the allocation of water … of streams flowing from one province into another." The board was given the responsibility of collating and analyzing whatever data it felt were required (though the actual collection was to be left to other branches of the federal and provincial bureaucracy). The expectation was that wide-ranging hydrological studies would be launched, followed by the formulation of a master plan for water use which would accommodate the needs of all concerned so far as possible. Yet the key point upon which the provinces insisted was that none of these recommendations would come into force until each of the four governments had ratified them by order-in-council. Only after the politicians had had their say would these findings acquire the force of law.[34]

The first meetings of the PPWB were devoted to reaching agreement on a set of priorities that would govern the approval of water-taking applications. The board agreed that its "general principles" should give the highest priority to allocations for projects already completed. Next in line would be water for partially finished developments. These guidelines implicitly recognized that certain undertakings were not using the full flow already allocated to them by a provincial government, which meant that the amount of water to be divided up in future was whatever remained once these pre-existing commitments had been fulfilled.

Once the PPWB was formed in 1948, Alberta formally requested a review of the existing allocations. Since the board agreed that these projects were to be accorded priority, approval for them was quickly granted. But the Saskatchewan government stalled on giving approval by order-in-council, claiming that Alberta was requesting more water than it was currently using. Agriculture minister I.C. Nollet advised the PPWB in 1949 that "any further allocation to Alberta at the present time will seriously affect this province, and, furthermore, will act as an effective barrier to the proper function of the Board." He was prepared to go no further than agreeing to the current levels of water-taking by Alberta.[35] The Alberta government was equally firm that the 1948 agreement had assumed that existing arrangements would be grandfathered. Though the PPWB finally passed a resolution in 1951 granting allocations for Alberta's existing uses, the Saskatchewan cabinet refused to pass the necessary order-in-council.

The interprovincial conflict became more acute in the spring of 1949 when Saskatchewan announced its intention to undertake the South Saskatchewan River Project (SSRP), building a vast dam on the river to irrigate up to 600,000 acres and generate hydroelectricity. The government requested from the PPWB an allocation of up to 5,500,000 acre-feet annually, more than the total flow of the river in dry years. In his formal request, agriculture minister Nollet admitted "that there may be a conflict between plans for the use of water ... in Saskatchewan and Alberta. Will your board report if there is a conflict, and if so, how far this conflict may be reconciled and by what means?"[36] In 1951 the PFRA endorsed the project, which would require substantial funding from Ottawa, but the following year a federal royal commission rejected the proposal. Saskatchewan then produced a formal reply which reiterated its support for the big dam.[37] The disagreement between the two prairie provinces, combined with the federal government's unwillingness to supply the required funds, made it impossible for the PPWB to produce any recommendations that could come into force through approval by orders-in-council.

The prospects for the South Saskatchewan River Project improved in 1957 when the Conservatives under John Diefenbaker won the national election, and Alvin Hamilton, another Saskatchewan MP who strongly favoured the SSRP, became an important federal minister. The plan for a huge dam near Outlook, Saskatchewan, to create a vast reservoir now seemed a real possibility. Yet Premier Ernest Manning's Social Credit government in Edmonton still had serious reservations about whether there was enough water available in the river to make the scheme feasible. In 1959 Manning wrote to his Saskatchewan counterpart, T.C. Douglas: "The concern of this province is that during some future drought period there will be a conflict between the use of water for irrigation in Alberta and the use of the same water for hydro power in Saskatchewan and Manitoba. It is our belief and one which is generally accepted that irrigation is a superior use of water to that of hydro, and, in fact, the Irrigation Act of Canada clearly sets forth this priority of use." Douglas was not prepared to give way; after consulting his cabinet, he replied: "Saskatchewan takes the position that no province can acquire a right to the waters of any interprovincial river which would prejudice the rights of another province to the use of a part of that river system, and that nothing can be done by the provinces, including Alberta, which would restrict the flow of the South Saskatchewan at any time in the future except to such an extent as has been or may be recommended by the [Prairie Provinces Water] Board and approved by the respective governments."[38]

Alberta's legal advisers became concerned in 1959 about the province's legal position in the event of such a dispute. The transfer of resources from Ottawa had left certain interprovincial issues unresolved. The Judicial Committee of the Privy Council's decision in the *St. Catherine's Milling* case (1888) seemed to confer absolute jurisdiction to each province over its own natural resources "without any need to answer to any adjoining province."[39]

At about the same time E.F. Durrant, who became the secretary of the PPWB in 1958, observed that under federal jurisdiction prior to 1930 a project on any stream in western Canada, interprovincial or otherwise, acquired priority according to the date of the issuance of the water-taking licence, the classic "first-come, first-served" principle. In 1931 the three provinces had accepted the already established priorities and subsequently followed the same principle. In addition, each province had adopted an order of precedence amongst competing uses similar to that in the North-West Irrigation Act, which ranked domestic requirements first, followed by municipal, industrial, and irrigation needs. Only Alberta's legislation referred specifically to waterpower use as being fifth in order of precedence. Yet the order of precedence for uses assigned by one province did not necessarily apply interprovincially. Wrote Durrant, "The transfer of water ownership to three provincial governments instead of one federal Crown has led to a paradox. On interprovincial streams all three provinces, through their legislation, can now claim ownership of the same waters."[40]

So long as the flow of an interprovincial river such as the South Saskatchewan was more than adequate to serve all competing uses, these problems remained theoretical, but by the 1960s there was talk of major hydroelectric and irrigation projects downstream which might absorb the full flow in the foreseeable future. Once the average volume had been allocated, a dry year would mean that not all users might be able to secure their full requirements. Without any established interprovincial priorities, it would be difficult to decide who should see their demands curtailed or even cut off entirely. In another memorandum Durrant asked what would happen to Alberta "if a downstream province should license a hydro plant to generate power from the full flow of the river today, how will consumptive uses in the upstream province be affected in the next 50, 100 or 200 years?"[41]

Durrant was loath to see such a dispute land up in the courts. Not only was this likely to create ill will between the governments, but it was risky because there had been little previous litigation concerning water law, making it difficult to know how the courts would rule on a case concerning an interprovincial dispute. If a resort to the courts was to be avoided, Durrant pointed out, the Prairie Provinces Water Board as presently constituted was not a satisfactory tribunal. In its first decade of existence the board had never attempted to fix either priority or precedence in interprovincial relations to the projects that had already been approved. As the demand for water increased in future, the board was likely to make allocations based on the average or median flow of the interprovincial rivers, but this would be insufficient in years of low flow. And even if the PPWB were to establish a clear scale, "there is still no machinery for satisfying these priorities in a dry year. That is, there is no board of control or water master to check the delivery of water across provincial boundaries. This type of control or information is necessary so that each province may be advised of its water supply in a dry year. Each province could then arrange to water master the projects within the province according to their license numbers."

As Durrant predicted, the early 1960s saw a number of new plans involving ever-larger diversions put forward by Saskatchewan and Manitoba, which aroused increasing concern for the government of Alberta. In 1964 Saskatchewan entered a claim to divert over 5,400,000 acre-feet annually from both the north and south branches of the Saskatchewan River. At that time Alberta, the only province then using significant quantities of water for irrigation, was still consuming only 1,000,000 acre-feet annually to irrigate about 650,000 acres. Alberta officials complained about an attempted grab by Saskatchewan, pointing out that it had taken their province fifty years to reach the current level of consumption and that it would probably take another fifty years before another 1,000,000 acre-feet were required.[42] As a result, the Albertans pressed the PPWB to approve quotas specifically for irrigation, which would ultimately allow them to double their diversion to 2,100,000 acre-feet annually, claiming that such uses should have priority over all others on the grounds that they were essential for food production. Saskatchewan and Manitoba resisted, arguing instead that the entire natural flow of the Saskatchewan River basin should be permanently allocated amongst the three jurisdictions on a percentage basis, so that the downstream provinces would not find themselves short of water in dry years once Alberta's needs had been met. Meanwhile, the newly created Saskatchewan Water Resources Commission put up stiff resistance to the idea of granting fixed apportionments for irrigation, and the idea was eventually dropped.[43]

By the mid-1960s the various competing claims for water from the South Saskatchewan had rendered the Prairie Provinces Water Board unable to arrive at any recommendations that would become effective through recognition by orders-in-council passed by the three provinces and the federal government. In particular, Alberta remained adamant that it should receive a higher allotment of water to be used for irrigation, while Saskatchewan would not accept such proposals for fear that the flow of the river would drop so low in dry years that the reservoir behind the SSRP dam would hold back insufficient reserves. On the other hand, if Saskatchewan received its desired allotment, Alberta feared that it would not have enough water to meet the needs of its irrigators.

The possibility of future water shortages was a particular concern of Alberta officials. The province still had plenty of untapped resources, but these were located well away from the population concentrated in the south. Two-thirds of the land area of the province lay in the watershed of the Mackenzie River, which flowed through sparsely peopled territory to the Arctic Ocean carrying about 85 per cent of Alberta's water. The southern third of the province, which consumed 80 per cent of its water supplies, possessed only 15 per cent of its total volume. In 1965 the chief engineer of Alberta's Water Resources Branch suggested a means of altering this situation by devising the Prairie Rivers Improvement and Management Evaluation Programme (PRIME), which proposed a huge series of dams and diversions to channel northbound Mackenzie water over into the Saskatchewan River watershed. Alberta engineers contended that in the long term the plan to divert water from the Arctic watershed into the Saskatchewan basin "will eventually cancel out our consumptive use of water for irrigation."[44]

The very scope and complexity of this scheme prevented any quick action, particularly after the Conservative opposition in Alberta joined in a chorus of criticisms of the plan as outdated and environmentally unsound.[45] Meanwhile, Saskatchewan kept up the pressure for an agreement on the apportionment of the flow of the Saskatchewan River. In 1966 PPWB member Harold Pope insisted to his Alberta counterpart that he sought no undue advantage for his province and was quite ready to give and take in an amicable spirit: "It is not a question of our province endeavouring to obtain something to the

Illus. 12.3 Bassano Weir, autumn 2002. (Authors' photo)

detriment of Alberta. I hope you feel that we are giving consideration to this matter on a realistic and practical basis with a view to developing our water resources in such a way as to benefit both our provinces as well as Manitoba and Canada as a whole."[46]

However, the hope that it might be possible to determine the "best use" of the waters of the South Saskatchewan proved unfounded. The Alberta government asserted that irrigation (to grow food) must be rated most highly, though with the additional unstated premise that non-consumptive uses (such as the generation of hydroelectricity) must also be accorded great worth. Having large irrigation projects, Alberta also insisted that their expansion represented the most efficient use of a finite flow. Saskatchewan wanted not only to expand its own irrigation and sanitary projects but also to build hydroelectric facilities within its borders. Lacking the capacity to arrive at binding decisions when provincial interests clashed, the PPWB collapsed.

By the end of the 1960s co-operative federalism as represented by both the Eastern Rockies Forest Conservation Board and the Prairie Provinces Water Board was no longer functional as far as Alberta was concerned. In watershed management and water apportionment, intergovernmental arrangements seemed to stand in the way of the province's best interests. It thus sought ways of disentangling itself from arrangements, which added new layers of complexity to dividing the waters. By this time the oil-rich province had the money and the administrative capability to go its own way, bearing, it might be said, a considerable chip on its shoulder over federal and extra-provincial interference in its jurisdiction.

Reunification: Provincial Self-Determination

The first step involved dismantling or modifying existing intergovernmental agreements to create a zone of exclusive provincial authority. Then the province had to put in place its own management plans and regulatory instruments within those boundaries. This change in provincial strategy roughly coincided with a change in government, in which a younger generation of statist Conservatives, led by Peter Lougheed, came to power in 1971. In a mood of affluent province-building, Alberta set out to reassert its authority over the water in the Bow River.

At the Prairie Provinces Water Board between 1964 and 1968 there had gradually emerged an alternative guiding principle. "Best use" of water might be replaced by the notion of "equitable division," under which each of the three provinces bordering the South Saskatchewan might receive a fair share of the total flow. If that proposition was accepted, then the flow could be allocated at the provincial boundaries, leaving it to each government to establish priorities amongst the competing users within its borders.

Translating these alternative propositions into a concrete agreement still posed a challenge. One attractive definition of an "equitable" division of the river's flow might be to grant Alberta the right to use half the flow of the river and to release the other half across the provincial boundary, while Saskatchewan in turn would allow half the river's flow to enter Manitoba. While this proposition might be attractive to the downstream provinces, Alberta remained insistent that it should be permitted to take a guaranteed minimum of 2,100,000 acre-feet annually to serve its irrigation projects. However, the Saskatchewan government was concerned that in years of low flow this requirement might mean that it would not receive enough water to meet its own commitments, particularly once the South Saskatchewan River Project was completed. Thus the principle of "equity" seemed to require some reciprocal commitment by Alberta to maintain a minimum flow in the river at the provincial boundary.

Gradually the details were hammered out and a consensus acceptable to all the parties eventually arrived at. In the autumn of 1969 ministers from the three provinces and the federal government signed a new "Master Agreement" to formalize the arrangements reached between them for "an equitable apportionment" of the waters "arising in or flowing through" Alberta which passed into Saskatchewan and thence onward to Manitoba. Schedule A permitted Alberta "to make a net depletion of one-half the natural flow of water" and to allow the remaining one-half to flow into Saskatchewan, which could then make a similar 50 per cent depletion from the flow of the Saskatchewan (including flows added within the province) before the water passed into Manitoba. Alberta was granted the right to make a minimum net depletion of 2,100,000 acre-feet from the South Saskatchewan River even in years of low flow, provided that the diversion did not reduce the water in the river to less than 1,500 cubic feet per second at the provincial boundary.[47]

A reconstituted Prairie Provinces Water Board assumed responsibility for measuring flows in order to carry out this agreement. Major monitoring stations would be established where the North and South Saskatchewan and Red Deer Rivers crossed the Alberta-Saskatchewan line and where the Churchill, the Saskatchewan, and the Qu'Appelle flowed into Manitoba. Alberta was permitted to divert water between its rivers so long as it met the commitment to its neighbour, and the flow of the Red Deer River, which joined the South Saskatchewan just east of the boundary between the two provinces, could be used to supplement the flows of the Bow and the Oldman, which were heavily used upstream for irrigation. As the reorganized PPWB turned its attentions to less controversial technical matters such as measuring flow and water quality, assessing demand, and preparing forecasts about possible future requirements, the two provinces were freed to regulate their own waters within the framework of the agreement.[48]

Saskatchewan officials considered this 1969 Master Agreement "the most important document ever signed on the prairies." The guaranteed water supply would enable that province to undertake planning for construction of the 1,800-foot-long, 600-foot-high Gardiner dam on the South Saskatchewan near Outlook, creating the 125-mile-long Lake Diefenbaker, which, with associated developments at Coteau Creek and Squaw Rapids, could turn out about 17 per cent of the hydroelectricity consumed in the province. Though irrigation was comparatively slow to develop around Lake Diefenbaker, the ability to store the flow of the South Saskatchewan made it a more reliable source of water for 40 per cent of Saskatchewan's citizens, including residents of its three largest cities.[49]

Saskatchewan was therefore particularly concerned to ensure that the volume of water in the river did not vary too much, even while recognizing that climatic conditions did not permit absolute regularity. Lower flow would not only affect hydro-power production but also be detrimental to recreation, fishing, and wildlife downstream. Relatively high volumes were needed to flush out riverbeds and prevent serious ice buildups in Lake Diefenbaker. Moreover, declines in water quality resulting from high mineral concentrations and nutrient loads from sanitary and irrigation uses were also of continuing concern to downstream users.[50]

Before any such projects could become a reality, the reconstituted Prairie Provinces Water Board had to conduct its stream-flow studies. From now on hydrological science would modulate political negotiations. Scientists were once more set to work gathering the necessary data and in this case doing the mathematical modelling required to arrive at a determination of an entirely imaginary value, the "natural flow" of the river. Natural flow was not something that could be observed in nature and measured. Rather, it had to be calculated after taking into account the many withdrawals and inputs, evaporation and infiltration, annual variations, and a host of other factors. The existing Data Network Planning Committee was charged with determining the natural flow of each stream, defined as "the quantity of water which would naturally flow in any watercourse had the flow not been affected by human interference or intervention." This committee, soon renamed the Committee on Hydrology, turned to the scientists of the Water Survey of Canada (WSC), who would do the actual work, for advice on methods of measuring natural flow. The WSC hydrologists also continued work begun for the PPWB in 1956 to gauge the return flow to the South Saskatchewan from Alberta's Eastern and Western Irrigation Districts and the Bow River development.

The ultimate aim of this data collection begun in 1971 was to allow stream-flow forecasting on each major interprovincial river, both of long-range water volume or water supply and real-time flow forecasts. By 1976 the data had been consolidated into computerized models based upon the US National Weather Service River Forecasting System and the University of British Columbia Basin Simulation Model. The following year the figures were reworked into stream-flow forecasts for each of the major river basins to allow forward planning.[51] These sophisticated calculations were necessary to the proper working of a regulatory regime that required half the natural flow of the river to be passed on.

The Master Agreement of 1969 broke the political deadlock between Alberta and Saskatchewan over various competing uses of the flow of the South Saskatchewan River. By accepting the principle of equitable allocation as between provinces, Alberta had secured the right to use half the flow of the Bow River for its own purposes according to its own priorities without interference from outside authorities, provided that a minimum flow was ensured in dry years. The agreement therefore left the issue of allocating uses among various competing purposes to be settled by the provincial government.

The Conservatives thus assumed power in 1971 at a moment when Alberta needed to develop a comprehensive and coherent provincial policy for water resource management. In meeting this challenge, the Lougheed government introduced not only a new way of thinking about water but also new ways of making water policy. First, the more urban-oriented Conservatives broadened the policy framework beyond water quantity to include water quality. Secondly, the new government tried to reconceptualize water and river management within an environmental, rather than a narrowly agricultural or economic, frame of reference. And thirdly, it sought to develop its policies in a participatory way, engaging interest groups, academics, businesses, social organizations, the media, and private citizens in a process that drew attention to the issues, educated a wider public, and generated some political support for action. In the argot of the day, the process was as important as the product: participation theoretically created the conditions for change. The new government thus set out to reunify the fractured responsibility for water management within provincial hands. Lougheed tapped a growing anxiety, mainly in urban areas, about water pollution and other forms of environmental degradation. In an era when major media attention focused on oil and gas, inflation, and the constitution, his government sought to refocus interest on other equally important strategic issues: water and the environment.

To signal this shift the government created a Ministry of the Environment in 1971 and passed Clean Air and Clean Water acts. As an indication of how fragmented responsibility for the environment had become, the new department had at least partial responsibility for administering thirty-two pieces of legislation, from the Agricultural Chemicals Act to the Wildlife Act.[52] As a keystone ministry, Environment was expected to coordinate the effective application of these statutes and bring to bear a concerted effort to enhance some vaguely defined notion of environmental protection. The ministry, made up of units drawn from other departments (e.g., Water Resources from Agriculture), was also to oversee the development of a coherent environmental strategy for the province which would reconcile the needs of agriculture, industry, towns and cities, recreation, tourism, and wilderness protection – a tall order. During the 1970s the Ministry of the Environment, along with the restructured Prairie Provinces Water Board, mobilized resources to do the basic science of sampling, analysis, measurement, and benchmarking upon which public policy could be based.[53]

In Alberta water was an issue that transcended even the remit of the revamped Ministry of the Environment. Dealing with water politics required more than simple bureaucratic management and scientific investigation because water issues were thoroughly interlaced with so many activities and because the water resource itself was not equally distributed across the province. The relatively thinly populated north had lots of water; the rapidly growing south had little water, and what it had was already heavily committed and its quality compromised. Moreover, the Master Agreement limited how much water Alberta could use. Because the agreement treated the three main tributaries of the South Saskatchewan as a single source, additional withdrawal in one river imposed limitations on withdrawals in the other two. Where the needs were greatest, water was scarce; conflicts over its use threatened to intensify as upper limits of use approached and water quality deteriorated. This was the new and potentially volatile politics of water the government had to address.

The pressures were greatest in the South Saskatchewan Basin, consisting of the Oldman, Bow, and Red Deer River watersheds. In opposition Peter Lougheed had strenuously contested large-scale inter-basin transfers of water from the north as a solution to southern shortages on environmental grounds. He and his party proposed instead to develop more efficient management of in-basin water resources.[54] This basic decision would guide the government in the years ahead: better management of what was available, rather than diversions from the north. It also meant the government would have to deal with in-basin conflicts over competing uses of a finite resource.

At the same time the government managed to disentangle itself from previous federal-provincial agreements. For example, Alberta extracted itself from its long-standing agreement with the federal government over the management of lands in the Saskatchewan River headwaters. The province quietly let the Eastern Rockies Forest Conservation Board agreement expire at the end of its twenty-five-year term in 1973 and then delegated these responsibilities to its own Environment Conservation Authority with a mandate to plan across a wider field of activities than simply water-flow maximization.[55] It should be noted, too, that in 1973, following negotiations, the federal government withdrew from agricultural rehabilitation programs in Alberta, primarily the construction and management of irrigation works, turning over all of its irrigation assets to the province for management. Thus between 1969 and 1973 the provincial government brought together under its own jurisdiction responsibilities that had been previously divided between the Prairie Provinces Water Board, the Eastern Rockies Forest Conservation Board, and the descendants of the Prairie Farm Rehabilitation Administration. Water resources and rivers had at last been brought under provincial control – or so it seemed.

Comprehensive policies thus had to be developed. A Water Advisory Committee was set up to inform cabinet decision-making. The committee set out to develop water management plans to address the quite different mix of problems (flooding, pollution, water-taking) in six major river basins (the Oldman, Bow, Red Deer, North Saskatchewan, Athabasca, and Peace).[56] During the early 1970s the Environment Conservation Authority conducted a major study of land and water use on the eastern slopes of the Rockies with particular emphasis upon the Bow valley corridor, which led to the opening up of the front range to park and recreational activities. Nevertheless, the authority also concluded that water should have the highest priority in the Bow basin and that the watershed should be managed to maximize the quality and quantity of water produced.[57] During this period the Ministry of the Environment focused much of its attention in southern Alberta on the Oldman and Red Deer Rivers, where water shortages were most acute.[58]

Eventually, however, mounting complaints from citizens, the media, and downstream users about water quality in the Bow River and the looming possibility of shortages and conflicts amongst competing users forced the government in the early 1980s to launch a major investigation of water use and needs in the entire South Saskatchewan River basin and afterwards to conduct an extended dialogue with stakeholders and interest groups to seek ways of improving quality, using water more efficiently, and resolving latent differences over priorities.[59] The South Saskatchewan River Basin Study (SSRBS), conducted between 1981 and 1984, had as its goal developing a strategic plan and providing citizens and decision-makers with information about options. The study dealt with the entire basin because under the Master Agreement the three southern Alberta rivers making up the South Saskatchewan (the Red Deer, the Bow, and the Oldman) were treated as one in measuring the flow at the border. Approvals for water-taking on one river necessarily limited choices elsewhere, as deficits must be remedied. Policy for the Bow therefore had to be coordinated with that on the other two rivers.

The SSRBS was a classic example of the bureaucratically led, scientifically based exercise in participatory policy formation. Environment Alberta scientific teams and academics gathered data about the river system,[60] planners drew up policy objectives and alternative regulatory scenarios, and for two years investigators consulted hundreds of individuals, scores of organizations, and more than two hundred agricultural, industrial, recreational, and municipal groups to determine and tabulate their water-use objectives.

The final report, succinctly written and attractively packaged in a glossy format, argued that not all desirable goals or interest group objectives could be achieved. Trade-offs would be necessary, such as limiting water-taking for irrigation to enhance the fishery; treating sewage more thoroughly to permit downstream water-taking; sacrificing scenic, wildlife, or recreational habitat for irrigation or hydroelectric storage; or limiting water-intensive development in one basin to permit expansion in other basins. The SSRBS sought to inform interested parties about the nature of the river systems, lay out the natural limits and policy constraints on development, reflect back the impact of historical demands upon the river, anticipate future demands, and identify explicit or latent community values. These goals it accomplished with some success. It was much less effective in moving towards an actual plan. It concluded with four overly complicated scenarios of development, each with five different mixes of outcomes for irrigation, fish protection, and recreation that were too technical, abstract, and confusing to be of practical use. Moreover, the opening observation in the report that Alberta typically passed on 83 per cent of the natural flow to Saskatchewan, rather then the 50 per cent minimum, seemed to suggest that there was a lot more water in the rivers for the taking before zero-sum trade-offs between sectors would be necessary.[61]

This study and its attendant public discussions emphasized the importance of establishing priorities in water management. The overarching message that resonated through the study and hearings was roughly as follows: for there to be enough water of good quality to sustain economic growth, the environment, and high living standards, water would have to be used more efficiently, according to agreed-upon priorities, in quantities that would have to be limited, and that it would have to be thoroughly treated before being returned to the rivers and passed on in good shape to Saskatchewan in the proportions agreed. But it was also evident that irrigated agricultural interests were driving the agenda. The main focus was on quantity. This is not surprising, given the fact that irrigated acreage in the basin had doubled between 1950 and 1980, most of the expansion coming in the 1970s.[62] The clear subtext of the discussion was that irrigation would need more water and that it should have it.

Meanwhile, as water became more of a contested issue, especially in the southern part of Alberta, the provincial government saw the wisdom of erecting another layer of bureaucratic armour-plating between itself and the frequently warring parties. As decision-making became more difficult, government sought buffering by intermediate quasi-judicial agencies to deflect and moderate criticism and, where possible, negotiate compromises. A prolonged drought in the Oldman River basin and erratic supplies of water in the upper reaches of the Red Deer River had raised fierce local storms of protest from the irrigation interests. On the Red Deer the government went ahead with a controversial irrigation dam, twice rejected by the Environment Conservation Agency, presumably on the grounds that voting farmers took priority over environmental issues.[63] On the Oldman irrigators were clamouring for a major dam to address current water shortages and allow for a doubling of irrigated acreage.

In 1983 the government established the Alberta Water Resources Commission to provide advice to cabinet "on long term planning in relation to agriculture, economic development, community and recreation, and environmental factors with respect to the water resource." The commission was also charged with supplying advice on specific projects, intergovernmental negotiations with respect to water, and water law reform.[64] As one of its first big responsibilities, the AWRC conducted an extensive series of public hearings and drafted a report aimed at the development of a comprehensive management plan for the South Saskatchewan basin. In this instance the commission was expected to pick up where the SSRBS had left off, in the hope that interest groups might saw off their differences in open discussions and a clearer zone of consensus might form before cabinet decisions had to be made.

The AWRC asked essentially the same questions about allocation priorities posed by the SSRBS. With fifteen days of public hearings in ten cities during November and December of 1984, the commission received 227 briefs and recorded about five thousand pages of testimony in transcripts. The very strong and forceful conflicting opinions expressed in the hearings stand in marked contrast to the rather anodyne report, which dissolved differences in platitudinous recommendations. Some witnesses detected a pro-irrigation bias in the selection of commissioners (three of the five had irrigation associations) and the disproportionate amount of time (eleven out of fifteen days) spent in irrigation-dependent centres.[65] Recent approval of the Oldman River dam by the Lougheed government over the objections of even the restructured Environmental Council of Alberta (which had replaced the Environment Conservation Agency) heightened suspicions. All told, in the course of the open sessions and submission of briefs the commission heard from fifty-nine municipalities, fifty-seven individuals, thirty-seven irrigation organizations, twenty-eight naturalist organizations, eighteen business groups, seven major industries, seven planning associations, five government agencies, three indigenous groups, and six unclassifiable delegations. An overwhelming majority of the submissions came from municipalities, chambers of commerce, farmers, ranchers, and officials from the irrigation districts.

The hearings did not really get to the heart of the matter of prioritizing trade-offs and therefore did not help the commission much in giving advice to government. Instead the 1984 AWRC hearings became the occasion for set-piece defences of interests and continued skirmishing by opponents over the recently approved Oldman dam. The notion that all the water in the Bow River had more or less been taken and that there was no new water to allocate did not really sink in, perhaps because even though the withdrawal permits already issued equalled most of the available water quota, water rights holders had yet to begin to withdraw water up to the limits of their permits. Thus to those with permits it seemed as if there was water enough. Those on the outside looking in took a different view, but they were not in the nature of things represented. Future shortages and the necessity for trade-offs were abstract concepts still, and until the reality of shortages began to bite, no real dialogue between the interest groups would occur.

Certainly, the commission heard from a very large number of irrigation advocates. In general, these spokespeople argued from the Jeffersonian-agrarian premise that agriculture was the best use of land and the highest calling, and therefore irrigation deserved to be the highest priority in use. In the words of Roy Jensen, the leading advocate for expanding irrigation, Albertans had a duty "to utilize our God-given resources of water and agricultural land to the best of our ability that we may feed the nations of the world."[66] The irrigators defended their water withdrawals against the criticisms of fishing and recreational organizations by stressing that irrigation added to the wildlife and recreational habitat. Indeed, they sought to shift the burden of responsibility for deteriorating water quality from their withdrawals to the effluent loading by upstream users, particularly Calgary. Irrigators also complained that reservoir storage for hydroelectric purposes reduced the amount of water available in the river at precisely the time the irrigators most needed it, mid to late summer. This existing hydroelectric priority needed to be changed. And they maintained that more efficient technology and better management practices meant that in future more acreage could be served by the same quantities of water.

The recent approval of the dam on the Oldman River haunted those representing recreational, naturalist, hunting, and fishing organizations. In their view, it presaged approval for a similar irrigation dam on the mainstem of the Bow between the Highwood River and the Carseland weir, which would effectively destroy a large portion of the renowned Bow River trout fishery. This proposal they set out in a collective fashion to resist.[67] Sharply critical of the wasteful methods and environmental indifference of the dominant irrigation interests, a broad coalition of naturalist organizations argued for the importance of maintaining minimum stream flows to preserve "a healthful and aesthetically pleasing natural environment." Government had a duty to protect not only economic interests but also the natural environment for future generations. Moreover, they insisted that the economic value of the scenery, recreation, boating, fishing, and hunting had been systematically overlooked in water management policy. The wildlife, recreational, and environmental groups were less exercised by upstream hydroelectric storage (which incidentally created new recreational and fishing possibilities) and Calgary nutrient loading (which they well understood supported the Bow River fishery) than they were about the slack water behind irrigation dams and the diminished flows and warmed waters that altered fish habitat for species deemed desirable by fishers. As Trout Unlimited insisted, the South Saskatchewan basin did not suffer from a crisis of supply but, rather, a crisis of water management.[68]

Mayor Ralph Klein deftly positioned Calgary within a regional context. The city benefited from the well-being of its entire region; thus the city should keep these broader interests in view, as well as its own narrow self-interest. Calgary had four objectives in water policy: a secure supply of water, a repository for waste-water effluent, recreational and visual amenities, and protection against floods. On the matter of the expensive measures taken to treat waste water, Klein cleverly reversed the implications of the city's "regional" view: the residents of the city alone should not have to pay the full shot when downstream users – primarily irrigators – reaped the benefit. Moreover, much of the diminished water quality in the lower reaches of the river, he said, was related to heavy water-taking for irrigation. He also chided his rural neighbours for their reluctance to provide access to the river in the same way the city of Calgary had with its river parks. He knew the river as a fisher and a canoeist; it was much easier to get to the river in Calgary than downstream. It was a politically astute, folksy, typically personal presentation intended to deflect anticipated criticism from the irrigators.[69]

The major economic interests in the region adopted a low-key, non-confrontational approach emphasizing that properly managed the water resources ought to serve all interests. The oil and gas industry, represented mainly by Dome Petroleum and the Canadian Petroleum Association, stressed the small amount of water used by industry compared with municipalities and irrigators, but it urged the commission to consider the rising future needs of the industry in extraction and processing.[70] Transalta, the electric utility, minimized the importance of its reservoirs as a recreational asset and defended itself against claims that its storage policies impinged upon irrigation. As a non-consumptive user whose dams regularized the naturally erratic flow of the river, the utility chipped away at the irrigators' presumed assumption of the moral high ground. In any event, the company argued, there was no reason why electricity customers should pay higher prices to allow farmers and agribusinesses to take more water for their own benefit free of charge.[71]

Among other things, the hearings revealed the intimacy of the personal experiences of many of the people who testified, from Mayor Klein in his canoe to the fishers wading and drifting in the river, the ranchers and residents who lived beside the river, and, of course, the irrigation district officials and ditch riders diverting its water into their canals and fields. They all knew the changing moods and flow of the river from extended first-hand experience and were sometimes eloquent in expressing their relationship with the river. But for the most part these public hearings simply provided an opportunity for various interests to be ventilated. It was a game in which the ball got batted back and forth between opposing courts without serious engagement with the issues. After almost two years of deliberation the commission issued a windy report in 1986 from which it was difficult to extract specific recommendations. But by its own estimation the commission advised the government to establish minimum in-stream flow requirements, stress multi-use management to maintain water quality, emphasize water conservation, and reserve some supplies for future and new uses. While, on balance, the irrigation interests received respectful consideration for their expansion plans, the recommendations for minimum flow, multi-use, and conservation represented major victories for the recreationists and the naturalists.[72]

But the South Saskatchewan River Basin Study and the Alberta Water Resources Commission's public hearings, which together chewed up a good deal of the 1980s, decided nothing. Rather, these two major inquiries positioned the key players for the future when actual or perceived shortages would force government action. The government was no better equipped with policies than when it started, but it did possess much better information about the river as a resource and the relative power of the interests configured around it.

The AWRC's public hearings did, however, reintroduce one voice into the debate. The Piikani, Kainai, and Siksika peoples presented briefs to the commission that stood apart from the dominant discourse. The Piikani, for example, who had lost a campaign to situate the Oldman dam on their reserve and thus obtain some degree of control over it, challenged the jurisdiction of the commission and the Alberta government over the rivers. They claimed that Treaty No. 7 gave them authority over the waters in their traditional territories.[73] The Siksika took a slightly different tack. Their spokespeople observed that the band had already surrendered a good deal of land to irrigation works, canals, and flooding by the Bassano dam but had received very little benefit from the diversions. In the past they had experimented with irrigation. In the future they expected to renew the effort with a better capitalized program. In any event they would be expecting water to be available to them when that time came. Beyond that, the Siksika representatives deliberately set out to shock the commission with the announcement of plans for a dam site and the construction of a coal-fired generating station on the reserve. These projects would require water allocations over and above those already granted by the government. As they claimed control over the water passing through their reserve, this water, the Siksika insisted, was theirs for the taking.[74] The commissioners and other observers likely regarded these claims as bluster, but along with Mayor Klein of Calgary, the commission was reminded of something that might have been forgotten with the passage of time: that a good portion of the river system, including the Indian reserves, fell under federal jurisdiction.[75]

In the absence of further development on the reserves, the arguments concerning the continuing power of the federal government and the residual rights of indigenous people either as riparian owners or under the treaty remained moot. That situation persisted until the coalition protesting the Oldman dam called upon the federal government to conduct an environmental assessment of the project, as required under its Navigable Waters Protection and Fisheries Acts. The Oldman, being a navigable river in law, they insisted, could not have an obstruction built which would hinder navigation without an environmental review as required by federal law. Similarly, the federal government had constitutional authority to protect inland fisheries. In the ensuing court case these claims were upheld by the Federal Court of Canada, and a full environmental assessment had to be conducted even as the dam was under construction. The courts found that an assessment was required under federal law whatever the Alberta government might think.[76] As the Oldman case and the indigenous claims before the AWRC demonstrated, other powers either claimed or actually exercised substantial powers over provincial waters.

To sum up, authority over the waters of the Bow initially became consolidated under the authority of the federal government, which in turn created numerous rights holders possessing licences of limited extent but often perpetual duration. Certain federal powers were devolved to the province of Alberta by stages after, but not immediately after, the Natural Resources Transfer Act of 1930. Financial difficulties and the need for interprovincial coordination on the prairies led Alberta to sustain a shared jurisdictional arrangement with the federal government through the 1960s. During the following two decades the province in theory gathered most of the powers over the river unto itself. But with jurisdiction came the responsibility for managing latent conflicts over what promised in the future to be an increasingly scarce resource, water. Notwithstanding this provincial ascendancy, authority to some extent remained divided. The full extent of residual riparian rights had never been tested in the courts, and the federal government exercised exclusive authority over water in the national park on the headwaters of the Bow and on the three Indian reserves bordering several long reaches of the river. The question of how much autonomous control indigenous people exercised over rivers passing through their reserves by virtue of their treaty rights or property rights associated with reserve lands remained unresolved. Beyond that, by virtue of its authority over navigable rivers and the environmental assessment provisions written into the Navigable Waters Protection Act, the federal government was potentially a decision-maker on every major project involving the river.

Whether power was concentrated under one jurisdiction or shared by several, rights in the river had nonetheless been widely dispersed to innumerable licensees, agencies, and corporations. The Master Agreement capped the amount of water that might be taken and specified what must be passed on to Saskatchewan, but up to that point individual licensees could exercise their rights up to the limits of their permits. For almost all of this period there was enough water available or licensees claimed more water than they actually used. Conflicts between users, to the extent they existed, were for the most part mediated by bureaucracy rather than the courts (as in log-driving and hydroelectric development) or tolerated (as in downstream water quality issues and irrigation dissatisfaction with hydroelectric storage practices). When bureaucrats and politicians opened discussions on the principles of management for the day when the river could no longer meet all needs, they encountered the entrenched resistance of significant vested interests.

But the Bow River, like any force of nature, was not a constant. Its waters fluctuated with the seasons and from year to year. It could flood capriciously or just as readily withhold water in years of drought. It faithfully carried downstream everything put into it. In that sense it created a community of users. But this was not a community capable of prioritizing and rationing needs because the day the river no longer carried enough usable water downstream to meet all the expectations placed upon it had not yet arrived.

13 · Who Has Seen the River?

Like a B-grade science fiction horror film, "the Blob" in the river appeared suddenly and without warning to threaten the city of Calgary and the entire lower Bow valley. Late in October 1989 an Alberta Environment official taking routine water samples from the Bow River in Calgary discovered "dinner plate–sized globules" on the bottom of the river. Subsequent tests proved that the tar-like substance, a toxic mixture of creosote and pentachlorophenol, had spread across forty square yards under the river and released a plume of carcinogenic benzopyrene that stretched downriver for 165 miles.

From 1924 to 1964 the Calgary Creosoting Company, later the Canada Creosoting Company, had operated a wood-preserving plant on the banks of the river at that location treating railway ties and bridge and construction timbers. Negligence, accidental spills, leakage from unlined waste ponds, and runoff from stored chemicals had contaminated the subsoil during its years of operations. Abandoned since 1964, the forty-acre site was eventually decontaminated in the late 1980s before being paved for roads and two auto dealerships. During the cleanup a cocktail of creosote residues had been found pooled on the bedrock thirty-four feet underground. The slope of the bedrock toward the river gave rise to the suspicion that there might be another nasty creosote deposit lurking somewhere under the river bottom ready to burst at any moment. However, water sampling had not detected any to that point. Discovery of the Blob, along with evidence it was slowly releasing its deadly cargo into the river, confirmed officials' worst fears, raised alarm in the media, and set off a mild panic in downstream communities.[1]

Apart from setting in motion a frantic cleanup, this apprehended environmental disaster marked the beginning of a new chapter in the relationship between the inhabitants of the Bow River valley and their river. Coming after a series of worrisome reports and unpleasant experiences with the river, the Blob melodrama, played out in full view between 1989 and 1996, focused public alarm and galvanized politicians at all levels. The Blob put the well-being of the river firmly at the top of the political agenda. But it also stimulated a renewed public appreciation of the importance of the river in the life of the community and provided the impetus to bring numerous interest groups and "users" together to rethink public and private policies. We conclude this examination of the environmental history of the Bow River with this dawning river awareness, which amounted to something of a cultural revolution in thinking if not behaviour. After the Blob, the river could no longer be taken for granted or its use expanded indefinitely. Nor could it be expected to clean itself up after public negligence. Through the different positions adopted after discovery of the Blob, the river began to speak back to the community, imposing a new regime of limitations and constraints on public policy.

The Ethic of Use

Without asking who would pay, Alberta Environment sprang into action, led by its ambitious new minister, Ralph Klein, the former populist mayor of Calgary. Immediately, workers removed the visible globules in the shallows manually. A convoy of dump trucks then pushed a 180-yard dike out into the river to contain the escaping toxins. A barbed wire–topped chain-link fence and two guards on twenty-four-hour duty secured the site. After an extensive study of the delicate situation, Alberta Environment eventually launched a systemic remedial operation, the cost, estimated to reach $50 million, to be shared by the province and the federal government. The successors to the company disclaimed any responsibility for the cleanup. For months and then years on end, huge vacuum trucks excavated the contaminated river bottom down to bedrock and slowly returned the cleansed gravel to the site. The cleanup of the Blob presented Calgarians with a long-running saga in plain view of the unintended consequences of ordinary industrial activities and the threat to public health posed by toxic pollutants long after operations had ceased and been forgotten.[2] The river did not forget; it carried its past with it. How many more creosote sites were still lining its banks waiting to burst?

Klein's ministry moved quickly and decisively to contain the environmental and the potential political crisis. The toxic spill was doubly embarrassing to the new minister since he had been mayor of the city during much of the eighties and a journalist before that, during which time he had made a name for himself as a crusader against pollution.[3] What is more, he had ambitions to become premier of the province. In classic proactive political fashion, Klein determined to get out in front of the issue. When a delegation of downriver communities and water users expressed concern about the safety and security of their water supply immediately after the Blob was discovered, Klein responded by striking a task force in May 1990 representing all of the stake-

Illus. 13.1 Cover of the Bow River Water Quality Task Force's 1991 report.

holders in the Bow valley. This independent investigative body was charged with identifying all sources of pollution in the Bow River, recommending measures to reduce or eliminate them, proposing means of preventing pollution in the future, and estimating the cost of these and alternative measures.[4] The Blob thus gave rise to a broadly based bureaucratic process to contain the much larger issue of the "pollution" of the Bow River, focus efforts aimed at a much delayed and long overdue cleanup of the river, and mobilize the political will to do something about it.

The Bow River Water Quality Task Force rested upon a clearly stated premise: that the river had been contaminated by pollution at many points; that the specific sources of that pollution could be identified and the problems rectified; and that the river, through these remedial measures, could be restored to health. After a year of extensive consultations, background studies, public hearings, and deliberations, the task force reported something quite different. The river was indeed the "lifeline" of southern Alberta, and preservation of its water quality was of paramount importance to the society and economy, the task force concluded. But simply cleaning up contaminated sites and addressing point sources of pollution could not achieve that goal. In his letter of transmittal to the minister in November 1991, the chair of the task force observed, employing the fashionable cant of the day: "One of the perhaps unstated, but certainly important, results coming out of this Task Force was an understanding by the various stakeholder members that the problems related to water quality in the Bow River are not singular and there is no simple solution. We have come to recognize that we are all part of the problem and in order to achieve our goals we must all be part of the solution."[5]

The task force discovered, first, that the river had become so intricately woven into all aspects of life in the valley and its character had been so dramatically changed through use that the notion of "preservation" required specification of what precisely was to be preserved. Secondly, it concluded that acceptable water quality required management of resources and regulation of human activity on a heretofore unprecedented scale. Preserving the river required searching personal and social introspection.

People spoke casually of the Bow River – singular. However, the task force decided that over its course of 619 kilometres the river passed through such varied geographies and climates that it made more sense to think of it as eight river segments, or reaches, each with its own unique ecological characteristics, river uses, and water quality challenges. The river in Reach 1, from the Bow Glacier to the Lake Louise townsite (35 kilometres), plunged down mountain slopes to a high valley in a national park. In this reach the relatively untouched stream served primarily as a wildlife habitat and recreational resource. From Lake Louise to the eastern boundary of Banff National Park (80 kilometres) in Reach 2, the river supported a sport fishery and was enjoyed by tourists primarily for its scenic character as it traversed a broad mountain valley. The river and its valley provided habitat for black and grizzly bears, elk, deer, moose, bighorn sheep, coyotes, wolves, cougars, beaver, mink, weasels, and muskrats. The river was used for sanitary purposes in the Lake Louise and Banff townsites.

In the foothills, Reach 3 (115 kilometres), from the Banff Park boundary to the Bearspaw dam, the river and its main tributary, the Kananaskis, had been extensively engineered for hydroelectric production, which had a significant impact upon the flow of the river. Impounded water stored for release to augment low winter flow created lakes used for seasonal recreation. Ranching, mainly on the Nakoda reserve, and municipal water supplies at Canmore and Cochrane exerted the main consumptive demands on the river in the foothills. In this reach sewage effluent and industrial users made their first significant impact upon water quality. As the river wound through the city of Calgary – Reach 4, from the Bearspaw dam to the Western Irrigation District Weir (28 kilometres) – domestic requirements, runoff, and industrial

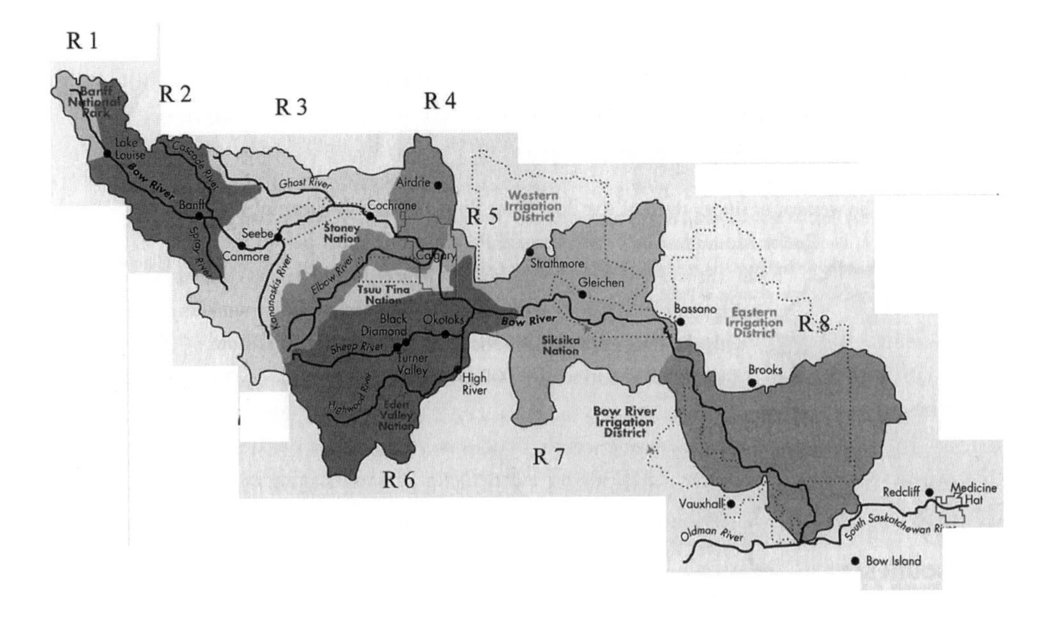

Map 13.1 Reaches of the Bow River. (Source: Bow River Water Quality Council, *Preserving Our Lifeline: A Report on the State of the Bow River* [1994], 11)

uses altered the character of the river as it flowed through a densely populated urban environment. Here the river was important, as well, as a site for riparian parks and recreation, mainly rafting, canoeing, kayaking, trout fishing, and hiking. In this reach the first major diversion of water for both sanitary purposes (the city of Calgary) and irrigation (the Western Irrigation District) subtracted substantial quantities of water from the flow.

The 25-kilometre stretch from the Western Irrigation District weir to Highway 22X, Reach 5, contained a major provincial park and bird sanctuary but also the city's two sewage-treatment plants, Bonnybrook and Fish Creek, where about 90 per cent of the water withdrawn by the city from Reach 4 returned in the form of treated effluent. In this section, intense biological and chemical loading had clearly stressed the capacity of the river to the breaking point. Over Reach 6, from Highway 22X to the Carseland weir (53 kilometres), the river slowed down and began to meander out onto the prairie in its deeply cut, broad valley. This reach was notable, in the words of the task force, because it supported "one of the best large, wild trout fisheries in North America."[6] Another large diversion of water to the Bow River Irrigation District occurred within this stretch.

In Reach 7, stretching 136 kilometres from the Carseland Weir to the Bassano dam, where the Eastern Irrigation District withdrew its water, the river wandered through ranchland and the Siksika reserve. Finally, in Reach 8, from the Bassano dam to the confluence with the Oldman River (147 kilometres), the Bow passed through rolling dryland prairie, where water was taken mainly for livestock. Warm-water fish species predominated in the shallow, sluggish waters, whose flow had been considerably reduced by irrigation withdrawals upstream. Ducks, geese, and pelicans congregated on the broad expanses of this reach during seasonal migrations and raised their young in the reed-protected backwaters and meanders.

The task force noted that the Bow River flowed out of Calgary more seriously compromised by day-to-day human activity than by any threats from the Blob or its other ninety-two potential counterparts lurking along the river. Briefly, in the body of its main report and then in exhaustive technical detail in the appendices, the task force documented the degraded state of water quality in the lower Bow.[7] In Reach 5, below the Calgary sewage-treatment plant outfalls, coliform bacteria levels made contact recreation (swimming, fishing, and boating) dangerous. Excessive weed growth caused by elevated phosphorus levels choked the channels and reduced dissolved oxygen levels at night to the point where fish kills were possible in periods of low flow. Heavy metals and silt from city streets and construction sites coated the bottom; noxious floating debris circled in backwaters. The pronounced oily smell of the water penetrated the flesh of sport fish, rendering them inedible. Weeds and algae affected the taste and odour of the water, and the brown foam frequently found floating on pools alarmed downstream communities that depended upon the river for their municipal water supply. Nutrient-rich water clogged the irrigation canals and storage reservoirs with massive, floating weed beds, entailing costly removal operations and restricting recreational use of the reservoirs in high summer. Intensive water testing during the summer and fall of 1991 showed that fecal coliform levels in Reach 5 always, in Reach 6 almost always, and in Reach 7 frequently exceeded safe levels for recreation. Concern extended as well to the upper reaches of the river, even though the levels of contamination were much lower. Coliform counts spiked upwards below the Lake Louise and Banff townsites, where seasonal demands occasionally overwhelmed the capacity of the sewage-treatment plants, and the communities of Canmore and Cochrane, where occasional raw sewage discharges or insufficiently treated effluent raised nutrient levels in the river. The rapid growth of these communities implied that levels would likely rise even further without vigorous remedial measures.[8]

The task force's terms of reference structured the Bow River story as a recovery narrative. Its report, by contrast, constructed the narrative as a story of the management of competing human desires within a complex system of flows, exchanges, feedback loops, trade-offs, and regulations. The river had different qualities and capacities depending upon the local ecology. In each reach diverse constellations of interest groups expected different things from the river and placed different demands upon it. Upstream activity, in turn, had a profound impact upon downstream users. In short, the river was a complex and diversified system that touched its bankside communities in a myriad of different ways. Geography changed the character of the river as it flowed from the mountains out through the foothills to the open prairie, but so, too, human intervention substantially altered the nature of the river as it passed through dams, turbines, municipal water systems, sewage-treatment works, storm sewers, and irrigation projects and ran off golf courses, city streets, fertilized fields, and intensive livestock operations. Seen from this perspective, the Blob was a serious but relatively minor threat to the integrity of the river and life in the valley. Almost everything about the river had been changed or altered through human intervention: its flow, vegetation, water quality and quantity, and aquatic life and, in some reaches, its smell and toxicity.

Political responsibility for the river was as fragmented and diverse as its geography and hydrology. Jurisdiction over the river lay primarily with the provincial government of Alberta, whose ministries of the Environment, Agriculture, Health, Labour, Forestry, Lands and Wildlife, Transportation, and Utilities and Crown agencies, such as the Natural Resources Conservation Board, the Energy Resources Conservation Board, and the Emergency Measures Organization, all administered statutes governing some aspect of the river. Even though natural resources fell under provincial jurisdiction after the Natural Resources Transfer Act of 1930, the federal government nevertheless still retained significant statutory powers over the Bow River through its responsibility for national parks, fisheries and oceans, Indian and northern affairs, and the environment. The third tier of government, urban and rural municipalities, operating under provincial authority, exerted considerable local control over the river. Finally, three First Nations communities and three organized irrigation districts had claims upon the river. Since many governments were to some extent responsible for the river, in effect no government could be held accountable for its condition. Because of this divided and overlapping jurisdiction, the first recommendation of the task force was the creation of a permanent Bow River Water Quality Council to promote public awareness and hold the responsible authorities accountable through regular "state of the river" reports.

The "lifeline" metaphor adopted by the task force (its report in 1991 was titled *Preserving Our Lifeline*) exerted its greatest power when cast in the negative. What did it mean for the community when its lifeline was near death? If the lifeline failed, that had ominous implications for those whose lives and livelihoods depended upon it, and in particular, it cast a dark shadow over future growth. The lifeline needed protection not simply to serve current needs but also because a promising future depended upon its continuing health. The task force report stressed the importance of determining the "finite assimilative capacity of the river" to establish regulatory limits for various types of biological and chemical loading. It recommended more effective data gathering, monitoring of water quality, and enforcement of loading limits. The task force underlined the urgent need to improve sewage treatment technology and sludge disposal. The flow of the river should be regulated by the electric utility dams not to maximize hydroelectric production but, rather, to maintain healthy flows in periods of low water to ensure optimum water quality. More needed to be done to control runoff from intensive livestock operations, golf courses, and urban storm sewers. The task force estimated that all of this would entail a one-time cost of about $4.5 million and additional annual operating costs of $350,000.

As the environmental historian Marc Cioc observed of changing perceptions of the Rhine, "It was not until the river was on the brink of death – and therefore unable to fulfill its duties to humans – that a new attitude about the river began to emerge."[9] Something similar happened in southern Alberta during the early 1990s. The Bow River Water Quality Task Force was essentially a body of concerned municipal, agricultural, industrial, and recreational users. Their experience, and now their carefully compiled scientific data, showed conclusively that in many reaches the river could no longer meet their various needs because of human impacts upstream. Nor could the declining health of the river sustain their ambitious hopes for economic growth and expansion in the future. Interest-group notions of utility animated the task force report. But by the same token, the power of the assembled interest groups, aggregated in the task force, meant that its report carried considerable political weight. Even in the midst of the fiscal crisis of the state, now under aggressive neo-conservative management, a sense of "something must be done" trumped fiscal concerns.

The government of Alberta, led now by its new premier, Ralph Klein, quickly established the Bow River Water Quality Council, an advisory body to the minister of Environmental Protection, "to promote awareness, improvement and protection of Bow River water quality, foster cooperation among agencies with water quality responsibilities, and provide communication links among governments, interest groups and the general public." This permanent advisory body was to maintain a continuous drumbeat of fact-based pressure on governments across jurisdictions and provide the media with the information necessary to keep river issues alive on a crowded political agenda. Consisting of seventeen stakeholder representatives, drawn from the rural and urban municipalities, recreation, wildlife, fisheries, agriculture, irrigation, health units, First Nations, parks, industry, and other interest groups, the council, in effect institutionalized the organized river users as the "river keepers."

The first "state of the river" report, published in 1994, measured water quality in each of the eight reaches of the river against recognized quantitative standards established for contact recreation, aesthetic enjoyment, cold-water ecosystem maintenance, drinking water supply, industrial uses, livestock watering, and irrigation. It was the intention of this report to identify specific objectives for improvement that the various responsible jurisdictions would then have to perform. Subsequent reports would, in turn, measure the effectiveness of the remedial program against these quantitative benchmarks and thereby hold various governments to account.

The council conducted extensive surveys of the river in 1992, testing for forty different water quality variables ranging from colour, temperature, turbidity, and dissolved oxygen to total dissolved solids, pH, chemical composition, coliform count, and organic composition. The "state of the river" report set these findings in the context of almost a decade of measurements (where the data were available) for filterable residues, total phosphorus, fecal coliform bacteria, nitrates and nitrites, benthic algae, and macrophytes in order to detect any measurable change over time in these key water quality indicators.

Overall, the council found the Bow River to be "in a reasonably good state," but it had some serious concerns and a broad agenda of improvement. "In 1992 the water quality in the Bow River ranged from being unaffected by local human activity and suitable for all uses in the headwaters, to impacted by pollutants and restrictive for some water uses in the lower reaches." The time series showed little consistent change over time – either improvement or deterioration – although tertiary treatment of Calgary's sewage effluent to remove phosphorus had reduced by two-thirds the mass of vegetation downstream that endangered nighttime oxygen levels in periods of low flow and choked irrigation intakes. Fecal coliform levels in the reaches below Calgary spiked upwards, sometimes exceeding safe limits, even though data for Reach 5, where the two Calgary sewage-treatment plants returned their effluent to the river, were not available or at least not published. Those data would have produced a much more alarming set of graphs. The council produced quantitative measures only for Reaches 1 through 4 and 6 through 8. The much lower but rising coliform counts in the reaches above Calgary, as a result of the growing urbanization of Canmore and Cochrane, were flagged as a cause for concern. The "state of the river" report specified an extensive itemized list of monitoring operations and proposals for integrated river basin planning and management aimed at controlling urban and rural runoff, reducing siltation and contamination, regulating flows, improving wildlife habitat, and enhancing sewage treatment. The matter of fact, uninflected tone of the document left the impression that the problems of the river were urgent but manageable with coordinated effort.[10]

The emphasis upon *water quality* in the "state of the river" report effectively shifted power on the river. Downstream water users now occupied the moral high ground. Upstream, mainly urban uses were deemed to have made the greatest impact upon the river. Apart from the obvious safety, health, and ecosystem concerns posed by pollution, urban source nutrients fed bottom-rooted aquatic plants that reduced the flow of the main irrigation channels and floating weedbeds that choked the irrigation reservoirs. The mass of stinking weeds entailed annual removal costs and also ruined the artificial irrigation lakes as recreational resources for prairie fishers, boaters, and swimmers. Agricultural interest groups complained, too, of heavy metals that were toxic to cattle in certain concentrations and contaminated irrigated fields.

The irony in all of this is that the focus on water quality represented by the Bow River Water Quality Council in its report shifted the onus of responsibility for changes in the river to cities that were, by comparison, rather minor consumptive users of the river compared to the irrigators. Cities withdrew quite a lot of water but also returned about 90 per cent of it. The downstream irrigators were the largest consumptive users of the river by far, accounting for 85 per cent of the water withdrawn from the river and 95 per cent of the water consumed. Irrigation withdrawals, saline returns, and runoff also had a major impact on the flow and character of the river below Calgary. For example, the combined withdrawals of the three major irrigation districts and other smaller users reduced the flow of the Bow River by almost 30 per cent. Moreover, the two hundred existing water-taking licences permitted the removal of almost double that amount.

The Blob floating to the surface in Calgary set off alarm bells, but subsequent scientific analysis and stakeholder investigation shifted attention from the minor matter of toxic industrial sites to the much larger problem of water quality degradation resulting from domestic use and its passage through an urban environment. It was not exactly news that the river flowed through the social system along its banks and that the future of those communities had been compromised by years of neglect, but the dramatic discovery of that long-repressed knowledge in the early 1990s created a new political equation in the Bow valley that privileged downstream interests.

The Ethic of Enjoyment

The political economy of "use" that animated the investigations of the Bow River Water Quality Task Force and the Bow River Water Quality Council can be largely understood as a function of who got to be defined as a stakeholder and the proportionate influence of these interest groups. The members appointed by the government represented downstream water users, agro-industry, health units, smaller municipalities in the watershed, the city of Calgary through its engineer, the irrigation districts, the Siksika nation, Transalta, and several recreational interests. The large downstream membership contingents (ten of twenty-one on the task force and eight of seventeen on the council) amplified the interests in the lower watershed. The underrepresentation of Calgary and urban users in both cases is most striking. In the course of their deliberations, the task force members interviewed and received representations from people largely like themselves: officials from Alberta Environment and Alberta Agriculture, health bureaucrats, academics from the University of Calgary, environmental consultants, managers of the irrigation districts, hydroelectric engineers, and organized recreation groups. Interest-group representatives and experts consulted with interest-group representatives and experts.

Nevertheless, between the first report in 1991 and the report released in 1994, it is possible to detect a subtle shift in underlying perceptions of the river and the place of humans in the environment. The first report had implicitly conceived of use in a narrow way as hydroelectric, domestic, industrial, sanitary, irrigation, agricultural, and recreational activity. The "aquatic eco-system" was a concern only in the reaches immediately below Calgary, where oxygen depletion, ammonia levels, and chemical residues periodically threatened fish. By contrast the "state of the river" report three years later evaluated the river ecosystem (nearshore vegetation, wetlands, and waterfowl, wildlife, and fish stocks) as one of three major categories of analysis (the other two being water quality and water quantity) in all eight reaches of the river. The executive summary explained: "The assessment of the state of the river is not considered complete without an evaluation of the riparian and aquatic aspects of the ecosystem because of their interdependence with the quality and quantity of water flowing in the river." Here, then, was the germ of a new conception of the river and an ethic of use beyond consumption, but related to it.

This shift in emphasis cannot be tied to any changes in membership. Other "users" of the river had participated in the task force and its successor, the council, in approximately similar numbers. Delegates from Ducks Unlimited, Trout Unlimited, the Bow Waters Canoe Club, and the Calgary Area Outdoor Council had been represented directly, and the Alberta Fish and Game Association and the Bow River Angling Outfitters Association indirectly.[11] For these groups, use meant enjoyment. By extension, enjoyment depended upon both the clean flow of the river and flourishing wildlife populations in and around

Illus. 13.2 Cover of the Bow River Water Quality Council's 1994 report.

it. Duck hunters valued ducks and busied themselves preserving and enhancing waterfowl habitat. Anglers heightened public awareness of the well-being of sport fish. For boaters and birders, aesthetic enjoyment depended upon the quality and quantity of water in the river and the forest, vegetation, and wildlife on the riverbanks. Together these interests articulated a view of the river based upon recreational use and aesthetic pleasure that received a more sympathetic hearing in the 1994 "state of the river" report, likely based upon the corroborating representations made by scientists from Alberta Environment and other agencies. This, to be sure, remained a human-centred environmental ethic, but one which nevertheless expressed greater concern for the nonhuman world than the narrower utilitarian ethic of consumption.

The "state of the river" ecosystem evaluation published in 1994 concluded that the total area of the riparian forest had not diminished over the past hundred years, but its composition had changed. Balsam poplar regeneration in the foothills and prairie reaches had declined because of the encroachment of urban development, the return of beaver, stabilization of river flow for hydroelectric storage, cattle grazing on the flood plain, and – by far the most important factor – the absence of regular floods that deposited silt which became the seedbed for trees popularly known as "cottonwoods." Fifteen kinds of fish inhabited the river, including several introduced species. While the river had been able to maintain "one of the best recreational trout fisheries in North America," the report noted that introduced species had interbred with or displaced some native fish, such as the bull, westslope, and cutthroat trout. Waterfowl in increasing numbers, mainly mallard ducks and Canada geese, used the river mainly as a staging ground for migrations, but some raised their broods in side channels, backwaters, and the wetlands created by irrigation works far from the river valley itself. In the various reaches different species of wildlife frequented the river on a seasonal basis, though beaver, deer, and coyotes were permanent residents in most reaches. This benchmark exercise found the river ecosystem to be in reasonably good shape, as it did the river generally. The importance of the "state of the river" report is that it broadened concern about the river to include its ecology and asserted the linkage between ecological diversity, water quality, and human use and enjoyment.

As the membership of the task force and the council indicated, several organizations had already established themselves in the region, asserting their special interests in the ecology of the Bow River. Outdoor recreation organizations, boaters, hunters and anglers, parks and wilderness advocacy groups, and individual citizens had in different ways attempted to bring the state of the river, as seen from their perspective, to wider public attention, with varying degrees of success. Representation on the two official bodies, however, brought some of their concerns into the mainstream, where they joined other convergent ideological currents.

Ducks Unlimited Canada, an affiliate of a US hunters' and conservation organization of the same name, had established its first chapter in Alberta in 1938. Since several of the main migratory duck flyways crossed the Canadian prairies, western Canada was densely populated with hunters, and it was a popular destination for hunters from all over the continent.[12] Ducks Unlimited focused its energies upon preserving waterfowl habitat, mainly wetlands on the prairie. Drainage projects associated with large-scale mechanized farming threatened many of the wetlands that sheltered ducks and geese during their migrations. Over the years, the organization sought, through publicity and a few demonstration projects, to create greater understanding of the importance of these wetlands. On the other hand, mechanization increased the availability of forage on harvested grain fields, and irrigation projects created new artificial wetlands along canals and on reservoirs. By the 1990s Ducks Unlimited Canada, employing about eighty organizers and with the support of almost twenty thousand supporters in Alberta, had sponsored more than a thousand projects that secured two million acres of waterfowl habitat.[13] Ducks Unlimited was less concerned with rivers than with preserving ponds, sloughs, and wetland, but it joined forces with other like-minded conservation organizations on various projects.

Trout Unlimited, a lineal descendent, also established a strong base in the province of Alberta. It would eventually form ten chapters located along the eastern front of the Rockies aimed at protecting the cold-water resources of Alberta, promoting sound conservation practices, fostering public awareness of these natural resources, and publicizing threats to trout streams. This organization flourished in southern Alberta with the rise of the Bow River trout fishery after the 1960s. Its concern for water quality matters was tempered somewhat on the Bow by the knowledge that the extraordinary trout-fishing conditions below Calgary depended in part upon nutrient loading from the city of Calgary. But there were limits. When overloading threatened to kill fish at night in periods of low flow, if the fish tasted of oily residues, or if pollution made it unsafe to fish in the river, then the organized anglers were eager to call attention to the problem and promote solutions.

In the 1990s Trout Unlimited launched a compelling and highly publicized campaign to save the Bow River trout not from pollution but, rather, from irrigation practices. Each winter when the irrigation districts closed their head gates and drained their canals for maintenance, thousands of stranded trout died. Beginning in 1996, Trout Unlimited organized annual rescue operations, usually during the first blasts of winter, mobilizing hundreds of supporters and schoolchildren to scoop up as many of the marooned rainbow trout, brown trout, and whitefish to be returned to the river. In 1997 this volunteer effort, prominently chronicled in the press, saved 17,000 fish. A year later the number rose to over 90,000. This campaign not only focused attention on the fishery in a dramatic way but also drew attention to one aspect of the adverse environmental impact of large-scale irrigation.[14]

These well-funded, permanently staffed subsidiaries of national and US organizations had established a high public profile as wildlife conservationists during the 1980s and were thus represented as stakeholders on the Bow River Water Quality Task Force and its successor bodies. But other local voluntary organizations had been stressing the importance of preservation and recreational use for many years. In 1979, for example, the Alberta Wilderness Society established a research team of canoeists, hikers, parks planners, biologists, and university professors to compile an inventory of "river landscapes" in southern Alberta to promote legislation intended "to protect and conserve our finest waterways as recreational or natural rivers." This team produced a handsome illustrated booklet in 1982, *Rivers on Borrowed Time: Eight Great Waterways and What We Must Do to Save Them*, the cover of which featured a wistful panoramic vista of the lower Bow valley with the Rockies barely visible

Illus. 13.3 Cover of the Alberta Wilderness Association's *Rivers on Borrowed Time* (1982).

on the horizon. Using the language of stewardship and trusteeship, this pamphlet sought to publicize the value of Alberta's rivers as recreational resources and to promote an action plan to preserve for future generations large sections of the natural environment of eight rivers: the Milk, Oldman, Bow, Upper and Lower Red Deer, North and South Ram, and North Saskatchewan. Hydroelectric and irrigation dams, riverbank residential development, and pollution from rapidly growing urban areas threatened the rivers of southern Alberta just as changing lifestyles and rising incomes revalued rivers for their aesthetic and recreation potential. Among its goals, the Alberta Wilderness Society sought to establish the Bow from the Bearspaw dam to the Siksika reserve as "a recreational river" where recreation – canoeing, fishing, hunting, scenic viewing, and nature watching – would be paramount in planning and regulatory considerations.[15]

As the title of its fundraising pamphlet suggested, time was of the essence. Even though these rivers offered tranquility and solitude to boaters, fishers, and naturalists and for the moment wildlife and fish were abundant, the organization argued that the Bow in particular was living on borrowed time. It counted more than 200,000 anglers on the river in 1977. Catch limits had to be strictly enforced or the fishery would perish at the hands of fishers. Sewage effluent overloading also threatened to wipe out the fish during periods of low flow. If recreational uses were raised to the highest levels of priority and conflicting uses regulated, it might still be saved. The society, though, had to admit that not all recreation was equal: high-speed jet boats and all-terrain vehicles disturbed wildlife and diminished the "aesthetic enjoyment of the Bow River" as well as posing safety hazards to canoeists and fishers.

Individuals also raised their own distinctive voices. During the late 1980s and early 1990s three talented and dedicated naturalists expanded public understanding of the Bow through their professional activities, television appearances, guidebooks, and art. Jim McLennan, a fly fisherman, outdoor writer, photographer, and conservationist, published a guide to the river that contained a wealth of information about its ecology, its fish, of course, and the various threats to them. Ted Godwin, a renowned abstract artist and ardent fly fisherman, marked his return to figurative painting with a major series of large-scale works inspired by fishing the lower reaches of the Bow. The handsome book-catalogue for a travelling exhibition in 1991–92, including a meditation by Godwin on rivers, fish, and art, gave the paintings much wider reach. Janice MacDonald, a mystery writer, children's author, and canoe enthusiast, with the help of several colleagues, in 1985 compiled a guidebook, *Canoeing Alberta*, a large section of which was given over to the Bow.

Each naturalist, author, and artist saw different qualities in the river. MacDonald, for example, marvelled at the power and beauty of fast moving water, especially "the tremendous echoing roar of tons of water incessantly tumbling over a rocky rapid." Her sport gave her a "respect" for the river and an ability to "read" its currents, velocity, eddies, morphology, and rapids – tips she passed on to her readers.[16] On a series of descriptive voyages downriver, she and her colleagues observed the wildlife and vegetation along the banks and pointed out important obstacles, hazards, and landmarks, but they were primarily interested in the changing flow of the river and the excitement of managing it. MacDonald captured the challenge to the canoeist presented by the river through a quotation from Rudyard Kipling about the Bow: "It does not slide nor rustle like Prairie rivers, but brawls across bars of blue pebbles, and a greenish tinge in its waters hints of the snows."[17]

For Jim McLennan, the fish of the Bow were its main, but by no means its only, attraction. His book, *Blue Ribbon Bow: A History of the Bow River: Canada's Greatest Trout Stream*, first published in 1987, was much more than a fisher's guide, though it was loaded with information about tackle, techniques, the character of the river, and its fish in its upper, middle, and lower reaches. The first edition, in large format and full of line drawings and maps, was as much an art book as a handbook. In the tradition of angling literature, McLennan's guide offered reflections on nature and humankind, the solitude and camaraderie of fishing, and the mystery of matching wits with wild creatures, and it contained the necessary quota of amiable fish tales – yes, a big one does jump right into the boat. A more lavishly illustrated second edition, published in smaller, easier-to-carry format, contained more practical advice on gaining access to the river, what flies to use at various time of the year on different parts of the river, and river etiquette. The most memorable passages of the book, however, reveal McLennan's joy in thinking like a fish. For him, the greatest pleasure of fishing the middle Bow – a part of the river more dear to him than any place on earth – came from "understanding what is going on it the fish's world."[18]

So much conspired against the fish that McLennan wondered if they could possibly survive much longer. Governments seemed indifferent to the economic value of the fishery and the importance of protecting it for the long term. Hydroelectric storage upstream held back water at critical times in midsummer. Thousands of mature fish perished each autumn with the draining of the irrigation canals. Jet skis and power boats, apart from the nuisance they created, interfered with spawning beds. The possibility of another large mainstem dam for irrigation purposes hovered like a death sentence over the fishery. McLennan wrote with an intimate concern for the fragility of the fishery he loved. While celebrating "Canada's greatest trout stream," his latter chapters are suffused with a sense of the transience of all things. Would human folly doom the world's greatest trout fishery as it had so many other trout streams? Or would recognition and wise self-regulation by governments, industries, irrigators, anglers, and recreational users keep this "magnificent accident" alive in perpetuity?[19]

Ted Godwin's *Lower Bow: A Celebration of Wilderness, Art and Fishing*, published in 1991, was at once a catalogue of the artist's luminous, luxuriant studies of reflections on the water and an essay of personal reflections on the restorative experience of fishing.[20] The Bow River had been inspiring and challenging artists since the mid-nineteenth century, a subject of sufficient interest and importance that we have devoted a separate book to it.[21] But in one sense Godwin could claim to be unique. In most of these large single panels, diptychs, and triptychs, the viewpoint is that of a fisherman looking from the middle of the river towards the shore. In every case there is a sharp contrast between the land in the background and the liquid in the foreground. Tangled, overgrown shoreline vegetation at the top of each picture encounters shimmering surfaces, rippled light and shadow, and fractured reflections on the riffles, pools, and swirls below. In reflection, water transforms, adds depth and mystery. The river is not so much a mirror as it is another shimmering world.

This series of paintings represented Godwin's re-encounter as an artist with landscape and a personal odyssey. He had grown up in Calgary fishing the Elbow and the Bow with his father. But his breathtaking images of the Bow represent not simply a nostalgic return to childhood, family, and the joys of fishing. Godwin had been one of Canada's most celebrated abstract painters in the 1950s and 1960s. These paintings of the lower Bow represent his abandonment of purely abstract forms and a return to landscape. Fly-fishing the Bow, closely studying its water and shoreline with the eye of a fisherman and a painter, Godwin perceived himself confronting landscape, the central problem of Canadian art historically. At one level his art records the river and its seasons and moods. In his prose the artist warns of the destruction of the river through pollution, dam-building, and neglect. For that reason there is a somewhat Arcadian, even archival, quality to these large-scale works. They document a golden age before corruption and decay irredeemably set in.

Throughout the 1980s and 1990s several organizations and numerous individuals discovered and publicized another Bow River, valued for the pleasure and outdoor recreation it provided. Seen from this perspective, the ethic of use had to be tempered by a higher ethic of human enjoyment through outdoor recreation. Fishing, bird watching, hunting, canoeing, camping, painting, and photography not only asserted sometimes competing claims upon the river vis-à-vis hydroelectric generation, sanitation, industrial usage, or irrigation, but these activities also necessarily heightened the public appreciation of nature and encouraged stewardship of it. Outdoor recreationists counterpoised aesthetic and recreational values with the consumption needs of industries, municipalities, and irrigators. This ethic sometimes conflicted directly with the ethic of use, pitting the trout fishers against the irrigators and hydroelectric production against canoeing, but more often than not these interests converged: for example, on the importance of improving water quality. Thus representatives of the two points of view could collaborate on the "state of the river" project.

This ethic of enjoyment, however, shifted political power on the river from downstream users, who claimed the high ground of food production and innocent victimhood of urban negligence, to the cultural needs of a leisured and largely urban society. In this view the cultural and therapeutic uses of the river trumped economic considerations. Seeing the river through the restorative, inspirational, physical, and intellectual pleasures it afforded, the outdoor enthusiasts argued, provided a broader, more humane, and more environmentally sensitive perspective on watershed management.

The Ethic of Environmental Integrity

The discovery of the Bow River during the 1980s and 1990s occurred mainly under the influence of two value systems: the ethic of use and the ethic of enjoyment. These two approaches differed in emphasis and agenda, but both were "instrumentalist" ways of looking at nature. Whether viewed as a lifeline or a playground, the river deserved to be restored, preserved, and protected primarily for human benefit. Human needs provided the measure of things. Both perspectives fragmented the river by emphasizing its different parts or attributes. People discovered the river through its utilities as an irrigator and sanitizer, as a habitat for ducks to shoot and fish to catch, and as a source of pleasure. They reacted and mobilized when other uses impaired "their river."

But outdoor recreation, when taken to the extreme, could be just as damaging in its own way as industrial, municipal, and agricultural uses. Fishing, hunting, jogging, skiing, bird watching, hiking, camping, cycling, and vacationing, when practised en masse, entailed hotels, highways, shops, power lines, campgrounds, golf courses, and subdivisions. Improving accessibility to the river and opening parks along its banks brought more and more people and their equipment – watercraft, recreational vehicles, hunting, fishing, and camping gear – into contact with the river.

Surveys confirmed the overwhelming importance of the automobile for the park environment and the direct impact of mass tourism. In the summer of 1972 the Socio-economic Research Division of Parks Canada decided to undertake a statistical survey of the users of eight national parks to discover who was visiting and why. Among the 27,700 respondents at Banff National Park, 55 per cent turned out to be Canadians, almost 45 per cent Americans, and fewer than one hundred parties from other countries; 40 per cent came from British Columbia and Alberta (nearly one-third of the overall total). Family groups were the overwhelming norm: more than 70 per cent of visiting parties included two adults, over 60 per cent with children along. Just over half listed the national park as their main destination, though one-third intended to stay only a single night, nearly a quarter two nights (though that same proportion did not answer this question). Almost 55 per cent opted for campgrounds for overnight stays, and more than 15 per cent motels, with less than 9 per cent intending to rent a cottage. Virtually none of these auto tourists intended to make use of what were usually considered the main draws of the park: playing golf, going boating or fishing, taking a bus tour, following a self-guided trail, camping in the backcountry, or visiting a mountain hut. More than three-quarters did not intend to visit the hot springs. Here, then, was the typical Banff National Park visitor in the auto age: part of a family group from the surrounding provinces or the United States aiming to stay but a day or two in fairly low-priced accommodation, while not straying far from town or campground to sample the wonders of nature. Presumably a short period spent in town, long enough to buy a souvenir or a postcard, look at the outside of a well-known hotel, and then back in the car and off to the next destination.[22]

Giving this volume of traffic and behaviour, parks authorities began to question what activities were appropriate within the boundaries of a national park and what might reasonably be excluded. A study from the mid-1970s determined that "high quality activities" included hiking, observing nature, photographing scenery, joining interpretive programs, sketching, attending campfire talks, and trail riding. Activities deemed inappropriate were listed as window-shopping, going to the movies, steam baths, disco dancing, waiting in traffic, roller skating, driving to people-watch, and drinking beer in taverns. The author recommended that the frowned-upon activities should be moved outside the boundaries of Banff National Park.[23]

A government-funded survey of Calgary households in 1987 determined that fully 70 per cent used the Bow valley for recreation at some point during the year. Projecting Calgary data onto the entire southern Alberta region, the investigators estimated 3.48 million user days spent in the valley over a twelve-month period. Fishing, along with picnicking and skating, they learned, were relatively minor attractions compared with walking (52 per cent of households), cycling (23 per cent), and jogging (18 per cent) in the urban river parks and trails. In all, according to this estimate, outdoor activity in the region accounted for expenditures in the region of $40 million annually.[24] The river had become the focal point of a leisure industry.

Upstream recreational use of the river intensified in the 1980s as well. A utilization survey of the Rocky Mountain national parks for the federal government in 1987–88 counted 3.2 million visitors to Banff and the upper Bow, most of them Canadians from Alberta and surprisingly few from Ontario and Quebec. Visitors sampled told the surveyors that they had been attracted by the natural features of the park, that the mountain scenery made the deepest impression upon them, and that on the whole they were satisfied with their visits, notwithstanding the shortage of public washrooms and parking places. Sightseeing, viewing wildlife, resting, relaxing, shopping, and taking photographs were the most popular activities, but 20 to 25 per cent of visitors also engaged in what were defined as strenuous activities, such as climbing, hiking, and downhill and cross-country skiing.[25]

Critics argued that the very popularity of Banff National Park threatened to destroy it, a claim made most forcefully by the Canadian Parks and Wilderness Society (CPAWS). Founded in 1962, this society was the foremost parks advocacy group and environmental lobbyist in the country. During the 1970s and 1980s, under the influence of the contemporary environmental movement, which was well represented amongst its membership, CPAWS moved away from its original instrumentalist "use and enjoyment" defence of parks towards an essentialist position which insisted that national parks were valuable in and of themselves as islands of autonomous, self-acting natural systems separate from human interference.[26] This "land ethic" put nature first and human needs second. In the parks and in the protected areas around them, nature could evolve on its own terms. Here humans needed to learn to respect and adjust to the complex interaction of ecosystems rather than dominate them. In this holistic view, everything was connected to everything else; no single part could be isolated for consumption without affecting the whole in some way. The mission of Parks Canada, CPAWS argued, was to preserve these natural ecosystems intact for posterity in as close to their natural condition as possible. Over the years Parks Canada had permitted inappropriate uses and management policies that threatened sensitive environments. But in a classic tragedy of the commons effect, overuse by visitors and the proliferation of facilities to service them had transformed the very ecology the parks had been created to preserve. CPAWS publicized the amount of environmental deterioration resulting from excessive use and commercial, residential, and recreational development in Banff National Park. It moved from being the leading advocate of national parks to being their most influential critic.

Lobbying by CPAWS had a profound impact upon park management policy in 1988, a year during which the Winter Olympics were held just outside the park boundaries, when parliament considered an amendment to the National Parks Act. It had long been recognized that the Parks Act needed to be strengthened in certain areas to permit more rigorous enforcement of regulations, to restrict development, and to allow for self-government for the anomalous municipalities of Banff and Jasper, located within national parks. First contemplated in the 1970s, these amendments were deferred by successive governments largely because of heavy lobbying by the government of Alberta, which opposed development restrictions on hotels and ski hills, especially within Banff National Park. In 1988, however, a Conservative government with a large majority, over the advice of the Alberta caucus, eventually, on a Friday afternoon, introduced a package of amendments that were defended mainly as a means of regulating poaching, creating new parks, and permitting certain aboriginal uses. The opposition parties criticized the government for the delay and tried to embarrass its Alberta members, but in large measure they approved the amendments.[27]

In committee, however, public consultations led to some important changes in the bill. Apparently, as a result of a CPAWS submission and lobbying to overcome government resistance, the text of the bill was amended to make ecological principles the first management priority for national parks.[28] The National Parks Act amendment, passed in 1988, declared: "Maintenance of ecological integrity through the protection of natural resources shall be the first priority when considering Park zoning and visitor use in a management plan."[29] Beginning formally in 1988, a new ethic of ecological integrity became the guiding principle for the management of Banff National Park and the Bow River within it.[30]

Nor, in the view of the Canadian Parks and Wilderness Society, was it enough simply to protect nature within the boundaries of parks. Banff National Park, for example, was but a small portion of a much larger mountain ecosystem, which was also coming under increasing pressure. That larger ecosystem needed to be protected as well if the ecology of the park within it was to remain vital. Restrictions on development within the park boundaries simply pushed commercial and residential services ancillary to tourism down the valley. Canmore, a dying coal-mining town, was enjoying a remarkable rebirth on the strength of the tourism boom, as a dormitory suburb of Banff, retirement community, and high-end mountain-living destination all its own. In its publication *The Bow Valley: A Very Special Place*, CPAWS pointed out that "the amount of undisturbed Montane in the valley had decreased by 50%" in recent years as a result of logging and urban development in the valley east of the park.[31] Fish species introduced to satisfy sports fishers had virtually extirpated the native bull trout in this reach. As urbanization increased, water and air quality deteriorated. Development put wildlife at risk, especially the large mammals that regularly migrated north and south across the corridor. To protect the valley from further degradation, CPAWS hoped that park protection would be extended to these key mountain sections of the river valley east from the national park to the Nakoda reserve. It regarded parks as beachheads from which ecological protection could be advanced into the interconnected regions beyond the gates.

Official policy may have changed to make ecological integrity the first priority of management, but events within and outside the park told a different story. By the early 1990s the number of visitors to Banff National Park surpassed five million, a surprising proportion of whom ventured into the backcountry in recreational pursuits. Applications for permits to develop hotels and services for this growing number of visitors and residential accommodation for park personnel mounted in Parks Canada files. Conflicts between pro- and anti-development groups around each proposal created a charged and deadlocked political atmosphere. The sheer number of applications and the rising tumult made approval on a project-by-project basis, the standard practice, unmanageable. In these circumstances the government of the day saw some virtue in declaring a moratorium on development, pending a report on a comprehensive management policy for Banff National Park. In 1994 the minister struck the Banff–Bow Valley Task Force to develop a plan for the park that would "integrate ecological, social and economic values" and provide guidance "on the management of human use and development in a manner that will maintain ecological values and provide sustainable tourism."[32]

The government asked for a squared circle, and the task force, with considerable skill and ingenuity, delivered it in a timely fashion. As usual, the task force set about its work by commissioning a number of research studies. It departed from the normal agenda of investigations in its approach to the problem. In parallel with the expert analysis, it also set in motion a consultative process remarkable for its breadth and intensity. In addition to public hearings, opinion surveys, and an active communications strategy, the task force went a step further towards giving the stakeholders some responsibility for fashioning a resolution. To break the pattern of a confrontation between opposing views from delegations, a round table of representative interest groups struggled through many meetings to work out a common vision for the park. Through its interactive and continuous consultations, the task force aimed to build a shared vision for the park among its members but, more importantly, through the process, to engender a political consensus among the stakeholders in the park regarding the principles that should guide policy and what they would entail in practice.[33]

Remarkably, out of this flurry of consultations and studies, a relatively stark and almost unanimous conclusion emerged. Banff National Park had arrived at a crossroads and was on the verge of losing its World Heritage Site designation. Present patterns of growth and development were unsustainable and, if allowed to continue, would result in "irreversible harm." The park was on the verge of being "lost" to destructive forces emanating from its own popularity as a tourist destination. The task force believed it could be saved, however, if following the amended National Parks legislation, the advice of its expert analysis, and the consensus emerging from its round table, "ecological integrity" became the touchstone of public policy. Parks management should make the preservation of biodiversity its first objective, a goal decreed not only by legislation but also on the higher ground of human survival: "National parks, as core protected areas of much larger ecosystems, are central to preserving the variety of life on earth. If we cannot respect the right of wildlife, plant species, fish, and yes, even insects, to live undisturbed in places set aside for their protection, what hope do we have for long term survival? What hope do we have for our own future?"[34]

An "Ecological Outlook Report" commissioned by the task force and an accompanying "Technical Report" documented the ways in which the ecological integrity of the park had already been compromised. Development in certain sensitive zones interfered with the seasonal migrations of wildlife. Aquatic habitat had been lost, as river flow patterns had been altered. Dams disrupted the movement and reproduction of fish. Fish and wildlife mortality inevitably rose with the extent of human contact. Fire suppression affected the evolution of vegetation and hydrology. Housing and highways occupied montane habitat, important to several intersecting ecosystems. Urbanization and transportation corridor crossings changed predator-prey balances. Sewage from the permanent and transient population of the park affected water quality. Non-native plants and species had been purposefully introduced and unintentionally imported.[35] Curiously, for such a crucial concept, the task force did not define precisely what it meant by the phrase "ecological integrity" in its report. It came closest by quoting a Parks Canada policy paper in which the term was used to mean "the perpetuation of natural environments essentially unaltered by human activity."[36]

The role of Banff National Park "as a public park and pleasure grounds" should continue, the task force concluded, but not at the expense of ecological integrity. Railway developments prior to the founding of the park obviously compromised this shibboleth somewhat. Some human intrusions in the park over long periods of time – the hotels and cottages, for example – had passed from being development to "heritage" worth preserving in its own right. The footprint of tourism, up to a certain point, could be tolerated without lasting damage. Some of these things could not be reversed;

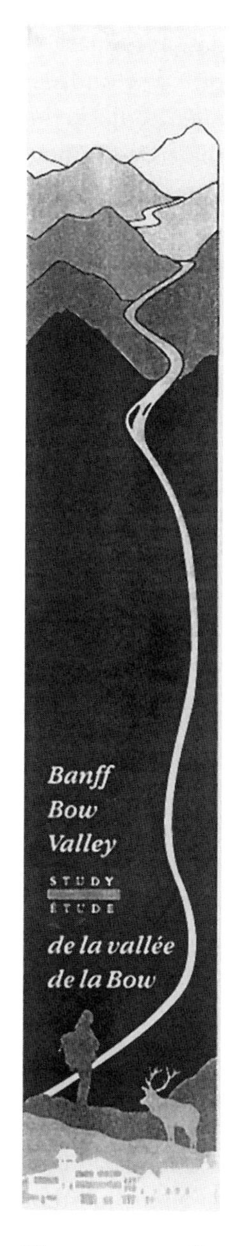

Illus. 13.4 Banff–Bow Valley Task Force logo, 1996.

nor should they be, in the eyes of the task force. But some inappropriate uses, such as the airstrip, must be phased out; others strictly regulated. Above all, future growth posed the potentially fatal threat. Access, activity, and development should be measured against their potential impact upon the ecology, which the park had a mission to pass on to future generations "unimpaired."

Broadly, this policy meant strict curbs on almost all forms of activity within the park and, in some respects, removal and restoration of prior conditions. The complete findings and recommendations of the task force lie outside the scope of this book. But specifically, with respect to the aquatic ecosystem, of which the Bow River was a central element, the task force determined that ecological integrity had been "compromised" in a number of ways: "Past and present human activities have substantially degraded aquatic resources within the Park, especially within the Bow Valley, where human use is most concentrated." The stocking of rivers and lakes with non-native species had extirpated some native species and substantially reduced the population of others. Introduced fish species from Europe and other parts of North America now predominated in the waters of the national park. As a result, the native cutthroat trout had disappeared completely from the Bow River below Lake Louise. The introduction of rainbow trout into previously fishless lakes had dramatically destabilized the original food chain.

Construction of the railway on one side of the Vermilion Lakes wetlands and roads on the other had altered their total area, changed water levels, reduced the habitat of some species, and dispersed large mammals such as moose. Channellization of the river during the construction of highways and railways had altered stream dynamics in places. The regulation of rivers within the park for hydroelectric and other purposes had greatly changed the patterns of stream flows, which in turn had implications for riverine ecosystems. Dams governed about 40 per cent of the flowing water in the Park. As Lake Minnewanka underwent successive developments to turn it into a storage reservoir between 1912 and 1942, its mean level was raised 25 metres and its area almost doubled. The annual filling up in the summer and drawing down in the winter created unstable shoreline zones in the reservoirs, which had wide-ranging ecological consequences. Storage doubled the wintertime flow of the river and greatly reduced summer peak flows, which changed scouring processes. Dams interfered with the migration of fish; reductions in stream flow or the complete elimination of some streams in the course of water storage interfered with breeding. The Spray River, for example, largely rerouted by diversion dams, had had its flow reduced by 80 per cent. In the process, 40 kilometres of angling were destroyed.

Regulation of flow reduced flooding, thereby nullifying those natural processes entailed by flooding that sustained aspen, poplar, and riparian shrubs and the life that depended upon them. Finally, increased nutrient loading from sewage-treatment plants, ski hills, golf courses, and septic fields had coated the river bottom in the park with a biofilm, increased algae growth, encouraged bacteria and certain insect populations, altered the oxygen available for fish, and changed the taste of the water.[37] In short, biodiversity had been lost, water flows altered, and water quality adversely affected by existing levels of activity.

Following its guiding principles, the task force recommended several measures to restore the ecological integrity of the park's aquatic ecosystem. It proposed major improvements in water quality through a reduction of fecal coliform bacteria and phosphorus loadings by the sewage-treatment plants in Banff and Lake Louise and numerous non-point sources. The task force proposed an extensive system of water quality monitoring against strict standards and public education to ensure compliance. To achieve the strategic goal of maintaining and restoring natural biodiversity of aquatic ecosystems, it recommended wetland rehabilitation projects, benchmark studies to identify precise objectives for each lake and stream, and then programs to restore native species, end stocking, remove non-native species on an experimental basis, and in the final analysis, eliminate fishing within the park. To return water levels and flows to their natural state, the task force recommended that priority in management be given to the needs of aquatic and riparian ecosystems over hydroelectric and sanitary water withdrawals. Ecological integrity required that the electric utility manage its storage reservoirs, dams, and hydroelectric stations to replicate as much as possible natural patterns of flow and to maintain aquatic life in its reservoirs. Some smaller dams could be removed immediately; otherwise, adjustment in operating policies could bring about the desired result. But in the long run, the task force implied, these non-conforming uses might be phased out completely in view of the minimal importance of hydroelectricity to the utility.[38]

The Banff–Bow Valley Task Force report, which was promptly accepted by the federal Heritage minister, dramatically changed the philosophy of management on the upper Bow River, first rhetorically and increasingly in practice. Its application of the concept of ecological integrity, critical observations on environmental degradation, and recommendations for remediation became the benchmark against which park management would henceforth be evaluated.[39]

Ideas do not necessarily flow downhill like rivers, but as we have seen with the concept of ecological integrity, its proponents intend that it should be sovereign in the much larger mountain ecosystem of which Banff National Park is but a part. The prevailing upstream ethic of ecological integrity may spill out of the park into surrounding regions, but as it flows downstream, it will join the two other reigning views of the river: the ethic of use and the ethic of enjoyment. In the swirling current these convergent but not wholly compatible ideas will struggle for supremacy. Ecological integrity may occupy the highest moral ground, but it may be the most unrealistic goal to achieve. An ethic whose benchmark is the absence of humanity will always find culture wanting.

The Return of the Blob

The Bow River acquired a constituency during the 1990s. A coalition of voices came forward purporting to speak for it: ministries, task forces, councils, advisory committees, academics, and a host of voluntary associations and individuals. This was something new. Previously, interest groups had been able to have their way with the river, negotiating differences with each other and largely compliant governments. But in the 1990s powerful organizations and governmental, academic, and popular pressure groups asserted a countervailing authority. In this chorus we have been able to discern three convergent but not entirely compatible environmental ethics, each with its own cohort of lobbyists, consultants, and scientists. For the purposes of exposition, each of these conceptions of the river has been dealt with separately, giving the misleading impression of sequential development. But the ideas and movements emerged more or less simultaneously. The "Preserving Our Lifeline" exercise took place at the same time as the work of the Banff–Bow Valley Task Force, the long-running cleanup of the Blob, Godwin's exhibition, and publication of the second edition of McLennan's fly-fishing guide.

These voices would subsequently be joined by others in the later 1990s and the first years of the new millennium. For example, under the title "A River Runs through Us," the *Calgary Herald* published a notable seven-part series on the Bow in 1998.[40] Descending the river through its various reaches, the series began with "Raw Beauty," centred on the scenic purity of the river from Bow Lake to Banff, and ended with two essays highlighting the tremendous water-consumption needs of irrigated agriculture along the lower Bow in "Nourishing the Country" and "Prairie Lifeblood." Journalism essentially echoed the structure, themes, and conclusions of the earlier "Preserving Our Lifeline" reports, with the inevitable human interest stories and featured personalities. The editorial concluding the series emphasized the importance of balancing conflicting needs to prevent "squandering our most precious resource."

The Biosphere Institute of the Bow Valley, located in Canmore, launched a program of public education in 1997 promoting ecological integrity. Led by Melanie Watt, this group gathered information, brought expert analysts into the public debate, facilitated research, and maintained a database of reference material to support its advocacy activities.[41] In 2003 another Canmore-based group of environmentalists formed themselves into the Bow Riverkeeper "to protect and restore the Bow River watershed in order to ensure a clean and sustainable water supply for all living communities that depend on the river now and in the future." Led by Danielle Droitsch and affiliated with a large US-based network of similarly named river protection organizations, the Bow Riverkeeper staged a well-publicized Big Bow Float, a twenty-nine-day, 650-kilometre canoe trip the length of the river, in 2005 to draw attention to the fragile state of the river.[42] The city of Calgary recruited a River Valleys Committee in the 1990s, and it maintains an Environmental Advisory Committee of sixteen stakeholder representatives to guide municipal action on river issues.[43] These new environmental organizations and educational activities reflect elements of the three environmental ethics discussed in this chapter. Drawing upon new financial resources, mainly charitable foundations, NGOs, and private-sector money, they have brought new methods of communication and institutionalized forms of surveillance to bear upon the river.

In recent years attention has shifted from rivers to water more generally, stimulated by heightened concerns about global climate change. During the United Nations International Year of Freshwater (2004), veteran environmental activist Robert Sanford directed a Wonder of Water program aimed at heightening public awareness of water vulnerability in the region. Scholarly research at the universities of Alberta, Calgary, and Lethbridge has recently fed directly into local discussions of water quality and potential water scarcity.[44] Professor David Schindler, a water specialist at the University of Alberta, has produced the most authoritative analysis of current water management problems and projections of the likely impact of climate change on prairie waters.[45]

Increasingly, these new voices are raising alarms that go well beyond the three ethics discussed in this chapter. Using scientific models of climate change, they predict that the supply of water is entering a long-term period of secular decline beyond human powers to alter.[46] The river is drying up, they warn, and that is a science-generated nightmare much more frightening than the Blob.

Conclusion

Books must have endings, even if the lives of rivers do not. The environmental history of the Bow River is obviously not over. In many respects it is just beginning. Major perceived trade-offs needed to accommodate economic growth to regulatory limitations on water use lie ahead. The ultimate fusion of nature and culture in a conscious, functional, sustainable river policy is yet to come. With that perspective in mind, we will conclude by looking ahead to the coming challenges posed jointly by cultural and environmental change on the river.

The Bow River has sustained a rich history of successive and overlapping socio-economic systems. For most of this period and over most of the length of the river, an instrumentalist attitude of appropriation for productive use has governed development of the river and its resources. The Bow has been seen as a resource to be used for human betterment by being brought to its highest use. In the earliest stages, this process involved displacing indigenous peoples and settling them on reserves. Over time, individuals and organizations with the necessary capital and technology extracted wealth and power from the river, sanctioned by the higher logic of progress through economic development under the permissive gaze of the developmental state.

This instrumentalist approach, by no means unique to the Bow River but characteristic of the extractive mentality governing most resource use in North America, fundamentally altered the character of the river. Logging temporarily changed its siltation and chemistry; electricity permanently shifted its seasonal flow; flow regulation, by cutting off the peaks of floods, stabilized vegetation on the banks and altered fish habitat; sanitation took vast quantities of water, putting much of it back badly soiled; fishing changed species and their numbers in the river, and nutrient loading bulked them up; irrigation consumed much of the river by redirecting its downstream flow onto fields and pastures, but it also extended the watery domain by spreading riverine habitat across the prairie far from the mainstem. Human use changed the nature of the river. The Bow River became a natural-cultural product without, at the same time, being destroyed – certainly altered but not destroyed.

Over time, technological change and economic development have shifted power among the competing users. For example, irrigators dethroned the power company as coal replaced hydro power as the main energy source and as irrigated acreage subsequently expanded under state patronage. Conflict arose at various points: between indigenous peoples and settlers, ranchers and homesteaders, loggers and dam operators. But many of the transitions occurred without major confrontations. Park-making proceeded relatively easily through the downtown core of Calgary, aided and abetted by deindustrialization that removed the logging booms, weirs, sawmills, power stations, and warehouses. When there was enough water to go around or readily available substitutes, life on the river was not a zero-sum game.

When conflicts between users over water have arisen, the interest groups themselves, sometimes mediated by government, arranged compromises. In general, interest groups rarely resorted to the law to settle disputes. There is remarkably little case law relating to water rights along the Bow. Other forms of resolution have proven more effective, such as regulation, legislation, or brokering amongst interests. For example, government brokered a three-way resolution when winter flooding, popularly associated with the release of stored water for hydroelectric generation, caused political friction. More recently, as drought reduced the flow of the Oldman River, threatening to trigger water rationing according to the "first in time, first in right" principle, the interest groups voluntarily entered into an ad hoc water-apportioning agreement to ensure that no one went without water.[1] The players have tended to avoid winner-take-all decision-making. Throughout our period the courts were surprisingly irrelevant as dispute settlement mechanisms, in part because the state gathered all powers relating to water unto itself. But why the law was not a tool to be used in these matters is a subject for more specialized deliberation.

Cultural and intellectual regimes change over time. Perceptions of the river are social constructs that do not remain permanent. A fundamental shift began to occur after World War II as notions of livelihood, commodity production, and beneficial use were challenged by a rising standard of living, increased levels of education, and an engagement with science and planning experts in government. A traditional emphasis on production was complicated by pressures born of consumption. Ultimately, conservation ideas stressing wise use would be challenged by environmental ideas that engaged a wider range of problems and concerns, from quality of life to leisure to ecological measures.[2] These broad shifts did not mean that the river came under less human intervention. Indeed, some of the newly realized potentialities of the river depended in part upon these prior human actions which stabilized banks, seemingly controlled flooding, and nurtured the fishery. River rehabilitation ideas involved direct intervention to "restore" the river to some imagined prior state. Materialist explanations can not be made to cover these fundamental shifts. In certain eras, power over the river is highly correlated with economic power, but not always. Other value systems – aesthetic, scientific, ecological, public safety – can vie with economic power for influence over policy, but they require legitimacy and efficacy through social expression. The past forty years have witnessed just such a fundamental social and ideological reorientation on the Bow.

The turning point seems to have been the 1960s, as a new public sensibility that elevated ideas of nature to a social value comparable with economic development began to alter the political economy on the river. That shift was marked, too, by a transition from private appropriation to public reassertion of ownership. This perhaps occurred because a relatively rich and leisured society growing prosperous on resource exploitation elsewhere could afford to displace marginal commercial activities from the river. The most visible manifestation of this change was the growth of park-making, which, over time, turned entire stretches of the river into a park.

Parks brought people closer to the river. Whether the closer contact or the prior ideological shift made the river more "their" river is an open question. But undoubtedly, something like collective possession occurred during the 1960s, bringing new social forces into discussions of use. Note that this change occurred as the river appeared to have become domesticated, having dispensed with its wilder ways of flooding in summer and then in winter. Apparently tamed and confined to its bed, the river posed less of a menace. At about the same time a web of less visible provincial regulation was spun over the river and its users.

By the 1990s the dialectic was firmly engaged as state bureaucracies, special-interest groups, and advocacy science brought countervailing power to bear upon established users. The reaction moved government from a permissive to a regulatory stance, establishing limits on withdrawals, waste-water loading, riparian development, and flow management and setting up vigilant monitoring processes to police human behaviour. In this the state was prodded, pushed, and supported by numerous advocacy groups and scientists purporting to speak for the river. The new rules have been codified in two major pieces of legislation, the Water Act (*Revised Statutes of Alberta*, 2000, c. W-3) and the Environment Protection and Enhancement Act (RSA, 2000, c. E-12). The policy document *Water for Life: Alberta's Strategy for Sustainability* (2003) spelled out the new framework within which comprehensive river basin management plans were to be hammered out amongst the interest groups.[3] Taken together, these statutes and regulations impose strict conditions upon water use on the Bow River. While, on the one hand, the new regime is permissive in that it recognizes existing rights to water-taking, on the other, it establishes household water use as the highest priority, prevents bulk water exports and inter-basin transfers, protects minimum in-stream flow requirements to preserve aquatic environments, and stiffens enforcement of regulations. The new laws create river basin councils to broker interests between users and also permit the transfer of water licences in situations where all available water has been allocated. In effect, the government of Alberta has replaced the "first in time, first in right" method of allocating water in times of shortage with a quasi-market system in which unused allocations can be bought and sold. At the same time it has held back a portion of unallocated water and made provision for water buybacks to meet inflow stream requirements and the anticipated future claims of First Nations to adequate water for activities on their reserves.[4]

A river plays different roles in the life of communities, depending upon their culture, social organization, and economic structures. The Bow, for example, is perceived in a different way as it flows through a national park, First Nation reserves, cities and towns, and the open range. Political economy has much to do with shaping these understandings, but it is not sufficient unto itself. The power of an idea linked to perceived group interests is one of the most potent driving forces in history.

Rivers have a history. They change under human intervention; they change on their own. However much a river has been used and abused, it is still very much a river, doing river things, which humans ignore or underestimate at their peril. Unplumbed reservoirs of hubris lie within the concepts of river management, environmental engineering, and ecological integrity. In our view, there is no more "pristine nature" out there to be preserved or restored. The history of humanity along the river is so bound up with the history of the river as to make the two inseparable.

What happened in Alberta on the Bow during the last forty years also occurred with local variations all over the Western world. Mark Cioc, an environmental historian of the Rhine, argues that during the 1990s governments "began to 'see' the river not simply as a navigational and commercial highway but also as a biological habitat in need of protection."[5] In the United States, river conservation, involving the removal of dams from the industrial era, restoration of riparian habitat, and mitigation of pollution, was one of the major accomplishments of the environmental movement.[6] In central Canada the Grand River in the 1990s emerged Cinderella-like from its squalid past as environmentalists, conservation authorities, and governments combined to redesign the river as a natural playground for hiking, canoeing, fishing, and ecotourism.[7] International information flows through the media, and environmental organizations made the Bow, in this respect, a tributary of all of the rivers in the developed world. Debates on the Bow drew upon and contributed to a collective discourse. But just as rivers are specific to particular geographies, human perceptions and experiences with rivers, while echoing universal archetypes, have their own distinctive logics, local social bases, and courses of resolution.

It might be said that the Bow River reasserted itself in the 1990s after having been the passive recipient of excessive use. But that conclusion would not be correct. The river no more spoke in the 1990s than it had in previous generations. It simply continued to flow; human perceptions fundamentally changed. This profound cultural shift, incorporated into legislation and policed by the state, has just begun to make its mark upon the river. Government imposed limits on what could be done to it, how much it could be consumed or diverted, what could be put back into it, and how its ecology should be regulated. These measures were, in turn, strongly supported (though often through criticism) by vigilant environmental watchdogs, special-interest groups, and independent scientists. Taken together, these political and extra-governmental forces have had their effect upon the river by regulating the behaviour of individuals, organizations, and municipalities. What was notable about this river redux regulatory regime is that most current practices could be accommodated within the prescribed limits. The upper limits on water-taking and diversion, for example, have yet to be reached. On the other hand, meeting nutrient loading targets required fairly aggressive steps to improve sewage treatment in cities and towns – including the national park – but these were doable things. What would happen when the society bumped up against these self-imposed regulatory ceilings? That is a question for the future that cannot be answered by drawing upon historical experience.

The twentieth-century west was perceived to offer limitless opportunity, inexhaustible resources, and bountiful harvests. By contrast, the operative phrase governing the outlook at the beginning of the twenty-first century would be the "limits on development." The myth of abundance is being gradually displaced by the grim prospect of scarcity. This is particularly the case with water, and most notably so with respect to the Bow River and the South Saskatchewan basin.[8] We sense that the really hard choices about water use along the Bow lie ahead and not too far into the future. For that reason we want to end by looking forward to those challenges to reaffirm the dynamic relationship of the nature-culture dialogue under changing social, technological, and environmental influences.

The limits are imposed by federal and provincial laws, interprovincial agreements, court rulings, and international treaties. These are social constructs subject, of course, to potential change. There is no telling what might happen as human behaviour confronts these constraints. Behaviour can be changed or the constraints loosened, or some combination of both. Where are the most likely points of conflict?

The Master Agreement apportioning water between Alberta and Saskatchewan will perhaps be among the first of the flashpoints. The amount of land under irrigation in Alberta doubled between 1971 and 1999.[9] Up until the present the limitations imposed upon Alberta by the agreement, a 50 per cent passage of the natural flow with a 1500 CFS minimum stream flow, have not yet significantly impinged upon the irrigators. Nevertheless, virtually all the available water in the basin has been allocated in water-taking licences, even though irrigators are not yet using water up to their full capability.[10] What will happen when all the licensed water has been taken up and the province is called upon in years of low flow or drought to meet its downstream obligations to Saskatchewan? Will the agreement be abrogated? Or will the irrigators be required to ration their water consumption to meet the requirements? If so, what type of water rationing regime will be negotiated in a notional "first in time, first in right" jurisdiction? Small steps have already been taken in the new Water Act towards creating markets for licensed but unused water for irrigation. As we have seen, an inherited water allocation doctrine enshrined in riparian rights was rejected in favour of a "first in time, first in right" licensing system. Under the pressure of possible wholesale water shortages, will this mode of allocation survive the inevitable social and political tensions associated with cutting off water in accordance with this arbitrary system? This problem will most likely arise if climate change leads to significant diminutions in total river flow.

The same question might be asked of municipal water-taking. Calgary, the main municipality, holds licences from the province permitting only so much diversion. The day when urban water-taking comes up against these provincial limits lies in the not too distant future. Calgary and other municipalities, following provincial guidelines but guided, too, by an internal politics, have begun to curb per capita water use so that more people can be accommodated within the prescribed limits. But even with these curbs on consumption, population growth alone will potentially push the city up against its water consumption ceiling. Will water-taking put a cap on urban growth? Or will domestic needs trump other uses?

At the other end of the pipe, the municipalities, especially Calgary, have made remarkable strides in reducing some aspects of river loading. Sewage-treatment facilities and new technologies have made it possible for Calgary to reduce its impact upon the river along some axes. However, on other measures the caps are of pressing concern. Inorganic substances, heavy metals, silt, and bacteria enter the water from stormwater runoff as well as the treatment process.[11] These multi-point sources of pollution are more difficult and expensive to control and are likely to multiply as the growing city spreads subdivisions ever farther across the prairie.

Humans seem particularly susceptible to forgetting when it comes to flooding. In popular perception, the river has been domesticated. In the 1970s something like a social contract was entered into by the interested and organized citizenry and the city over assessment of flood risk. The population preferred to accept the risk of floods rather than undergo the tree removal, diking, and zoning changes necessary to minimize that danger. The devastating 2004 flood below Calgary on Fish Creek may have served as a reminder that the city itself was not immune.[12] Had that storm struck a few kilometres farther northwest, Calgary might have been inundated. What will happen to the social contract when the hundred-year flood returns, as it eventually will?

Up to this point, water use by First Nations has not figured prominently in river management thinking. First Nations, on the other hand, have been affected by hydraulic development, hydroelectric plants on the Nakoda reserve, and irrigation canals on the Siksika reserve. The precise nature of aboriginal water rights under treaties and as riparian owners remains unresolved. Domestic requirements have up to this point placed few demands upon the river. However, should major irrigation, residential, or industrial development occur on reserve lands and if First Nations demand appropriate water allocations from the river that conflict with other users' rights or agreements or legislation in force, what would likely happen?[13]

The hydroelectric dams on the Bow are more or less redundant from a base-load perspective in an era when most electricity is produced in thermal generating plants. Theoretically, the dams could be removed and the flow of the river returned to its "natural state," as has happened on some rivers in the United States. All other things being equal, the Bow is a likely candidate to be turned into a "wild river" by removing its dams, as some environmental groups might insist. But a coalition of recreational users prefer the river in its modified, regulated form. The dams provide outdoor recreation opportunities valued by campers, boaters, fishers, and tourists. Moreover, international agreements and deregulation of utilities markets give the power company a new interest in its hydroelectric assets. The Kyoto Accord values hydroelectric capacity as a "green" offset for a company heavily dependent upon coal. Similarly, the opening of a continental electricity market – hydroelectric facilities that can be rapidly turned on and off – makes the company more nimble in responding to spot-market sales opportunities. Decades ago the dams and turbines were depreciating assets; a new regulatory regime has imparted new value to them.

Most troubling, however, are the implications of global warming for the region and the river. Analyses of the implications of climate change in Alberta predict a general rise in temperatures that will drive higher levels of evaporation, as well as significant reductions in stream flow associated with glacier melt. Glaciers in the Bow's headwaters have already shrunk by about 25 per cent in the last century.[14] If these predictions turn out to be accurate, there will be considerably less water in the Bow to be divided up, greater need to preserve water from consumption to maintain in-stream needs to protect aquatic environments, and much less loading absorptive capacity during minimum flow periods. A smaller river will perhaps require a recalibration of regulatory limits.

In brief, will the Master Agreement be stretched and broken when irrigators come up against its caps? Political power has shifted towards the irrigators. Will they trump environmental and recreational influence? Will water shortages lead to a new system of apportionment? Will the caps on water-taking and river-loading prevail against urban population growth? First Nations' rights to water and the relation of those as yet undetermined rights to existing regulations remain open questions. And overall, how will the political system cope with the social and economic conflicts during an impending era of less flow in the Bow?

Historians cannot predict. We offer this environmental history of the Bow River as a contribution towards a better understanding of the partnership between nature and culture and the shifting power relations between the partners as ideologies, technologies, and the river change. Our history offers as much hope to those who look forward to a continuation of river rehabilitation under robust state regulation as it does evidence that regulations will likely not withstand serious economic pressures. The environmental history of the Bow River, under something like zero-sum game conditions in which one group's interests can only be maximized at the expense of others' losses, is, in a sense, just beginning. But in this new regime the nature-culture dialogue we have explored in these pages will go on. The Bow will continue to be in part our creation, a product of human choice, as we join with natural forces to negotiate the ways in which it will change.

Appendix: Calgary Power-Generating Capability in the Bow Watershed, 1911–1960

Hydroelectric Capability in 1951

Station	Capability (kilowatts)
Horseshoe (1911)	13,900
Kananaskis (1914)	10,000
(1951)	8,900
Ghost (1929)	28,000
Cascade (1942)	18,000
Barrier (1947)	12,900
Spray (1951)	49,900
Rundle (1951)	17,000
Three Sisters (1951)	3,000
Total	161,600

Hydroelectric Capability Added after 1951

Station	Capability (kilowatts)
Ghost (1954)	22,900
Bearspaw (1954)	16,900

Pocaterra (1955)	14,900
Interlakes (1955)	5,000
Cascade (1957)	17,900
Spray (1960)	52,900
Rundle (1960)	32,900
Total, 1954–60	163,400

Notes

Chapter One

1 Richard Glover, ed., *David Thompson's Narrative, 1784–1812* (Toronto: Champlain Society, 1962), lxxiii, lxxxvi, 46–9, 234–70. For an attempt to answer the question "Who were the first white men to visit the valley of the Bow River?" see J.W. Porter, *Nineteenth Century Bow Valley Exploits*, a reprint from an American Society of Petroleum Geologists Field Guide Manual for a symposium in 1971, Banff, Alberta, (Glenbow Library, PAM 971.238 B784p).

2 Nathaniel Rutter, "Geomorphology and Multiple Glaciation in the Area of Banff, Alberta," *Geological Survey of Canada Bulletin* 206 (Ottawa: Department of Energy Mines and Resources, 1970); Mel Reasoner and Ulrike M. Huber, "Postglacial Palaeoenvironments of the Upper Bow Valley, Banff National Park, Alberta, Canada," *Quaternary Science Reviews* 18 (1999): 475–92; E.C. Pielou, *After the Ice Age: The Return of Life to Glaciated North America* (Chicago: University of Chicago Press, 1991), 227–48; L.E. Jackson and M.C. Wilson, eds., *Geology of the Calgary Area* (Calgary: Society of Petroleum Geologists, 1987), chapter 1; and M.C. Wilson, *Geology and Landscape Evolution, Bow River Valley Field Trip* (Calgary: Society of Petroleum Geologists, 1960).

3 BP = before present. Present = 1950.

4 Daryl W. Fedje, James M. White, Michael C. Wilson, D. Erle Nelson, John S. Vogel, and John R. Southon, "Vermilion Lakes Site: Adaptations and Environments in the Canadian Rockies during the Latest Pleistocene and Early Holocene," *American Antiquity* 60 (1995): 81–108; Barbara Huck and Doug Whiteway, *In Search of Ancient Alberta* (Winnipeg: Heartland Publishers, 1998), 92–4.

5 These cultural sequences are mapped and illustrated in plates 5 to 10 in R.C. Harris, ed., *Historical Atlas of Canada*, vol. 1 (Toronto: University of Toronto Press, 1987); see also E.J. Hart, *The Place of Bows: Part I, To 1930: Exploring the Heritage of the Banff–Bow Valley* (Banff: EJH Literary Enterprises, 1999), 9–29.

6 Stephen Pyne, *Vestal Fire* (Seattle: University of Washington Press, 1997) and *Awful Splendour: A Fire History of Canada* (Vancouver: UBC Press, 2007); and Reasoner and Huber, "Postglacial Palaeoenvironments," 486. Reasoner and Huber conclude:

"Fire has been an important disturbance agent in the vicinity of Crowfoot Lake since ca. 9000 C year BP."

7 See S.A. Haslam, *The Historic River: Rivers and Culture Down through the Ages* (Cambridge: Cobden of Cambridge Press, 1991), for a comprehensive, illustrated treatment of the different relations human cultures, mainly European, have enjoyed with rivers.

8 Simon Schama, *Landscape and Memory* (New York: Random House, 1996).

9 Hugh MacLennan, "Thinking Like a River," in *Rivers of Canada* (Toronto: Macmillan, 1974), viii, 7, 71, 127, and *Seven Rivers of Canada: The Mackenzie, the St. Lawrence, the Ottawa, the Red, the Saskatchewan, the Fraser, the St. John* (Toronto: Macmillan, 1960), vii. For the context of the writings, see Elspeth Cameron, *Hugh MacLennan; A Writer's Life* (Toronto: University of Toronto Press, 1981), 306–7, 346–7.

10 W.H. New, "The Great River Theory: Reading MacLennan and Mulgan," *Essays in Canadian Writing*, no. 56 (1995): 168–71. See also Neil S. Forkey, "Thinking Like a River: The Making of Hugh MacLennan's Environmental Consciousness," *Journal of Canadian Studies* 41 (2007): 42–64.

11 MacLennan here is under the thrall of another great river novel. His St Lawrence played the same role as the Mississippi for Mark Twain; *Two Solitudes* presented a similar elemental social divide as Huck Finn encountered on the river's flow, one that could like the river be bridged by the human imagination. To complete the analogy, MacLennan's *Seven Rivers* could be seen as the counterpart to Twain's *Life on the Mississippi*. MacLennan refers directly to Mark Twain and *Life on the Mississippi* in his "Thinking Like a River," 9.

12 McLennan, *Rivers of Canada*, 12. For this book MacLennan reordered the chapters of *Seven Rivers* and composed new essays on the Miramichi, the streams of Nova Scotia, the Niagara, the rivers of Ontario, the Saguenay, and the Hamilton and added an epilogue, "And On to the Eternal Sea."

13 MacLennan, *Rivers of Canada*, 9–10.

14 Donald Creighton, *Harold Adams Innis: Portrait of a Scholar* (Toronto: University of Toronto Press, 1957), quoted 58.

15 H.A. Innis, *The Fur Trade in Canada* (New Haven: Yale, 1930), 392.

16 H.A. Innis, "Transportation in Canadian Economic History," in M.Q. Innis, ed., *Essays in Canadian Economic History* (Toronto: University of Toronto Press, 1956), 74. For another statement of this position in a comparative Canada-US perspective, see "Recent Trends in Canadian-American Relations," in Daniel Drache, ed., *Staples, Markets and Cultural Change: Harold Innis, Selected Essays* (Toronto: University of Toronto Press), 262–3.

17 Innis, *The Fur Trade in Canada*; D.G. Creighton, *The Commerical Empire of the St. Lawrence* (Toronto: Ryerson, 1937); C. Berger, *The Writing of Canadian History* (Toronto: Oxford, 1976).

18 Creighton, *The Commercial Empire of the St. Lawrence*, 6–7.

19 Born in 1877 in the Cariboo region of British Columbia, daughter of a Hudson's Bay Company fur trader, Skinner left the province about the turn of the century for Los

Angeles, Chicago, and eventually New York, where she made a living as a writer and journalist. After World War I she wrote and edited volumes for Yale University's Chronicles of America series, which made her reputation as a popular historian. Thereafter she managed to eke out a living as a freelance writer-editor in New York. See Jean Barman's touching portrait of this fiercely independent writer, "'I walk my own track in life & no mere male can bump me off it': Constance Lindsay Skinner and the Work of History," in Beverly Boutilier and Alison Prentice, eds, *Creating Historical Memory: English-Canadian Women and the Work of History* (Vancouver: UBC Press, 1997), 129–63, and *Constance Lindsay Skinner: Writing on the Frontier* (Toronto: University of Toronto Press, 2002).

20 Skinner's "vision" of rivers in American history is presented in a posthumous essay appended to J. Dana's *The Sacramento: River of Gold* (Farrar & Rinehart, 1939).

21 Henry Beston, *The St. Lawrence* (New York: Rinehart, 1942); Bruce Hutchison, *The Fraser* (New York: Rinehart, 1950); Marjorie Wilkins Campbell, *The Saskatchewan* (New York: Rinehart, 1950); Leslie Roberts, *The Mackenzie* (New York: Rinehart, 1949).

22 The "cradle" metaphor is most fully developed in Robert Coffin's *The Kennebec* (New York: Rinehart, 1937), 3–5.

23 Ibid., 9.

24 Brion Davis's *The Arkansas* (New York: Rinehart, 1940), Arthur Tourtellot's *The Charles* (New York: Rinehart, 1941), Walter Hard's *The Connecticut* (New York: Rinehart, 1947), and Carl Carmer's *The Hudson* (New York: Rinehart, 1939) are classic examples of the genre.

25 Donald Davidson, *The Tennessee* (New York: Rinehart, 1946), 6.

26 Stanley Vestal, *The Missouri* (New York: Rinehart, 1945), 5.

27 Hutchison, *The Fraser*, 5.

28 Mabel Dunham, *Grand River* (Toronto: McClelland and Stewart, 1945), and William Toye, *The St. Lawrence* (Toronto: Oxford, 1959).

29 Ted Lewis, *The Hudson: A History* (New Haven: Yale University Press, 2005), and Peter Ackroyd, *Thames: Sacred River* (London: Chatto & Windus, 2007).

30 See the website: www: greatcanadianriver.com for the twenty-seven rivers profiled thus far.

31 Coffin, *The Kennebec*, 271.

32 Stewart Holbrook, *The Columbia* (New York: Holt, Rinehart, Winston, 1974).

33 Frank Waters, *The Colorado* (New York: Rinehart, 1974 ed), xvii and xx.

34 Samuel P. Hays, *Beauty, Health and Permanence: Environmental Politics in the United States, 1955–1985* (New York: Cambridge University Press, 1987).

35 Hugh Aitken, *The Welland Canal Company* (Cambridge: Harvard, 1954); John Dales, *Hydroelectricity and Industrial Development – Quebec, 1898–1940* (Cambridge: Harvard, 1957).

36 H.V. Nelles, *The Politics of Development* (Toronto: Macmillan, 1974). N. Hundley Jr wrote a comparable and contemporary book in the United States: *Water and the West* (California: University of California Press, 1975).

37 C. Armstrong, *The Politics of Federalism* (Toronto: University of Toronto Press, 1981).
38 William H. McNeill, *Plagues and Peoples* (New York: Anchor, 1976).
39 Both quotations from Donald Worster, *The Wealth of Nature: Environmental History and the Ecological Imagination* (New York: Oxford, 1993), vii.
40 Aldo Leopold, *A Sand County Almanac* (New York: Ballantine, 1966), 137–41. This edition also contains most of the material from *Round River.*
41 Aldo Leopold, *Round River: From Aldo Leopold's Journal* (New York: Oxford, 1953).
42 Donald Worster, "Thinking Like a River," in *The Wealth of Nature: Environmental History and Ecological Imagination* (New York: Oxford, 1993), 124.
43 Donald Worster, *Rivers of Empire: Water, Aridity and the Growth of the American West* (New York: Oxford, 1985).
44 See, in addition to *Rivers of Empire*, Worster's collections of essays *The Ends of the Earth: Perspectives on Modern Environmental History* (New York: Cambridge University Press, 1988) and *The Wealth of Nature: Environmental History and the Ecological Imagination* (New York: Oxford, 1993).
45 See, for example, Ted Steinberg, *Nature Incorporated: Industrialization and the Waters of New England* (New York: Cambridge University Press, 1991), and Ari Kelman, *A River and Its City: The Nature of Landscape in New Orleans* (California: University of California Press, 2003).
46 B. Harden, *A River Lost* (New York: Norton, 1996). For a variation on this theme, see Ellen Wohl, *Virtual Rivers: Lessons from the Mountain Rivers of the Colorado Front Range* (New Haven: Yale, 2001).
47 Fred Pearce, *The Dammed: Rivers, Dams and the Coming World Water Crisis* (London: Bodley Head, 1992). See also his more recent *When the Rivers Run Dry* (Toronto: Key Porter, 2006).
48 R.C. Bocking, *Mighty River* (Vancouver: Douglas & McIntrye, 1997).
49 John McPhee, *The Control of Nature* (New York: Farrar, Straus and Giroux, 1989).
50 John M. Barry, *Rising Tide: The Great Mississippi Flood of 1927 and How It Changed America* (New York: Simon and Shuster, 1997).
51 J.M. Bumsted, *Floods of the Centuries* (Winnipeg: Great Plains, 1997); see also T. Thomson, *Faces of the Flood* (Toronto: Stoddart, 1998). Hurricane Katrina has tragically revived the genre.
52 R. White, *The Organic Machine: The Remaking of the Columbia River* (New York: Hill & Wang, 1995), x–xi.
53 Ibid., 59; the quotation that follows is from 108.
54 For another fine example, see John O. Anfinson, *The River We Have Wrought: A History of the Upper Mississippi* (Minneapolis: University of Minnesota Press, 2003). Alfinson shows how both the Army Corps of Engineers and the Izaac Walton League together and in different ways shaped the river.
55 William Cronon, "The Trouble with Wilderness," in *Uncommon Ground* (New York: Norton, 1996). See also Cronon's contribution to the round table on environmental history in *Journal of American History* 76 (1990), where he argued: "We can no longer

assume the existence of a static, benign climax community in nature that contrasts with dynamic but destructive human change. Rather than benign natural stasis and disruptive human change we need to explore differential rates and types of change" (1182).

56 Matthew Evenden, *Fish versus Power: An Environmental History of the Fraser River* (New York: Cambridge, 2004), 267.

57 Clarence Glacken, *Traces on the Rhodian Shore: Nature and Culture in Western Thought from Ancient Times to the End of the Eighteenth Century* (Berkeley: University of California Press, 1967).

58 Keith Thomas, *Man and the Natural World* (Harmondsworth: Penguin, 1984).

59 Donald Worster, *Nature's Economy: A History of Ecological Ideas*, 2nd ed. (New York: Oxford, 1994); see also Peter Coates, *Nature: Western Attitudes Since Ancient Times* (Berkeley: University of California Press, 1998).

60 In 1995 Alberta Fish and Wildlife captured and destroyed forty "nuisance beavers" in Calgary and another ten in Canmore on the grounds they were cutting down trees, damming streams, blocking pathways, and interfering with fish breeding. See "Calgary's Beavers Get Chopped," *Calgary Herald*, 4 September 1995.

61 *The Painted Valley*, published by the University of Calgary Press in November 2007.

62 Donald Worster defined the new field in his contribution to the *Journal of American History* round table, "Transformation of the Earth: Toward an Agroecological Perspective in History" (1087–147). Canadian scholars have recently made important contributions to this field reflected in special issues of the journals BC *Studies*, nos. 142–3 (2004), ed. Graeme Wynn; *Urban History Review* 34, no. 1 (2005), ed. Stephen Bocking; *Revue d'histoire de l'Amérique française* 60, nos. 1–2 (2006); *Globe* 9 (2006), ed. Stéphane Castonguay; and *Environmental History* 12, no. 4 (2007), ed. Matthew Evenden and Alan MacEachern; and the volumes appearing in the Nature/History/Society series from UBC Press edited by Graeme Wynn; see http://www.ubcpress.ca/books/series_nhs.html. Other notable recent book-length contributions include Colin Coates, *The Metamorphoses of Landscape and Community in Early Quebec* (Toronto: University of Toronto Press, 2000); Alan MacEachern, *Natural Selections* (Montreal and Kingston: McGill-Queen's University Press, 2001); Clinton Evans, *The War on Weeds* (Calgary: University of Calgary Press, 2002); George Colpitts, *Game in the Garden* (Vancouver: University of British Columbia Press, 2002); and Graeme Wynn, *Canada and Arctic North America: An Environmental History* (Santa Barbara: ABC Clio Press, 2007).

Chapter Two

1 On the Bad River: this designation may be found, for example, in Peter Fidler's maps referred to later in the text. In Fidler's hand the Bad River referred to both the Bow and the South Saskatchewan to the confluence with the Red Deer River.

2 For a general discussion of some of the considerations in describing the problems of cultural encounter or contact or intersections in the context of western Canada, see

Sarah Carter, *Aboriginal People and Colonizers of Western Canada to 1900* (Toronto: University of Toronto Press, 1999), 31–46; Richard White, *The Middle Ground: Indians, Empires, and Republics in the Great Lakes Region, 1650–1815* (New York: Cambridge University Press, 1991).

3 Alice B. Kehoe, "Blackfoot and other Hunters on the North American Plains," in *Cambridge Encyclopaedia of Hunters and Gatherers*, ed. Richard B. Lee and Richard Daly (Cambridge: Cambridge University Press, 1999), 36. Ted Binnema argues this point over the very long term in *Common and Contested Ground: A Human and Environmental History of the Northwestern Plains* (Norman: University of Oklahoma Press, 2001).

4 Ian A.L. Getty and Erik Gooding, "Stoney," in *Handbook of North American Indians: Plains*, vol. 13, ed. Raymond J. DeMallie (Washington, DC: Smithsonian Institution, 2001), 596–603; Bill B. Brunton, "Kootenai," in *Handbook of North American Indians*, vol. 12 (Washington, DC: Smithsonian Institute, 1998), 223–37; Hugh Dempsey, "Blackfoot," in *Handbook of North American Indians*, vol. 13: 604–28 (on the shifting territories, see 608); Hugh Dempsey, "Sarcee," in *Handbook of North American Indians*, vol. 13: 629–37; The Blackfoot Gallery Committee, *Nitsitapiisinni: The Story of the Blackfoot People* (Calgary: The Glenbow Museum, 2001), 4–5.

5 Richard G. Forbis, *Cluny: An Ancient Fortified Village in Alberta*, Occasional Papers No. 4 (Calgary: Department of Archaeology, University of Calgary, 1977); David Thompson, *Travels in Western North America, 1784–1812*, ed. Victor G. Hopwood (Toronto: Macmillan, 1971), 200.

6 Thompson, *Travels in Western North America*, 193, 196; John C. Ewers, *The Horse in Blackfoot Indian Culture* (Washington, DC: Smithsonian Institution, 1955), 17–18. Thanks to Gerald Conaty for explaining the different names applied to horses.

7 Binnema, *Common and Contested Ground*, 102; quotation, 103.

8 Arthur J. Ray, *Indians in the Fur Trade* (Toronto: University of Toronto Press, 1974), 170. Jody F. Decker, "Tracing Historical Diffusion Patterns: The Case of the 1780–82 Smallpox Epidemic Among the Indians of Western Canada," *Native Studies Review* 4, no. 1–2 (1988): 1–24; Maureen Lux, *Medicine That Walks: Disease, Medicine and Canadian Plains Native People, 1880–1940* (Toronto: University of Toronto Press, 2001), 14–15. See also Thompson's reference to rivers and smallpox in *Travels in Western North America*, 199; quotation, 199.

9 Decker, "Tracing Historical Diffusion Patterns"; Lux, *Medicine That Walks*, 106–7; quotation, 16.

10 HBCA, Chesterfield House (Bow River) Post Journal, microfilm reel no. 1M20; Eugene Y. Arima, *Blackfeet and Palefaces: The Pikani and Rocky Mountain House* (Ottawa: Golden Dog Press, 1995), 117–19. Mackenzie quotation from David J. Fairfield, "Chesterfield House and the Bow River Expedition" (MA thesis, University of Alberta, 1970), 143. See also J.E.A. Macleod, "Piegan Post and the Blackfoot Trade," *Canadian Historical Review* 24 (1943): 273–79.

11 See E.J. Hart, *The Place of Bows: Part I, To 1930: Exploring the Heritage of the Banff-Bow Valley* (Banff, Alta: EJH Literary Enterprises, 1999), 35–49, for the search for mountain passes.

12 Doug Owram, *Promise of Eden: The Canadian Expansionist Movement and the Idea of the West, 1856–1900* (Toronto: University of Toronto Press, 1980); W.L. Morton, *Henry Youle Hind, 1823–1908* (Toronto: University of Toronto Press, 1989).
13 On the growth of inventorial sciences in Canada, see Suzanne Zeller, *Inventing Canada: Early Victorian Science and the Idea of a Transcontinental Nation* (Toronto: University of Toronto Press, 1987); Irene Spry, ed., *The Papers of the Palliser Expedition, 1857–1860* (Toronto: Champlain Society, 1968), xv–cxxxviii.
14 Hugh Dempsey, "A—CA—OO—MAH—CA—YE (Ac ko mok ki, Ak ko mock ki, A'kow—muk—ai, known as Feathers and Ak ko mokki (Old Swan)), Blackfoot Chief; d. 1859 or 1860," in *Dictionary of Canadian Biography.* Available online at http://www.biographi.ca; accessed 21 August 2006. A copy of Hector's map may be found on the back leaf of Spry, *The Papers of the Palliser Expedition.*
15 Binnema has provided a wide-ranging analysis of this map, to which we are indebted; see Theodore Binnema, "How Does a Map Mean? Ak ko mokki (Old Swan)'s Map of 1801 and the Blackfoot World," in *From Rupert's Land to Canada: Essays in Honour of John E. Foster*, ed. Theodore Binnema, Gerhard Ens, and R.C. Macleod (Edmonton: University of Alberta Press, 2001). Binnema translates *mistakis* as "backbone" (212). Barbara Belyea throws cold water on the intriguing possibility (repeated by Binnema) that Ak ko mokki (Old Swan)'s map influenced Arrowsmith's. See Belyea, *Dark Storm Moving West* (Calgary: University of Calgary Press, 2007), 46–7.
16 John Nicks, "Thompson, David," in *Dictionary of Canadian Biography*; accessed 16 August 2006. Derek Hayes describes Thompson's map as "defining the west"; see Hayes, *Historical Atlas of Canada* (Vancouver: Douglas & McIntyre, 2002), 166–7. See also D'arcy Jenish, *Epic Wanderer: David Thompson and the Mapping of the Canadian West* (Toronto: Doubleday, 2000).
17 Spry, *The Papers of the Palliser Expedition*, lxiii.
18 Ibid., 289.
19 Binnema, "How Does a Map Mean?" 215; quotation from 223n31; Ewers, *The Horse in Blackfoot Indian Culture*, 40; Hugh A. Dempsey, ed., "Alexander Cuthbertson's Journey to the Bow River," *Alberta Historical Review* 19, no. 4 (1971): 17. The Montana Historical Society holds the original version of Cuthbertson's narrative. See also Spry, *The Papers of the Palliser Expedition*, 260.
20 Hugh A. Dempsey, ed., *The Rundle Journals, 1840–1848* (Calgary: Alberta Historical Society and Glenbow Alberta Institute, 1977), 261; quotation, 273; Spry, *The Papers of the Palliser Expedition*, 288.
21 James Bradley cited in Ewers, *The Horse in Blackfoot Indian Culture*, 144; Weasel Tail quotation from 105.
22 On the social role of rendezvous points, see David Meyer and Paul C. Thistle, "Saskatchewan River Rendezvous Centers and Trading Posts: Continuity in a Cree Social Geography," *Ethnohistory* 42 (1995): 403–44; Gerald T. Conaty, "The Bow River Downstream," in *The Bow: Living with a River*, ed. Gerald T. Conaty (Toronto: Key Porter Books, 2004), 119; Dempsey, ed. "Alexander Cuthbertson's Journey to the Bow River," p. 18; Gerald T. Conaty, "Calgary," in *The Bow*, ed. Conaty, 90.

23 David Thompson, "Journey to the Bow River," in *Columbia Journals / David Thompson*, ed. Barbara Belyea (Montreal and Kingston: McGill-Queen's University Press, 1994), 18; HBCA, Chesterfield House (Bow River) Post Journal, microfilm reel no. 1M20, 25 October entry; Fairfield, "Chesterfield House and the Bow River Expedition," 104.

24 Quotation from George W. Arthur, *An Introduction to the Ecology of Early Historic Communal Bison Hunting Among the Northern Plains Indians*, Mercury series (Ottawa: National Museum of Man, 1975), 74. See also R.E. Morlan and M.C. Wilson, "Bison Hunters of the Plains," in *Historical Atlas of Canada*, vol. 1, ed. Cole Harris (Toronto: University of Toronto Press, 1987), plate 10. In this case the Bad River would appear to be the Oldman River, identified as "Mookoo wans." This detail can be read well in Derek Hayes's reproduction of Fidler's map in Hayes, *Historical Atlas of Canada*, 153.

25 Quotation from Treaty 7 Elders and Tribal Council with Walter Hildebrandt, Dorothy First Rider, and Sarah Carter, *The True Spirit and Original Intent of Treaty 7* (Montreal and Kingston: McGill-Queen's University Press, 1996), 107; Thompson quotation from Thompson, "Journey to the Bow River," 17. See also Spry, *The Papers of the Palliser Expedition*, 313.

26 Ewers, *The Horse in Blackfoot Indian Culture*, 41–2; Alison Landals, "Horse Heaven: Change in late Precontact to Contact period Landscape Use in Southern Alberta," in *Archaeology on the Edge: New Perspectives from the Northern Plains*, ed. Brian Kooyman and Jane Kelley (Calgary: University of Calgary Press, 2004), 244–5; Treaty 7 Elders and Tribal Council with Hildebrandt, First Rider, and Carter, *The True Spirit and Original Intent of Treaty 7*, 105; Getty and Gooding, "Stoney," 599–600; Chief John Snow, *These Mountains are our Sacred Places* (Toronto: Samuel Stevens, 1977), 19.

27 John C. Hellson, *Ethnobotany of the Blackfoot Indians*, Mercury series (Ottawa: National Museum of Man, 1974), 19. Hellson notes that more recently willows have been used to construct sweat lodges. On pines as a building material, see ibid., 116–17; on Saskatoon berries as a staple food, ibid., 100–1; on medicines, ibid., 62–85. An accessible version of the Fidler map is in Hayes, *Historical Atlas of Canada*, 154. See also J. Rod Vickers and Trevor R. Peck, "Islands in a Sea of Grass: The Significance of Wood in Winter Campsite Selection on the Northwest Plains," in *Archaeology on the Edge*, ed. Kooyman and Kelley, 95–124.

28 Fidler quotation from Alice M. Johnson, ed., *Saskatchewan Journals and Correspondence: Edmonton House, 1795–1800; Chesterfield House, 1800–1802* (London: Hudson's Bay Record Society, 1967), 268; Mackenzie quotation from HBCA, Chesterfield House (Bow River) Post Journal, microfilm reel no. 1M20, 6 October entry. See also Spry, *The Papers of the Palliser Expedition*, 263.

29 Getty and Gooding, "Stoney," 597; Arapooash quotation in Binnema, *Common and Contested Ground*, 17; Hector quotation in Spry, *The Papers of the Palliser Expedition*, 317.

30 Thompson, *Travels in Western North America*, 246–49.

31 Spry, *The Papers of the Palliser Expedition*, 287, 290, 296.

Chapter Three

1 A good start has been made in Theodore Binnema, *Common and Contested Ground: A Human and Environmental History of the Northwestern Plains* (Toronto: University of Toronto Press, 2004); George Colpitts, *Game in the Garden: A Human History of Wildlife in Western Canada to 1940* (Vancouver: University of British Columbia Press, 2002); Clinton Evans, *The War on Weeds* (Calgary: University of Calgary Press, 2002); and Alwynne B. Beudoin, "What They Saw: The Climatic and Environmental Context for Euro-Canadian Settlement in Alberta," *Prairie Forum* 24 (1999): 1–40.

2 The classic texts are Chester Martin, *"Dominion Lands" Policy* (Toronto: Macmillan, 1938); W.A. Macintosh, *Prairie Settlement: The Geographical Setting* (Toronto: Macmillan, 1934); and A.S. Morton, *History of Prairie Settlement* (Toronto: Macmillan, 1938). On the broader context of land taking in settler societies, see John Weaver, *The Great Land Rush* (Montreal and Kingston: McGill-Queen's University Press, 2004).

3 Lucien M. Hanks and Jane Richardson Hanks, *Tribe under Trust: A Study of the Blackfoot Reserve of Alberta* (Toronto: University of Toronto Press, 1950); Hana Samek, *The Blackfoot Confederacy, 1880–1920: A Comparative Study of Canadian and U.S. Indian Policy* (Albuquerque: University of New Mexico Press, 1987); Sarah Carter, *Aboriginal People and the Colonizers of Western Canada to 1900* (Toronto: University of Toronto Press, 1999); Binnema, *Common and Contested Ground*). On the NWMP, see R.C. Macleod, *The NWMP and Law Enforcement, 1973–1905* (Toronto: University of Toronto Press, 1976), 22–5.

4 In addition to the books by Hanks and Hanks, Samek, and Carter cited above in note 3, the circumstances surrounding the negotiation of Treaty No. 7 and the differing perceptions of what was being negotiated are discussed in the essays by Chief Snow, Hugh Dempsey, and Stan Cuthand in Ian A.L. Getty and Donald B. Smith, eds, *One Century Later: Western Canadian Reserve Indians Since Treaty 7* (Vancouver: UBC Press, 1978), and in greater detail from a Native perspective in Treaty 7 Elders and Tribal Council with Walter Hildebrandt, Sarah Carter, and Dorothy First Rider, *The True Spirit and Original Intent of Treaty 7* (Montreal and Kingston: McGill-Queen's University Press, 1996); Hugh A. Dempsey, *Red Crow: Warrior Chief* (Saskatoon: Western Producer Prairie Books, 1980) and *Crowfoot: Chief of the Blackfeet* (Edmonton: Hurtig Publishers, 1976).

5 A.C. Isenberg, *The Destruction of the Bison* (New York: Cambridge University Press, 2000).

6 Carter, *Aboriginal People and the Colonizers of Western Canada to 1900*, final chapter.

7 Hanks and Hanks, *Tribe under Trust*, 3–52; Samek, *The Blackfoot Confederacy, 1880–1920*, 56–69; Hugh Dempsey, "One Hundred Years of Treaty Seven," in Getty and Smith, *One Century Later*, 20–30.

8 Open-range ranching, the iconic moment in Alberta history and public memory, is amply documented in popular literature beginning with L.V. Kelly, *The Rangemen* (Toronto: 1913) and continuing with R.D. Symons, *Where the Wagon Led* (Calgary: Fifth House, 1997); Sherm Ewing, *The Range* (Missoula: Mountain Press, 1990);

Edward Brado, *Cattle Kingdom: Early Ranching in Alberta* (Vancouver: Douglas and McIntyre, 1984); Hugh Dempsey, *The Golden Age of the Canadian Cowboy* (Calgary: Fifth House, 1995); Sheilagh Jameson, *Ranches Cowboys and Characters* (Calgary: Glenbow Museum, 1987); Charlie Russell, *The Canadian Cowboy* (Toronto: McClelland and Stewart, 1993); and Candance Savage, *Cowgirls* (Vancouver: Douglas and McIntyre, 1996). The academic literature begins with David H. Breen, *The Canadian Prairie West and the Ranching Frontier, 1874–1924* (Toronto: University of Toronto Press, 1983) and Simon Evans, "The Origins of Ranching in Western Canada," in *Essays on the Historical Geography of the Canadian West*, ed. L.A. Rosenvall and Simon Evans (Calgary: Department of Geography, University of Calgary Press, 1987), 70–92. More recent work includes Simon Evans, Sarah Carter, and Bill Yeo, eds, *Cowboys and Ranchers and the Cattle Business* (Calgary: University of Calgary Press, 1999). A. Warren Elofson in *Cowboys, Gentlemen and Cattle Thieves* (Montreal and Kingston: McGill-Queen's University Press, 2000) disputes Breen's interpretation about the peacefulness of the frontier but upholds most of his other conclusions. On the transnational ranching experience, see Terry Jordan's *North American Cattle Ranching Frontiers* (Albuquerque: University of New Mexico Press, 1993), and Richard Slatta, *Cowboys of the Americas* (New Haven: Yale University Press, 1990). On the business of ranching, see Max Foran, *Trails and Trials: Markets and Land Use in the Alberta Beef Cattle Industry, 1881–1948* (Calgary: University of Calgary Press, 2003), and Henry Klassen, *Eye on the Future: People and Business in Calgary and the Bow Valley, 1870–1900* (Calgary: University of Calgary Press, 2002), chapter 7.

9 The economic weakness of large leasehold ranching is discussed in Breen, *The Canadian Prairie West*, 59–68; Foran, *Trails and Trials*, 1–27; and Klassen, *Eye on the Future*, 149–84. Elofson, *Cowboys, Gentlemen and Cattle Thieves*, 71–98, focuses upon the adverse impact of the occasionally harsh northern environment on ranching.

10 Breen, *The Canadian Prairie West*, 47–61. William Pearce, the de facto federal land agent in Alberta, who sided firmly with the leaseholders, believing grazing to be the only viable activity in an arid landscape, deployed all fair means and some foul to keep squatters off the leases, especially out of the critical river bottoms and watering places. See UAA, William Pearce Narrative, vol 1, 34ff; see also ibid., 9/2/5/1/14, Ejection of Squatters, 1894–96, for Pearce's memoranda on proper procedures for removing illegal homesteaders and tearing down their houses. The Mounted Police were understandably reluctant to carry out these instructions and had to be regularly stiffened by Pearce. A partial view of Pearce's career is to be found in E. Alyn Mitchner, "William Pearce and Federal Government Activity in Western Canada, 1882–1904" (PhD thesis, University of Alberta, 1971).

11 On the small ranchers, see Klassen, *Eye on the Future*, 171ff, and Foran, *Trails and Trials*, 29–56. Klassen's essay "A Century of Ranching at the Rocking P and Bar S," in *Cowboys and Ranchers and the Cattle Business*, ed. Evans, Carter, and Yeo, 101–22, presents a case study of the evolution of two successful small ranches.

12 There is a large literature on the building of the CPR, beginning with the memoirs of some of those involved: J.H.E. Secretan, *Canada's Great Highway: From the First Stake to the Last Spike* (Toronto, Longmans, 1924); Charles Aeneas Shaw, *Tales of*

a Pioneer Surveyor (Toronto, Longman, 1970); and P. Turner Bone, *When the Steel Went Through: Reminiscences of a Railroad Pioneer* (Toronto: Macmillan, 1947). The best-known account is Pierre Berton, *The National Dream* (Toronto: McClelland and Stewart, 1970) and *The Last Spike* (Toronto: McClelland and Stewart, 1971). See also Omer Lavallée, *Van Horne's Road* (Montreal: Railfare Books, 1974). The underlying impetus behind the building of the CPR is examined in A.A. den Otter, *The Philosophy of Railways* (Toronto: University of Toronto Press, 1997).

13 Much less has been written about the operation of the railway and its effects; that analysis is just beginning. See John Eagle, *The Canadian Pacific Railway and the Development of Western Canada* (Montreal and Kingston: McGill-Queen's University Press, 1989), and the essays collected in Hugh Dempsey, ed., *The CPR West: The Iron Road and the Making of the Nation* (Vancouver: Douglas and McIntyre, 1984).

14 On homesteading, the land rush, and the wheat boom after the turn of the century, see, among others, Lewis H. Thomas, "A History of Agriculture on the Prairies to 1914," reprinted in *The Prairie West*, ed. R. Douglas Francis and Howard Palmer (Edmonton: University of Alberta Press, 1985), 221–36; Ken Norrie, "The National Policy and the Rate of Prairie Settlement," reprinted ibid., 237–56; Gerald Friesen, *The Canadian Prairies* (Toronto: University of Toronto Press, 1984), 301–38; and Ronald Rees, *New and Naked Land: Making the Prairies Home* (Saskatoon: Western Producer Books, 1988). David C. Jones's account of the settlement of the dry belt at the eastern end of the Bow River in the first decade of the twentieth century and its abandonment by settlers in the 1920s captures both the exhilaration of acquisition and the agonizing heartbreak of failure. See Jones, *Empire of Dust: Settling and Abandoning the Prairie Dry Belt* (Calgary: University of Calgary Press, 2002). There is also a rich local history and memoir literature; notable examples dealing with this region would include *Lachlin McKinnon, 1865–1948, Pioneer* (Calgary: Privately printed, 1956?) and J. Angus McKinnon, *The Bow River Range, 1898–1974* (Calgary: Privately printed [1970s]).

15 Chester Martin, *"Dominion Lands" Policy* (Toronto: McClelland and Stewart, 1973), 171.

16 A. Mitchener, "The Bow River Scheme: CPR's Irrigation Block," in Dempsey, *The CPR West*, 259–74; John Eagle, *The Canadian Pacific Railway and the Development of Western Canada* (Montreal and Kingston: McGill-Queen's University Press, 1989), 173–212.

17 Although controversial, the Siksika decision added substantially, over several years, to the band fund to ensure their long-term survival. See Samek, *Blackfoot Confederacy*, 112–13; Hanks and Hanks, *Tribe under Trust*, 35–53; Valerie K. Jobson, "The Blackfoot Farming Experiment, 1880–1945" (MA thesis, University of Calgary, 1990), chapter 2. For the documents, see LAC, RG 10, vol. 3702, file 17537-3, part one, c10123, Blackfoot Agency – correspondence, reports, and petitions re: the surrender of 6 townships on the Blackfoot reserve.

18 Carter, *Aboriginal People and the Colonizers of Western Canada to 1900*, 58–9, quoting the deputy minister of the Interior's instructions to the surveyors locating the western reserves; from LAC, RG 10, vol. 3622, file 5007, 16 July 1875.

19 Hanks and Hanks, *Tribe under Trust*, 32, 47, 57.

20 Canon H.W. Gibbon Stocken, *Among the Blackfoot and Sarcee* (Calgary: Glenbow Alberta Institute, 1976), 62.
21 See chapter 10 and John Warkentin, *The Western Interior of Canada: A Record of Geographical Discovery, 1612–1917* (Toronto: McClelland and Stewart, 1964), as well as Doug Owran, *The Promise of Eden: The Canadian Expansionist Movement and the Idea of the West, 1856–1900* (Toronto: University of Toronto Press, 1980), chapters 5–7.
22 Duncan MacEachern, *Notes of a Trip to the Bow River, North-West Territories and A Journey over the Plains from Fort Benton to the Bow River and Back* (pamphlets reprinted from the Montreal *Gazette*, November 1881, in Glenbow Library).
23 Alexander Begg, "Stock Raising in the Bow River District Compared with Montana," in John Macoun, *Manitoba and the Great North-West* (Guelph: The World Publishing Company, 1882), 269–81.
24 AA, G.M. Dawson's Field Notebooks, microfilm C 4844, no 2748, GMD no. 4 (1881), McLeod to Bow Pass & Blackfoot Crossing.
25 See Suzanne Zeller and Gale Avrith-Wakeam's excellent biography of Dawson in the *Dictionary of Canadian Biography*; available online at http://www.biographi.ca.
26 The virtual absence of wildlife, even insects, in their transit of the prairie from Battleford to Montana contributed to the surreal experience of the British visitors. To their great disappointment, they encountered only one small group of bison near the Red Deer River, three of which had to be shot, of course, for sport. See Reverend James MacGregor, "Lord Lorne in Alberta," ed. Hugh Dempsey, *Alberta Historical Review* 12 (1964): 1–14.
27 Ibid., 1–14; see also GA, Marquis of Lorne Fonds, microfilm newspaper reports of Lord Lorne's tour of Western Canada, 1881.
28 Marquis of Lorne, *Canadian Pictures Drawn with Pen and Pencil*, engraved by Edwin Whymper (London: Religious Tract Society, 1885), 158–72.
29 MacGregor, "Lord Lorne in Alberta," 10.
30 Alwynne Beaudouin, "What They Saw: The Climatic and Environmental Context for Euro-Canadian Settlement in Alberta," *Prairie Forum* 24 (1999): 33ff.
31 See Breen, *The Canadian Prairie West*, 31–51; Elofson, *Cowboys, Gentlemen and Cattle Thieves*, 31–2, and especially Rees, *New and Naked Land*, 136ff, for the orientations of ranchers to the valley.
32 *Lachlin McKinnon, 1865–1948*, 23–54. The McKinnons obviously prospered as ranchers and farmers, but a close reading of their memoirs reveals how many of their kin and neighbours did not.
33 GA, J. Angus McKinnon Fonds, M4070, Autobiography, 15. A privately published version with additions appeared in the 1970s: J. Angus McKinnon, *The Bow River Range, 1898–1974* (Calgary: Privately printed, [c. 1975]). J. Angus was one of Lachlin's sons. See *Lachlin McKinnon, 1865–1948*, 40–50, for haying on the bottom lands. Putting up hay accomplished two purposes: it provided fodder for the stock in winter, and it was an excellent means of breaking in new horses.
34 Sue Anne Donaldson, "William Pearce: His Vision of Trees," *Journal of Garden History* 3 (1983): 233–44.

35 F. Girard, the medical doctor on the Blood Reserve, writing to Sir Hector Langevin, 18 November 1884, quoted in Breen, *The Canadian Prairie West*, 48.
36 GA, Amelia Brown Banister Fonds, M8364, three-volume unpublished novel in manuscript titled "Chinook Medley." Amelia Brown, born in Palmerston, Ontario, in 1873, moved west to Alberta to teach school, switched to Montana for a time, and then in 1903 moved back to Alberta, where she married John Bannister. On a ranch named "Grotto," near Davisburg on the Bow River, she raised four children before moving to Vancouver Island in 1919. After a financial setback in 1929, they returned to Ontario to farm near Port Hope. Her novel-memoir, compiled in 1950, a story of Fredina ("Freddy") Milne, a schoolteacher who marries a rancher, unfolds in three volumes: "Lower Than the Angels," "Daddy Man," and "Nostalgia."
37 Carter, *Aboriginal People and the Colonizers of Western Canada to 1900*, 27; Elofson, *Cowboys, Gentlemen and Cattle Thieves*, chapter 4, "Nature's Fury."
38 GA, J. Angus McKinnon Fonds, M4070, Autobiography, 12.
39 Elofson, *Cowboys, Gentlemen and Cattle Thieves*, 43, Shaw, *Tales of a Pioneer Surveyor*, 143–4.
40 *Lachlin McKinnon, 1865–1948*, 29.
41 Ibid., 53; Elofson, *Cowboys, Gentlemen and Cattle Thieves*, 138; R.D. Symons, *Where the Wagon Led* (Calgary: Fifth House, 1997 ed), xvii–xxxi.
42 Hanks and Hanks, *Tribe under Trust*, 43.
43 Breen, *The Canadian Prairie West*, 53, for the Pearce quotation from 1884 and 124 for the die-off of 1904. See *Lachlin McKinnon, 1865–1948*, 28, 97, 107, for an account of the closing of the range through fencing, a activity in which he actively participated.
44 *Lachlin McKinnon, 1865–1948*, 54ff; Breen, *The Canadian Prairie West*, 34.
45 *Lachlin McKinnon, 1865–1948*, 35, 42–3. His son Angus remembers hauling the building materials for the new homestead from the station at Dalemead; see GA, J. Angus McKinnon Fonds, M4070, Autobiography, 33.
46 AA, G.M. Dawson's Field Notebooks, microfilm C 4844, no. 2748, GMD no. 4 (1881) McLeod to Bow Pass & Blackfoot Crossing.
47 Robert M. Stamp, *Royal Rebels: Princess Louise & the Marquis of Lorne* (Toronto: Dundurn Press, 1988), 176.
48 GA, Isaac Kendall Kerr Fonds, M629, The Diary of I.K. Kerr and Eau Claire Lumber Company Fonds, box 3, file 32.
49 Shaw, *Tales of a Pioneer Surveyor*, 147.
50 Dawson made this observation near the Highwood: "The river today unsuited for steamboats. In many of the little rapids where spreads widely but 3' of water, though river now high for the season" (AA, G.M. Dawson's Field Notebooks, microfilm C 4844, no 2748, GMD no. 4 [1881] McLeod to Bow Pass & Blackfoot Crossing).
51 *Lachlin McKinnon, 1865–1948*, 51.
52 Berton, *The Last Spike*, 16.
53 For Macoun's influence, see W.A. Waiser, *The Field Naturalist: John Macoun, the Geological Survey, and Natural Science* (Toronto: University of Toronto Press, 1989).
54 Shaw, *Tales of a Pioneer Surveyor*, 105. Routine correspondence between 20 June 20 and 5 December 1883 relating to the location of this section of the line may be found

in Canada, House of Commons, *Sessional Papers*, 1884, no. 31: 7–17. Oral tradition among the Siksika recorded by the Hankses in the 1930s suggested that the CPR bribed Chief Crowfoot to obtain the right of way, but also that no one objected to the railway. See Hanks and Hanks, *Tribe under Trust*, 22. The railway did not acquire a route with better grades into Calgary across the Siksika reserve until 1912, when the main line was rebuilt following an agreement with the band.

55 LAC, RG 10, vol. 3583, file 1084, part A, C10103, Hayter Reed to A.E. Forget, 3 July 1895.

56. Max Foran, "The CPR and the Urban West," in Dempsey, *The CPR West*, 89–105, and Foran, *Calgary: An Illustrated History* (Toronto: Lorimer, 1978).

57 Shaw, *Tales of a Pioneer Surveyor*, 138.

58 Rogers's report to Van Horne dated 10 January 1883 indicates that F.P. Davis, one of his subordinates, had responsibility for surveying the line from the summit eastwards; see Canada, *Sessional Papers*, 1884, no. 31: 241–4. He reported: "The 40 miles of line located from the summit of the Rocky Mountains, eastward, is very easy work, affording light grades and easy alignment. The descent from the summit eastward is at the rate of 75 feet per mile for the first five miles, and for the remainder of the distance, the maximum is 37 feet to the mile, and while the surveys eastward from Fort Calgary have not yet been completed, I have reason to believe that the maximum gradients may be confined within the figures last named" (243).

59 Shaw, *Tales of a Pioneer Surveyor*, 126–7, for his first exploration of the valley, and 143–8 for his experiences during the 1883 survey.

60 Secretan, *Canada's Great Highway*, 106–7.

61 Shaw, *Tales of a Pioneer Surveyor*, 146. What irked Shaw most about this episode is that he found coal in the gravel of Devil's Head Creek, later known as the Cascade River. An experienced mining surveyor, Shaw immediately filed mining claims for the location only – in his view – to have corrupt civil servants and politicians in Ottawa steal the properties from him during the application process.

62 *Lachlin McKinnon, 1865–1948*, 59.

63 GA, J. Angus McKinnon Fonds, M4070, Autobiography, 17.

64 Surveyors boarding with the McKinnons calculated the cutbank opposite their ranch house to be 282 feet high; see ibid.

65 GA, Amelia Brown Banister Fonds, M8364, Banister, "Lower Than the Angels," 201.

66 Ibid., Banister, "Daddy Man," 25.

67 Ibid., Banister, "Lower Than the Angels," 58.

68 *Lachlin McKinnon, 1865–1948*, 51.

69 See ibid., 47, for the winter party to mark the opening of their second house.

70 GA, Amelia Brown Banister Fonds, M8364, Banister, "Lower Than the Angels," 202–3.

71 *Lachlin McKinnon, 1865–1948*, 51.

72 GA, Amelia Brown Banister Fonds, M8364, Banister, "Daddy Man," 66.

73 Hanks and Hanks, *Tribe under Trust*, 31.

74 P. Turner Bone served as the engineer for several of these bridges, see his memoir, *When the Steel Went Through*, 57–61, for the bridges near Calgary. When Lachlin

McKinnon came out from Ontario in the mid-1880s, he worked first as a logger in the Eau Claire Lumber Company camp and then as a labourer and a carpenter on the CPR replacing the original pile bridges with truss bridges on the line above Canmore. They framed the bridges in a yard at Canmore and moved them in pieces onto the site. See GA, J. Angus McKinnon Fonds, M4070, Autobiography, 11.

75 GA, George W. Craig Fonds, Personal and Business Papers of City Engineer, Calgary, 1913–23, Typewritten report on the construction of the Centre Street Bridge, January 1917, 2: "The problem from the view of the designer may be stated as follows: Given a height of 29 feet from the water line to grade at the south bank, 47 feet, or 18 feet more at the north bank, with a skew of 600, to obtain a structure which shall provide for all probable traffic requirements on a high and low level during its life, and at the same time present an appearance in keeping with the dignity which should characterize public works of a monumental character, and this for a predetermined amount of money."

76 Elofson, *Cowboys, Gentlemen and Cattle Thieves*, 54.

77 *Lachlin McKinnon, 1865–1948*, 30–7, 41.

78 Shaw, *Tales of a Pioneer Surveyor*, 144–7.

79 When records began to be kept in the 1920s, drowning accounted for 11 per cent of the average four hundred accidental deaths each year in Alberta. See AA, Department of Health, *Annual Reports*, 1919–1927.

80 We owe these newspaper references to research done on our behalf by Harry Sanders. In his reading of the Calgary press from the 1890s to the 1930s, he came across twenty-five drownings, a half-dozen bodies found floating in the river, and thirteen suicides by drowning. The river was also a convenient dumping place. Even as late as 1967 enterprises and individuals were being reprimanded for throwing refuse into the river; see CCA, Engineering and Environmental Services Records, series III, box 133, file 2990.

81 GA, Amelia Brown Banister Fonds, M8364, Banister, "Lower Than the Angels," 35–7.

82 Ibid., Banister, "Daddy Man," 87–96.

83 Hanks and Hanks, *Tribe under Trust*, 18, 57, 113, 126, 155; Binnema, *Common and Contested Ground*, 35.

84 Shaw, *Tales of a Pioneer Surveyor*, 102, 109, 132.

85 GA, Amelia Brown Banister Fonds, M8364, Banister, "Lower Than the Angels," 195.

86 *Lachlin McKinnon, 1865–1948*, 48, 102.

87 GA, J. Angus McKinnon Fonds, M4070, Autobiography, 9–10.

88 GA, Alberta Ice Company Fonds, three clippings: Pat Wiltshire, "The Ice Man Is a Big box These Days," *Calgary Herald*, 8 September 1973; C.W. Mowers, "Harvesting Ice for the West," *Western Business and Industry* 19, no 6 (June 1945); and Denny Layzell, "Pioneer Firm Helps Keep Calgarians Cool," *Herald Magazine*, 3 July 1965. The company cut ice from the river beginning in 1900 but found shallow artificial pools beside the river worked better. The water froze more solidly and could be chlorinated, and harvesting from Christmas to March would not be interrupted by the variable flow of the river. See also AA, Department of the Environment Fonds, Acc. 65.44, box 22, item 664, Alberta Ice Company

89 We are indebted to Don Smith for a file folder of documents and clippings related to the ghost town of Bow City. See also GA, Charles Westgate Fonds, pamphlets advertising Bow City, the future "Pittsburgh of Canada."

90 Ben Gadd, *Bankhead: The Twenty Year Town* (Banff: Friends of Banff National Park and the Canadian Parks Service, 1989); and John D.R. Holmes, "The Canmore Corridor: The Historical Geography of a Pass Site, 1880–the Present" (MA thesis, University of Calgary, 1978), the essential contents of which are reported in his essay "The Canmore Corridor, 1880–1914: A Case Study of the Selection and Development of a Pass Site," in *Essays on the Historical Geography of the Canadian West*, ed. Rosenvall and Evans, 27–47.

91 Hanks and Hanks, *Tribe under Trust*, 57, 74–5.

92 *Lachlin McKinnon, 1865–1948*, 59.

93 "From Prairie to Park: Green Spaces in Calgary," in *At Your Service: Calgary's Library, Parks Department, Military, Medical Services, and Fire Department* (Calgary: Century Publications, 1975), 121–256.

94 William Pearce's 1925 memo, "Reservation of Land at Calgary," was published in *Alberta History* 27 (1979): 22–8.

95 McKinnon, *The Bow River Range 1898–1974*, 3–8.

96 *Lachlin McKinnon, 1865–1948*, 51; GA, J. Angus McKinnon Fonds, M4070, Autobiography, 9.

97 GA, Amelia Brown Banister Fonds, M8364, Banister, "Daddy Man," 47.

98 For the environmental impact of farming, see Reese, *New and Naked Land*, especially the chapters "The Cult of the Tree," "Plant Propagation and Culture," and Gardens and Shelterbelts." Environmental historians would dispute his contention that ranchers did not want to change the landscape so much as live in it (136–55). For a counterblast whose subtitle explains it all, see Debra L. Donahue, *The Western Range Revisited: Removing Livestock from Public Lands to Conserve Native Biodiversity* (Norman: University of Oklahoma Press, 1999), 114–60, which relies heavily upon the analysis of Thomas L. Fleischner, "Ecological Costs of Livestock Grazing in Western North America," *Conservation Biology* 8 (1994): 629–44. Clinton L. Evans, *The War on Weeds in the Prairie West: An Environmental History* (Calgary: University of Calgary Press, 2001), 92–9, shows how eleven native grasses and plants resurfaced in wheat monoculture as weeds, supplemented by sixteen alien invaders such as stinkweed, tumble mustard, and Russian thistle. Reese notwithstanding, the broader question of evaluating the environmental change associated with prairie agriculture has yet to be tackled in Canada.

Chapter Four

1 GA, I.K. Kerr Fonds, M629. This paragraph is based upon material contained in a typescript autobiography written sometime in the early 1920s and the diary of I.K. Kerr kept in the summer of 1883.

2 GA, Eau Claire and Bow River Lumber Company Fonds, M1564 (hereafer Eau Claire Fonds), Timber Limit Maps, box 3, file 29; Canada, House of Commons, *Sessional Papers*, 1883, Department of the Interior.

3 GA, Eau Claire Fonds, box 1, file 1, K.N. MacFee to O.H. Ingram and Associates, Eau Claire, 23 June 1883; box 1, file 2, Agreement, 25 June 1883, O.H. Ingram, Putnam & Bayless, J.G. Thorp, W.A. Rust, I.K. Kerr, H.M. Stocking, Alex McLaren, McDonnell & Irvine, F. Holman, A.C. Bruce, L.E. Waterman, John B. Kehl, H.D. Davis, Daniel Donellan and K.N. MacFee.

4 GA, Eau Claire Fonds, box 1, file 1, MacFee to Ingram, 23 June 1883.

5 GA, Eau Claire Fonds, box 3, file 21, Department of the Interior, 3 July 1883, Memorandum of the Opening of Certain Tenders for Timber Berths Received up to this Date at the Department of the Interior, in Accordance with Notifications Issued to Various Parties Tendering. The Eau Claire bids averaged approximately $100 per acre; the next highest bids averaged $42. Limit B, on which the parties did not bid, was awarded to the railway contractor James G. Ross for an $81 bid. Members of the syndicate – Thorpe, Ingram, McLaren, Bayless, MacFee, Kehl, Holman, Rust, and Kerr – bid as unrelated individuals. The applications of Ross, Thorpe, McLaren, and Ingram can be found in LAC, RG 15-D-V-1, Department of the Interior, Timber and Grazing Branch Records, now reorganized as R190-138-X-E.

6 GA, Kerr Fonds, M629, Licence no. 88, Ref. no. 8888, 1 February 1886, Thomas White, Minister of the Interior to the Eau Claire and Bow River Lumber Co. of Eau Claire, Wisconsin, Licence to Cut Timber on Dominion Lands situate on the Bow River and Its Tributaries, in the Provisional District of Alberta, 478 36/100 Square Miles, Berths A, C, D, E, F, G, H, I, J, and K on map in Timber, Mineral, and Grazing Land office.

7 Max Foran, *Calgary: An Illustrated History* (Toronto: Lorimer, 1978), 11–24, and Omer Lavallée, *Van Horne's Road* (Montreal: Railfare, 1974), 70–104.

8 E.J. Hart, *The Place of Bows: Part I, To 1930: Exploring the Heritage of the Banff–Bow Valley* (Banff: EJH Literary Enterprises, 1999), 72–3, 85.

9 GA, I.K. Kerr Fonds, M629, Diary, 16 August–2 September 1883 (hereafter Kerr Diary).

10 Ibid., 18 August, 1 September.

11 Ibid., 18 August, 2 September.

12 GA, Eau Claire Fonds, box 3, file 27, Field Notes and Reports: No. 1, Berths A, C, D; No. 2, Berths E, F; No. 3, Berths G, H, I, J; addressed to the Minister of the Interior following instructions of 8 January 1884, written in April 1884.

13 Hart, *The Place of Bows*, 103–4; GA, Eau Claire Fonds, box 3, Manuscript History of the Eau Claire Lumber Company.

14 Canada, *Sessional Papers*, 1892, vol. 9, no. 13, Department of the Interior, "Annual Report 1892," Pt. V, Report of the Superintendent of Rocky Mountains Park, 5.

15 LAC, RG 10, vol., file 1084, part 1, Blackfoot Crossing Agency, 1885–1892. A letter from Hayter Reed, the deputy superintendent general of Indian Affairs to the Indian commissioner, A.E. Forget, 9 April 1889, starts an extended correspondence with J.C. Nelson of Calgary, Magnus Begg, and A.W. Ponton, local agents of the department,

which runs to 1896. See LAC, RG 10, vol. 3583, file 1084, file A, John C. Nelson to Indian Commissioner, 7 August 1891 quoting Begg on the hundred-year supply. Later it became clear that local agents had underestimated the difficulty of logging this terrain. Nelson reported in 1892 to the Indian commissioner that "it would be difficult even for whites to get the wood out, and consequently, impossible for Indians." Moreover, it would be expensive to transport the timber to the Blackfoot reserve. See ibid., Copy of a Report of a Committee of the Hon of Privy Council approved 5 April 1893; agreement re. Castle Mtn timber limit, 24 December 1892. The Blackfoot timber limit file peters out in 1897 as agents begin looking for land near Sheep Creek; see LAC, RG 10, vol. 3570, file 102, part 18, Blackfoot Agency – Land Sale, A.E. Forget, Indian commissioner, to A.W. Ponton [a local surveyor], 10 August 1897; Ponton to Forget, 15 November 1897; Forget to Ponton, 24 November 1897; Ponton to Forget, 29 November 1897. A quite different and terse account of the timber limit and its abandonment has been provided by the committee who organized the Siksika exhibit at the Glenbow Museum. In *Nitsitapiisinni: the Story of the Blackfoot People* (Toronto: Key Porter Books, 2001), 78–9, the authors write: "Early in the twentieth century a woman died at the Siksika timber limit and the area was avoided for many years. When the Sikisika tried to return, they discovered that the timber reserve had become part of Banff National Park." The area surveyed for a limit was indeed enclosed within an expanded park in 1902, but just what use was made of this surveyed (but not granted) limit is difficult to judge.

16 See the chronology of the claim provided by the Indian Claims Commission; available at http://www.indianclaims.ca/claimsmap/siksikacastlemed-en.asp; accessed 6 January 2008

17 LAC, RG 84, series A-1-a, T11019, vol. 973, file B206-318E, Eau Claire Timber Limits, 1928–1930, Summary memo, 17 March 1937, resolving trespass issues from the late 1920s and reporting all the merchantable timber has been removed and the timber berth cancelled.

18 Canada, House of Commons, *Sessional Papers*, 1887, no. 7, Department of the Interior, "Annual Report," Report of the Calgary Crown Timber Agent, 69; 1888, no. 14, 71; GA, Eau Claire Fonds, box 3, file 29, Map of property owned by the Calgary Water Power Company.

19 Hart, *The Place of Bows*, 103.

20 Theodore Strom, "With the Eau Claire in Calgary," *Alberta Historical Review* 12 (1964): 2–3. Strom arrived in 1886 to help build the mill and stayed on many years as an employee. He dictated these memoirs in the 1930s.

21 Strom, "With Eau Claire in Calgary," 4–8; W.E. Hawkins, *Electrifying Calgary: A Century of Public and Private Power* (Calgary: University of Calgary, 1987), 33–57.

22 GA, Calgary Power Company Fonds, box 1, Calgary Water Power Company Minutebook, June 1890–July 1920; box 4, Ordinance of Incorporation, Calgary Water Power Company, 1889–94; box 8, Department of the Interior, Confirmation of Operations on Bow River, land leases; GA, Eau Claire Fonds, box 1, Leases, licences, agreements, and permits, 1883–1954; box 1, Agreement, 31 March 1891, Calgary Water Power Com-

pany and Eau Claire and Bow River Lumber Company (Price of $2.50 to drive 1,000 bf from Limits G and H, $3.00 from all the others); box 1, Licence, Department of the Interior, 15 December 1897, for right to divert water for use of steam engine; box 3, Map of the Mill and its water system, undated; GA, M1565, Incorporation Document and Plan for the proposed Bow River Dam at Calgary, 11 November 1889; GA, Peter A. Prince Fonds.

23 Strom, "With Eau Claire in Calgary," 7–8.

24 Hawkins, *Electrifying Calgary*, 51–71.

25 J.P. Dickin McGinnis, "Birth to Boom to Bust: Building Calgary, 1875–1914," in A.W. Rasporich and H. Klassen, eds., *Frontier Calgary* (Calgary: McClelland and Stewart, 1975), 6–19; Foran, *Calgary*.

26 GA, Col. James H. Walker Fonds, M986, Letterbook of the Bow River Saw and Planing Mills, 1883–1903; see also Grant MacEwan, *Colonel James Walker: Man of the Western Frontier* (Saskatoon: Western Producer, 1989), 97–101, 107–8, 112–13. See also Calgary City Council, *The Historical Development of the Downtown and the Inner City* (Calgary: n.p., 1979), on the growth of the city and 15 for an illustration of Colonel Walker's mill.

27 GA, Col. James H. Walker Fonds, M986, Letterbook, letter to Robert Kerr, of the CPR, January 1887, about freight rates charged; Walker to Parish & Lindsay, Brandon, 18 February 1887; Walker to Robert Kerr, 8 March 1887; Walker to Robert Kerr, 17 March 1893; Walker to Mr Whyte of the CPR, 6 April 1893; Walker to Robert Kerr, 5 February 1896. In 1891, when the Eau Claire Lumber Company made an arrangement to fold its river operations into the Calgary Water Power Company, the Power Company charged the Lumber Company $2.50 to drive 1000 bf from Limits G and H and $3.00 from all the others. But these transfer prices were intended in part to convey some of the potential profits of the lumber company, subject to the profit-sharing agreement with MacFee, to the power company, where the principals were not similarly encumbered. See GA, Eau Claire Fonds, box 1, Leases, licences, agreements and permits, 1883–1954; box 1, Agreement, 31 March 1891, Calgary Water Power Company and Eau Claire and Bow River Lumber Company.

28 GA, Eau Claire Fonds, box 1, file 4, Amos Carr, Statutory Declaration, 26 September 1886; box 1, file 1, documents this trespass; box 3, file 21, for the lawsuit. The CPR claimed it took the timber under section 19 of its charter before the land had been licensed. The government agreed. The courts appear to have upheld this view.

29 GA, Eau Claire Fonds, box 12, file 1, Lougheed, McCarthy & Beck to P.A. Prince, 10 May 1890, and same to I.K. Kerr, 21 July 1891; box 1, file 3, Legal Papers, Opinion of Mr. Griffin as to the rights of K.N. Macfee; box 1, file 4, Agreement K.N. Macfee et al and Eau Claire, 12 January 1903. MacFee sold out for $8,000.

30 GA, Eau Claire Lumber Company Fonds, box 2, file 15, and Canada, *Sessional Papers*, Department of the Interior, "Annual Report," Returns of the Calgary Crown Timber Agency, 1887–1911. The Crown timber returns shadow the company figures but typically at lower levels of output, and the two data points are not easily reconciled. First, the two organizations maintained different year-ends, and those dates changed over

time. Secondly, the company and the government probably measured different things, the Eau Claire number being an all-inclusive record of output. Finally, the possibility of systematic under-reporting of output on which duties had to be paid to the government should not be discounted.

31 For a general discussion with illustrations, see Foran, *Calgary*, 25–117; Calgary City Council, *The Historical Development of the Downtown and Inner City* (n.d.); and especially B.P. Melnyk, *Calgary Builds: The Emergence of an Urban Landscape* (Saskatoon: Canadian Plains Research Centre, 1985).

32 GA, Eau Claire Fonds, box 2, file 15, Ground Rent, Stumpage, Royalty and Production statistics.

33 Hawkins, *Electrifying Calgary*, 51–7.

34 Ibid., 63–9; GA, I.K. Kerr Fonds, M629, Clipping, Calgary *Albertan*, 1916.

35 GA, Eau Claire Fonds, box 1, file 1, Sale and Reincorporation documents; see in particular the prospectus; financial details of the transaction; W.H. McLaws to Carr, 1 November 1928; McLaws to Mr. Tanner, Tull and Arden, 4 February 1929; and Carr to G.F. Tull & Arden, 17 September 1928. Carr appears to have bought the company for about $350,000. The new company issued 1,500 7 per cent preferred shares at $100, creating a liability for dividends of $15,750 against a profit calculated to be $62,655 for nine months.

36 Ibid., Preferred Shareholders as at 29 December 1928.

37 GA, Eau Claire Fonds, box 1, file 7, Income Tax and Wages, 1929 Income Tax Return.

38 See Foran, *Calgary*, 117–56, on the Great Depression in Calgary.

39 LAC, RG 84, series A-1-a, T8104, vol. 628, file 577722 Water Power Branch, Kananaskis Site, 1912–27; file contains documents about the dispute between Eau Claire Lumber Company and Calgary Power concerning floating logs on Bow River. See also GA, Eau Claire Fonds, box 3, file 32, Eau Claire Sawmills to Ben Russell, Director, Water Resources, Alberta, 6 October 1944, setting out the history of the situation.

40 GA, Eau Claire Fonds, box 3, file 32, Timber Cruise, C. Edwards, 30 September 1941.

41 GA, Eau Claire Fonds, box 2, files 12–20, document the failing fortunes of the company.

42 See AA, Acc. 69.289, Premiers' Papers, "Forests," file 487, federal Interior minister T.G. Murphy to Premier J.E. Brownlee, 24 November 1932.

43 AA, Acc. 74.49, ERFCB records, box 1, item 1, "A Report to the Minister of Lands and Forests concerning the Conservation of the Eastern Slopes Watershed in Alberta by Eastern Rockies Forest Conservation Board, January 1972," by W.R. Hanson.

44 AA, Acc. 74.49, ERFCB records, box 12, item 221, Crerar to Alberta Lands and Mines Minister N.E. Tanner, 29 January 1938.

45 AA, Acc. 69.289, Premiers' Papers, "Forests," file 1496B, federal Mines and Resource minister J.A. Glen to N.E. Tanner, 4 April 1947; Premier E.C. Manning to Glen, 7 May 1947.

46 AA, Acc. 74.49, box 12, item 221, contains the House of Commons debates on this bill. On 28 February 1947 Mines and Resources minister J.A. Glen reviewed the events leading up to the Legislation (889ff). The correspondence for 1937–38 is in the same file; see especially T.A. Crerar to N.E. Tanner, 28 January 1938.

47 Canada, House of Commons, *Debates*, 28 February 1947, 889 (Glen); 23 June 1947, 4917 (Gardiner). The enabling legislation was *Statutes of Canada*, 1947, c.59, and *Statutes of Alberta*, 1948, c. 21, and the initial appointees to the board were Major General Howard Kennedy and J.M. Wardle of the Department of Mines and Resources, along with Alberta's H.G. Jensen.

48 AA, Acc. 74.49, box 12, item 221, House of Commons *Debates*, 23 June 1947, 4534–5, 4912–4, 4917–18; July 3, 1947, 5065. See also ERFCB, "Annual Reports," 1947–48 in AA, Acc. 74.49.

49 AA, Acc. 74.49, ERFCB, Annual Reports, 1949/50 to 1954/55; a complete run of bound copies of these annual reports from 1947/48 to 1972/73 can be found in AA, "Provincial Archives Annual Report file." On fire-fighting cost-sharing see AA, Acc. 74.49, ERFCB records, box 12, item 221, federal Northern Affairs and Natural Resources minister Alvin Hamilton to Alberta minister of Lands and Mines minister N.A. Willmore, 2 October 1958, and Willmore to Hamilton, 30 October 1958.

50 AA, Acc. 74.49, box 12, item 221, "History of the Eastern Rockies Forest Conservation Board, 1947–73 A Report by W.R. Hanson, Chief Forester in Collaboration with Members of the Eastern Rockies Forest Conservation Board" (mimeo). Only one dam was ever built on Kline Creek in 1963 for use as a fish hatchery.

51 AA, Acc. 74.49, ERFCB, Annual Reports (mimeo), 1954/55 to 1956/57.

52 Ibid., ERFCB, Annual Report (mimeo), 1964/65.

53 Ibid., ERFCB records, box 12, item 222, "The Conservation of a Canadian Watershed – A Case Study," prepared by Chief Forester W.R. Hanson for United Nations Symposium on the Environment, May 1971.

54 Ibid., item 221, "History of the Eastern Rockies Forest Conservation Board, 1947–73: A Report by W.R. Hanson, Chief Forester in Collaboration with the Members of the Eastern Rockies Forest Conservation Board" (mimeo).

55 AA, Acc. 74.49, ERFCB records, Finding Aid, 5 September 1974.

56 Inferences drawn from Paul Japson, "A Report on the Effects of Log Driving on the St. Joe River, Idaho," available at https://research.idfg.idaho.gov/Fisheries, and Laurie Cooper and Douglas Clay, "A Historical View of Logging and River Driving in Fundy National Park," Greater Fundy Ecosystem Research Project, UNB Faculty of Forestry and Environmental Management; available at www.unbf.ca/forestry/centers/cwru/soe/logging.html.

57 We are grateful for the guidance given us on these matters by Professor Edward Johnson, director of the Kananaskis Forest Research Station, University of Calgary, and for sharing with us papers he has written over the years with colleagues explaining changes in the forest. See especially E.A. Johnson and G.I. Fryer, "Historical Vegetation Change in the Kananaskis Valley, Canadian Rockies," *Canadian Journal of Botany* 65 (1987): 853–8, which uses the Louis B. Stuart timber cruise notes of 1884 as a baseline to measure changing fire patterns in the twentieth century; Fryer and Johnson, "Reconstructing Fire Behaviour and Effects in a Subalpine Forest," *Journal of Applied Ecology* 25 (1988): 1063–72; and Johnson and C.P.S. Larsen, "Climatically Induced Change in Fire Frequency in the Southern Canadian Rockies, *Ecology* 72 (1991): 194–201.

58 The available evidence for and against on this point of Native use of fire is marshalled by A.R. Byrne, "Man and Landscape Change in the Banff National Park Area Before 1911" (MA thesis, University of Calgary, 1968), 33–40. For railway fires, see 90–106. For a dramatic description of one of these fires, see Charles Aeneas Shaw, *Tales of a Pioneer Surveyor* (Toronto: Longman, 1970), 147–8.
59 I.S. MacLaren, in "Captured Wilderness in Jasper National Park," *Journal of Canadian Studies* 34 (1999): 37–8, discusses the increases in forest fires in the 1880s in making his main point about the remarkable suppression of fire subsequently. For the changing areas burned in Banff and for the elimination of changing weather as a factor in the changing fire regime, see U. Fuenekes and C.E. Van Wagner, "A Century of Fire and Weather in Banff National Park" (Parks Canada, 1995).

Chapter Five

1 General information about the drainage basin and stream flow of the Bow River can be found most readily in the Bow River Water Quality Council document *Preserving Our Lifeline: A Report on the State of the Bow River* (Calgary, 1994) and Alberta, Water Resources Branch, *South Saskatchewan River Basin Summary Report* (Edmonton, 1984).
2 Stream-flow data are taken from Environment Canada, Inland Water Directorate, Water Resources Branch, *Historical Streamflow Summary: Alberta to 1973* (Ottawa: Water Survey of Canada, 1973). This document has been updated and metrified in Environment Canada, Inland Waters Directorate, Water Resources Branch, Water Survey of Canada, *Historical Streamflow Summary: Alberta to 1990* (Ottawa: Supply and Services, 1992). One of the goals of the South Saskatchewan River Basin Study was the calculation of the "natural flow" of the Bow River making adjustments for all of the twentieth-century storage reservoirs, diversion projects, water withdrawals, dams, and other engineering works that interfered with that flow. For their calculations, see J.R. Card, F.D. Davis, and R.A. Bothe, *South Saskatchewan River Basin Historical Natural Flows, 1912 to 1978* (Edmonton: Hydrological Branch, Alberta Environment, 1982).
3 GA, I.K. Kerr Fonds, Diary, text and sketch.
4 CCA, Montreal Engineering Company and Water Resources Division, Department of Agriculture, Alberta, Schematic Profile of the Bow River, Banff to Calgary, 1 February 1968, plate 2.
5 We hope ultimately to publish a history of the hydroelectric development on the Bow River which exists as a manuscript authored by Armstrong and Nelles.
6 This chapter relies on our manuscript (see note 5), which is based upon the Calgary Power Company papers, the records of the Department of Interior, the Beaverbrook Papers, and a large collection of other public and private papers. References here will be to the underlying documentation in that manuscript. On Max Aitken in this stage of his career, see A.J. P. Taylor, *Beaverbrook* (London: Widenfeld, 1976), which does not have a firm grasp of Aitken's business affairs, and Gregory P. Marchildon, *Profits*

and Politics: Beaverbrook and the Gilded Age of Canadian Finance (Toronto: University of Toronto Press, 1996), which does. Aitken as a promoter of overseas utilities features as well in C. Armstrong and H.V. Nelles, *Southern Exposure: Canadian Promoters in Latin America and the Caribbean, 1896–1930* (Toronto: University of Toronto Press, 1986).

7 Details of this wheeling and dealing, and especially the extended negotiations with the city, can be found in W.E. Hawkins, *Electrifying Calgary: A Century of Public and Private Power* (Calgary: University of Calgary Press, 1987), chapters 5 and 6.

8 LAC, MG 28, III 7, Kerry and Chace Papers, vol. 14, Report No. 839, 2 November 1907.

9 Ibid.; LAC, RG 85, vol. 734, Arthur L. Ford to J.T. Johnston, 10 March 1921.

10 LAC, MG 28, III 7, Kerry and Chace Papers, vol. 14, Report No. 838, on Construction of Hydro-Electric Power Plant at Horseshoe Falls on the Bow River, 50 Miles West of Calgary, Alberta, n.d. [1910?]; GA, Calgary Power Collection, MB, 23 December 1910, 2 May 1911.

11 C. Armstrong and H.V. Nelles, "Competition vs Convenience: Federal Administration of Bow River Waterpowers, 1906–13," in Henry Klassen, ed., *The Canadian West* (Calgary: University of Calgary Press, 1977), 163–80.

12 LAC, MG 30, E 169, Harkin Papers, Memorandum, 20 March 1914; for more on the doctrine of "usefulness" of the resources in the national parks, see chapter 8 below.

13 LAC, RG 84, file B39-5, P.M. Sauder to R.H. Campbell, 6 February 1911; A.B. Macdonald to Campbell, 20 February 1911.

14 Ibid., John Standly to R.H. Campbell, 15 May 1911; Campbell to Standly, 22 May 1911.

15 LAC, RG 85, vol. 737, E.F. Drake to J.B. Challies, 18 January 1911; House of Commons, *Debates*, 28 April 9, 17 May 1911, cols. 8084–6, 8606–41, 8650–74, 9345–7; Canada, *Statutes*, 1–2 Geo V, 1911, ch. 10, s. 17.

16 LAC, RG 84, file B39-8, Memorandum from Harkin to W.W. Cory, 24 December 1912; Memorandum from J.B. Challies to Harkin, 26 March 1913; Memo from Harkin to Challies, 3 April 1913.

17 LAC, RG 84, file B39-5, "Montreal Engineering Company, Limited, Proposed Cascade Development," 28 August 1922; file R39-8, Memorandum from J.B. Harkin to W.W. Cory, 22 May 1926.

18 Ibid., Memorandum from M.C. Hendry to J.B. Challies, 3 December 1914; Memorandum from J.T. Johnston to Challies, 7 December 1914; W.W. Cory to W.A. Found, 30 December 1914; G.F. Desbarats to Cory, 11 January 1915; Memorandum from Cory to Challies, 20 January 1915.

19 LAC, RG 85, vol. 734, Memorandum re "The Power Situation in Calgary," March 1920; A.L. Ford to J.T. Johnston, 10 March 1921; RG 84, file B39-5, Memorandum from J.T. Johnston to F.H.H. Williamson, 3 March 1921.

20 LAC, RG 84, file B39-5, Montreal Engineering Company (per Gaherty) to Interior Minister Charles Stewart, 31 July 1922.

21 Ibid., Harkin to W.A. Found, 5 March 1921; R.S. Stronach to Harkin, 5 March1921; memorandum from Harkin to J.B. Challies, 9 March 1921; memorandum from Challies to Harkin, 14 March 1921.

22 Ibid., Montreal Engineering Company to Minister of the Interior, 31 July 1922; "Montreal Engineering Company Limited, Proposed Cascade Development," 28 August 1922.
23 Ibid., Drury to the Minister of Interior, 12 April 1923.
24 Ibid., Parks Commissioner J.B. Harkin to Rocky Mountains Park Superintendent R.S. Stronach, 13 January 1922. The parks bureaucrats infuriated the company by announcing at that very moment that they intended to build a 1,000-horsepower station on the Cascade River (through which Lake Minnewanka drained into the Bow) to supply the Banff townsite since the CPR was closing its thermal generating station at Bankhead; see ibid., Harkin to Drury, 23 January 1922.
25 Ibid., memorandum to W.W. Cory, 13 October 1922; Memorandum from Harkin to W.W. Cory, 7 October, 1922; clipping from Calgary *Morning Albertan*, 29 January 1923; Drury to Minister of the Interior, 12 April 1923.
26 LAC, RG 84, file B39-8, memorandum from M.C. Hendry to J.B. Challies, 14 December 1912; RG 85, vol. 737, K.H. Smith to Challies, 2 October 1912.
27 LAC, RG 84, file B39-5, Montreal Engineering Company (per Gaherty) to Charles Stewart, 31 July 1922 (quoted); V.M. Drury to Stewart, 12 April 1923; file B39-8, memorandum from J.M. Wardle, 3 February 1923 (quoted); RG 85, vol. 733, memorandum from J.T. Johnston to J.B. Challies, 16 March 1923; Gaherty eventually became president of Calgary Power.
28 LAC, RG 84, file B 39-8, memo from Harkin to DM W.W. Cory, 28 March 1923
29 LAC, RG 84, B39-5, T.B. Moffatt to Stewart, 27 March 1923; file B39-8, Minutes of the first meeting of the National Parks Association of Canada, 2 August 1923; see letters ibid., B39-8, and R39-8, which includes memorandum re Proposed Spray Lakes Project, 4 December 1926, "List of Organizations protesting against the above project."
30 GA, Western Canada Coal Operators Association records, box 17, file 110, W.J. Selby Walker (vice-president of the CNPA) to R.M. Young, 11 December 1923, 5 January, 22 April 1924; Young to Walker, 5 January 1924.
31 UAA, Pearce Papers, file 421.1, Pearce to Bennett, 12 May 1923 (quoted); Pearce to editor, *Calgary Herald*, 19 May 1923, Personal; Memorandum by Pearce, 22 May 1923; Pearce to Interior Minister Charles Stewart, 7 June 1923. Pearce could not make quite the impact that he wished upon this debate because, as an employee of the CPR, he had to operate largely behind the scenes since his activities soon generated a subtle rebuke from his superiors on the grounds that he was causing the company "some embarrassment in Ottawa. Now that he has done his duty by announcing his views would regard it as a personal favor if he would desist from further controversy" (see ibid., file 421.1, Pearce to D.C. Coleman, vice-president, CPR, Winnipeg, 25 June 1923, quoting a cipher telegram that had been handed to him).
32 Canada, House of Commons, *Debates*, 10 April 1924, 1259–60; LAC, RG 84, Premier Herbert Greenfield to Stewart, 17 April 1923; Stewart to Greenfield, 3 May 1923.
33 AA, Acc. 69-289, file 467, Greenfield to Sir Adam Beck, 5 December 1923; Acc. 70.414, Sessional Paper no. 15, 1925, *Report on the Development and Distribution of Hydro-Electric Power in the Province of Alberta by the Hydro-Electric Power Commission of Ontario* (2 vols).

34 LAC, RG 84, file B39-8, Report to the Alberta legislature by Greenfield, 8 April 1925; AA, Acc. 69.289, file 467, Greenfield to Director of Reclamation and Water Power, 24 April 1925; Canada, House of Commons, *Debates*, 26 June 1925, 5008-11.

35 LAC, RG 84, file R39-8, George Webster to Calgary Mayor Fred. E. Osborne, 2 February 1927; resolution of the National Council of Women, n.d., and Henrietta L. Wilson and Lydia M. Parsons to Charles Stewart, 19 January 1927; "*Attention! Most Important!* The Spray Lakes in Banff National Park" (circular from CNPA Executive Secretary Arthur O. Wheeler, 15 February 1927); Calgary City Council, *Spray Lakes, the Need and ... the Answer*, enclosed in J.T. Johnston to W.W. Cory, 5 February 1927.

36 LAC, RG 84, file B39-8, Brownlee to Stewart, 2 March 1927; AA, Acc. 69.289, file 465, Brownlee to Stewart, 14 February 1928; Stewart to Brownlee, 28 February 1928.

37 H.B. Neatby, *William Lyon Mackenzie King, 1924–1929: The Lonely Heights* (Toronto: University of Toronto Press, 1963), 294–5.

38 LAC, R.B. Bennett Papers, series E, vol. 48, Canadian National Parks Association, *Bulletin*, no. 8, "Support the Parks Bill," 1 January 1930 (quoting Stewart); series F, vol. 469, CNPC, *Bulletin*, no. 9 (1 July 1930); RG 84, file B39-5, clipping from Calgary *Daily Herald*, 5 July 1930; Canada, *Statutes*, 20–21 Geo V, 1930, c. 33. The act also removed sizable areas on the Ghost, Red Deer, and Clearwater Rivers and divided the Rocky Mountains Park into Banff, Jasper, Yoho, and Glacier.

39 LAC, RG 84, B39-8, memorandum re Spray River flow, 28 February 1928; memorandum from J.M. Wardle, 13 April 1928 (quoted); file B39-5, J.O. Apps, CPR, to Stewart, 17 April 1930 (quoted). The normal summer flow of the Spray averaged 556 CFS, which would require less than 7 per cent of the water stored in the reservoir.

40 Hawkins, *Electrifying Calgary*, 168–9; AA, Premiers' Paper, box 43, Stewart to Brownlee, 14 April 1928; GA, Calgary Council Minutes, 12 May 1928, and proposed agreement with Calgary Power; 15 May 1928, memorandum of agreement; GA, Calgary Power Papers, box 1, file 1, directors' minutebook, 31 July 1928.

41 LAC, RG 84, file B39-5, C.M. Walker to T.S. Mills, 9 September 1940.

42 Ibid., Walker to Mills, 9, 16, September 1940; G.A. Gaherty to R.A. Gibson, 10 October 1940.

43 Ibid., Jennings to F.H.H. Williamson, 7 September 1940; memorandum from Byrne to Williamson, 2 October 1940; memorandum from Wardle to R.A. Gibson, 30 September 1940; Gaherty to Gibson, 10 October 1940.

44 Ibid., Howe to Crerar, 12 November 1940; Crerar to Howe, 13 November 1940; Walker to James Smart, 4 September 1940; R.A. Rooney to R.A. Gibson, 22 November 1940; mimeographed letter from Banff Advisory Council to Alberta MPs, 27 November 1940. The company wanted Howe to write to Crerar so that its director, Power Controller H.J. Symington, would not be placed in a conflict of interest.

45 Ibid., V.M. Meek to J.M. Wardle, 26 November 1940; clipping from Calgary *Albertan*, 26 November 1940; Crerar to Howe, 30 November 1940.

46 Ibid., Deputy Minister of Justice W.S. Edwards to DM of Mines Charles Camsell, 26 November, 10 December 1940; Howe to Crerar, 3 December 1940; Aberhart to Crerar, 5 December 1940; Crerar to Gaherty, 16 December 1940, enclosing PC 7382, 13 December 1940.

47 Ibid., R.A. Gibson to James Smart, 7 December 1940; Gaherty to Camsell, 2 January 1941; Memorandum to Gibson, 3 January, 27 February 1941. Departmental officials were concerned, however, when the Nakoda men brought along their families, fearing they might seek to establish permanent residence in the town of Banff, and eventually the band's Indian agent had to assume responsibility for supervising the workers and ensuring that they did not remain once the work was completed; see RG 22, vol. 238, file 33-4-4, memorandum from Gibson to J.M. Wardle, 3 March 1941.

48 LAC, RG 84, file B39-5, press release, 10 April 1941; Walker to James Smart, 25 April 1941; Canada, House of Commons, *Debates*, 4 June 1941, 3578; Canada, *Statutes*, An Act to Amend the Alberta Natural Resources Act, 4–5 Geo. VI, c. 22, 1941.

49 LAC, RG 84, file B39-5, James Smart to R.A. Gibson, 21 October 1941; memorandum from J.E. Spero to file, 9 July 1943; RG 22, vol. 238, file 33-4-4, Charles Camsell to Gaherty, 21 January, 4 December 1941; memorandum from J.M. Wardle to Camsell, 18 December 1941; Gaherty to Wardle, 24 April 1944; memorandum from Wardle to Camsell, 28 April, 3, 20 May 1944.

50 LAC, RG 84, file B39-5, Smart to Professor F. Webster, 12 April 1942; memorandum from Smart to R.A. Gibson, 30 April 1942; memorandum from Smart, 6 October 1942; J.M. Wardle to G.A. Gaherty, 14 October 1942.

51 Ibid., memorandum from James Smart to R.A. Gibson, 21 August 1946; memorandum to Smart, 9 September 1946; J.A. Mackinnon to Smart, 8 April 1947; A.G. Mackinnon to Smart, 8 April 1947; Gibson to Smart, 2 May 1947; Licence for storage and development of water power, Lake Minnewanka and Cascade River, Banff National Park, Alberta, 14 May 1947. The rental payments varied according to how much the value of the lands was depreciated by commercial use from that which they would have as parkland.

52 In 1945 the company considered a new dam lower on the Bow near Cochrane as well as on the upper Kananaskis River and also studied the Kootenay River in the Columbia watershed; see GA, Calgary Power Company records, finding aid, "Highlights in Company's Growth," n.d..

53 The Social Credit government considered a public scheme at the end of World War II and rejected the idea, but it remained under pressure from supporters of rural electrification, who demanded a plebiscite on the issue at the same time as a provincial election on 17 August 1948. Manning (who was easily re-elected as premier) campaigned actively against any takeover of existing private utility companies, and public power was very narrowly defeated by a vote of 139,991 to 139,840. See John Richards and Larry Pratt, *Prairie Capitalism: Power and Influence in the New West* (Toronto: McClelland and Stewart, 1979), 81–2, and AA, Premiers' Papers, Acc. 69.289/1615, Memorandum re "Electrification Plebiscite, August 17th, 1948."

54 LAC, RG 84, file B39-8, Gaherty to J.A. Mackinnon, 26 May 1948.

55 Ibid., Manning to Mackinnon, 21 May 1948; memorandum from Gibson to James Smart, 27 May 1948; Howe to Mackinnon, 15 June 1948.

56 Ibid., Gaherty to Mackinnon, 26 May 1948.

57 Ibid., memo from Ben Russell to Ernest Manning, 12 June 1948.

58 LAC, RG 22, vol. 238, file 33-4-10, Manning to Prime Minister W.L.M. King, 23 June 1948; King to Manning, 28 June 1948; RG 84, file B39-8, A.D.P. Heeney to Mackinnon, 23 June 1948.
59 LAC, RG 22, vol. 238, file 33-4-10, Fraser Duncan to Mackinnon, 31 August 1948, Personal; vol. 89, Mackinnon to Walker, 22 November 1948; RG 84, file B39-8, memorandum from R.A. Gibson to James Smart, 30 September 1948 (quoted); Smart to Devereux Butcher, 9 June 1949.
60 Canada, House of Commons, *Debates*, 23 March 1949, 1910–11; Canada, *Statutes*, The National Parks Amendment Act, 13 Geo. VI, c. 5 1949; LAC, RG 84, file B39-5, R.A. Gibson to A.C.L. Adams, 30 August 1949; James Smart to Gibson, 14 September 1949; T.D. Stanley to Superintendent J.A. Hutchison, 22 November 1949; Hutchison to Smart, 23 November 1949; Norman Marr to Smart, 2 December 1949.
61 LAC, RG 22, vol. 316, file 33-4-10, memorandum from James Smart to J.M. Wardle, 22 May 1951; memorandum from Wardle to Deputy Minister, 25 July 1951; Harold Riley to Deputy Minister of Justice, 24 October 1951; memorandum to Deputy Minister, 19 January 1952; memoranda from Deputy Minister H.A. Young to file, 30 January, 21 April 1952; memorandum from Smart to Young, 4 March 1952; Gaherty to Young, 9 May 1952; Young to Gaherty, 27 May 1952.
62 LAC, RG 84, file B39-8, R.A. Mackie to J.A. Hutchison, 23 October 1952; memorandum from H.A. deVeber to Hutchison, 4 November 1952; Hutchison to James Smart, 14 November 1952.
63 See appendix in this book and AA, Acc. 90.618, Public Utilities Board records, Calgary Power, Water power leases, 1909–72; on the lack of challenge to expansion from the Parks Service over Cascade and Spray expansion, see LAC, RG 84, file B39-5, J.A. Hutchison to file, 9 September 1955; T.D. Stanley to Hutchison, 22 September 1955.
64 Calgary Power Ltd., *Annual Report*, 1954, report to shareholders, 24 March 1955.
65 These data are taken from GA, J.Angus McKinnon Fonds, M1863, Submission of the Calgary Power Ltd to the Royal Commission Inquiring into Ice Conditions on the Bow River, 22 April 1952, p. 12–18.
66 Data from Environment Canada, *Historical Streamflow Summary: Alberta to 1973*.
67 Stewart B. Rood, Kristina Taboulchanas, Cheryl E. Bradley, and Andrea R. Kalischuck, "Dynamics of Riparian Cottonwoods Along the Bow River, Alberta," *Rivers* 7 (1999): 33–48; Stewart B. Rood and John M. Mahoney, "Collapse of Riparian Poplar Forests Downstream from Dams in Western Prairies: Probable Causes and Prospects for Mitigation," *Environmental Management* 14 (1990): 451–64. See also the Bow River Basin Council, *The 2005 Report on the State of the Bow River* (Calgary, 2005), 28, 56, 70, 87, for a discussion of the specific effects of hydroelectric operations on the upper reaches of the Bow River.

Chapter Six

1 F.H. Newell, *Report on Agriculture by Irrigation in the Western part of the United States at the Eleventh Census: 1890* (Washington, DC: Government Printing Office,

1894). A few years earlier, the US government published a document tabulating irrigation abroad but reported no activity in Canada; see Official Consular Reports, *Canals and Irrigation in Foreign Countries* (Washington, DC: Government Printing Office, 1891).

2 David H. Breen, *The Canadian Prairie West and the Ranching Frontier, 1874–1924* (Toronto: University of Toronto Press, 1983), chapters 1–3.

3 A. Mitchner, "The Bow River Scheme: The CPR's Irrigation Block," in Hugh A. Dempsey, ed., *The CPR West: The Iron Road and the Making of the Nation* (Vancouver: Douglas and McIntyre, 1984), 259.

4 The concern with bad publicity is discussed in Canada, *Sessional Papers*, 1895, Annual Report of the Department of the Interior, 1894, A.M. Burgess, "Report of the Deputy Minister," xxi. See also A.A. Den Otter, "Irrigation in Southern Alberta, 1882–1901," *Occasional Paper*, no. 5 (Whoop-up County Chapter Historical Society of Alberta, 1975), 7.

5 Pearce's role in early irrigation promotion is covered in detail by E. Alyn Mitchner, "William Pearce and Federal Government Activity in Western Canada, 1882–1904" (PhD thesis, University of Alberta, 1971); Mitchner, "The Development of Western Waters" (Unpublished MS, University of Alberta, 1973), 77–106 (available at the Glenbow Library, Calgary).

6 Sue Anne Donaldson, "William Pearce: His Vision of Trees," *Journal of Garden History* 3 (1983): 233–44.

7 Patricia Wood has recently documented that, in the course of this project, a canal was dug across the Tsuu T'ina reserve, causing confusion and problems for years to come; see Patricia Wood, "Winding 'in ways most devious': The Calgary Irrigation Company and the Tsuu T'ina Nation" (unpublished paper). We thank Patricia Wood for allowing us to read this paper in advance of publication.

8 Mitchner, "The Bow River Scheme," 260.

9 UAA, William Pearce Papers, Acc. 74-169, M9272, file 6, Pearce to Burgess, 7 January 1891; M9271, file 2, Schumacher to Pearce, 18 July 1893.

10 On contending legal traditions in the United States, see Donald Worster, *Rivers of Empire: Water, Aridity and the Growth of the American West* (New York: Pantheon, 1985), and Donald Pisani, *From Family Farm to Agribusiness: The Irrigation Crusade in California and the West, 1850–1931* (Berkeley: University of California Press, 1984).

11 Canada, *Sessional Papers*, 1895, Annual Report of the Department of the Interior, 1894, William Pearce, "Report of the Superintendent of Mines," 33.

12 For a statement of one such expert's views, see Elwood Mead, *Irrigation Institutions* (New York: Arno Press, 1972 [1903]).

13 Breen, *The Canadian Prairie West*, chapters 2 and 3.

14 "Irrigation in the Territories," *Calgary Tribune*, 1892 (available at GA).

15 The North-West Irrigation Act and its connections to international developments in water law has been the subject of several papers. See Charles S. Burchill, "The Origins of Canadian Irrigation Law," *Canadian Historical Review* 29 (1948): 353–62; David R. Percy, "Water Law in the Canadian West: Influences from the Western States," in *Law*

for the Elephant, Law for the Beaver: Essays in the Legal History of the North American West, ed. John McLaren, Hamar Foster, and Chet Orloff (Regina: Canadian Plains Research Center, 1992), 274–91. For Pearce's retrospective account of the framing of the legislation and its intent, see UAA, William Pearce Papers, Acc. 74-169, M9272, file 12, Pearce to F.H. Peters, 16 April 1913.

16 AA, Acc. 65.44, Department of the Environment, box 1, item 5, Letter to Magnus Beggs, Esq., Indian Agent, Blackfoot Indian Reserve, Gleichen, 11 May 1895; Magnus Beggs to J.S. Dennis, Chief Inspector of Surveys and Irrigation, 11 July 1896; LAC, RG 10, vol. 3852, file 988, A.W. Ponton to Hayter Reed, 7 October 1893; Ponton to Indian Commissioner, 24 October 1893. "Ponton's folly" references may be found in AA, Acc. 65.44, Department of the Environment, box 1, item 5, F.R. Burfield to Commissioner of Irrigation, 29 September 1921. When a later Indian agent proposed to reconstruct the irrigation ditch, he was met by "opposition from the Indians" (AA, Acc. 65.44, Department of the Environment, box 1, item 5, G.H. Gooderham to V. Meek, Commissioner of Irrigation, 29 August 1922). On the broader policy of turning Siksika into agriculturalists, see Valerie K. Jobson, "The Blackfoot Farming Experiment, 1880–1945" (MA thesis, University of Calgary, 1990). See also Canada, *Sessional Papers*, 1895, Annual Report of the Department of the Interior, part I, 1894, William Pearce, "Report of the Superintendent of Mines," 33.

17 Mitchner, "The Bow River Scheme," 260.

18 D.J. Hall, *Clifford Sifton: The Young Napoleon*, vol. 1 (Vancouver: University of British Columbia Press, 1981), 255.

19 GA, M2269, CPR Land Settlement and Development, box 196, file 1937, J.D. McLean, Sec., Dept of Indian Affairs, to J.S. Dennis, Asst. to Second VP, CPR, 30 May 1910; A.S. Dawson to J.D. McLean, 23 March 1915; Chief Engineer to Secretary, Department of Indian Affairs, 3 February 1915.

20 For a description of the surveys, see Canada, *Sessional Papers*, 1896, Department of the Interior Annual Report, 1895, J.S. Dennis, "Canadian Irrigation Surveys." See also John A. Eagle, *The Canadian Pacific Railway and the Development of the West, 1896–1914* (Montreal and Kingston: McGill-Queen's University Press, 1989), 196.

21 Mitchner, "William Pearce," 245. Mitchner reports figures from part III of the Department of the Interior Annual Report, published in 1896.

22 LAC, RG 89, vol. 52, file 29(1), J.S. Dennis, Chief Inspector of Surveys Memorandum, "The Bow River Irrigation Canal" to Surveyor General, Department of the Interior, 27 March 1896.

23 Ibid., Copy of Canadian Gazette notice, 1897, which sets out a ten-year reservation for a water right on the Bow River canal scheme.

24 Mitchner, "The Bow River Scheme," 262.

25 Ibid., 264.

26 "The Irrigation Project of the Canadian Pacific Railway," *Engineering News* 53 (27 April 1905): 429–31; "Progress on the Canadian Pacific Railway Co.'s Irrigation Works, at and Near Calgary," *Engineering News* 68 (12 August 1912): 247; H.B. Muckleston, "The Design and Construction of the Bassano Dam – Part II," *Engineering News* 72

(3 September 1914): 484–8; David Finch, *Much Brain and Sinew: The Brooks Aqueduct Story* (Brooks: Eastern Irrigation District, 1993), 12–23.

27 These anecdotal points are drawn from the autobiographical account of Hew Grant Cochrane contained in GA, Hew Grant Cochrane Fonds, M 235; and Muckleston, "The Design and Construction of the Bassano Dam – Part II."

28 "Biggest Irrigation Plant in America," *New York Times*, 1 October 1911.

29 Z.E. Black, "America's Greatest Irrigation Project, the Bassano Dam," *Scientific American* 113 (18 September 1915): 252–3.

30 GA, Western Irrigation District papers, M2273, box 18, file 156, Western Section Irrigation Block report of August 1914, p. 2a; M6388, EID Papers, box 1, Annual Report, 1915, 6; box 2, Annual Report, 1920, 240; Annual Report, 1922, 153.

31 Mitchner, "The Bow River Scheme," 264.

32 GA, EID Papers, box 3, Annual Report, 1923, 12; box 2, Annual Report, 1919, 101; Annual Report, 1921, 67.

33 GA, EID Papers, box 1, Annual Report, 1915, 1–20.

34 On the business of organizing land sales, see the still valuable James B. Hedges, *Building the Canadian West: The Land and Colonization Policies of the Canadian Pacific Railway* (New York: Russell & Russell, 1939), especially 294–319. Glenbow Library, CPR Pamphlets, undated, but c. 1910: "Public Opinion Concerning the Bow River Valley of Southern Alberta"; "Starting A Farm in the Bow River Valley Southern Alberta Canada"; "Animal Husbandry in the Kingdom of Alfalfa Bow River Southern Alberta"; "Back to the Land The Famous Bow River Valley Southern Alberta Canada."

35 Glenbow Library, CPR Pamphlets, undated, but c. 1910: "Facts Concerning the Bow River Valley"; "The Staff of Life: A Story of Winter Wheat Production in Southern Alberta."

36 GA, RCT-90-11, Carl Anderson Interview, n.d. Anderson, a prominent Swedish-American settler, commented upon the relatively cheap land compared to equivalent opportunities in the American West.

37 GA, EID Papers, box 1, Annual Report, 1915, 4.

38 Mitchner, "The Bow River Scheme," 267.

39 GA, EID Papers, box 2, Annual Report, 1919, 5–6; box 3, Annual Report, 1928, 2; Annual Report, 1929, 3.

40 GA, EID Papers, box 1, Annual Report, 1915, 15; Annual Report, 1916, 1; Annual Report, 1917, 34; and Annual Report, 1918, 56.

41 UAA, William Pearce Papers, Acc. 74-169, M 9721, file 2, Burgess to Pearce, 3 November 1894. In this letter Burgess raises concerns, regarding malaria and rheumatism. See also ibid., Pearce to A.M. Burgess, Deputy Minister of the Interior, 19 April 1893 (copy). Here Pearce responds to claims from a Mr. Klotz to Burgess that alkali could prove to be a problem. Also ibid., M 9274, file 1, Draft of Paper read to the American Society of Irrigation Engineers, 1895–96, by John Titcomb on behalf of William Pearce on "Irrigation in Alberta, Canada"; and M 9721, file 2, Pearce to A.M. Burgess, Deputy Minister of the Interior, 19 April 1893 (copy).

42 UAA, William Pearce Papers, Acc. 74-169, M 9721, file 2, Pearce to Burgess, 17 November 1894.
43 Ibid., Pearce to A.M. Burgess, Deputy Minister of the Interior, 19 April 1893 (copy).
44 GA, M2260, CPR Land Settlement and Development Papers, box 10, file 18, P.L. Naismith to Mr. Walker, 18 January 1918.
45 AA, Acc. 79.363, Department of the Environment, Water Resources Management Branch, box 11, item 94, C.S. Clendening, District Manager, Lethbridge Northern Irrigation District to P.M. Sauder, Official Trustee, LNID, 15 December 1941. Clendening wrote from further south but assessed the situation of all of the irrigation districts in southern Alberta.
46 Quoted in Finch, *Much Brain and Sinew*, 29. No date is given for the Yeats interview.
47 "Prairie Lake Provides Living for 43 Fishermen Near Brooks," *Calgary Herald*, 15 February 1938; "Fishermen to Wind Up Years' Operations," *Brooks Bulletin*, 23 March 1939.
48 T.E. Randall, "Birds of the Eastern Irrigation District," *Canadian Field Naturalist*, November–December 1946, 124–31. The current check list of birds on offer from the Eastern Irrigation District lists 237 different species.
49 On the loss of bird habitat in the prairies and the use of agricultural landscapes by birds, see Robert M. Wilson, "Seeking Refuge: Making Space for Migratory Waterfowl and Wetlands Along the Pacific Flyway" (PhD thesis, University of British Columbia, 2003).
50 GA, M2269, CPR Land Settlement and Development Papers, box 12, file 152, P.L. Naismith to J.B. Harkin, Commissioner, Canadian National Parks, 20 October 1921; M6588, Lea N. Kramer, "Ducks Unlimited Canada in the Eastern Irrigation District: A History of Co-operation" (Brooks, 1983), 3–4.
51 GA, EID Papers, box 2, Annual Report, 1919, 98.
52 GA, EID Papers, box 1, Annual Report, 1917, 12–13; box 2, Annual Report, 1919, 104. The quotation is from 1917, 12.
53 GA, EID Papers, box 2, Annual Report, 1919, 97; Annual Report, 1920, 229–31; Annual Report, 1922, 115. The quotation is from 1920, 119.
54 Ibid., Annual Report, 1921, 81.
55 GA, EID Papers, box 3, Annual Report, 1926, 159–60.
56 Ibid., Annual Report, 1923, 153–4.
57 Ibid., Annual Report, 1925, 170.
58 Ibid., Annual Report, 1923, 24.
59 For a brisk analysis, see the still useful W.A. Mackintosh, *Economic Problems of the Prairie Provinces* (Toronto: Macmillan 1935), especially chapters 2 and 3.
60 GA, M6193, Ernest Corbett Papers, Corbett report on soil drifting, 24 August 1931.
61 GA, EID Papers, box 3, Annual Report, 1923, 18–19.
62 D.C. Jones, *Empire of Dust: Settling and Abandoning the Prairie Dry Belt* (Edmonton: University of Alberta Press, 1987).
63 In 1931 it was estimated that the ratio of debt to farm values in Census Division 3 of Alberta (containing the Eastern Section as well as some dry land farming areas) was

68.52:100. This figure compared with a provincial average of 38.75:100. See *Census of Canada, 1931*, vol. 8, table 23, Mortgage Indebtedness, 1931, and Farm Expenses, 1930, by Census Division, 678. In the mid-1930s, agricultural economist Charles Burchill wrote of the dire circumstances in the Eastern Section: "The 780 land contracts still in existence in 1934 had an average age of fourteen years and represented an original cost of $4,423,800. But the present debts on these same contracts now totalled $5,164,240. Fewer than one-fifth of the farmers owed less than they had originally agreed to pay. The average man had wasted fourteen years of labor." See Burchill, "Alberta's Great Experiment in Community Co-operation: History of the Eastern Irrigation District," *Western Farm Leader*, 3 June 1938, 3–4.

64 GA, EID Papers, box 3, Annual Report, 1925, 19.

65 GA, EID Papers, box 4, Annual Report, 1926, 158.

66 R. Gross and L.N. Kramer, *Tapping the Bow* (Brooks, Alta: Eastern Irrigation District, 1985), 160.

67 GA, EID Papers, box 4, Annual Report, 1928, 178.

68 GA, EID Papers, box 3, Annual Report, 1924, 27.

69 GA, EID Papers, box 4, Annual Report, 1928, 2.

70 For a discussion of Trego's role as an agitator and organizer, see D.C. Jones, *Feasting on Misfortune: Journeys of the Human Spirit in Alberta's Past* (Edmonton: University of Alberta Press, 1998), 81–102.

71 GA, M6959, Carl Anderson Papers, box 6, file 61, contains minutes of the UFA local no. 904, Scandia, 1920–28.

72 The Bolivia threat was more than rhetoric. A settlement company provided advice to prospective settlers, and Trego promoted the venture. See Jones, *Feasting on Misfortune*, 95–101.

73 GA, M6959, Carl Anderson Papers, box 6, file 61, contain minutes of the UFA local no. 904, Scandia, 1920–28.

74 Irrigation officials noted a shift towards sheep production in the section as early as 1923; see GA, EID Papers, box 3, Annual Report, 1923, 67.

75 GA, M6959, Carl Anderson Papers, box 5, file 54, Notes, n.d.

76 Ibid.

77 Mitchner, "The Bow River Scheme," 271.

78 The events and terms of the transfer may be traced in Gross and Kramer, *Tapping the Bow*.

79 On the forgiving of debt, see C.S. Burchill, "The Eastern Irrigation District," *The Canadian Journal of Economics and Political Science* 5 (1939): 213. The rate of mortgage debt in Census Division 3 (containing the Eastern Irrigation District) per farm reporting decreased astonishingly from $5,380.20 in 1931 to $1,601.66 in 1936. See *Census of Canada*, 1931, "Table 23, Mortgage Indebtedness, 1931 and Farm Expenses, 1930 by Census Division," 678; *Census of the Prairie Provinces*, 1936, "Table 103, Farm Values, Mortgage Debts covered by Liens, 1936, Farm Expenses and Value of Farm Products, 1935, by Census Division," 1152.

80 Gross and Kramer, *Tapping the Bow*, 202.

81 Betty Crook, "This is Another Country: EID Bright Spot on the Bald Prairie," *Calgary Herald*, 20 August 1938, 19.
82 PFRA, "The Prairie Farm Rehabilitation Act" (Dominion Department of Agriculture, n.d.).
83 GA, M3761, box 42, file 367, "Confidential Resolutions Agreed to by the Executive to the Advisory Committee on Land Utilization, PFRA," Meeting held at Regina, 18 and 19 October 1937; "Gardiner Plans to Develop Community Pastures," *Calgary Herald*, 3 August 1937, 3; "Plan to Set Up Drouth Areas, Move Farmers," *Calgary Herald*, 13 November 1937, 1; "Community Pastures Are Being Planned for the Prairie Provinces," *Calgary Herald*, 22 January 1938, 20.
84 LAC, RG 17, vol. 3290, file 5467, Premier Aberhart to George Spence, PFRA, 26 April 1938.
85 Ibid., file 559-13-30, E.L. Gray to J.G. Gardiner, 17 August 1937.
86 Ibid., Article of Agreement between the Eastern Irrigation District and the Water Development Committee of the Prairie Farm Rehabilitation Administration, 12 September 1935.
87 AA, Acc. 73.307, Department of Agriculture, file 7, Dominion Department of Agriculture Prairie Farm Rehabilitation branch, Expenditures in Alberta from Inception to March 31st, 1942.
88 Eastern Irrigation District Archives, Rolling Hills file, 1943, 3-R-1, box 11, file 89, Mark Mann, "Report on Rolling Hills," Regina, PFRA, 14 July 1943.
89 GA, M3761, box 11, file 89, EID District, *Annual Report*, 1938, 1.
90 Eastern Irrigation District Archives, Rolling Hills file, 1939, 3-R-1, Mike Heisler to Eastern Irrigation District, 13 February 1939.
91 Ibid., E.L. Gray to George Spence, PFRA, 28 June 1938.
92 Ibid., "Application for Land" (blank template, n.d.). Farmers requesting an application were informed that they must deposit $175 in advance: see E.L. Gray form letter, 30 September 1938.
93 "Albertans Ineligible for Rehabilitation: Re-settle Saskatchewan Farmers in Alberta Area Because No Government Agreement Made," *Calgary Herald*, 19 July 1938, 3.
94 Eastern Irrigation District Archives, Rolling Hills file, 3-R-1, 1938, E.L. Gray to George Spence, 15 September 1938.
95 LAC, RG 17, vol. 3290, file 559-13-30, H. Barton, Deputy Minister of Agriculture, to E.S. Archibald, 18 December 1937.
96 AA, Acc. 73.307, Department of Agriculture, box 91, T. Lhjar to Minister of Agriculture, 19 October 1938.
97 Ibid., J. Schaufele to Minister of Agriculture, 23 June 1938.
98 AA, Acc. 69.289, Premier's Office, file 676, unsigned to Premier Aberhart, 7 July 1938. The next letter in the file is a response from Aberhart to Wilson E. Cain on the matter, 21 July 1938.
99 "Gardiner Refutes Mullen 'Land Grab' Charges," *Calgary Herald*, 15 January 1938, 3.
100 Eastern Irrigation District Archives, Rolling Hills file, 1939, 3-R-1, E.L. Gray to N.E. Tanner, 16 January 1939.

101 AA, Acc. 80.168, Department of Environment, Water Resources Branch, box 13, file 427, "A Group of EID farmers per J. Eade" to D.B. Mullen, Minister of Agriculture, 21 January 1939.
102 "Judge Jackson Named the Commissioner to Probe 'Irregularities' in EID," *Calgary Herald*, 21 April 1939, 1.
103 "Government Interference Strongly Protested by EID Residents," *Calgary Herald*, 22 March 1939, 1–2; "Mass Meeting at Brooks Demand EID Board Reinstated at Once until Government Charges Proved," *Calgary Herald*, 25 March 1939, 1 and 5.
104 The articles appeared in the *Calgary Herald* between March and June 1939.
105 AA, Acc. 73.307, Department of Agriculture, box 92, "Commissioner's Report" by J.A. Jackson, into management of the EID, 22 July 1939; Ian Clarke, "Social Credit Overreaction to Innovative Business Practices in the Eastern Irrigation District: 1935–1940," *Canadian Papers in Rural History* 7 (1990): 293–308.
106 Eastern Irrigation District Archives, Rolling Hills file, 1939–41, 3-R-1. This and various other jurisdictional issues arose in the early years of the project.
107 Ibid., 1940, Petition of Rolling Hills Water Users, undated and untitled; Report re: petition, September 1940.
108 Quoted in *The History of Rolling Hills, 1939–1964* [1964], 81. No author or place of publication is given on this publication, which is available at the Glenbow Library.
109 Eastern Irrigation District Archives, Rolling Hills file, 1940, Report re: petition, September 1940.
110 *PFRA: A Record of Achievement* (Ottawa: Department of Agriculture, 1943), 45.
111 *Report of the Royal Commission on the South Saskatchewan River Project, 1952* (Ottawa: Queen's Printer, 1952), 152.
112 Ibid.
113 GA, M2273, Western Irrigation District Papers, box 1, file 1, P.M. Sauder, Director of Water Resources to N.E. Tanner, Minister of Dept of Lands and Mines, 18 March 1942.
114 John F. Gilpin, *Prairie Promises: History of the Bow River Irrigation District* (Vauxhall, Alta: Bow River Irrigation District), 155–80.
115 GA, Alberta Irrigation Projects Association Fonds, M7804, "Resolutions from Lethbridge meeting of the Alberta Irrigation Association which forwarded to Ministers of Agriculture in Edmonton and Ottawa" [c. April 1947].
116 Gilpin, *Prairie Promises*, 134–6.
117 Len Ring, director, Irrigation Secretariat, Alberta Agriculture, Food and Rural Development, "Irrigation Infrastructure Rehabilitation in Alberta: 38 Years of Government/Industry Cooperation" (paper delivered to the Canadian Society for Bioengineering, Edmonton, Alberta, 16–19 July 2006); available at http://asae.frymulti.com/request.asp?JID=5&AID=22091&CID=csbe2006&T=2; accessed 18 January 2008.
118 GA, Carl Anderson Papers, M6959, box 5, file 53, Carl Anderson to Ben McEwan, Deputy Minister of Agriculture, 20 February 1987; Anderson to H.B. McEwen, 27 March 1987.

119 Ibid., Anderson to Mrs Connie Osterman, Minister of Social Services, 16 June 1987.
120 Gilpin, *Prairie Promises*, 164–8.
121 Ring, "Irrigation Infrastructure Rehabilitation in Alberta."
122 AA, Acc. 79.363, Department of Environment, Water Resources Management Branch, EID Board of Trustees, "Brief Describing Irrigation Districts' Water Needs" January 1952.
123 K. Cannon and D. Went, "Determination of Historical Changes in Salinity," (paper delivered at 37th Annual Alberta Soil Science Workshop, 22–24 February 2000, Medicine Hat, Alberta); available at http://www1.agric.gov.ab.ca/$department/deptdocs.nsf/all/aesa8421; accessed 21 August 2008.

Chapter Seven

1 The Thames provides a notorious example of how far this neglect can go before remedial action seems necessary in self-defence. See Alwyn Wheeler, *The Tidal Thames: A History of a River and Its Fishes* (London: Routledge, 1979), and Bill Lukin, *Pollution and Control: A Social History of the Thames in the Nineteenth Century* (Bristol: Hilger, 1986). See also Marc Cioc, *The Rhine: An Eco-Biography, 1815–2000* (Seattle and London: University of Washington Press, 2002), and David Blackbourn, *The Conquest of Nature: Water, Landscape and the Making of Modern Germany* (New York: Norton, 2006).
2 See C. Armstrong and H.V. Nelles, *Monopoly's Moment: The Organization and Regulation of Canadian Utilities, 1830–1930* (Philadelphia: Temple, 1986), 11–33, for a discussion of waterworks politics in Canada's major cities in the nineteenth century.
3 See CCA, City Council Minutes, 4 March 1885, for the report of the Fire, Water and Light Committee and 11 March, 6 May, 20 May, and 7 October for construction. The history of Calgary's waterworks is presented in greater detail and splendidly illustrated in Harry Sanders, *Watermarks: One Hundred Years of Calgary Waterworks* (Calgary: The City of Calgary, 2000).
4 T.C. Keefer consulted on virtually every major waterworks project in Canada, including Montreal, Toronto, Hamilton, Dartmouth, and Quebec. See CCA, City Council Minutes, 6, 27 October, 16 November, 7 December 1887. For Keefer's report see Calgary *Tribune*, 25 November 1887.
5 CCA, City Council Minutes, 23 May 1888, call for tenders; 29 August, Calgary Gas and Water Tender; 19 September, Special Committee Report on the Waterworks Contract; Letters of Inquiry to various municipalities, 8, 15, 16 August 1888.
6 GA, Orr Papers, Letterbook, Orr to W.H. Wolf, 31 October 1888. Orr tried to sell the water company's bonds on commission apparently without success; see Orr to C.A. Hanson, Montreal, 15 April 1899; Orr to Osler, Hammond & Nanton, 13, 22, 25 October, 5 November 1890. He was also the local agent for the credit-rating agency Dunn Wiman. See his report on George Alexander, "an Irish barrister" whose assets included ranch interests, considerable Calgary real estate, and the Gas and Water

Company: Orr to Dunn Wiman, 8 December 1891. See also Sanders, *Watermarks*, chapter 1.

7 CCA, City Council Minutes, 9 September 1881, 8 October 1892, 4 June 1895, for complaints about water pressure and service.

8 GA, James and Mary Russell Fonds, M1087, James Russell Waterworks Diaries and Timebooks, 1891–96.

9 CCA, City Council Minutes, 14 February 1893, for resolution; 22 October 1895 for special committee to deal with George Alexander; GA, Cross Papers, letter from George Alexander to A.E. Cross, the local MLA, 12 April 1899, complaining about the city threat of competition. See also ibid., Charles Macmillan, Calgary City Clerk to Cross, 20 March 1899, enclosing draft bill and asking him to co-operate with R.B. Bennett in getting it passed.

10 CCA, City Council Minutes: 4 May 1899, the company was asked for its selling price; 18 May 1899, it replied asking $110,000; Col. Ruttan, the city engineer of Winnipeg, was retained to advise the city in these discussions; 31 July 1899, a return offer of $65,000 was submitted by the city; 5 October 1899, plans were submitted for a new system; throughout October and November negotiations continued for the purchase of the system; Memorandum of Agreement between Bondholders and the City, 8 February 1900; 8 March 1900, amendments to city charter approved; 19 April 1900, agreement with Water Works Company following 30 March agreement with bondholders was approved. See also CCA, Engineering and Environmental Services Records, series VI, Waterworks Department Records, 1899–1928, box 1, file 1, Correspondence between Charles McMillan, the city clerk, and Willis Chipman, civil engineer, James Mackenzie, chair of the Fire, Water and Light Committee, and the Calgary Gas and Water Company, 30 October, 20 July, and 30 June 1899 respectively; box 1, file 3, Agreement between City of Calgary and the Calgary Gas and Water Company, letter Lougheed and Bennett to City, 31 January 1900, proposing a sum of $85,000.

11 CCA, Engineering and Environmental Services Records, series VI, Waterworks Department Records, Undated letter, City Engineer to R.H. Ruttan, City Engineer, Winnipeg, asking for advice on what should be done in Calgary. The prime requirement of the system, he emphasized in upper-case letters, was "That it will be able to supply sufficient water at high pressure for all FIRE REQUIREMENTS."

12 CCA, Engineering Department Fonds, series I, file 1, Scrapbook of Newspaper Clippings, 1904–08, "Will Lower the Insurance: Fire Underwriters of Toronto Say Gravity System is Very Good Thing," *The Albertan*, 31 May 1907; "Opposes the Gravity By-Law: Ratepayers' Reasons For Asking for the Rejection of the Bill," *The Albertan*, 3 June 1907; Editorial, The Gravity System, *The Albertan*, 5 June 1905; letter from the Water Works Committee (thanks for support, ends with "A Word to the Ladies": "What a good supply of pure water means to the household is perhaps better understood and more appreciated by the ladies than anyone else, therefore we are hoping for their support tomorrow, which if we have we are assured the by-law will carry"); "Gravity By-Law by a Big Majority By a Vote of 249 to 40 Ratepayers Authorize Big Expenditure," *The Albertan*, 6 June 1907. See CCA, Council Minutes, 11 March 1907,

for consideration of the gravity water system from the Elbow River; Calgary Council Papers, box 18, files 108–9, for the technical details of negotiating the easement for the conduit, purchase of materials, and construction. See also Sanders, *Watermarks*, chapter 2.

13 CCA, City Council Minutes, 6 April 1909, for the report of the Special Committee on Water, approval of bylaws 385 and 386 to borrow for waterworks reconstruction.

14 CCA, City of Calgary *Annual Report*, 1909, Medical Health Department Report, 34–5; 1912, Health Department Report, 206ff (documents a typhoid epidemic in summer traced to waste supply and improper drainage); 1913, 175ff (Dr Mahood's medical health officer's report focused on water: "The agitations for an improvement in our water supply have aroused the public to the very great need and have undoubtedly had their effect in crystallizing opinion as to the necessity of a filtration plant or a gravity system. The installation of one or the other cannot be much longer delayed if the community would preserve its self respect as well as offer its members reasonable security as to health. We know that many diseases are transmitted by water-borne germs and it is clearly the City's duty to protect the citizens against their ill effects"); 1914, 202 (medical health officer reports close supervision needed during times of year "when from tests made the [water] supply becomes dangerous").

15 City Chemist F.C. Field's findings were reported in the Calgary *News-Telegram*, 29 September 1916 (clipping in the CCA, Engineering and Environmental Service Records, series VI, Water Works Department Records, box 6, file 35). Newspapers discovered manure and oil from nearby storage tanks seeping into manholes at the pumping station; see *News-Telegram*, 15 June 1917.

16 See John Duffy, *The Sanitarians: A History of American Public Health* (Urbana: University of Illinois Press, 1990); Heather MacDougall, *Activists and Advocates: Toronto's Health Department, 1883–1983* (Toronto: Dundurn Press, 1990), chapter 4.

17 G.W. Craig, *Report re: Improved Water Supply for Calgary* (1916; available in the Glenbow Library); CCA, Engineering and Environmental Services, series VI, Waterworks Department Records, box 6, file 35, Water Works Clippings: "Engineer Submits Schemes for New Water Supply," Calgary *Albertan*, 14 October 1916; Sanders, *Watermarks*, chapter 3.

18 CCA, Engineering and Environmental Services, series VI, Waterworks Department Records, box 6, file 35, Water Works Clippings, *News-Telegram*, 26 October 1916; *Albertan*, 3 November 1916; *Albertan*, 4 November 1916; letter to the editor, *News-Telegram*, 22 November 1916; *News-Telegram*, 6 December 1916. See also the *Calgary Herald*, 15 June 1917, for the marvellous screed from Mrs C.R. Edwards, chair of the People's Pure Water Committee, in favour of filtration as opposed to sedimentation reservoirs, which she likened to stagnation. As a member of the Local Council of Women and later on the National Council of Women, she campaigned relentlessly for children's health; see her obituary, *Calgary Herald*, 13 May 1938. See also GA, Arthur G. Graves Papers, M450, file 9.

19 For a discussion of public health issues in Calgary, see J.P. Dickin McGinnis, "A City Faces an Epidemic," *Alberta History* 24 (1976): 1–11. There is a small historical litera-

ture dealing with public health and water in Canada, focusing mainly upon epidemics and the social disparities in water service. See, for example, Sheila Lloyd, "The Ottawa Typhoid Epidemic of 1911," *Urban History Review* 8 (1979): 66–89; Chris Warfe, "The Search for Pure Water in Ottawa," *Urban History Review* 8 (1979): 90–112; Rosemary Gagan, "Mortality Patterns and Public Health in Hamilton, Canada, 1900–14," *Urban History Review* 17 (1989): 161–76; François Guerard, "Ville et santé au Québec: Un bilan de la recherche historique," *Revue d'histoire de l'Amérique française* 53 (1999): 19–45; John S. Hagopian, "Debunking the Public Health Myth: Municipal Politics and Class Conflict during the Galt, Ontario, Waterworks Campaigns, 1888–1890," *Labour/Le Travail* 39 (1997): 39–68; Hagopian, "Would the Benefits Trickle Down? An Examination of the Paris, Ontario, Waterworks Campaign of 1882," *Ontario History* 87 (1995): 129–53; Hagopian, "The Political Geography of Water Provision in Paris, Ontario, 1882–1924," *Urban History Review* 23 (1994): 32–51; Collen MacNaughton, "Promoting Clean Water in Nineteenth-Century Public Policy: Progressors, Preachers, and Polliwogs in Kingston, Ontario," *Histoire sociale/Social History* 32 (2000): 49–61; Claire Poitras, "Construire les infrastructures d'approvisionnement en eau en banlieue montréalaise au tournant du XXe siècle: le cas de Saint-Louis," *Revue d'histoire de l'Amérique française* 52 (1999): 507–31.

20 Martin Melosi, *The Sanitary City: Urban Infrastructure in America from Colonial Times to the Present* (Baltimore: Johns Hopkins University Press, 2000), 123–6. After we finished writing this chapter, Jamie Benidickson published *The Culture of Flushing: A Social and Legal History of Sewage* (Vancouver: UBC Press, 2007). The core chapters of that book provide excellent European and North American context against which to read the Calgary episode treated in this chapter. Calgary conformed to the broad outlines of municipal behavour presented by Benidickson, differing only in the effectiveness of its clean up activities after the 1970s.

21 CCA, Water Supply, Glenmore file, L. Simpson and M. Fedori, "Glenmore Water Treatment Plant, Dam Heritage Evaluation Report," May 1991.

22 CCA, City Council Papers, file 1484, Gore, Naismith and Storrie report, which begins with a useful historical summary of Calgary's water problems. According to newspaper clipping in the Water Supply, Glenmore file, in November 1929, 4,279 (72 per cent) ratepayers voted in favour of the expenditure and 1,679 (28 per cent) against.

23 See CCA, Water Supply, Glenmore file, for clippings and other documents pertaining to the construction of the Glenmore plant, especially files 1479, 1511, and 1517-30 regarding the Ewing Inquiry into cost overruns. See also Water Supply General in this file and article by the plant designer, William Storrie: "Calgary's New Water Works System," *Journal of the American Water Works Association* 26 (1934): 977–1014; all conveniently surveyed in Simpson and Fedori, "Glenmore Water Treatment Plant, Dam Heritage Evaluation Report," May 1991. See also Sanders, *Watermarks*, chapter 4.

24 LAC, RG 84, reel T12668, vol 45, file Bennett Personal Parts 1–5, 1912–21, Bennett to J.B. Harken [*sic*], 14 December 1914.

25 GA, Robert T. Hollies Fonds, M533. Hollies, a retired engineer with the City Waterworks Department, was retained as a consultant to do feasibility studies on various

projects; see CCA, Engineering Department, Waterworks Division, box 1, file 12, R.T. Hollies, P Eng., City of Calgary, Waterworks Department, Report on the Proposed Use of the Bow River for Water Supply, 1959; box 2, file 13, R.T. Hollies, Report on Diversion of Water from Bow River to Elbow River for Water Supply Purposes, June 1961; see also CCA, Engineering and Environmental Services Records, series III, box 126, file 2738, Bearspaw.

26 CCA, City of Calgary, Engineering and Environmental Services, Annual Reports, Waterworks section, 1959–98; Sanders, *Watermarks*, chapters 5 and 6.

27 South Saskatchewan River Basin, *Summary Report* (Edmonton: South Saskatchewan River Basin Planning Program, 1984), 35–9. Copy in University of Calgary Library.

28 George Fuller, *Sewage Disposal* (New York, 1912), quoted in Melosi, *The Sanitary City*, 163; GA, G.W. Craig Fonds, Estimate of Cost of Labor and Material for intercepting Sewer along the Elbow River … to the Bridge at 25th Avenue, 9 April 1914.

29 If the calculation were based on the minimum stream flow, approximately 1,123 CFS during this period, the river could accommodate a city of only 224,000 rather than 800,000.

30 Jack Peach, *100 Years of Connections 1890–1990* (Calgary: City of Calgary, Sewer Division, 1990), 1–24. There is a growing literature on sewage disposal in Canadian cities. See, for example, Arn Keeling, "Urban Waste Sinks as a Natural Resource," *Urban History Review* 35 (2005): 58–70, and "Sink or Swim: Water Pollution and Environmental Politics in Vancouver, 1889–1975," *BC Studies* 142/3 (2004): 69–101; Robert Gagnon, *Question d'égouts: Santé publique, infrastructures et urbanisation à Montréal au XIXe siècle* (Montreal: Boréal, 2006); Catherine Brace, "Public Works in the Canadian City: The Provision of Sewers in Toronto, 1870–1913," *Urban History Review* 23 (1995): 33–43; Michel Dagnais and Caroline Durand, "Cleaning, Draining and Sanitizing the City: Conceptions and Uses of Water in the Montreal Region," *Canadian Historical Review* 87 (2006): 621–51.

31 Bassano *News*, 25 July 1912; Bassano *Mail*, 7 December 1916.

32 Peach, *100 Years of Connections*, 31; CCA, City Council Minutes, 513, 27 October 1917; Commissioners Report; 515, 12 November 1917, Communication from Town of Bassano re Sewage Disposal Plant; 534, 10 November 1917; Commissioners Report, Report of the City Engineer, Waste Water Survey; 502, 29 October 1918; 143, 31 March 1919, approval of Commissioners Report re Sewage Disposal Plant; 151, 29 March 1919, Commissioners request city engineer to submit estimates of the cost of building a sewage disposal plant; 269, Commissioners report, 15 May 1919, recommends by-law be submitted to ratepayers for $350,000 for the construction of a sewage disposal plant; 273, estimate; City Engineer's Report, 15 May 1919: Mechanical equipment and sterilization $200,000, buildings $25,000, grit chamber and digestion tanks $85,000, outfall, miscellaneous $40,000.

33 Brooks *Bulletin*, 23 September 1922; Bassano *Mail*, 4 October 1928; "Bassano Scores Victory in Sewage Dispute," Bassano *Mail*, 17 January 1929; Bassano *Mail*, 25 September, 9 November 1930.

34 Peach, *100 Years of Connections*, 37–9; CCA, City Council Minutes, 30 January 1929, 42, Finance Committee; 30 September 1929, 337, Sewage Plant by-law borrowing

$50,000 read first time; 12 November 1929, 394, 395, two more $50,000 borrowing bylaws introduced; 23 November 1929, Screening house tenders let for $27,000; 2 August 1930, 497, Commissioners Report.

35 LAC, RG 84, reel T11017, vol. 975, file B80, Division of Sanitary Engineering, Department of Public Health, Province of Alberta, *Stream Pollution Survey of Bow River and Tributaries* (Edmonton, August 1952). The report is not clear as to whether imperial or US gallons were used. This analysis assumes imperial measures in its calculations. The report is neither comprehensive nor consistent; Calgary effluent per capita data have been used to calculate missing data from other communities. The Turner Valley refineries and gas-processing plants daily effluent volumes were not included in the report and were not estimated, but they would have been small relative to the BA Refinery in Calgary, for example.

36 AA, Acc. 79.150, box 16, arch 116, Environmental Records, Standards and Approvals Division, 1954–69, Dr A. Somerville, chair, Provincial Board of Health, to Calgary City Council, 4 June 1953 conveying orders and attaching the 1952 report; quotation from 12–14 of the report; clipping, Calgary *Albertan*, 1 November 1952. See also CCA, Engineering and Environmental Servies Records, series III, box 20, file 168a.

37 AA, Acc. 79.150, box 16, arch 116, Environmental Records, Standards and Approvals Division, 1954–69, Memo re "Sewage Treatment – City of Calgary," March 1959; K.G. Evans, Calgary Sewer Engineer, to Dr A. Somerville, Alberta Board of Health, 30 December 1954, reporting Calgary's response to the orders.

38 For an excellent brief account deeply informed by over thirty years of fly-fishing, see Jim McLennan, *Blue Ribbon Bow: A Fly-Fishing History of the Bow River – Canada's Greatest Trout Stream* (Red Deer: Johnson Gorman Publishers, 1998), 55–64.

39 CCA, Engineering and Environmental Service Records, series III, ox 116, file 2441, Bow River 1965. A glowing piece in *Outdoor Life* drew responses from Harry Fairbridge, a lay minister, which enraged Calgary officials. This file documents growing public concern over pollution.

40 AA, Acc. 79.150, box 16, arch 116, Environmental Records, Standards and Approvals Division, 1954–69, J.H. Crichton, Acting Chief Inspector, Calgary Dept of Health, to Harvey Hogge, Provincial Sanitary Engineer, 9 August 1962; Hogge to Crichton, 16 August 1962.

41 CCA, Engineering and Environmental Service Records, series III, box 116, file 2441, Bow River 1965, R.H. Ferguson, Summary Report Bow River, 1962 to 1965, Department of Public Health, Division of Sanitary Engineering, Alberta, 23 July 1965 (the appendices provide a detailed chemical analysis of water quality in the river circa 1961–65), and P.H. Bouthillier, "A Report on Fish in the Bow River below Calgary," Department of Public Health, Division of Sanitary Engineering, Alberta, August 1965, and addendum.

42 City of Calgary Engineering Department Library, Class TD 365.B68 1968, Bow River Monitoring Surveys, 1950s and 1960s.

43 CCA, Engineering and Environmental Service Records, series III, box 116, file 2441, Bow River, 1965, Minutes of Provincial Board of Health Meeting, 25 October 1965;

AA, Acc. 79.150, box 16, arch 116, Environmental Records, Standards and Approvals Division, 1954–69, H.L. Hogge, chair, Provincial Board of Health, to Mayor and council of Calgary, 5 May 1966.

44 AA, Acc. 71.311, box C, Department of the Environment, P.G. Shewchuck, Pollution Survey Summary, Bow River, 1970–71; City of Calgary Engineering Department Library, Class TD 365 .U41 1972, Water Quality, Bow River, 1971–72, P. Ullman, Division of Pollution Control, Department of the Environment, Alberta, 23 August 1972.

45 AA, Acc. 79.150, Environmental Records, Standards and Approvals Division, arch 102, PRC 79/39 Bow River Water Pollution 1966–1968, Subjects Bow River Fish Kill, 1967, Bow River Fish Taste, Brooks complaint, Calgary Sewage Treatment Facilities; AA, Acc. 87.327, file 146, clipping, *Calgary Herald*, 2 September 1971, reporting Bow pollution levels.

46 AA, Acc. 78.110, Deputy Minister Environment file, 1971, box 1.

47 Interview with Wolf Keller, acting director, Wastewater, Utilities and Environmental Protection, 8 December 2003.

48 Ralph Klein, "The Filthy Bow: What's Bad For Us Is Even Worse Downstream," *Calgary Magazine*, June 1980, 70–1; "Klein Wants Help to Clean Up Bow River," *Calgary Herald*, 26 October 1982; *Calgary Herald*, 27 June 1986.

49 Samuel P. Hays's *Beauty, Health and Permanence* (Pittsburgh: University of Pittsburgh Press, 1987) and *A History of Environmental Politics since 1945* (Pittsburgh: University of Pittsburgh Press, 2000) provide a reliable guide to environmental politics in the United States. J.R. McNeill examines this phenomenon in a global perspective in the last chapter of his *Something New Under the Sun: An Environmental History of the Twentieth Century World* (New York: Norton, 2000).

50 City of Calgary Engineering Department Library, Class PR83002, City of Calgary, "The Bow River: A Wastewater Management Study for the City of Calgary," Reid Crowther & Partners, January 1983.

51 Peach, *100 Years of Connections*, 39–56. A brief history and description of Calgary's waste-water treatment program is conveniently located on the Water Services Business Unit website, which features as well recent recognition of innovation in sewage treatment.

52 Data from CCA, Engineering and Environmental Services, Annual Reports, various years and reports on waste-water treatment plants; Engineering and Environmental Services Records, series II. Annual reports for the 1990s can be found in the Calgary Public Library.

53 City of Calgary Engineering Library, Class SH 328 .R425 1994, "Instream Flow Needs Investigation of the Bow River for Alberta Environment, Fish and Wildlife Division," Environmental Management Associates and WER Engineering, December 1994.

54 Interview with Wolf Keller, acting director, Wastewater, Utilities and Environmental Protection, 8 December 2003.

55 Alberta Environment, "Alberta Surface Water Quality Guidelines."

56 This shift in civic public policy can be read in documents filed in the City of Calgary Engineering Library, such as City of Calgary Engineering Department, Waterworks

Division, *The City of Calgary Water Conservation Study* (Acres Consulting Services in association with Gore & Storrie Limited, February 1980), PR80002 c.3; Reid Crowther & Partners, *The Bow River: A Wastewater Management Study for the City of Calgary*, January 1983, PR 8302 1983; *Future Water Use and Management in the Bow River Basin from the City of Calgary's Point of View*, Class CED 1989 14; Paul Do, Barry Kobryn, Garry Lamb, Red T. Seidner, Engineering Department, Presented to CWRA/AIPA 1989 Conference; and Reid Crowther & Partners, *City of Calgary Wastewater Master Plan Update 2000–2006*.

57 CCA, Annual Reports of the Engineering and Environmental Services Division. See also *The City of Calgary Water Conservation Study* for consumption data and recommendations.

58 City of Calgary, Year-End Water Conservation Report, 2006. The objective is to reduce consumption levels further to 350 by 2033.

59 For a study of the impact of nitrogen and phosphorus as factors in the growth of algal biomass, see S.E.D. Charlton et al., *The Limnological Characteristics of the Bow, Oldman and South Saskatchewan Rivers, 1979–82* (Edmonton: Alberta Environment, March 1986), and S.E.D. Charlton et al., *A Summary of Ecological Characteristics of the South Saskatchewan River Basin with Specific Reference to the Bow River, 1979–82* (Edmonton: Alberta Environment, March 1986). Studies such as these provided the basis for the provincial government placing limits on the quantity of chemicals permitted in waste-water effluent as a condition of the treatment plant licence.

60 Karen A. Saffran, *Temporal and Spacial Trends For Coliform Bacteria in the Bow River, 1951–1994* (Edmonton: Surface Water Assessment Branch, Technical Services and Monitoring Division, Alberta Environmental Protection, January 1996), and A.J. Sosiak, *An Evaluation of Nutrients and Biological Conditions in the Bow River, 1986 to 1988* (Edmonton: Environmental Quality Monitoring Branch, Environmental Assessment Division, Alberta Environment, 1990); Calgary Public Library (CPL), Engineering and Environmental Services, Annual Reports, 1995, 1996.

61 CPL, Engineering and Environmental Services, Annual Report, 1996; City of Calgary Engineering Library, "Deerfoot Trail Extension: Deerfoot Trail–Bow River Pond Stormwater Management Report," Reid Crowther & Partners for Calgary Engineering and Environmental Services Department, February 2000, Class SWM2000040. See also Bow River Basin Water Council, *Preserving Our Lifeline: Survey of Urban Water Use and Management in the Bow River Basin* (Calgary, July 1998) and *Preserving Our Lifeline: Solutions for Management of Urban Stormwater in the Bow River Basin* (Calgary, September 1999); "Calgary Storm Sewer Run-off Putting Bow Fish in Peril: City Official," Canadian Press Newswire, 26 May 2004.

62 Jim McLennan, "The Bow Revisited," *Canadian Fly Fisher*, Fall 1999, 17–20; we are indebted to our colleague Richard Hoffmann for this reference. Interview with Wolf Keller, acting director, Wastewater, Utilities and Environmental Protection, 8 December 2003.

63 For the most recent analysis, see Bow River Basin Council, *The 2005 Report on the State of the Bow River Basin* (Calgary, 2005), especially chapters 7, 8, and 12.

Chapter Eight

1 Jim McLennan, *Blue Ribbon Bow: A Fly-Fishing History of the Bow River – Canada's Greatest Trout Stream* (Red Deer: Johnson Gorman, 1998). McLennan quotes an unnamed writer on 55.

2 Richard B. Miller, *A Cool Curving World* (Toronto: Longmans, 1962), 210–11. The book was completed in 1956 and published posthumously. For a recent indictment of the management of rivers and lakes for fish in Banff National Park, see David W. Schindler, "Aquatic Problems Caused by Human Activities in Banff National Park, Alberta, Canada," *Ambio* 29 (2000): 401–7.

3 Joseph S. Nelson and Martin J. Paetz, *The Fishes of Alberta* (Edmonton: University of Alberta Press, 1992), 47–55.

4 Brian J. Smith, "The Historical and Archaeological Evidence for the Use of Fish as an Alternate Subsistence Resource among Northern Plains Bison Hunters," in *Aboriginal Resource Use in Canada: Historical and Legal Aspects.* ed. Kerry Abel and Jean Friesen (Winnipeg: University of Manitoba Press, 1991), 35–50.

5 Hugh Dempsey, *Red Crow: Warrior Chief* (Saskatoon: Western Producer, 1980).

6 LAC, RG 10, vol. 3908, 107297-1, part O, C10159, Agent to Secretary, Department of Indian Affairs, 20 January 1898; Agent to Secretary, Department of Indian Affairs, 25 January 1898.

7 Archives of Ontario, F709, M11 757, series 3, Crawford Family Papers, Writings of Colin Crawford, Diary entry, 23 June, 1883. Thanks to Graeme Wynn, who brought this reference to our attention.

8 *Fishing and Shooting along the Line on the Canadian Pacific Railway* (Montreal: CPR, 1893).

9 Whyte Museum and Archives, M265/V394, Acc. 2961, Guest Book Lake Minnewanka Hotel, 24 June 1890–14 October 1906; Rodney Touche Papers, M377, Beach House Register, 1887–1892, Lake Minnewanka, for the quotations.

10 Evidence of this workers' fishery derives mainly from reports critical of it. They must be read with caution; see LAC, RG 23, vol. 365, file 3216, part 3, Final Report Alberta and Saskatchewan Fishery Commission 1911, 97; RG 84, vol. 70, file B296, S.C. Vick, Fishery Officer, to R.S. Stronach, Superintendent, Rocky Mountains Park, 8 September 1922; Commissioner (unsigned copy) to Superintendent, Rocky Mountains Park, 15 September 1919; J.M. Wardle, Superintendent to J.B. Harkin, Commissioner, Dominion Parks, 24 September 1918; A.C. Finlayson, "Memorandum Re Fishing in Spray Lakes," 6 July 1918, forwarded to J.B. Harkin by W.A. Found (Department of Marine and Fisheries); "F.H.W." for Commissioner, Memo to Mr Scott, 16 June 1913.

11 LAC, RG 23, vol. 1001, file 721-4-37[26], W.D. Elliot, President, and Frank Watt, Secretary, Highwood Angling and Protective Association, to Deputy Minister of Fisheries, 14 May 1925.

12 LAC, RG 23, vol. 365, file 3216, part 3, Final Report Alberta and Saskatchewan Fishery Commission 1911, Summary of Public Meetings, 8–13. For a brief history of the commission, see George Colpitts, "Science, Streams and Sport: Trout Conservation

in Southern Alberta, 1900–1930" (MA thesis, Department of History, University of Calgary, 1993), chapter 1.

13 George Colpitts, *Game in the Garden: A Human History of Wildlife in Western Canada* (Vancouver: UBC Press, 2002), chapter 5.

14 LAC, RG 23, vol. 1001, file 721-4-37[19], Memorandum re Closing Tributary Streams to Highwood River to Minister of Fisheries, 23 May, 1922, n.a.

15 W.H. Bell, *Bow River Days: Growing Up in Calgary, 1927–1951* (Nanaimo: Bell Enterprises, 2000), 33–5.

16 "Bassano Becoming Popular with Visiting Fisherman," *Bassano Recorder*, 27 July 1939.

17 The title of this section was taken from LAC, RG 84, vol. 70, file B296, "Banff Fish Hatchery Angler's Best Friend," *Calgary Herald*, 18 June 1925.

18 Thomas R. Dunlap, *Nature and the English Diaspora: Environment and History in the United States, Canada, Australia, and New Zealand* (New York: Cambridge University Press, 1999).

19 The activities of Alberta fish and game groups in importing game birds may be traced in GA, Austin De B. Winter Fonds, M1327.

20 Joseph E. Taylor III, *Making Salmon: An Environmental History of the Northwest Fisheries Crisis* (Seattle: University of Washington Press, 1999).

21 Matthew Evenden, "Locating Science, Locating Salmon: Institutions, Linkages and Spatial Practices in Early British Columbia Fisheries Science," *Environment and Planning D: Society and Space* 22 (2004): 355–72.

22 On fish introductions, see Nelson and Paetz, *The Fishes of Alberta*, and McLellan, *Blue Ribbon Bow*, 57–8. Some of the fish escapes from the Banff hatchery ended up in the Calgary water supply reservoirs; see CCA, Engineering and Environmental Services Records, Waterworks Records, box 3, file 17.

23 Lake Newell was stocked with whitefish sometime in the 1920s: "Prairie Lake Provides Living for 43 Fishermen Near Brooks," *Calgary Herald*, 15 February 1938.

24 G. Allen Mail, "The Mosquito Fish *Gambusia Affinis* (Baird and Girard) in Alberta," *Mosquito News* 14 (1954): 120. Mail claimed to be responsible for an introduction in 1924, but other sources imply earlier introductions. See W.E. Round, "The Mystery of the Mystery Fish Is Now Explained," *Crag and Canyon*, 11 May 1928. Round suggests an introduction date of 1918, citing E. Hearle, dominion entomologist, as his source. Another document would support this latter claim: LAC, RG 84, vol. 71, file B296-12, Dr H.M. Rogers to Hoyes Lloyd, 18 December 1940.

25 See LAC, RG 84, vol. 71, file B296-12, Dr H.M. Rogers to Hoyes Loyd, 18 December 1940, for the history of the mosquito fish (*Gambusia affinis*) in the Cave and Basin.

26 D.E. McAllister, "Introduction of Tropical Fishes into a Hotspring near Banff, Alberta," *Canadian Field Naturalist* 83 (1969): 31–5; "Hot Water and Mosquito Fish," *Crag and Canyon*, 22 May 1974.

27 LAC, RG 23, vol. 395, file 3737, Charles Webster, Secretary, Calgary Board of Trade, to Hon. Louis Brodeur, Minister of Marine and Fisheries, 5 October 1910.

28 GA, Austin De B. Winter Fonds, M1327, box 2, file 21, Alberta Fish and Game Protective Resolutions, nd. Attached to a letter from M.K. Reiter to F. Darker, 11 February 1910.

29 LAC, RG 23, vol. 578, file 704-16-1, "Government Hatchery at Banff Attracts Thousands Daily," *Edmonton Journal*, 6 June 1929.

30 Colpitts, "Science, Streams and Sport"; see, in particular, chapter 4, "The 'Angler's Best Friend': The Banff Hatchery and the Ethics of Conservation" (84–107).

31 D.B. Donald, "Assessment of the Outcome of Eight Decades of Trout Stocking in the Mountain National Parks, Canada," *North American Journal of Fisheries Management* 7 (1987): 545–53; V.E.F. Solman, J.P. Curruer, and W.C. Cable, "Why Have Fish Hatcheries in Canada's National Parks?" *Seventeenth North American Wildlife Conference*, 1952, 226–34.

32 LAC, RG 84, vol. 109, file U125-19, Alberta Fish and Game Assoc., Report of the President, 1941.

33 On the development of the hatchery, see GA, Molson Brewery Papers, M8581, file 597, "The Calgary Fish Hatchery, Calgary Aquarium, and Horseman's Hall of Fame," n.d. [possibly 1967].

34 David W. Schindler, "Aquatic Problems Caused by Human Activities in Banff National Park, Alberta, Canada," *Ambio* 29 (2000): 402.

35 Elizabeth Tough, "An Investigation of the Potential to Restore Native Trout Populations: Goat Creek, Banff National Park, Alberta" (Master of Environmental Design thesis, University of Calgary, 2000).

36 LAC, RG 23, vol. 825, file 719-8-12, part 1, D.A. Richardson, Fishery Overseer to R.T. Rodd, Inspector of Fisheries, Edmonton, 31 July 1923; Richardson to Rodd, 31 October 1923; Richardson to Rodd, 30 September 1923; Rodd to W.A. Found, Department of Marine and Fisheries, 8 December 1923; Rodd to Found, 23 February 1924; Richardson to Rodd, 27 October 1924; Found to Rodd, 27 December 1924; Richardson to Rodd, 29 May 1925; Richardson to Rodd, 27 August 1925; Richardson to Rodd, 28 September 1925.

37 LAC, RG 23, vol. 825, file 719-8-12, part 1, D.A. Richardson to J.D. Willson, Inspector of Fisheries, 20 June 1923.

38 Ibid., D.A. Richardson to Inspector of Fisheries, Edmonton, 31 October 1923.

39 Ibid., W.D.E. Elliot, President, Highwood River Angling Protective Association, to G.S. Davidson, Chief Inspector of Fisheries, Indian Head, Saskatchewan, 21 April 1920.

40 Ibid., D.A. Richardson, Fishery Overseer, to J.D. Willson, Inspector of Fisheries, 30 September 1921; quotation from Richardson to Willson, 31 October 1922.

41 LAC, RG 23, vol. 1501, file 769-11-18, J.A. Rodd, Director of Fish Culture, Alberta to D.S. Rawson, 16 January 1936.

42 Ibid., "Report on the Biological Examination of the Kananaskis Lakes, Alberta," D.S. Rawson, Department of Biology, University of Saskatchewan, February 1937; prepared for Fisheries Service, Department of Lands and Mines, Alberta.

43 J.S. Nelson, "Effects of Fish Introductions and Hydroelectric Development on Fishes in the Kananaskis River System, Alberta," *Journal of the Fisheries Research Board of Canada* 22 (1965): 721–51. For a related study on the Barrier reservoir, see J.R. Nursall, "The Early Development of a Bottom Fauna in a New Power Reservoir in the Rocky Mountains of Alberta," *Canadian Journal of Zoology* 30 (1952): 387–409.

44 Richard B. Miller and Martin Paetz, "The Effects of Power, Irrigation, and Stock Water Developments on the Fisheries of the South Saskatchewan River," *Canadian Fish Culturist* 25 (1959): 14. See also Jean-Paul Cuerrier, "The History of Lake Minnewanka with Reference to the Reaction of Lake Trout to Artificial Changes in the Environment," *Canadian Fish Culturist* 15 (1954): 1–9.

45 Miller and Paetz, "The Effects of Power, Irrigation, and Stock Water Developments," 25.

46 L.A. Langmore and C.E. Stenton, "The Fish and Fisheries of the South Saskatchewan River Basin: Their Status and Requirements" (Edmonton: Alberta Energy and Natural Resources, Fish and Wildlife Division, 1981), 214.

47 LAC, RG 23, vol. 365, file 3216, part 3, Final Report, Alberta and Saskatchewan Fishery Commission, 1911, 99,

48 Two instances of sawdust pollution are chronicled in LAC, RG 23, vol. 439, file 702-7-5, S.H. Fox to Department of Marine and Fisheries, 17 January 1916, and R.T. Rodd to W.A. Found, 1 November 1929.

49 LAC, RG 23, vol. 439, file 702-7-5, R.T. Rodd, Inspector of Fisheries, Alberta, to W.A. Found, Director of Fisheries, 10 December 1923.

50 Ibid., D.A. Richardson, "The investigation of alleged pollution of the Bow River by Imperial Oil Refinery," extracts submitted 14 August 1926.

51 Ibid., D.A. Richardson, Fishery Overseer, to R.T. Rodd, Inspector of Fisheries, Edmonton, 14 August 1926.

52 Ibid., W. Boote to Mayor of Bassano, 6 March 1940 (copy).

53 Ibid., "Anon." to Mayor of Bassano, 9 November 1937 (copy).

54 LAC, RG 84, reel T11017, vol. 975, file B80, Division of Sanitary Engineering, Department of Public Health, Province of Alberta, *Stream Pollution Survey of Bow River and Tributaries* (Edmonton, August 1952).

55 UAA, Acc. 74-169, William Pearce Papers, M9271, file 4, Pearce to Secretary, Department of the Interior, 23 August 1894.

56 See http://www.bowriver.org/index.html
and http://www.tucanada.org/alberta/fishrescue.htm.

57 "Prairie Lake Provides Living for 43 Fishermen Near Brooks," *Calgary Herald*, 15 February 1938.

58 "Fishermen to Wind Up Years' Operations," *Brooks Bulletin*, 23 March 1939.

59 Miller and Paetz, "The Effects of Power, Irrigation and Stock Water Developments," 22.

60 "Ask Fishing in E.I.D. Lakes Be Stopped," *Brooks Bulletin*, 23 March 1939.

61 Miller and Paetz, "The Effects of Power, Irrigation and Stock Water Developments," 19.

62 "The Sport Fishery in Alberta: Facts and Figures for 1975 and 1980," Fisheries Management Report No. 28 (Edmonton: Alberta Energy and Natural Resources, Fish and Wildlife Division, 1983).
63 L.A. Langmore and C.E. Stenton, "The Fish and Fisheries of the South Saskatchewan River Basin: Their Status and Requirements" (Edmonton: Alberta Energy and Natural Resources, Fish and Wildlife Division, 1981).
64 Jim McLennan, "The Bow Revisited," *Canadian Fly Fisher*, Fall 1999, 18. Thanks to Richard Hoffmann for bringing this article to our attention.
65 Alberta Wilderness Association, *Rivers on Borrowed Time*, River Study Team: Hans Buhrmann, Brian Forestall, and Cliff Wallis; Study Coordinator: Ron Hooper (Calgary, 1982).
66 S.A. Telang, "Effects of Reservoir-Dam, Urban Industrial, and Sewage Treatment Run-off on the Presence of Oxygen and Organic Compounds in the Bow River," *Water, Air and Soil Pollution* 50 (1990): 77–90.

Chapter Nine

1 GA, J. Angus McKinnon Fonds, M 1683 (hereafter McKinnon Fonds), Submission of the Calgary Power Company to the Royal Commission on the Bow River, 22 April 1952, Appendix A, History and Description of Ice Jamming and Flooding on the Bow River at and above Calgary, 8–10, clipping from the *Calgary Herald* and *Albertan*.
2 We have been guided in our understanding of the science of how rivers work by Jeffrey F. Mount, *California Rivers and Streams: The Conflict Between Fluvial Processes and Land Use* (Berkeley: University of California Press, 1995).
3 Information about Bow River floods is conveniently collected in two documents: GA, McKinnon Fonds, Submission of the Calgary Power Company to the Royal Commission on the Bow River, 22 April 1952, 5–10; and CCA, Rivers General file, Montreal Engineering Company, *Flood Plain of Bow River in the City of Calgary*, 2 vols (February 1968), 1: 1–13.
4 LAC, RG 89, vol. 53, file 29(3), Report on flood conditions on Bow compiled by P.M. Sauder, Chief Hydrographer, to F.H. Peters, Commissioner of Irrigation, not dated but 1912.
5 Ibid.
6 GA, George W. Craig Fonds, Report on Construction of the Centre Street Bridge, 10.
7 LAC, RG 89, vol 226, file 2677, Flood on the Bow and Elbow, 1915–27, "Bow River Drainage Basin," G.H. Whyte on the floods in the Bow basin, 1915, 15 pp.
8 Inland Waters Directorate, Water Resources Branch, *Historical Streamflow Summary, Alberta* (Ottawa, 1991), 60.
9 *Lachlin McKinnon, 1865–1948, Pioneer* (Calgary: Privately Printed, 1956?), 41–4.
10 N. Smith, *Man and Water: A History of Hydro Technology* (London: P. Davies, 1976).
11 LAC, R653-0-8-E, Inland Waters Directorate sous fonds, Biography/Administrative History; Inland Waters Directorate, *Historical Streamflow Summary: Alberta to 1990* (Ottawa: Water Resources Branch, Water Survey of Canada, 1991), v. The Water Man-

agement Division of Alberta Environment has more recently revised, computerized, and posted online streamflow data for the entire South Saskatchewan River Basin.

12 M.C. Hendry, *Bow River Power and Storage Investigations*, Water Resources Paper no. 12, Sessional Paper 25e (1914; copy in Glenbow Library).

13 Leo G. Dennis and Arthur V. White, *Water Powers of Canada* (Ottawa: Commission of Conservation, 1911); Leo G. Dennis and J.B. Challies, *Water Powers of Manitoba, Sakatchewan and Alberta* (Commission of Conservation, 1916); for the Bow, see 194–225.

14 LAC, RG 89, vol. 560, file 501, Arthur L. Ford, "Report on the Flood Prediction Surveys," 18 February 1921 (Water Resources Branch, Ottawa); vol. 632, file 1913, O.H. Hoover, Engineer in Charge, "Bow River Snow Survey, Lake Louise," 6 October 1936.

15 CCA, Rivers file, Montreal Engineering, *Flood Plain of Bow River in the City of Calgary*; see 1: 17 for a table showing periods of heavy rainfall not accompanied by flooding.

16 Ibid., 1: 18.

17 AA, Acc. 72.302, Minister of Agriculture Papers, file 56, box 1, J.C. Watson, Mayor of Calgary to Premier Ernest Manning, 13 January 1947.

18 CCA, Environment and Engineering Records, series III, box 9, file 74, Report on the Hillhurst Flood, typescript, 16 May 1946.

19 LAC, RG 84, series A-2-a ,vol. 225, file EC3-1-22-1, Ice Conditions on the Bow, 1941–47, reel T 12952 for letters of complaint and newspaper clippings.

20 CCA, Environment and Engineering Records, series III, box 37, file 318, Calgary Power Company to the City Commissioner, 19 October 1948. Calgary Power offered to pay 75 per cent of the cost up to a maximum of $75,000 for raising dikes along the river to 12½ feet. See also box 14, file 125a, Extracts from Council Meetings re Flooding, 1946–48.

21 LAC, RG 84, series A-2-a, vol. 225, file EC3-1-22-1, Ice Conditions on the Bow, 1941–47, reel T 12952, P.M. Sauder, Water Resources Branch to Charles Camsell, Deputy Minister, Department of Mines and Resources, 8 March 1941.

22 Ibid., H.J. McLean, Production Superintendent, Calgary Power, to P.M. Sauder, Director of Water Resources, 4 March 1941. The Water Resources Branch had on file accounts of earlier winter floods; see, for example, LAC, RG 89, vol. 226, file 2677, Flood on the Bow and Elbow, 1915–27, A.W.P Lowrie member of Waterpower and Reclamation Service, to Commissioner of Irrigation, 21 November 1927, describing ice jamming and flooding on the Bow River upstream of Calgary.

23 AA, Acc. 72.302, Minister of Agriculture Papers, box 4, file 197, Ben Russell to Hon. D.B. MacMillan, 21 February 1948, "The Bow River Problem."

24 Ibid. This file is full of letters from municipal officials demanding the provincial government do something; see, for example, the Mayor of Bowness to the Director of Water Resources, 3 April 1948. See also ibid., box 1, file 56, J.C. Watson, Mayor of Calgary, to Premier E.C. Manning, 4 November 1947. And see the city of Calgary putting pressure on the federal government to build Bearspaw, even as it campaigned in support of the Spray Lakes development to alleviate the power shortage: LAC, RG 84, vol. 8, file B39-8, V.A. Newhall to Prime Minister W.L.M. King, 6 May 1948.

25 AA, Acc. 72.302, Minister of Agriculture Papers, box 1, file 56, Minister of Agriculture to Mayor of Calgary, 24 January 1947: "I do not think that either the Dominion, the Province or the Calgary Power Company is prepared to admit any legal responsibility for property damage at Calgary, and that certainly the City of Calgary is much more morally responsible for damage to the property of its citizens than any of the three parties referred to"; also box 4, file 197, Minister of Agriculture to the City Clerk of Calgary, 19 March 1948.

26 GA, McKinnon Fonds, Evidence and Hearings of the Bow River Flood Commission (hereafter E&H), vol. 9, 8 May 1952, for Stanley and vol. 14, 15 May for Teske. Others proposed variations on a log boom theme to trap the ice. Professional engineers readily shot down these ideas as ineffective based on experience, but they could not counter with something that would work.

27 CCA, Engineering and Environment Records, series III, box 5, file 34, letters to the Mayor from concerned citizens re Bow River flooding: J.L. Davey, 6 December 1950; Lorne Saunders, 5 December 1950 (two letters); John Boyd, 14 December 1950; and box 14, file 125a, City Engineer Thomas to Wing Commander Border, 19 December 1952: "These bombs I presume would have a minimum charge, but that of course is your field rather than mine."

28 Ibid., Olive Clare to the Mayor, 4 December 1950.

29 GA, McKinnon Fonds, E&H, vol. 1, 22 April 1952.

30 CCA, Engineering and Environment Records, series III, box 9, file 73, Minutes, Permanent Flood Committee, November–December 1950, 28 March, 3 October 1951; box 14, file 125a, Report of the Permanent Flood Committee, 16, 17 January 1952.

31 Ibid., box 14, file 125a, Bruce Nesbitt to J.I. Strong, City Engineer, 14 February 1952. This letter reports a story from a trustworthy woman that she and her husband were told on the afternoon of 30 December 1951 to be prepared for the worst as the Calgary Power Company were going to flush out the Ghost Dam about 11 o'clock that night.

32 Ibid., box 9, file 73, correspondence between V.A. Newhall and Ben Russell, February 1951; AA, Acc. 72.302, Minister of Agriculture Papers, box 11, file 494, correspondence with Calgary re Ice Jams, 1950; box 13, file 628, correspondence for 1951 with Calgary mainly over the sharing of flood costs and the construction of a dam at Bearspaw. The political context of the royal commission's creation can been gleaned from box 17, file 799, which also contains F.J. Graham's report of 28 March 1952 on the flooding the previous December. After the province appointed the commission on 11 March 1952, the city deferred its motion regarding legal liability pending the outcome. The province moved with remarkable speed. The chairman of the commissioner was out of town and not consulted before being appointed; another commissioner was simply telephoned by the minister the day before the announcement.

33 Andrew Baxter in GA, Mackinnon Fonds, E&H, vol. 14, 15 May 1952.

34 Ibid., vol. 3, 25 April 1952, examination of Ivor Strong. See also CCA, Engineering and Environment Records, series III, box 14 file 125a, Brief of the City to the Royal Commission.

35 GA, McKinnon Fonds, Submission of the Calgary Power Company to the Royal Commission on the Bow River, 22 April 1952, part II: 2–4.

36 Ibid., part III: 17–18, part IV: 1–4, 17–18.

37 GA, McKinnon Fonds, E&H, vol. 17, 4 June 1952, examination of Ben Russell; AA, Acc. 72.302, Minister of Agriculture Papers, box 17, file 800, Submission of the Director of Water Resources, 20 May 1952.

38 GA, McKinnon Fonds, Submission of the Calgary Power Company, 22 April 1952, part IV: 8–11. With excess capacity in the system, the company could afford to run the Ghost plant at a constant volume as an experiment. "There was no apparent improvement in the winter flooding situation and, insofar as one year's results may be taken as a guide, it can now be said that variable flow is not the cause of ice jamming in the river. This phenomenon remains fundamentally a result of low temperatures and turbulent river flow" (part IV: 9).

39 Ibid., E&H. For examples, see vol. 15, 2 June 1952, examination of Mr Sexton, and vol. 16, 3 June 1952, examination of Sexton and Stanley.

40 Ibid., vol.. 2, 23 April 1952, examination of Leonard Cooper.

41 Ibid., vol. 16, 3 June 1952, examination of Mr Stanley.

42 Glenbow Library, Commission to Conduct an Inquiry into Causes and Conditions Contributing to Floods in the Bow River at Calgary, *Report* (typescript, 1952).

43 GA, McKinnon Fonds, "The Bow River Royal Commission 1952." The flavour of this poem, running to about 350 lines, may be sampled in the following:

> Calgary Power presented a brief,
> Crammed full of facts on every leaf.
> The expert, Jack Sexton, was brought from Montreal,
> Engineers from McGill graced the Legal Hall.
>
> Sexton prepared Calgary Power's submission,
> The most valuable material presented to the Commission.
> It contained charts, and maps, and theories no end,
> With a Royal Air Force Mosaic of the land-slide at River Bend.

44 Glenbow Library, Commission to Conduct an Inquiry, *Report*, 41–55, Remedial Works Recommendations, Summary, and Recommendations.

45 AA, Acc. 72.302, Minister of Agriculture Papers, box 14, file 645, B. Russell to G. Gaherty, 2 May 1951.

46 Ibid., box 17, file 783, Minister of Agriculture Ure to G. Gaherty, 29 September 1952. This file contains correspondence before and after this key letter invoking the clause setting forth the company's concerns about financing the project and its quite different interpretation of its legal obligations. The province took the view that it was time for the company to co-operate. Nevertheless, things proceeded reasonably swiftly; a draft agreement was negotiated by November.

47 Ibid., box 28, file 1533, contains memoranda from 1957 summarizing the provincial assistance; see also CCA, Engineering and Environmental Services Records, series III, box 14, file 124, 1951 Dyking Costs; box 62, file 856, Report, 1959.

48 AA, Acc. 72.302, Minister of Agriculture Papers, Grant MacEwan's question in the Legislative Assembly, 22 March 1956; box 32, file 1792, 1959 correspondence; AA, Acc.

70.414, box 74, item 2961; box 69, item 2816; box 69, item 2825; and box 71, item 2889, Questions in the Legislative Assembly re flooding and seepage, 1955–56.

49 Data from South Saskatchewan River Basin Historical Natural Flows, 1912–95, CD-ROM version 2.02, Water Management Division, Alberta Environment.

50 In fact, the city engineer, more familiar with flood hazard, expressed alarm at the Parks Department program of construction for Prince's Island. Citizens, consultants, and Parks employees should be told that the island would be under water at a flow of 50,000 CFS. See CCA, Engineering and Environment Records, series III, box 144, file 3359, Bow River, 1969, 2 October 1969, to Commissioner of Operations.

51 CCA, MOO 3967, Montreal Engineering Company, *Flood Plain of Bow River in Calgary*, vol. 1; *Report*, vol. 2, twenty air photos covered with mylar transparencies showing the flood plain and much wider floodway in detail.

52 GA, Calgary Local Council of Women Fonds, box 10, file 104, Bow River Beautification, 1973–78, Brief to the Lombard Group.

53 University of Calgary Library, GB 1230, A5B66, Lombard North Group, *The Bow River Impact Study* (Calgary: Lombard North Group, 1973), part I, summary and recommendations; part II, overview; part III, analysis and impact assessment; see especially part III: 69–78 and part IV on public participation.

54 Glenbow Library, *The Bow River Study Committee Report*, January 1974, W. Solodzuk chair; *Bow River Community Survey*, January 1975; and Debbie Bathory, Chris Bray, Rita Gore, Gary Johannesson, Rose Kolibaba, Rose Lamoureux, and Lyn Tyler, mimeograph report (PAM 307.33 B 784).

55 CCA, Parks and Recreation Papers, Calgary River Management Committee, box 2864, file PL0012d; Bow River Park System, box 2864, file PL0012, 1975.

Chapter Ten

1 The extensive literature on the origins of parks includes, among many others, Donald Edgar, *The Royal Parks* (London: W.H. Allen, 1986); Erik Orsenna, *André Le Notre: Gardener to the Sun King* (New York: George Braziller, 2001); and Dorothy Stroud, *Capability Brown* (London: Faber and Faber, 1975).

2 Judith Adams, *The Great Amusement Park Industry: A History of Technology and Thrills* (Boston: Twayne, 1991); David Nye, *Electrifying America: Social Meanings of the New Technology* (Cambridge, MA: MIT Press, 1990), 122–3.

3 Alfred Runte, *National Parks: The American Experience*, 3rd ed. (Lincoln: University of Nebraska Press, 1997); Marguerite S. Shaffer, *See America First: Tourism and National Identity, 1880–1940* (Washington, DC: Smithsonian Institution, 2001), 43–5; Mark D. Barringer, *Selling Yellowstone: Capitalism and the Construction of Nature* (Lawrence: University of Kansas Press, 2002), 15–27.

4 David Blackbourn, "'Taking the Waters': Meeting Places of the Fashionable World," in Martin H. Geyer and Johannes Paulmann, eds, *The Mechanics of Internationalism* (Oxford: Oxford University Press, 2001).

5 This account is drawn from a paper read by Pearce to the Historical Society of Calgary on 16 December 1924, which was published in the *Calgary Herald*, 27 December 1924; a copy is in file 473 of Pearce's papers at UAA, Acc. 74-169.

6 Pearce's views of the springs may be found in UAA, Pearce Papers, Acc. 74-169, files 473–4, and in vol. 51, Letterbooks, 1886, Land Claims, memorandum re Hot Springs Banff, February 1886.

7 Ted Binnema and Melanie Niemi, "'Let the Lines Be Drawn Now': Wilderness Conservation, the Exclusion of Aboriginal People from Banff National Park in Canada," *Environmental History* 11 (2006): 724–52.

8 Pearce's 1924 article, cited in note 5 above, is quoted; the act creating the park is Canada, *Statutes*, 50–1 Vic., c. 32, 1887. The complicated story of the early development of Banff is reviewed in L.A. Taylor, "The Cave and Basin – Birthplace of National Parks," Background Information Document, 31 March 1978 (Whyte Museum, M317, Canadian Parks Service Papers, file F 17).

9 Macdonald is quoted in E.J. Hart, *The Place of Bows: Part I, To 1930: Exploring the Heritage of the Banff-Bow Valley* (Banff: EJH Literary Enterprises, 1999), 119. R.C. Brown has contended that Macdonald's park policy showed little evidence of enthusiasm for wilderness preservation, preferring "usefulness"; see Brown, "The Doctrine of Usefulness: Natural Resources and National Park Policy in Canada, 1887–1914," in *The Canadian National Parks: Today and Tomorrow*, vol. 1, ed. J.G. Nelson and R.C. Scace (Calgary: National and Provincial Parks Association and University of Calgary, 1969), 94–110. Alan MacEachern, however, makes the point that Macdonald and many contemporaries like Pearce understood that creating a park reserve was necessary to ensure that the area around the hot springs was not simply ruined by unbridled commercial exploitation; see MacEachern, *Natural Selections: National Parks in Atlantic Canada, 1935–70* (Montreal and Kingston: McGill-Queen's University Press, 2001), 16–18.

10 The story of that hotel and its successors is delightfully told in Bart Robinson, *Banff Springs: The Story of a Hotel* (Banff: Summerthought, 1988).

11 Hart, *The Place of Bows*, 141–58; PearlAnn Reichwein, "Beyond the Visionary Mountains: The Alpine Club of Canada and the Canadian National Park Idea, 1906 to 1969" (PhD thesis, Carleton University, 1995), chapters 1–3, which quotes the ACC constitution from the *Canadian Alpine Journal*, 1907, at 79, as well as an article in that journal in 1986 by Margaret Johnston and J. Marsh, entitled "The Alpine Club of Canada, Conservation and Parks, 1906 to 1930," which claimed, "Many of the statements regarding the Club's philosophy coincide with statements used contemporaneously to promote the national parks" (80).

12 See especially E.J. Hart, *The Selling of Canada: The CPR and the Beginnings of Canadian Tourism* (Banff: Altitude Publishing, 1983), and one of the early guidebooks produced by local newspaperman and promoter E.J. Luxton and revised annually, *50 Switzerlands in One: Banff the Beautiful, Canada's National Park* (Banff: *The Crag and Canyon*, 1912).

13 Robinson, *Banff Springs*, 37–54; Hart, *The Place of Bows*, 246–7, records the first of several additions in 1904 to Brett's original Sanitarium Hotel, along with the opening

of the King Edward (1904), subsequently expanded four times; the Alberta (1904), originally the National Park later the Cascade; the Banff, later the Mount Royal (1908), subsequently expanded; the Homestead (1910); the Alpine (1912); at the Upper Hot Springs there was a rebuilding of the Upper Hot Springs Hotel (1904–05) and the reconstructed Grand View Villa (1905) and Hydro Hotel (1909).

14 Hart, *The Place of Bows*, 247–8; Robinson, *Banff Springs*, 37–54.

15 A collection of postcards covering the first half of the twentieth century acquired by the authors includes many views of the hotel, most often showing its setting with the view eastward down the Bow but also looking up from the valley with Sulphur Mountain behind.

16 Sladen's *On the Cars and Off* (London: Ward, Lock and Bowden, 1895) is quoted in Robinson, *Banff Springs*, 23; McEvoy, *From the Great Lakes to the Wide West* (Toronto: William Briggs, 1902), is quoted in John A. Jakle, *The Tourist, Travel in Twentieth-Century North America* (Lincoln: University of Nebraska Press, 1985), 9.

17 *Canadian National Park (Rocky Mountains)* (copy in the Glenbow Library).

18 McEvoy's *From the Great Lakes* is quoted in Robinson, *Banff Springs*, 26.

19 Hart, *The Place of Bows*, 168, 302–3. A less successful experiment in 1929 saw forty Nakoda "boys," recruited to add "local colour" by acting as caddies on the golf course, housed "in a village by themselves" near the first tee. See LAC, RG 84, reel 104915, series A-2-a, vol. 71, file R313, part 7, Park Superintendent R.S. Stronach to CPR General Manager, Western, H.F. Matthews, 28 March 1929; Associate Banff Superintendent Arthur L. Ford to J.B. Harkin, 4 April 1929; Ford to Matthews, 9 April 1929; Ford to J.A. Angus, CPR, 24 August 1929.

20 By 1908, however, local officials were reporting, "The Bow River has been pretty well fished out, and there is no very good fishing ground in the vicinity of Banff," so the Department of the Interior agreed to pay to import 10,000 yearling rainbow trout (along with 500 mature fish paid for by the CPR) to be planted in the two lakes that fed Lake Minnewanka, since "these small lakes form one of the best and most convenient places for propagating rainbow trout, which is one of the best species of game fish." See LAC, RG 84, reel T8100, series A-1-a, vol. 1126, file 550756, memorandum from R.H. Campbell to Deputy Minister, 29 May 1908.

21 LAC, RG 84, reel T8100, vol. 127, file 549669, Banff Waterworks, 1904–06, H. Douglas, Commissioner, Rocky Mountains Park, to Secretary, Department of Interior, 21 August 1905; Memorandum, R. Rothwell to Secretary, Department of Interior, 19 May 1905; John Galt, Consulting Engineer, to W.W. Cory, Deputy Minister, 16 February 1905.

22 LAC, RG 84, reel T8100, vol. 126, file 562966, Memorandum, 16 February 1909, re complains of the residents of Banff – Commissioner Campbell to the Minister of the Interior re: complaints of Banff Liberal Association, 28 December 1908; reel T12668, vol. 45, file Bennett Personal, Parts 1–5, 1912–21, Bennett to Harkin re lowering water rates, 25 March 1914; Harkin to Bennett, 2 April 1914 ("I may say that to date the revenue derived from the Banff water and sewer system do not compare at all favourably with the capital expenditure and the cost of maintenance"); reel T10438, vol. 198, file B3-9, Banff Sewer and Plumbing Regulations, 1917–55; ibid., file B3-31, Banff Water

and Sewer Regulations, 1942–57; see *Calgary Herald*, 17 January 1957; Jean Lesage to R.E.W. Edwards, Secretary, Banff Advisory Council, 5 December 1956; Public Meeting, Wednesday, 26 September 1956, to discuss proposed new rates; *Canada Gazette*, 6 March 1946, Regulations for the Control and Management of Waterworks Systems for Townsites in the National Parks of Canada (rates specified for each park); reel T10439, vol. 198, file B80, Banff Water Supply General, 1956–57, re new sewer and water rates, October 1956; reel T12872, vol. 95, file U3-31, Water Regulations General, pt 2, 1947–56.

23 LAC, RG 84, reel T10438, vol. 198, file B3-31, Banff Water and Sewer Regulations, 1942–57, Memorandum, C.H.E. Powell, re Water Rates at Banff, 4 October 1948.

24 LAC, RG 84, reel T10451, vol. 216, file R80-13, Rocky Mountains Park – Water Supply, 1905–30, Banff Springs Hotel Contract; reel T11017, vol. 975, file B80-15, Water Supply Banff Springs Hotel – CPR 1906–49.

25 LAC, RG 84, reel T11017, vol. 975, file B80, Banff National Park Water Supply, 1949–53, pt 9, Memorandum, 28 June 1951, Turbidity – Banff Water Supply.

26 LAC, RG 84, reel T8100, vol. 127, file 549669, Banff Waterworks, 1904–06, pt 1, Memo H. Douglas to Secretary, Department of the Interior, 18 October 1904.

27 LAC, RG 84, reel T8101, vol. 128, file 549669, Banff Water Works, 1906–10, pt 3, for details of the construction of the system and John Galt's report on its successful operation, 15 July 1909.

28 LAC, RG 84, reel T10438, vol. 198, file B80, Banff Water Supply, Part 11 General, January 1956–December 1957, Report of the Municipal Fire Fighting Facilities and Structural Conditions, Town of Banff, Ottawa, January 1957.

29 E.J. Hart, *Golf on the Roof of the World, A History of the Banff Springs Golf Club and Course* (Banff: EJH Literary Enterprises, 1999).

30 MacEachern, *Natural Selections*, 160, notes that up to World War II officials remained keen on providing the same sort of recreational facilities in Atlantic Canada too.

31 Hart, *The Place of Bows*, 9–12, notes that later explorations would reveal that the same area held sites for pit houses erected by the first Banff residents, Native peoples moving up and down the valley from BCE to the late pre-contact period, that an archaeologist from the Geological Service of Canada was able to secure that area in 1913, making it the first protected archaeological site in the country, and that recent study suggests the pit sites date from as early as 10,700 BP.

32 LAC, RG 84, reel T10422, vol. 137, file B272, Banff, Investigation of the Beaver, 1938–45, Medical Officer of Health, 3 November 1938; reel T12975, series A-2-a, vol. 512, file 272, Banff National Park, Beaver, 1946–58, J. Smart to P.J. Jennings, Superintendent, 12 August 1946; reel T10422, series A-2-a, vol. 138, file B212-2, Report on Beavers by Green, 1944, pp. 1773–97; reel T12975, series A-2-a, vol. 512, file 272, parts 2, 3, National Parks Controller James Smart to Banff Superintendent P.J. Jennings, 12 August 1946. Beaver were blamed again in the spring of 1982, when the now outdated water supply system became infected with giardia ("beaver fever"), which was contracted by 160 local residents; see E.J. Hart, *The Battle for Banff: Exploring the Heritage of the Banff Bow Valley*, Part 2, 1930 to 1985 (Banff: EJH Literary Enterprises, 2003), 317.

33 LAC, RG 84, reel T12975, series A-2-a, vol. 512, file 272, parts 2, 3, A.F. Banfield to Dr Harrison Lewis, 8 January 1948; Lewis to Smart, 14 January 14 (quoted); Smart to Superintendent J.A. Hutchison, 6 September 1951 (quoted); J.R.B. Coleman to Smart, 14 December 1951.

34 This museum is now a historic site; see the Banff Park Museum National Historic Site website, www.pc.gc/lhn-nhs/ab/banff. It is also an splendid example of the almost vanished Banff style of cross-log decorative architecture. See Edward Mills, *Rustic Building Programs in Canada's National Parks, 1887–1950* (Ottawa: Parks Canada, 1994).

35 Hart, *The Place of Bows*, 114–34, 159–82, 205–29; see also Hart, *Golf on the Roof of the World*.

36 Whyte Museum, M317, Canadian Parks Service Papers, F3, S.F. Kun, Director, National Parks Branch, to Maryalice Stewart, Director, Archives of the Canadian Rockies, 9 October 1974, re boundary changes at Banff; see also Donald Kerr and Deryck Holdsworth, eds, *The Historical Atlas of Canada* (Toronto: University of Toronto Press, 1990), vol. 3, plate 36; and Nelson and Scace, *The Canadian National Parks: Today and Tomorrow*, vol. 1.

37 Maps of the shifting boundaries of Banff National Park (though hard to read) can be found in Kerr and Holdsworth, *The Historical Atlas of Canada*, vol. 3, plate 36, and Robert C. Scace, "Banff Townsite: An Historical-Geographical View of Urban Development in a Canadian National Park," in *Canadian National Parks*, ed. J.G. Nelson and R. Scace (Calgary: University of Calgary Press, 1969), 775. For the relevant legislation, see Whyte Museum, Parks Canada Papers, M317, F3, Changes in Boundaries and Areas of National Parks, S.F. Kun, Director, National Parks Branch, to Maryalice Stewart, Director, Archives of the Canadian Rockies, 9 October 1974, and Walter Hildebrandt, *An Historical Analysis of Parks Canada and Banff National Park, 1968–1995* (Banff: Banff–Bow Valley Task Force, 1995), 5–14. See also Ben Gadd, *Bankhead: The Twenty Year Town* (Banff: Friends of Banff National Park, 1989); R.W. Sanford, *Lake Minnewanka: The Spirit of the Waters* (Banff: Lake Minnewanka Boat Tours, 1999); Hart, *The Place of Bows*, 230–52.

38 Lougheed set off from Calgary with his family for Banff as soon as he acquired his first car in 1912. Once the MacKinnon family achieved some success on their farm near Dalemead, they piled into their Ford for a brief vacation in Banff.

39 Hart, *The Place of Bows*, 256–63, quoting Harkin at 263; UAA, Pearce Papers, file 474.

40 UAA, Pearce Papers, file 479.1, memorandum, n.d. (quoted), apparently sent to Pearce by Calgary MP R.B. Bennett, which seems to have been written by Harkin, 24 January 1914; file 479.2, Pearce to F.W. Lent, 2 November 1917.

41 This map is in UAA, Pearce Papers, file 479.3. Hart, *The Place of Bows*, 263–8, includes the quotation from Harkin.

42 The tendency of motoring tourists to move quickly from place to place with few stops in between was noted from an early date: "driving seemed to impose its own special momentum. Even with the many new sights, drivers could not help but want to go faster, to move on, regardless of immediate attractions"; see Warren J. Belasco, *Amer-*

icans on the Road: From Autocamp to Motel, 1910–1945 (Cambridge, MA: MIT Press, 1979), 86.

43 Hart, *The Place of Bows*, 268–9; Canadian Pacific Railway, *Resorts in the Rockies* (circa 1931 ed.); Hart, *The Battle for Banff*, 64–7, 81–2.

44 Hart, *The Battle for Banff*, 93–7. The Banff National Park Information Bureau recorded the following visitor registrations: 1940, 18,430; 1941, 22,371; 1942, 10,255; 1943, 11,068: 1944, 12,219; 1945, 13,703; 1946, 26,679; 1947, 27,934; 1948, 36,801; see LAC, RG 84, reel T11998, series A-2-a, vol. 5, file B109, part 3, Report on Registrations of Visitors at Banff National Park information Bureau, 1940–46, by R.E.W. Edwards, 30 September 1946; Edwards to Banff National Park Superintendent J.A. Hutchison, 31 October 1947, noting that Canadian registrants came from every province and Americans from every state, 13,855 of the former, 13,170 of the latter; Edwards to Hutchison, 31 October 1948, recording a 34 per cent increase over the previous year, with Canadians now numbering 20,018 compared with 16,321 Americans.

45 LAC, RG 84, reel T10407, series A-2-a, vol. 60, file TC 60, part 1, memorandum from Mills to J.M. Wardle, 2 July 1947; reel T11998, series A-2-a, vol. 5, file B109, part 3, Report on Banff National Park by Wilson, 15 October 1949.

46 LAC, RG 2, series B-2, vol. 96, file R-16-6, Dominion-Provincial Conference on the Trans-Canada Highway, 14–15 December 1948, Secret. The federal government also agreed to pay the provinces half the cost of any roads constructed between 1928 and 1949 which could be incorporated into a transcontinental route.

47 Ibid.; report by the Joint Planning Committee to the Chiefs of Staff Committee on Strategic Implications of Trans-Canada Highway, 11 February 1949; copy of Yellowhead Route Letter, vol. 2, no. 3, May 1949; memorandum from Reconstruction and Supply Minister R.H. Winters to cabinet, 25 October 1949. Even though the entire cost of the road through the national parks was to be borne by Ottawa, Manning tried to extract an agreement to pay two-thirds of the province's costs for selecting the southerly route but was turned down by Winters.

48 LAC, RG 84, reel T10408, series A-2-a, vol. 60, file TC60, part 3, memorandum from C.V.F. Weir to J.M. Wardle, 24 January 1952. For a more detailed account of the impact of the Trans-Canada Highway upon the national park, see Christopher Armstrong and H.V. Nelles, "Car Park: The Influence of Auto Tourism on Banff National Park," paper presented to the T2M Conference, Mobility and the Environment, Ottawa, Museum of Science and Technology, September 2008.

49 LAC, RG 84, reel 10411, series A-2-a, vol. 63, file TC7-8-1-1, part 1, memorandum from J.M. Wardle to Deputy Minister H.G. Keenleyside, 29 March 1950; reel T10407, series A-2-a, vol. 59, file TC64, part 1, memorandum from J.A. Hynes to Chief Engineer C.V.F. Weir, 14 June 1950; vol. 60, file TC60, part 2, Camp to Hutchison, 3 October 1950; memorandum from Hutchison to J. Smart, 20 October 1950 (quoted); reel 10408, series A-2-a, vol. 60, file TC60, part 3, memorandum from J.M. Wardle to Smart, 31 January 1952.

50 Hart, *The Battle for Banff*, 33–51, 183–206.

51 LAC, RG 84, reel T13012, series A-2-a, file B121, part 6, E.H. Davis, president, Calgary Olympic Development Association, to Northern Affairs and Natural Resources Minister Walter Dinsdale, 22 June 1961.
52 Ibid., file B121-13, part 6.
53 LAC, RG 84, reel T10981, series A-2-a, vol. 1938, file TCH60-2, part 5, memorandum from J.R.B. Coleman to Deputy Minister, 21 January 1964.
54 The most influential preservationist lobbies were the Alpine League of Canada, headed by A.W. Wheeler, and Calgary resident W.J. Selby Walker's Canadian National Parks Association; see Reichwein, "Beyond the Visionary Mountains."
55 The most nuanced analysis of Harkin's thought and policies during this period is to be found in MacEachern's *Natural Selections*, 28–33. The Harkin quotation is from Hart, *The Place of Bows*, 223.
56 For the debate see Canada, House of Commons, *Debates*, vol. 2, 1930, 9 May, 1930–4; Canada, *Statutes*, 20–21 George V, c. 33, 1930.
57 LAC, RG 84, reel T10438, vol. 198, file B80, Banff Water Supply, pt 11 General, January 1956–December 1957, Banff Townsite Report on Water Supply and Sewerage, pt 1, 1 November 1955, Canadian-British Engineering Consultants; reel T10439, vol. 198, file B80, Banff Water Supply General 1956–57, T.C. Main, Canadian-British Engineering Consultants, to J.A. Hutchison, Director, National Parks Branch, 24 January 1956.
58 Hart, *The Battle for Banff*, 201–4, 214–15, 232–52, 333–5.
59 Ibid., 339–42.

Chapter Eleven

1 There is a small but growing literature on the history of parks and the variations in their forms and functions. See, for example, Clare Latimer, *Parks for the People: Manchester and Its Parks, 1846–1926* (Manchester: Manchester City Council, 1987); Peter Bailey, *Leisure and Class in Victorian Britain* (New York: Methuen, 1987); and Martin Daunton, *House and Home in the Victorian City* (London: Edward Arnold, 1983), which in quite different ways emphasize the controlling aspects of parks over the working class. On urban parks in North America, see David Schuyler, *The New Urban Landscape* (Baltimore: Johns Hopkins University Press, 1986); Galen Cranz, *The Politics of Park Design* (Cambridge: MIT Press, 1982); and Roy Rosenzweig and Elizabeth Blackmar, *The Park and the People* (Ithaca: Cornell University Press, 1992). On the transition from the Arcadian to romantic natural form under the influence of Calvert Vaux and Frederick Law Olmsted, see Witold Rybczynski, *A Clearing in the Distance* (New York: Harpers, 1999); Laura Roper, *FLO: A Biography of Frederick Law Olmsted* (Baltimore: Johns Hopkins University Press, 1973); and Charles Beveridge and Paul Rocheleau, *Frederick Law Olmsted: Designing the American Landscape* (New York: Rizzoli, 1995). On parks more generally, see William Irwin, *The New Niagara* (University Park: Pennsylvania State University Press, 1996); Patricia Jasen, *Wild Things: Nature, Culture, and Tourism in Ontario, 1790–1914* (Toronto: University of

Toronto Press, 1995); and Linda Martin and Kerry Segrave, *The City Parks of Canada* (Oakville: Mosaic Press, 1983). For the great variety in contemporary city parks, see Alan Tate, *Great City Parks* (London: Spon Press, 2001).

2 N.C. Landrum, *The State Park Movement in America* (Columbia: University of Missouri Press, 2004).

3 Gerald Killan, *Protected Places: A History of Ontario's Parks System* (Toronto: Dundurn Press, 1993).

4 Sue Anne Donaldson, "William Pearce: His Vision of Trees," *Journal of Garden History* 3 (1983): 233–44.

5 UAA, William Pearce Papers, file 92, Islands in the Bow River 1908, copy of a seven-page letter to the city, 3 April 1908, documenting the unsatisfactory city response and enclosing his plan.

6 *At Your Service: Calgary's Library, Parks Department, Military, Medical Services, and Fire Department* (Calgary: Century Publications, 1975); see "From Prairie to Park: Green Spaces in Calgary" (72–4) for the evolution of the zoo.

7 Ibid., 121–256. Pearce also transferred a prominent parcel of land in the middle of the city, land which became Victoria Park, the main civic space and floral garden in the city. See Donaldson, "William Pearce," 238.

8 UAA, William Pearce Papers, file 74, 169–87, letter to his son Seabury, 28 February 1913, instructing him what to show Mawson in the event he might be out of town and unable to conduct a guided tour himself.

9 John Crosby Freeman, "Thomas Mawson: Imperial Missionary of British Town Planning," *Revue d'art canadienne/Canadian Art Review* 2 (1975): 41. See also Walter Van Nus, "The Fate of City Beautiful Thought in Canada, 1893–1930," in *The Canadian City*, ed. G. Stelter and A. Artibise (Ottawa: Carleton Library, 1979), 162–85. For the influence of these ideas in the United States, see also W.H. Wilson, *The City Beautiful Movement* (Baltimore: Johns Hopkins University Press, 1989).

10 The plan is usually treated as a footnote, a sidebar, or a remarkable curiosity in popular histories of Calgary; see, for example, T. Ward, *Cowtown: An Album of Early Calgary* (Calgary: McClelland and Stewart, 1975), 382–3.

11 For perceptive critiques of the Mawson plan, see CCA, E. Joyce Morrow, *"Calgary Many Years Hence": The Mawson Report in Perspective* (Calgary: City of Calgary and the University of Calgary, 1979), and Freeman, "Thomas Mawson," 37–47. Mawson's drawings toured Alberta's provincial fairs for several years until they remained unclaimed in the lost luggage department at one of the railway stations. At some point a workman appropriated the neglected artwork for use as wallboard in his garage. Mawson's drawings resurfaced in the 1970s during the demolition of the garage. Art historians rescued the remaining work and restored to its former glory, forming the subject of the exhibition that accompanied Morrow's booklet.

12 Terry Bullick, *Calgary Parks and Pathways* (Calgary: Rocky Mountain Books, 1990), 12–14, and *At Your Service*, 186.

13 Bullick, *Calgary Parks and Pathways*, 82–8; *At Your Service*, especially 174–205. On the difficulties of dealing with the prickly owner of the Inglewood property, see GA,

Austin de B. Winter Fonds, M1327, box 3, file 41, Confidential Memorandum Re: Inglewood Bird Sanctuary, February 1941.

14 AA, Acc. 83.92, Provincial Parks Board Records, 1929–58, box 6, Deputy Minister Robertson to District Engineers, 3 July 1929, the responses from various district engineers, and the report on possible provincial park sites, 7 November 1929, by Seymour, Smith and Robertson.

15 AA, Acc. 71.212, Provincial Parks and Reserves, Report of W.T. Aiken, secretary, Provincial Parks Board, to Chairman C.A. Davidson, 30 September 1935, on Provincial Parks for Alberta.

16 AA, Acc. 69.289, Premier's Papers, file 319A, Provincial Parks; contains letters and documents from 1919–32 regarding the creation of Aspen Lake, Ghost River, Gooseberry Lake, and Elk Lake provincial parks, as well as letters from organizations and individuals supporting the policy.

17 AA, Acc. 83.93, Provincial Parks Board Records, 1929–58, box 6, Report on Alberta Provincial Parks, 1946; Report on Parks in Canada by Province and Area, 1950; Chairman of the Parks Board to Mrs Janssen, 4 August 1950, regretting the lack of literature on parks.

18 AA, Acc. 83.92, Provincial Parks Board Records, 1919–58, box 6, R.A. Wileman to Fish and Game Commissioner E.S. Huestis, 3 September 1953.

19 AA, Acc. 83.92, Provincial Parks Board Records, 1959–72, box 7, Memo, D. Williams to Chair of the Provincial Parks Board, 29 October 1954.

20 AA, Acc. 83.92, Provincial Parks Board Records, 1919–58, box 6, Parks Board to Canadian Economic Research Associates, 13 December 1957; Doug Morgan, Calgary Fish and Game Association, to Norman Willmore, Minister, 17 May 1957; Morgan to Willmore, 3 July 1957; Willmore to Morgan, 5 July 1957; D.V. Hicks, Calgary Downtown Business Association, to Willmore, 28 April 1959. The province of Ontario, among others, also had to respond to the postwar outdoor recreation boom by developing more parks and different kinds of parks; see Killan, *Protected Places*, 74–119.

21 AA, Acc. 83.92, Provincial Parks Board Records, 1919–58, box 6, Memo, F.A. Jones, Forest Officer Canmore, to F.V. Keats, Superintendent, Bow Valley Forest, 23 June 1957.

22 See AA, Acc. 74.49, Eastern Rockies Forest Conservation Board, for the main file; see also Acc. 74.169, Eastern Rockies Forest Conservation Board, Minutes, 1968.

23 AA, Acc. 83.498, Parks Records, box 2, Bow Valley Park, file 54A, Report on Proposed Provincial Park in the Calgary Area, 1959; Order-in-Council, 23 July 1959, declares provincial park in Bow Valley area; file 55A, Warden to E.P. Shaver, Parks Commissioner, 25 January 1961; same to same, 3 June 1961; referred 11 April 1962 to Provincial Sanitary Engineer; file 56, 1963 Lime plant to be moved; Acc. 83.467, Provincial Parks: Supervision Reports, 1967–69, box 1, Bow Valley Park Supervisor's Reports; Acc. 83.92, box 14, Provincial Parks Reports; Acc. 83.92, Parks Records, box 14, *Calgary Herald*, 27 July, 14 September 1959. On the jamboree, see Acc. 85.166, GS Department of Parks and Recreation Records, box 16; Acc. 83.467, Provincial Parks: Supervision Reports, 1967–69, box 1, Bow Valley Park Supervisor's Reports; Acc. 86.116, Parks Records – Contracts, Services, box 1, Bow Valley, 1961–77.

24 On the space devoted to residential and commercial uses, see the excellent map in CCA, Board of Commissioners Papers, series V, box 227, file 6200.4, circa 1963.
25 Marjorie Norris, *A Leaven of Ladies: A History of the Calgary Local Council of Women* (Calgary: Detselig, 1995), 198.
26 CCA, Engineering and Environmental Services, series III, box 62, file 856, 170–2, Memorandum on The River Bank Development Scheme, 3 December 1959; box 62, file 855, 15 December 1959, report of the River Bank Development Committee.
27 CCA, Council Minutes, box 58, file 441, Urban Renewal Report, no. 2, December 1961.
28 For a fuller treatment of this incident, see H.V. Nelles, "How Did Calgary Get Its River Parks?" *Urban History Review* 34 (2005): 28–45. On the continuing role of the women's movement, see Norris, *A Leaven of Ladies*, 197–212; the political agitation of Alderman Jack Leslie is discussed in Jean Leslie, *Three Rivers Beckoned: Life and Times with Calgary Mayor Jack Leslie* (Bragg Creek: Fay-Mar Publishing, 2004), 165–218.
29 On the history of planning in Canada, see Jill Grant, "Planning Canadian Cities: Context, Continuity and Change," in *Canadian Cities in Transition*, ed. T. Bunting and P. Filion, 2nd ed. (Toronto: Oxford University Press, 2000), 443–61.
30 See CCA, RG 1501, Town Planning Commission, box 1, City of Calgary, *The Future of Downtown Calgary* (Calgary: Planning Advisory Committee, 1966), 1, 9, for the critique and the tables and maps purporting to show the "health" of the downtown. There is a brief account of the history of zoning and planning in Calgary on 11 and 76. For redevelopment of the Eau Claire site and the riverbank, see 65–6.
31 See GA, Colin Campbell McLaurin Fonds, M 6072, box 1, file 15, Bow River Beautification Committee, for correspondence regarding the Underwood McClellan study and the study itself; box 2, file 16, for correspondence on his unsuccessful attempt to turn this committee into a more permanent, incorporated Calgary Beautification Foundation.
32 CCA, Parks and Recreation Papers, Prince's Island files 1969–79, box 1056, file PG0009, 1968–70.
33 CCA, 800.004, City of Calgary, *The Calgary Plan: A General Plan Prepared on Behalf of the Civic Administration by the City of Calgary Planning Department* (May 1973). The black and gold design symbolized the oil and wheat foundations of the local economy. Section 5 deals with parks and recreation.
34 CCA, Engineering and Environmental Services, box 144, file 3359, Bow River, 1969, 2 October 1969, City Engineer to Commissioner of Operations.
35 CCA, Parks and Recreation, Calgary River Management Committee, box 2864, file PL0012d; Bow River Park System, box 2864, file PL0012, 1975.
36 Personal communication from Bill Robinson, the retired Parks Department official who oversaw the development of the river parks, April 2003.
37 CCA, Parks and Recreation, Bow River Park System, box 2864, for the key 1975 documents; box 2087, file PL0012b, for expansion of the program in 1976. Land acquisition

and trail development continued on into the early 1980s. See also Leslie Beck and Bill Robinson, *Trail Systems* (Calgary: The City of Calgary, Parks and Recreation, 1975); copy in University of Calgary Library.

38 See AA, Acc. 70.417, Highways Correspondence, 1886–1905, Bow Bottom Trail maps.

39 AA, Acc. 81.312, Department of the Environment, Subject files, Provincial Parks Act file, 1963–64, Policy Document Relating to the Provincial Parks System, March 1967, Department of Lands and Forests. A classification system established separate management policies for six different kinds of parks each of which involved a different degree of natural protection: (a) wilderness; (b) historical, ethnological, and archaeological; (c) unique natural area; (d) natural environment recreation area; (e) outdoor recreation area; and (f) parkway.

40 For Alberta, John Richards and Larry Pratt's landmark book *Prairie Capitalism: Power and Influence in the New West* (Toronto: McClelland and Stewart, 1979) provides an excellent beginning whose approach has not been extensively followed up by other scholars.

41 AA, Acc. 83.92, Parks Board Records, box 1, General Correspondence files. The ERFCB is discussed further in an earlier chapter, but Alberta likely terminated this federal provincial agreement at the end of its twenty-five-year term because the board resisted provincial pressure for an aggressive park development policy on the eastern slopes of the Rockies. See AA, Acc 74.169, Eastern Rockies Forest Conservation Board, Minutes, 1968, where the board resists the recommendations of "The Real Need in Alberta – Parks and Recreation," a report of an independent agency urging province-wide planning. See also Acc. 74.345, Environment Conservation Authority Papers, Interim Report No. 3, Bow River Basin, May 1974, 17–20.

42 AA, Acc. 74.345, GSE Environment Conservation Authority, "Land Use and Resource Development in the Eastern Slopes," Interim Report No. 3, Bow River Basin, May 1974, 3.

43 Ibid. Interim Report No. 3 dealt specifically with the Bow basin. The previous two reports dealt more broadly with policy for the entire eastern slope. Chapter 2, 17–41, surveyed opinion; chapter 3, 45–69, summarized what was said in the hearings and proffered some general conclusions.

44 AA, Acc. 93.91, Eastern Slopes Planning Committee, box 1, file 400/16, Eastern Slopes Interdepartmental Planning Committee Progress Report, 2 April 1976, details the process and recommendations. See also Acc. 92.468, Environment Records, Canmore Corridor Study, Calgary Regional Planning Commission, "A Proposed General Plan Study of the Canmore Corridor and the Eastern Slopes," September 1976 (mimeo), and "The Canmore Corridor Integrated Land Management Plan," Prepared for the Eastern Slopes Interdepartmental Planning Committee" by R.W. Benfield, May 1977; Acc. 85.401, Lougheed Papers, Executive Correspondence series IXa, 54, files 972–4, Kananaskis Country, 1975–78, Material on park planning, Canmore corridor study, Resource Management policy for the Eastern Slopes, Summary of Draft Report "The Canmore Corridor Integrated Land Management Plan" for the Premier, 27 June 1977.

45 AA, Acc. 96.524, Provincial Parks Records – Kananaskis and Fish Creek, box 4, 1978, official opening of the Kananaskis Country Park; Acc. 85.401, Lougheed Papers, Executive Correspondence series IXa, box 54, files 972–4, Kananaskis Country, 1975–78.

46 AA, Acc. 83.498, Provincial Parks Records, box 2, Bow Valley Park, file 54A, Report on Proposed Provincial Park in the Calgary Area, 1959; Acc. 85.116, Provincial Parks Records, box 14, Fish Creek Provincial Park, 1973–81, Progress Report, November, 1977; box 15, Assistant Deputy Minister of the Environment H.W. Thiessen to Minister J.W. Cookson, 13 December 1979, setting out some of the early steps toward creation of the park.

47 AA, Acc. 85.401 Lougheed Papers, box 54, file 970, Fish Creek Park, 1972–75, Press Conference, 19 February 1973; Request for Cabinet Decision, 31 October 1972; Acc. 85.116, Parks Records, box 14, Provincial Parks Policy for Alberta, Position Paper No. 13, May 1973.

48 GA, Burns Fonds, M7771, file 49, P. Burns Ranches Limited; file 49b, The Setup as from 1928; file 30, Rights and Obligations of Adjoining Property Owners Respecting the Water in Fish Creek, 1968. On Hull, see Henry Klassen, *Eye on the Future: Business People in Calgary and the Bow Valley* (Calgary: University of Calgary Press, 2002), 247–50.

49 AA, Acc. 90.242, Parks Records, box 1, Fish Creek Provincial Park, 1978–82, Report of T. Ward and J. Saunders, Proposals for the Development and Interpretation of Bow Valley Ranch, Fish Creek Provincial Park, an historic site within a natural park, 8 September 1978.

50 Ibid., Minutes of Park Management Committee, 14 September 1978, 22 October 1981; box 2, Memo, William C. Pearce, 8 December 1981; Barry Potyondi to Wendy Martindale, 14 December 1981.

51 AA, Acc. 96.523, Provincial Parks Records, box 3, Fish Creek file, Memo on Flow, 30 July 1975; Acc. 85.401, Lougheed Papers, Executive Correspondence series IXa, box 54, file 970, Dedication Ceremony, 29 June 1975; Press Conference, 21 April 1976, on Fish Creek Development; Acc. 85.166, Parks Records, box 15, file 3100, Fish Creek Provincial Park, M.R. Brower to K. Bowsma, 30 July 1975; John Filby to Bill Payne, MLA, 3 October 1979; Payne to the Minister, 11 October 1979.

Chapter Twelve

1 David R. Percy, the leading Canadian legal scholar in respect of water rights, has written, "The courts have rarely addressed fundamental questions of constitutional jurisdiction over water except in some leading cases which occurred at the end of the nineteenth century. The absence of a definitive body of constitutional law with regard to water means that many issues, particularly concerning inter-jurisdictional waters, remain uncertain and their discussion, in the absence of any clear legal precedent, tends to be clouded by the preferences of the commentator." See Percy, "Federal-Provincial Jurisdictional Issues," Appendix D, 83, in *Water Law and Policy Issues in Canada*, ed. Harriet Rueggeberg and Andrew R. Thompson (Vancouver: Westwater Research Centre, 1984).

2 Discussion of water law in a Canadian context begins with Bora Laskin's "Jurisdictional Framework for Water Management," in *Resources for Tomorrow Conference*, Background Papers, vol. 1 (Ottawa, 1961): 211–25, which opens with a lucid exposition of the law of riparian ownership. On riparian rights, see also Rueggeberg and. Thompson, *Water Law and Policy Issues in Canada*, 6–10, and Appendix B by D. Chesman, "Memorandum on Riparian Rights," 61–70, which focuses upon the issue of the extent to which latent riparian rights continue to exist in western water law despite legislative attempts to extinguish them. David Percy has published a succinct compendium of three distinct water law regimes in eastern Canada, the north, and western Canada in *The Framework of Water Rights Legislation in Canada* (Calgary: Canadian Institute for Resource Law, 1988); riparian rights within the three regimes are discussed on pages 17–21, 58–60, and 73–6. A useful bibliography compiled by Maureen Bell, of the University of Calgary, "Water Rights in Alberta: Annotated Bibliography" (available at www.law.ucalgary.ca/system/files/b7.pdf) also covers material relating to Canada generally, to environmental law, and selectively to international water law. See also the memorandum by the secretary of the Prairie Provinces Water Board in AA, Acc. 90.159, Deputy Minister of the Environment, box 15, E.F. Durrant, "Some Notes On Problems Arising From The Lack of Interprovincial Water Law," n.d. [c.1959]. For the more recent accounts of the development of riparian law and the law of navigable rivers, see Tristan M. Goodman, "The Development of Prairie Canada's Water Law, 1870–1940," in *Laws and Societies in the Canadian Prairie West, 1670–1940*, ed. Louis Knafla (Vancouver: University of British Columbia Press, 2005), 266–79, and Jamie Benedickson, *The Culture of Flushing* (Vancouver: University of British Columbia Press, 2007), chapters 1–2.

3 See the government of Canada's Indian and Northern Affairs website, www.ainc-inac.gc.ca/pr/trts/trty7_e.html, for a copy of Treaty and Supplementary Treaty 7 made 22 September and 4 December 1877 between Her Majesty the Queen and the Blackfeet and other Indian Tribes, at Blackfoot Crossing of Bow River and Fort MacLeod. For the context, see Ian A.L. Getty and Donald B. Smith, eds, *One Century Later: Western Canadian Reserve Indians Since Treaty 7* (Vancouver: UBC Press, 1978). A modern reading of the treaty from an indigenous perspective can be found in Treaty 7 Elders and Tribal Council with Walter Hildebrandt, Sarah Carter, and Dorothy First Rider, *The True Spirit and Original Intent of Treaty 7* (Montreal and Kingston: McGill-Queen's University Press, 1996).

4 Glenbow Library, Map Collection, GR3503 IR146 1883, C212, Treaty No. 7, "Survey of the boundaries of the Blackfoot Indian Reserve on the Bow River, Nor'West Territories and settled by amended Treaty, June 20, 1993."

5 Glenbow Library, Map Collection, G3503 IR142 1890, M297, "Map of the Stoney Reserve," and a slightly later version, G3503 IR142 1889, C212; see note 3 above.

6 Glenbow Library, Map Collection, G3503 IR146 1883 C212, Treaty No. 7, "Survey of the boundaries of the Blackfoot Indian Reserve on Bow River Nor'West Territories and settled by amended Treaty, June 20, 1993"; LAC, RG 89, vol. 605, file 1514, Deputy Minister of Justice F.P. Varcoe to Deputy Minister of Mines and Resources, 15 September 1942, noting the 1933 opinion of his predecessor, W.C. Edwards, and continu-

ing, "The reason relied on for advising that it was to be presumed that the riparian title included the bed of the stream *ad medium filum* still seen to be valid. I have discovered no good reason to vary this opinion."

7 On navigation and regulation of navigable rivers, see Laskin, "Jurisdictional Framwork for Water Management," and Rueggeberg and Thompson, *Water Law and Policy Issues in Canada*, 18.

8 LAC, RG 89, vol. 509, file 1514, Department of Mines and Resources, Water Power Branch Memorandum "Re Jurisdiction, Bow River Licenses," 2 March 1944, unsigned, initialed by J.M. W[ardle of the Water Power Bureau].

9 This judgment is quoted in an Ontario Ministry of Natural Resources legal opinion, "Ownership Determination – Beds of Navigable Waters," Lands and Water Branch, PL 2.02.02, 11 February 1997. Nor do obstructions such as rapids or low water invalidate potential navigability. The Ontario legal opinion stated the tests of navigability, as determined by the courts, as follows: a stream must be usable by small craft or floatable for purposes of river log-driving. Ontario fortified its claim to the bed of navigable rivers as against federal claims with a 1911 statute, the Beds of Navigable Waters Act.

10 On the history of federal-provincial conflict over water control, see Christopher Armstrong, *The Politics of Federalism* (Toronto: University of Toronto Press, 1981), and Neil Forkey, *Shaping the Upper Canadian Frontier* (Calgary: University of Calgary Press, 2003).

11 We have relied heavily upon David R. Percy, "Water Rights in Alberta," *Alberta Law Review* 15 (1977): 142–65, which contains a critical analysis of the licensing process and the rights granted under it, focusing upon the permanent nature of the rights conferred. See also Percy's *Framework for Water Rights Legislation in Canada*, 3–47. We have benefited as well from a memorandum on Alberta water rights drafted by Eric Strikwerda of York University in 1999.

12 This claim was made in "General Report on Irrigation in the Northwest Territories (1894)" in *Annual Report of the Interior Department*, 1895, 28, but as David Percy points out, subsequent commentators have noted that neither the Interior minister nor MPs who supported the bill understood its import; see Percy, "Seventy-Five Years of Alberta Water Law: Maturity, Demise & Rebirth," in *Alberta Law Review* 35 (1996–97): 224n9. Percy also notes at 223n5 that riparian owners probably continued to enjoy the right to "reasonable use" of waters flowing over or beside their lands even if that must necessarily have some effect upon the flow downstream.

13 Percy, "Seventy-Five Years of Alberta Water Law," 223n7, notes that the original act (Canada, *Statutes*, 1894, c. 30) merely followed the Australian precedent in granting control of water use, while the 1895 amending act (c. 33, s.2) vested ownership of all water in the Crown.

14 Voluminous records generated by this process can be found in several series in AA, Acc. 92.415 and 93.121, Water Rights files, 1895–1985, and Acc. 80.166, Environment Department, Water Rights Ledgers. See chapter 6 above for the history of irrigation in the Bow valley.

15 GA, Eau Claire Lumber Company Fonds, M1564, box 1, file 2, Licence, Dept of the Interior, 15 December 1897, to divert water.

16 The CPR needed water at its divisional points and main stations, in this case Gleichen, Canmore, and Calgary. Water-taking at Gleichen was complicated by the need to lay a pipe across the Siksika reserve, which required permission of the band, who demanded compensation for horses killed by railway operations on the northern border of the reserve. The water-taking itself was federally licensed. See LAC, RG 10, vol. 3583, file 1084, part A, C10103, for correspondence 1888–95, regarding this matter.

17 LAC, RG 10, vol. 3563, file 82, part 16, C10099, Frank Oliver to Clifford Sifton, 1 July 1903.

18 LAC, RG 89, vol. 605, file 1514, Memorandum "Re Jurisdiction, Bow River Licenses," 2 March 1944, unsigned but initialled by J.M. W[ardle of the Water Power Branch].

19 LAC, RG 10, vol. 3686, file 13119-2, 3, 4, C10120, Memorandum, 14 July 1903; H.E. Sibbald to Indian Commissioner, 17 August 1903; Deputy Superintendent General of Indian Affairs to Frank Oliver, 31 August 1903; Secretary of the Calgary Board of Trade to Department of the Interior, 10 October 1903; Guardian Insurance Company to Department of Indian Affairs, 30 September 1904; Johnson and Johnson, Barristers on behalf of E.R. Wood to Department of Indian Affairs, 15 September 1905.

20 LAC, RG 85, vol. 737, file 1436-3-5 (1), J.B. Challies to Mr Young, copy to Wm Cory, Deputy Minister, 4 January 1907, recapitulates the matter.

21 LAC, RG 10, vol. 3686, file 13119-2, 3, 4, C10120, Frank Pedley, Deputy Superintendent General of Indian Affairs, to Rev. John McDougall, 23 May 1906; reply, 22 May 1906; McDougall to Pedley, 25 May 1906; Pedley to McDougall, 6 June and 12 June 1906; McDougall to Pedley, 6 June 1906, reporting the asking price. The minister, Frank Oliver, replied to McDougall himself, 24 July 1906, stating the price to be "unjustified by the value of the power in its undeveloped state."

22 LAC, RG 85, vol. 737, file 1436-3-5 (1), W.M. Alexander and W.J. Budd to the Minister of Interior, 21 December 1906; J.B. Challies to Mr Young, copy to Mr Cory (Deputy Minister), 4 January 1907. Creation of a provincial government in 1905 created confusion: the bureaucrats thought the land under the river was "under the jurisdiction of the local government," that is, the province. Authority to grant diversion or water power permits, however, rested with the Department of the Interior, but the authorities were not certain and added a phrase about possible jurisdiction by "local government if rights have been passed on to the province." See ibid., Lands and Timber Branch to Deputy Minister Cory, 30 January 1907.

23 LAC, RG 85, vol. 737, file 1436-3-5 (1), Charles Mitchell (Engineer for Alexander and Budd) to Department of the Interior, 26 February 1907; R.E. Secretary, Department of the Interior, to J.B. Challies, 1 March 1907.

24 Ibid.

25 LAC, RG 85, vol. 737, file 1436-3-5 (1), Surrender Agreement with T.I. Fleetham, agent, Moses Bearspaw, Chief, Peter Wesley, Chief, Jonas Two Youngman, Chief, and James Swampy, Amos Big Stony, John Mark, Hector Crawler, George McLean, Councillors,

22 March 1907; Draft Lease, 1 April 1907. See also GA, Calgary Power Papers, M1546, vol. 4, file 11, Material Relating to the development of the Horseshoe Falls, and file 12, Quit Claim, Stony Indians, Morley, 1907, sale of 1,000 acres adjacent to Horseshoe Falls.

26 This position was adopted despite the fact that the office of superintendent general was usually, though not always, occupied by the minister of the Interior.

27 LAC, RG 85, vol. 737, file 1436-3-5 (1), J.B. Challies to R.E. Young, Memoradum, 13 March 1909; J.D. McLean, Secretary, Department of Indian Affairs, to R.E. Young, 6 April 1909; Water Power Regulations under the Dominion Lands Act, passed by order-in-council, 2 June 1909.

28 The amendments to the Dominion Lands Act and the regulations respecting water allocation and development are printed in Leo G. Denis and Arthur V. White, *Water-Powers of Canada* (Ottawa: Commission of Conservation, 1911), 275 ff, and again in Leo G. Denis and J.B. Challies, *Water-Powers of Manitoba, Saskatchewan and Alberta* (Ottawa: Commission of Conservation, 1916), 302 ff. This latter volume contains detailed hydrographical analysis of the Bow River and its flow conducted by J.B. Challies, superintendent of the Water Power Branch (178–226). The records of the Water Power Branch are to be found in the LAC, RG 89.

29 LAC, RG 85, vol. 737, file 1436-3-5 (1), H. Fitsimmons to W. Cory, Deputy Minister, 8 July 1909; Andrew Haydon and T.R.E. McInnes, Calgary Power, to Frank Oliver, Minister, 28 August 1909; C.F. Adams to Secretary, Department of the Interior, 15 October 1909. The lease included the phrase "subject to the regulations governing water power rights then or thereafter in force."

30 LAC, RG 89, vol. 605, file 1514, "Reference to the Department of Justice re Ownership and Control of Developed Power Sites on the Bow River in Alberta," by M.F. Cochrane, April 1931.

31 We benefited greatly from Roderick I. Stutt, "Water Policy-Making in the Canadian Plains: Historical Factors that Influenced the Work of the Prairie Provinces Water Board (1948–1969)" (PhD thesis, University of Regina, 1995), 73–4. We also read the Alberta collection of documents relating to this body in various series in the AA and the annual reports of the PPWB in the Alberta Government Library. Our research assistant, Liza Piper, examined the main body of PPWB papers in the Saskatchewan Archives.

32 Stutt, "Water Policy-Making," 79.

33 This unpublished document, dated 1947, is discussed in Stutt, "Water Policy-Making," 86–9, from which the following is drawn.

34 Stutt, "Water Policy-Making," 81–5. Alberta's activities are documented in AA, Acc. 72.303, Agriculture Department Deputy Minister's Records, file 353, box 7, Special Committee on Water, 1949.

35 Stutt, "Water Policy-Making," 107.

36 Nollet to L.B. Thomson, 4 May 1949, in Stutt, "Water Policy-Making," 108.

37 Stutt, "Water Policy-Making," 105.

38 AA, Acc. 78.110, Deputy Minister of Environment records, box 2, Manning to Douglas, 30 June, 9 November 1959; Douglas to Manning, 23 December 1959.

39 Ibid., box 15, memorandum from Desharpe on "The Law of Water as Relating Between Provinces," 14 October 1959.
40 Ibid., "Some Notes On Problems Arising From The Lack of Interprovincial Water Law," n.d.
41 Ibid., memorandum from Durrant re "Hydro-Power Projects and the Prairie Provinces Water Board," n.d.
42 Ibid., box 2, "A New Look at Alberta's Long Range Water Plan," 1964.
43 Ibid., Director of Water Resources F.L. Grindley to Alberta Deputy Minister of the Agriculture R.M. Putnam, 2 July 1964; memorandum from Grindley to Premier Harry Strom, 23 April 1965.
44 Ibid., box 17, Proposal for the Alberta Water Plan, The Prairie Rivers Improvement and Management Evaluation Programme (PRIME) by R.E. Bailey, 1965; ibid., box 2, "A New Look at Alberta's Long Range Water Plan," 1964.
45 AA, Acc. 85.401, Peter Loughheed papers, Executive Correspondence, series IXa, box 38 (offsite), file 675, pamphlet prepared by University of Alberta students entitled "PRIME: is it a CRIME?" A Saskatchewan–Nelson Basin Board was also created in Regina in 1968, which prepared a massive report in 1972 noting fifty-five possible storage sites in the watershed involving twenty-five dams for diversions and storage; see ibid., file 674, "The SNBB Study and the People Who Do It," 1972.
46 AA, Acc.90.159, Deputy Minister of Environment records, box 15, Pope to W.S. Russell, 31 October 1966.
47 Schedule A of the Master Agreement provided that "Alberta shall be entitled in each year to consume, or to divert or store for its consumptive use, a minimum of 2,100,000 acre-feet net depletion out of the flow of the watercourse known as the South Saskatchewan River even though its share for the said year ... would be less than 2,100,000 acre-feet, provided however Alberta should not be entitled to consume ... more than one-half the natural flow of the said ... watercourse, if the effect therefore at any time would be to reduce the actual flow ... at the common boundary ... to less than 1,500 cubic feet per second." In 1973 the PPWB approved the following gloss: "The foregoing is interpreted to mean that Alberta may ... consume 2,100,000 acre-feet net depletion even though this in excess of its 50 percent share, provided the result does not reduce the flow of the South Saskatchewan River at the ... boundary to less than 1,500 cfs. In the case when the natural flow at the boundary is less than 3,000 cfs, Saskatchewan's share shall be 50 percent of the natural flow." See PPWB, Annual Report, 1972–73, which notes that in the calendar year 1972 the recorded flow of the South Saskatchewan and Red Deer at the boundary was actually 7,948,600 acre-feet, or 91.7 per cent of the calculated natural flow that year.
48 The parties bound themselves to submit any disputes to the Exchequer Court of Canada for adjudication. In such cases the PPWB could be directed to prepare a "factual report" on any dispute prior to a reference to the court. The provinces would submit proposals to the board, which would review them and make recommendations after requesting additional information on water quantity and quality where needed. See Alberta Government Library, PPWB, "Master Agreement on Apportionment" between Canada, Alberta, Saskatchewan, and Manitoba, 30 October 1969, with

Schedules A and B concerning Alberta-Saskatchewan and Saskatchewan-Manitoba agreements respectively. See also correspondence leading up to this agreement in AA, Acc. 90.159, Deputy Minister of the Environment records, box 15.

49 Legislative Library of Alberta, Edmonton, AWRC, Hearings re South Saskatchewan River Basin Planning Programme (hereafter AWRC, Hearings), Calgary, vol. 1, 7 November 1984, testimony of Dave MacLeod, vice-president of Saskatchewan Water Corporation, 343–66 (transcript).

50 Ibid., 347–50.

51 By 1977 the Streamflow Forecasting Report Package for the North and the South Saskatchewan (as well as for the Qu'Appelle) involved five documents each – a Main Report, a Combined Use Manual, a User Manual Water Supply, a User Manual Routing, and a User Manual Basin Simulation – and similar work was proceeding on the Saskatchewan and the Churchill; this study is briefly described in Government of Alberta Library, PPWB, Annual Report, 1976–77 (typescript). The current mathematical model of the river with actual flows and calculated natural flows is available on CD: Alberta Environment, Water Management Division, South Saskatchewan River Basin Historical Natural Flows, 1912–95, Version 2.02. It is also available online through the department's website.

52 AA, Acc. 78.110, Deputy Minister of Environment records, 1971, box 1.

53 Ibid., box 3, Memo, Alberta Water Resources Commission, which sets out several possible changes in administrative structures to deal with water issues. This memo led up to the creation of a ministry. For an example of early monitoring activities, see AA, Acc. 71.311, Department of the Environment, box C, Pollution Survey Summary, Bow River, 1970–71, P.G. Shewchuk. See also Acc. 91.123, Environment records, Environment files relating to the Committee on Hydrology, Prairie Provinces Water Board, 1976–83; Acc. 90.159, Deputy Minister of the Environment records, box 15, Water Agreements, E.W. Allison, "A Summary of Prairie Provinces Water Board Water Quality Activities, 1969 to 1980," October 1980.

54 AA, Acc. 85.401, Peter Loughheed papers, Executive Correspondence, series IXa, box 38 (offsite), file 675, pamphlet "PRIME: is it a CRIME?" Years later, when one of his cabinet ministers proposed a version of an interbasin transfer scheme, Lougheed had a study done to ensure that it did not contradict his position while in opposition; see ibid., Meeting, Water Resources, 1 August 1979, Peter Lougheed to Collen Collins, Research Officer, 23 June 1979. The proposal involved a small transfer into the Red Deer River.

55 AA, Acc. 83.92, Parks Board records, box 1, General Correspondence files; Acc. 74.169, Eastern Rockies Forest Conservation Board, Minutes, 1968, where the board resists the recommendations of "The Real Need in Alberta – Parks and Recreation," a report urging province-wide planning. See also Acc. 74.345, Environment Conservation Authority Papers, Interim Report No. 3, Bow River Basin, May 1974, 17–20.

56 AA, Acc. 85.401, Lougheed Papers, Executive Correspondence series IXa, box 38, file 674, Water Advisory Committee, Aug. 1981, Memo: River Basin Planning Status Report.

57 Ibid., box 138, Eastern Slopes Report. Environment Conservation Authority, Land Use and Resource Development in the Eastern Slopes, Report and Recommendations, September 1974 (Environment Conservation Authority, 1974), 54–8.

58 Alberta Environment and Alberta Agriculture, *Water Management for Irrigation Use* (Alberta Sessional Paper 502/75, May 1975).

59 The South Saskatchewan Basin *Summary Report on Water Use Objectives* (1984, mimeo, University of Calgary Library) concluded from its public opinion survey that the main pressing concerns were "water quality problems on the Bow River and the desire for increased irrigation" (23).

60 Among the studies was a further attempt to estimate the "natural flow" in the basin; see J.R. Card, F.D. Davies, and R. Bothe, *South Saskatchewan River Basin Historical Natural Flows, 1912 to 1978: Main Report* (Hydrological Branch, Technical Services Division, Alberta Environment, 1982), 6 vols. (copy in University of Calgary Library, GB 1230 A5S67).

61 South Saskatchewan River Basin Planning Program, *Water Use Objective Report, Bow River Sub-Basin*, (Alberta Environment, Water Resources Management Services, Planning Division, 1984) (copy in University of Calgary Library, CA2AL EN51 84S57A).

62 Ibid., 9.

63 Jack Glenn, *Once upon an Oldman: Special Interest Politics and the Oldman River Dam* (Vancouver: UBC Press, 1999), 25–42, provides the best available overview. When its sunset clause kicked in, the Environment Conservation Agency was subsequently restructured as the Environmental Council of Alberta with the intent of bringing it more closely into line with government policy.

64 Alberta Water Resources Commission, *Annual Report*, 1984–85.

65 AWRC, Hearings, Calgary, vol. 1, Stan Ainslie, Calgary Hook and Hackle Club, 96; vol. 7, Red Deer, Ed Wolf, private citizen, 449–53.

66 AWRC, Hearings, Lethbridge, vol. 1, 137–40, quoted in Glenn, *Once upon an Oldman*, 44.

67 See most emphatically Neil Jennings, Bow River Protection Society, AWRC, Hearings, High River, vol. 4, 44.

68 In addition to more than a dozen local fish and game associations, the commission received briefs and heard from the following organizations with broadly similar approaches – Ducks Unlimited, Trout Unlimited, the Alberta Wilderness Association, the Sierra Club, the Federation of Alberta Naturalists, the Bow Waters Canoe Club, the Oldman River Canoe and Kayak Association, and the Bow River Protection Society – and the well-known fishing guides R. Harding and J. McLennan. The quotation is from Doug Kariel, chair, Alberta Section of the Sierra Club, AWRC Public Hearings, Calgary, vol. 1, 368. The commission provided a complete list of the participants as an appendix to its report. The Trout Unlimited statement is to be found in AWRC, Hearings, Calgary, vol. 1, 28.

69 See Mayor Klein presentation, AWRC, Hearings, Calgary, vol. 1, 4–25, 19–20, for his observations as a fisher on the river.

70 AWRC, Hearings, Calgary, vol. 1, 310–4, 328–34, Canadian Petroleum Association, and Dome Petroleum.
71 AWRC, Hearings, Strathmore, vol. 2, 138–224.
72 Alberta Legislative Library, AWRC, *Water Management in the South Saskatchewan River Basin: Report and Recommendations* (Edmonton, 1986; typescript), and AWRC *Annual Report*, 1986/87.
73 AWRC, Hearings, Peigan Band presentation, Red Deer, vol. 3, 433–46.
74 AWRC, Hearings, Brooks, vol. 2, 224–41. Russell Wright, presenting the brief of the Blackfoot Tribal Council, introduced his discussion of the dam with the phrase, "Okay, this is a biggy here" (224).
75 AWRC, Hearings, Calgary, vol. 1, 8. Klein was concerned about upstream residential and industrial development on indigenous lands, which would have an impact on Calgary's downstream water quality. In a legal monograph published in 1988 Richard H. Bartlett argued, drawing heavily upon US jurisprudence, that aboriginal people in Canada had title to water by virtue of their treaties and secondly as reserve property; see *Aboriginal Water Rights in Canada: A Study of Aboriginal Title to Water and Indian Water Rights* (Edmonton: Faculty of Law, Canadian Institute of Resources Law, 1988).
76 For a full account of this legal battle, which sustained authority in western navigable waters, fisheries, and indigenous reserves, see Glenn, *Once upon an Oldman*. See also the Supreme Court Decision that settled the issue, *Friends of the Oldman River Society v. Canada (Minister of Transport)*, 1992.

Chapter Thirteen

1 "The Blob from the Bow River," *Alberta Report*, 20 November 1989, 31; "Klein checks on toxic cleanup," *Calgary Herald*, 5 November 1989.
2 "Post-industrial poison: Calgary begins cleaning up at Canada Creosote," *Alberta Report*, 22 April 1991, 30; "Creosote cleanup set: new method to get tryout on Bow River," *Calgary Herald*, 21 March 1990; "River cleanup costly 'nightmare:' toxic creosote bill to suck up millions from taxpayers' purse," *Calgary Herald*, 31 May 1991; "Bow River oil slick nearly contained," *Calgary Herald*, 7 July 1991; "Bow creosote cleanup begins," *Calgary Herald*, 27 September 1991; "Engineering nightmare carries costly price tag," *Globe and Mail*, 12 June 1991.
3 Ralph Klein, "The Filthy Bow: What's Bad For Us Is Even Worse Downstream," *Calgary Magazine*, June 1980, 70–1; "Klein wants help to clean up Bow River," *Calgary Herald*, 26 October 1982; "Klein, Kowalski as Ottawa to aid in Bow River cleanup," *Calgary Herald*, 27 June 1986.
4 *Report of the Bow River Water Quality Task Force, Preserving Our Lifeline* (Calgary: November 1991), "Terms of Reference," 61.
5 *Bow River Water Quality Task Force, Preserving Our Lifeline*, preface, John. W. Donahue, Chairman, to The Honourable Ralph Klein, Minister of the Environment, 26 November 1991.
6 Ibid., 11.

7 Ibid., passim, and *Preserving Our Lifeline, Appendix VI*, a separate volume of tables, graphs, and compilations of technical data.
8 "Urban users have big impact on Bow River," *Calgary Herald*, 7 December 1991.
9 Marc Cioc, *The Rhine: An Eco-Biography, 1815–2000* (Seattle: University of Washington Press, 2002), 199.
10 *Preserving Our Lifeline: A Report on the State of the Bow River* (Calgary: Bow River Water Quality Council, October 1994); "Bow River fails to make grade downstream," *Calgary Herald*, 15 October 1994.
11 Ducks Unlimited, which sent a representative to the Task Force, did not participate directly in the Water Quality Council, though it would be represented on the later Bow River Basin Water Council. Al Taylor, representing Calgary outdoor recreation organizations, and Jim Rouse, from Trout Unlimited, whose recreational pursuits linked them more directly to the river, participated in both bodies.
12 In the 1970s and 1980s Nelles's neighbour Tom Harling made annual fall pilgrimages to Alberta and always returned with a plump bag of wildfowl to be plucked, trussed, and roasted for memorable dinners.
13 See, for details, the Ducks Unlimited Canada, Canada's Conservation Company website at http://www.ducks.ca/province/ab/.
14 See the Trout Unlimited Canada Alberta Council website, http://www.tucanada.org/alberta/ with its emphasis on its "Fish Rescue" operations. *Canadian Geographic* magazine did a photo essay on the annual fish rescue by these "Anglers of Mercy."
15 John Eisenhauer, Bozena Kolar, and Alberta Wilderness Society, *Rivers on Borrowed Time* (Calgary, 1982). The study team included Hans Buhrmann, Brian Forestall, Cliff Wallis, and Ron Hooper. See also the beautiful handbook published by the same group, *Eastern Slopes Wildlands: Our Living Heritage* (Calgary: Alberta Wilderness Association, 1986), a proposal for a vast wildlands reserve on the front range.
16 Janice MacDonald, *Canoeing Alberta* (Red Deer: Lone Pine Publishers, 1985).
17 Ibid., 21.
18 Jim McLennan, *Blue Ribbon Bow: A History of the Bow River – Canada's Greatest Trout Stream* (Edmonton: Lone Pine Publishing, 1987), and the second edition, titled *Blue Ribbon Bow: A Fly-Fishing History of the Bow River – Canada's Greatest Trout Stream* (Red Deer: Johnson and Gorman Books, 1998). Quotations are from the 1998 edition: 55, 63, and 70.
19 Ibid.; see in particular "The State of the Bow River Trout Fishery" (148–63).
20 Ted Godwin, *Lower Bow: A Celebration of Wilderness, Art and Fishing* (Calgary: Hard Art Moving and Storage Company, 1991). The seventy-eight-work exhibition, sponsored by PetroCanada, was displayed at galleries in Regina, Calgary, Vancouver, and Toronto between November 1991 and March 1992. Paintings from this show have found their way into corporate and private collections across the country. One graces the Canadian embassy in Kiev; two panels brighten the bar of the Bank of Montreal's Institute for Learning in Toronto.
21 For a more detailed understanding of the art history of the Bow River, see C. Armstrong and H.V. Nelles, *The Painted Valley: The Artists Along Alberta's Bow River, 1845–2000* (Calgary: University of Calgary Press, 2007).

22 LAC, RG 84, Parks Canada, G0000527, Park visitor user survey – Banff National Park, 1972, Printout Job D923576W, 79–139. See also University of Alberta Library, Survey and Analysis of Resident and Non Resident Travel, 1971, Kates Peat Marwick for Alberta Industry and Commerce, October 1972 (Gov Doc CA2 AL IC50 72S71).

23 Donald W. Irwin, *Banff: Visitors' Activities and Regional Growth* (Calgary: University of Calgary, Faculty of Environmental Design, n.d. [c. 1975]), 18.

24 John H. Thompson et al., *Household Survey*, vol. 4 of the Alberta Forestry, Lands and Wildlife Department, *Bow River Recreation Survey, 1987* (University of Calgary Library, GV 181.46 B69 1987) vol. 4; the other three volumes appear to be missing.

25 AA, Acc. 90.145, Rocky Mountain National Parks Utilization Study, 1989, Executive Summary, General Summary – volume titled "Value, Trip Characteristics, General Summary – Activities, Sites and Attitudinal Information." The federal government conducted the survey to provide data for a marketing campaign to attract more tourists. Parks Canada learned, for example, that it should target older travellers without children to maximize tourist revenue, give Japanese tour groups a more satisfying experience to get them to stay longer, provide a relaxing atmosphere to the American tourists who valued that the most, and market the Rocky Mountain parks in eastern Canada as "Canada's Parks" to overcome "national frictions over regionalism" that turned off Ontario and Quebec visitors.

26 For the evolution of environmental science and the rise of the environmental movement based upon a John Muir–Aldo Leopold styled "land ethic" of varying degrees of asceticism, see Donald Worster, *Nature's Economy: A History of Ecological Ideas*, 2nd ed. (New York: Cambridge University Press, 1994), parts 4, 5, and 6. See also Carolyn Merchant's account, which adds chaos theory and feminism to the mix, in *Reinventing Eden: The Fate of Nature in Western Culture* (New York: Routledge, 2003), and the thumbnail account in her handbook *The Columbia Guide to American Environmental History* (New York: Columbia University Press, 2002), 159–90.

27 See Canada, House of Commons, *Debates*, 15 April 1988, 14523–35, and 19 April 1988, 14593–607, for the most substantial discussion. Curiously, though the minister, Mr McMillan, was present in the chamber he neither introduced nor defended his bill. The government's unusually passive approach to this legislation probably could be explained by an unwillingness to court diversionary controversy during debate of its highest policy priority, the Canada-US Free Trade Agreement.

28 A Liberal critic, Don Boudria, drew attention to the CPAWS intervention in committee; see ibid., 18 July 1988, 17704.

29 "Ecological and historical integrity" had been included as management objectives in an internal Parks Canada policy document in 1979. A CPAWS member and ecologist, Stephen Woodley, had been instrumental as an academic in developing the concept ecological integrity. For the evolution of the idea see inter alia, *"Unimpaired for Future Generations"? Conserving Ecological Integrity with Canada's National Parks* (Ottawa: Parks Canada), 2: 1-1–8.

30 The concept of ecological integrity is developed and applied in a collection of scholarly papers edited by Stephen Woodley, James Kay, and George Francis, of the Cana-

dian Parks Service and the University of Waterloo respectively, *Ecological Integrity and the Management of Ecosystems* (n.p.: St. Lucie Press, 1993). James R. Karr, an ecologist who initially developed this idea for the evaluation of streams, is represented in the volume. In a more recent collection of papers, Karr defines ecological or biological integrity as follows: "Biological integrity refers to the condition of places at one end of a continuum of human influence, places that support a biota that is the product of evolutionary and biogeographic processes with little or no human influence from industrial society. This biota is a balanced, integrated, adaptive system having its full range of elements (genes, species, assemblages) and processes (mutation, demography, biotic interaction, nutrient and energy dynamics, and metapopulation processes) expected in areas of minimal human influence." See David Pimental, Laura Westra, and Reed F. Noss, eds, *Ecological Integrity: Integrating Environment, Conservation and Health* (Washington: Island Press, 2000), 212. In this collection the logician Mark Sagoff revisits his devastating philosophical critique of the concept on 61–78.

31 Canadian Parks and Wilderness Society, *The Bow Valley: A Very Special Place* (Calgary, n.d. [early 1990s]) (copy in Glenbow Library, PAM 333.73 L498b).

32 The task force was chaired by Dr Robert Page, then dean of the Faculty of Environmental Engineering at the University of Calgary. The members were academics Dr Suzanne Bayley, a biologist, Dr Brent Ritchie, a business professor with a specialty in tourism, and two environmental consultants, J. Douglas Cook from the business side and Jeffrey E. Green, a wildlife ecologist. The task force adopted for its identifying logo a vertical banner of a river valley winding out of a mountain range into a deep valley.

33 See Banff–Bow Valley Task Force, *Banff–Bow Valley: At The Crossroads Summary Report* (Ottawa, 1996), 9–11, 20–4, for the vision statement arrived at.

34 Ibid., 4–5.

35 Ibid., 26, 33.

36 Ibid. 12. The more detailed *Technical Report* used Stephen Woodley's definition as follows: "Ecological integrity is defined as a state of ecosystem development that is optimized for its geographic location, including energy input, available water, nutrients and colonization history. For parks and protected areas, this optimal state has been referred to by such terms as natural, naturally-evolving, pristine and untouched. It implies that ecosystem structures and functions are unimpaired by human-caused stresses, that native species are present at viable population levels and, within successional limits, that the system is likely to persist. Ecosystems with integrity do not exhibit the trends associated with stressed ecosystems. Parks and protected areas are part of larger ecosystems and determinations of integrity in national parks must consider these larger ecosystem." The task force expanded upon this definition by enumerating the characteristics of ecosystems with integrity, namely, health, or the ability to develop and evolve; biodiversity, or the functioning of evolving ecological interdependencies; stability, or the ability to resist stress; and sustainability, the

capacity of the ecosystem to maintain itself in perpetuity (Banff–Bow Valley Task Force, *Technical Report*, 94).

37 Banff–Bow Valley Task Force, *Technical Report*, 101–10.

38 Ibid., 158–74.

39 See, for example, Parks Canada, *"Unimpaired for Future Generations"?* 2 vols. (Ottawa, 2000), and Parks Canada, *First Priority: Progress Report on Implementation of Recommendations of the Panel on Ecological Integrity of Canada's National Parks* (Ottawa, 2001). Canada's National Park Act was amended in February 2001 to strengthen the ecological integrity clause.

40 *Calgary Herald*, 7, 8, 9, 10, 11, 12, 13, and 14 August 1998.

41 http://www.biosphereinstitute.org/

42 http://www.bowriverkeepr.org. http://www.riverkeeper.org/ is the link to the central organization. For a history of this movement, see John F. Kennedy Jr and John Cronin, *The Riverkeeper* (New York: Scribner, 1997), and a Princeton BA thesis copied on this site, Charles Scribner, "The History of Waterkeeper Alliance: An International Grassroots Movement Flows from the Hudson" (Princeton University, Department of History, April 2005).

43 http://www1.calgary.ca/bcc/BCC35.asp.

44 Water research programs are directed by Dr David Schindler, Dr Leland J. Jackson, and Dr James Byrne at the three universities respectively. In addition, the Canadian Institute of Resources Law at the University of Calgary and a research consortium, the Canadian Water Network, centred in the west provide additional supporting research on water matters.

45 See, for example, David Schindler and W.F. Donahue, "An Impending Water Crisis in Canada's Western Prairie Provinces," *Proceedings of the National Academy of Sciences*, Environmental Sciences Inaugural Article, 10 April 2006 (10.1073/pnas.0601568103).

46 Andrew Nikiforuk, "Water Shock: Smaller snowfalls, shrinking glaciers, and a growing population could mean Calgary's well is about to run dry," *Avenue Magazine*, April, 2003, 41–9; Robert William Sanford, *Water, Weather and the Mountain West* (Vancouver: RNB Alliance, 2007), 45–50; "Glaciers shrinking that feed Calgary water supply: Alta environment minister," Canadian Press Newswire, 21 April 2004.

Conclusion

1 Rob de Loe and Reid Kreutzwiser, "Challenging the Status Quo: The Evolution of Water Governance in Canada," in *Eau Canada: The Future of Canada's Water*, ed. Karen Bakkar (Vancouver: UBC Press, 2007), 90.

2 For a broad statement of this transition, see Samuel P. Hays, *Beauty, Health and Permanence: Environmental Politics in the United States, 1955–1985* (New York: Cambridge University Press, 1987). A more focused and highly suggestive case study of this process is explored in Gregory Summers, *Consuming Nature: Environmentalism in the Fox River Valley, 1850–1950* (Lawrence: University Press of Kansas, 2006).

3 For the policy document and legislation, see http://www.waterforlife.gov.ab.ca/.

4 David R. Percy, "Responding to Water Scarcity in Western Canada," *Texas Law Review* 83 (2005): 2091–107; Bakker, *Eau Canada*, chapters 10, 11, and 12.
5 Marc Cioc, *The Rhine: An Eco-Biography, 1815–2000* (Seattle and London: University of Washington Press, 2002), 179.
6 See, for example, the organization American Rivers, which champions dam removal and the re-wilding of rivers: http://www.amr.convio.net.
7 For information on the rehabilitation of the Grand River and its status as a Heritage River and benchmark performance over the past decade, see the website of the Grand River Conservation Authority, http://www.grandriver.ca/index.cfm.
8 John B. Sprague, "Great Wet North? Canada's Myth of Water Abundance," in *Eau Canada*, ed. Bakkar, 23–35.
9 Alberta, Agriculture and Rural Development, *Irrigation in Alberta* (Edmonton, 2000), 2; available at http://www1.agric.gov.ab.ca/$department/deptdocs.nsf/all/irr7197.
10 In 2002, for example, actual water use by irrigators in Reach 7 of the Bow equalled 74.3 per cent of the total water withdrawals permitted under existing licences; see Bow River Basin Council, *The 2005 Report on the State of the Bow River Basin*, 145.
11 Ibid., 115–16, 130–2, for evaluation of water quality below Calgary. Data gathering in Reach 5, where Calgary's effluent re-enters the river, curiously has not been adequate to establish measurements in relation to Canadian Water Quality Standards.
12 For a recent response see the report on the 2004 flood: http://www.calgary.ca/docgallery/bu/water_services/emergency_planning/2005_flood/full_report_on_flooding.pdf.
13 Ardith Walkem, "The Land is Dry: Indigenous Peoples, Water, and Environmental Justice," in *Eau Canada*, ed. Bakkar, 303–19, and Richard Bartlett, *Aboriginal Water Rights in Canada: A Study in Aboriginal Title to Water and Indian Water Rights* (Calgary: Canadian Institute of Resources Law, 1986).
14 D.W. Schindler and W.F. Donahue, "An Impending Water Crisis in Canada's Western Prairie Provinces," *Proceedings of the National Academy of Sciences* 103 (2006): 7210–16.

Acknowledgments

It took more than the three of us to write this book. In our research we drew heavily upon the collections and the staff of the many archives and libraries identified in our notes, but none more so than the estimable Glenbow Archives in Calgary. Archivist Doug Cass, librarian Lindsey Moir, and their staffs guided us towards otherwise overlooked documents and cheerfully put up with our urgent requests for help.

Our research expenses were primarily funded by a grant from the Social Sciences and Humanities Research Council of Canada. We are particularly indebted to the SSHRCC for its support of our work over the years. Additional research support came from McMaster University, York University, and the University of British Columbia.

During the course of writing, some parts of these chapters took shape as papers and talks delivered at York, McMaster, UBC, Simon Fraser, Calgary, the University of Western Ontario, and Queen's and as conference papers for meetings of the International Water History Association, the Canadian Historical Association, the American Society for Environmental History, the Association of American Geographers, the T2M Conference on Mobility and the Environment, and a NiCHE (Network in Canadian History and Environment) summer workshop.

Portions of some chapters previously published in scholarly journals appear in substantially revised form here. Part of chapter 1, "Discovery," began as a Snider Lecture by Nelles at Carleton University and was subsequently posted on the NiCHE website as a discussion paper. Part of chapter 6, "Watering a Dry Country," started out as an article by Evenden: "Precarious Foundations: Irrigation, Environment, and Social Change in the Canadian Pacific Railway's Eastern Section, 1900–1930," *Journal of Historical Geography* 32, no. 1 (January 2006): 74–95. An earlier version of chapter 7, "The Sanitary Imperative," was published by Evenden and Nelles as "The Bow: Calgary's Sanitary River," in a special issue of the *Journal of the West* 44, no. 3 (Summer 2005): 30–7. A section of chapter 11, "Greening Alberta," was greatly expanded by Nelles following a public presentation at the Nickle Art Gallery in Calgary into "How Did Calgary Get Its River Parks?" in the *Urban History Review* 34 (2005): 27–44.

Who did what? We developed the overall research plan collectively. We then divided up the research between us according to our special interests; in other cases we worked together on topics and pooled our sources. Responsibility for the first cut at chapters was divided between us as follows: Nelles drafted "Discovery," "Home on the Range and River," "The Wooden River," "The Sanitary Imperative," "Overflow," "Greening Alberta," and "Who Has Seen the River?" Evenden drafted "Homeland and Margin," "Watering a Dry Country," and "The Fishing River." Armstrong took responsibility for "Power and Flow," "Building Banff," "Water Powers," and an art chapter that in the end became a separate book, *The Painted Valley* (Calgary: University of Calgary Press, 2007). During the revision stage we worked on each other's chapters to create a collaborative text for which we take joint responsibility. We drafted the "Introduction" and "Conclusion" collectively through an exchange of drafts. Finally, Evenden planned all the maps, which Eric Leinberger, UBC Geography, prepared.

Several people read draft chapters for us and assisted us with sources. Gerry Conaty read "Homeland and Margin"; Pat Myers read "Greening Alberta" (and also provided geomatics data on provincial park boundaries); Diane Coleman read "The Sanitary Imperative"; George Colpitts, Cole Harris, and Graeme Wynn read an early version of "The Fishing River"; Clint Evans, Royden Loewen, Jamie Murton, and Graeme Wynn commented on an earlier version of "Watering a Dry Country"; James Muir read "Water Powers"; Don Smith provided a vigorous critique of the parks material, as did Sean Kheraj; Ruth Sandwell reviewed the "Home on the Range and River" chapter. We extend a special thanks to Graeme Wynn for engaging (particularly with Evenden) in the ideas behind this project and for his practical support offered as head of the department of geography at UBC.

Joseph Nelson and the University of Alberta Press kindly allowed us to use some photographs of fish in a map included in "The Fishing River," taken from Nelson and Martin Paetz's *Fishes of Alberta*.

Several research assistants helped us at various stages: at UBC John Thistle, Kayla Pompu, Jonathan Peyton, and Laurie Dickmeyer; and at York Eric Strickwerda, who worked on water law; Liza Piper, who read the Prairie Provinces Water Board Papers in Regina; and Doug Parker, who searched the City of Calgary Council minutes.

At McGill-Queen's University Press Philip Cercone responded eagerly to the prospect of publishing this book, Susanne McAdam ably managed production, and Elizabeth Hulse, as copy editor, did her best with our prose and inconsistencies.

When we first started this project, several people in Calgary supplied practical help and encouragement. The Department of History at the University of Calgary provided Nelles with space and library access. The late Henry Klassen encouraged us in this endeavour. Don Smith gave us a host of good suggestions. David Johnston took Evenden canoeing and fishing on the river. Not only was Bob Page a generous host, but he also helped us gain access to Transalta records. John Gilpin talked to us about irrigation. With his customary enthusiasm and ingenuity, Harry M. Sanders guided us with his unparalleled mastery of Calgary historical sources. Several people hosted us at different stages, including Diane Roulson and Helen MacRae, David Johnston, Kirstin Evenden, Ben and Chris Heazell, Don Smith, and Nancy Townshend.

We would especially like to thank our wives, Diane, Val, and Kirsty, who accompanied us at various times during our research, listened to our half-baked ideas, and tolerated a good deal of dam tourism over several years. Finally, we would like to dedicate this book to our children in the hopes that they might enjoy this river in the future too.

Separately and together we have enjoyed prowling the Bow River from its mountainous glacial origins to its meandering, deeply entrenched course across a dusty prairie. As can be seen from these pages, throughout our explorations the machinery of the river – the storage reservoirs, hydroelectric stations, plumbing, and especially the majestic weir at Bassano – has enchanted us as much as its glorious geography and supposed natural beauty.

Index